The Drone Chronicles 2001 – 2016 Catalogue

ÉDITIONS CENTRE DE LA PHOTO— GRAPHIE GENÈVE

Spector Books

2001

Aeronautics Ltd.

Aerostar (RQ-26A)

© Aeronautics

Release Date: June 2001
Price: $1.5 million
Number of Rotors: 1 (Fixed Wing)
Control System: Ground Control Station, Remote Control
Applications: Military Purposes, Maritime Patrol, Reconnaissance, Surveillance
Dimensions: 4.5 × 8.7 × 1.2 m
Wingspan: 8.7 m
Weight: 230 kg
Max. Payload: 50 kg
Frame Material: Composite
Rotor Material: Wood
Available Colors: Grey
Max. Flight Time: 12 hours
Max. Speed: 200 km/h (124 mp/h)
Max. Operating Range: 250 km
GPS Compatible: √
Altitude Hold: √
Position Hold: –
Flight Planning: √
Autonomous Flight: √
Target Tracking: √
Automatic Landing: √
Return To Home: √
Obstacle Avoidance: √
Headless Mode: –
Built-in Camera: √

Korea Aerospace Industries (KAI)

RQ-101 Night Intruder 300

© Korea Aerospace Industries

Release Date: October 2001
Price: Unknown
Number of Rotors: 1 (Fixed Wing)
Control System: Ground Control Station
Applications: Military Purposes, Reconnaissance, Surveillance
Dimensions: 4.7 × 6.4 × 1.47 m
Wingspan: 6.4 m
Weight: 215 kg
Max. Payload: 45 kg
Frame Material: Composite
Available Colors: Grey
Max. Flight Time: 6 hours
Max. Speed: 185 km/h (115 mp/h)
Max. Flying Altitude: 4,500 m (14,764 ft)
Max. Operating Range: 200 km
GPS Compatible: √
Altitude Hold: √
Position Hold: –
Flight Planning: √
Autonomous Flight: √
Target Tracking: √
Automatic Landing: √
Return To Home: √
Obstacle Avoidance: √
Headless Mode: –
Built-in Camera: √

Northrop Grumman

RQ-4A Global Hawk

Release Date: November 2001
Price: $61 million
Number of Rotors: – (Fixed Wing)
Control System: Ground Control Station
Applications: Military Purposes, Reconnaissance, Surveillance
Dimensions: 13.4 × 35.3 × 4.6 m
Wingspan: 35.3 m
Weight: 11,612 kg
Max. Payload: 907 kg
Frame Material: Composite
Motors: 1 × Allison Rolls Royce AE3007H turbofan engine
Available Colors: Grey
Max. Flight Time: 35 hours
Max. Speed: 650 km/h (404 mp/h)
Max. Flying Altitude: 19,800 m (64,960 ft)
Max. Operating Range: 22,779 km
GPS Compatible: √
Altitude Hold: √
Position Hold: –
Flight Planning: √
Autonomous Flight: √
Target Tracking: √
Automatic Landing: √
Return To Home: √
Obstacle Avoidance: √
Headless Mode: –
Built-in Camera: √

Turkish Aerospace Industries (TAI)

Keklik

Release Date: November 2001
Price: Unknown
Number of Rotors: 1 (Fixed Wing)
Control System: Ground Control Station, Remote Control
Applications: Military Purposes, Target Acquisition
Dimensions: 1.36 × 1.61 × 0.32 m
Wingspan: 1.61 m
Weight: 7 kg
Max. Payload: 3 kg
Frame Material: Composite
Motors: 1 × Graupner OS-MAX 91 Fx two-stroke engine
Available Colors: Red
Max. Flight Time: 30 minutes
Max. Speed: 150 km/h (93 mp/h)
GPS Compatible: √
Altitude Hold: √
Position Hold: –
Flight Planning: √
Autonomous Flight: √
Target Tracking: √
Automatic Landing: –
Return To Home: √
Obstacle Avoidance: –
Headless Mode: –
Built-in Camera: √

2002

Leonardo S.p.A.

DRS RQ-15 Neptune

© US Navy/Daniel J. McLain

Release Date: January 2002
Price: Unknown
Number of Rotors: 1 (Fixed Wing)
Control System: Ground Control Station
Applications: Military Purposes, Reconnaissance, Surveillance
Dimensions: 1.83 × 2.13 m
Wingspan: 2.13 m
Weight: 36 kg
Frame Material: Composite
Motors: Two-stroke piston engine
Available Colors: Grey
Max. Flight Time: 4 hours
Max. Speed: 161 km/h (100 mp/h)
Max. Flying Altitude: 2,440 m (8,000 ft)
Max. Operating Range: 74 km
GPS Compatible: √
Altitude Hold: √
Position Hold: –
Flight Planning: √
Autonomous Flight: √
Target Tracking: √
Automatic Landing: √
Return To Home: √
Obstacle Avoidance: √
Headless Mode: –
Built-in Camera: √

Boeing Insitu

X-45A

© NASA/Jim Ross

Release Date: May 2002
Price: Unknown
Number of Rotors: – (Fixed Wing)
Control System: Ground Control Station
Applications: Military Purposes, Armed Reconnaissance, Surveillance, Target Acquisition
Dimensions: 8.08 × 10.3 × 2.14 m
Wingspan: 10.3 m
Weight: 3,630 kg
Frame Material: Composite
Motors: 1 × Honeywell F124-GA-100 turbofan
Available Colors: White
Max. Speed: 919 km/h (571 mp/h)
Max. Flying Altitude: 13,200 m (43,307 ft)
Max. Operating Range: 2,400 km
GPS Compatible: √
Altitude Hold: √
Position Hold: –
Flight Planning: √
Autonomous Flight: √
Target Tracking: √
Automatic Landing: √
Return To Home: √
Obstacle Avoidance: √
Headless Mode: –
Built-in Camera: √

AeroVironment

RQ-14 Dragon Eye

© National Air and Space Museum

Release Date: July 2002
Price: $75,000
Number of Rotors: 2 (Fixed Wing)
Control System: Ground Control Station
Applications: Military Purposes, Reconnaissance
Dimensions: 0.91 × 1.14 × 0.28 m
Wingspan: 1.14 m
Weight: 2.7 kg
Frame Material: Carbon Fiber, Styrofoam
Rotor Material: Plastic
Available Colors: White
Max. Flight Time: 60 minutes
Max. Speed: 35 km/h (21,7 mp/h)
Max. Flying Altitude: 152 m (500 ft)
Max. Operating Range: 5 km
GPS Compatible: √
Altitude Hold: √
Position Hold: –
Flight Planning: √
Autonomous Flight: √
Target Tracking: –
Automatic Landing: –
Return To Home: √
Obstacle Avoidance: –
Headless Mode: –
Built-in Camera: √

Raytheon

Silver Fox

© Raytheon

Release Date: September 2002
Price: Unknown
Number of Rotors: 1 (Fixed Wing)
Control System: Ground Control Station
Applications: Military Purposes, Reconnaissance, Surveillance
Dimensions: 1.47 × 2.4 × 0.28 m
Wingspan: 2.4 m
Weight: 11 kg
Max. Payload: 2.2 kg
Frame Material: Composite
Available Colors: Grey
Max. Flight Time: 10 hours
Max. Speed: 93 km/h (58 mp/h)
Max. Flying Altitude: 3,657 m (12,000 ft)
Max. Operating Range: 37 km
GPS Compatible: √
Altitude Hold: √
Position Hold: –
Flight Planning: √
Autonomous Flight: √
Target Tracking: √
Automatic Landing: √
Return To Home: √
Obstacle Avoidance: √
Headless Mode: √
Built-in Camera: √

Textron Systems

RQ-7A Shadow

© US Marine Corps

Release Date: October 2002
Price: $15.5 million
Number of Rotors: 1 (Fixed Wing)
Control System: Ground Control Station
Applications: Military Purposes, Reconnaissance, Surveillance
Dimensions: 3.4 × 3.89 × 1 m
Wingspan: 3.89 m
Weight: 148 kg
Frame Material: Composite
Motors: 1 × AR741-1100 engine
Available Colors: Grey
Max. Flight Time: 5.5 hours
Max. Speed: 194 km/h (120,5 mp/h)
Max. Flying Altitude: 3,000 m (9,842 ft)
Max. Operating Range: 50 km
GPS Compatible: √
Altitude Hold: √
Position Hold: –
Flight Planning: √
Autonomous Flight: √
Target Tracking: √
Automatic Landing: √
Return To Home: √
Obstacle Avoidance: √
Headless Mode: –
Built-in Camera: √

2003

Northrop Grumman

X-47A Pegasus

Release Date: February 2003
Price: Unknown
Number of Rotors: – (Fixed Wing)
Control System: Ground Control Station
Applications: Military Purposes, Research & Development, Armed Reconnaissance, Surveillance
Dimensions: 8.5 × 8.5 × 1.86 m
Wingspan: 8.5 m
Weight: 2,500 kg
Frame Material: Composite
Motors: 1 × Pratt & Whitney Canada JT15D-5C turbofan
Available Colors: Grey
Max. Speed: 900 km/h (560 mp/h)
Max. Flying Altitude: 12,192 m (40,000 ft)
Max. Operating Range: 2,778 km
GPS Compatible: √
Altitude Hold: √
Position Hold: –
Flight Planning: √
Autonomous Flight: √
Target Tracking: √
Automatic Landing: √
Return To Home: √
Obstacle Avoidance: √
Headless Mode: –
Built-in Camera: √

AeroVironment

RQ-11 Raven

Release Date: May 2003
Price: $200,000
Number of Rotors: 1 (Fixed Wing)
Control System: Ground Control Station
Applications: Military Purposes, Reconnaissance, Surveillance
Dimensions: 0.91 × 1.37 m
Wingspan: 1.37 m
Weight: 1.9 kg
Frame Material: Composite
Motors: 1 × Aveox electric motor
Available Colors: Grey
Max. Flight Time: 90 minutes
Max. Speed: 97 km/h (60 mp/h)
Max. Flying Altitude: 4,600 m (15,092 ft)
Max. Operating Range: 10 km
GPS Compatible: √
Altitude Hold: √
Position Hold: –
Flight Planning: √
Autonomous Flight: √
Target Tracking: √
Automatic Landing: √
Return To Home: √
Obstacle Avoidance: √
Headless Mode: –
Built-in Camera: √

Sagem

Sperwer MK-II

© Wikipedia/Tangopaso

Release Date: June 2003
Price: $3 million
Number of Rotors: 1 (Fixed Wing)
Control System: Ground Control Station
Applications: Military Purposes, Reconnaissance, Surveillance
Dimensions: 3.5 × 4.2 m
Wingspan: 4.2 m
Max. Payload: 50 kg
Frame Material: Composite
Available Colors: Grey
Max. Flight Time: 6 hours
Max. Speed: 166 km/h (104 mp/h)
Max. Flying Altitude: 4,600 m (15,092 ft)
Max. Operating Range: 200 km
GPS Compatible: √
Altitude Hold: √
Position Hold: –
Flight Planning: √
Autonomous Flight: √
Target Tracking: √
Automatic Landing: √
Return To Home: √
Obstacle Avoidance: √
Headless Mode: –
Built-in Camera: √

Turkish Aerospace Industries (TAI)

Pelikan/IHA-X2

© TAI

Release Date: September 2003
Price: Unknown
Number of Rotors: 2 (Fixed Wing)
Control System: Ground Control Station
Applications: Military Purposes, Reconnaissance, Surveillance
Dimensions: 3 × 3.6 × 0.42 m
Wingspan: 3.6 m
Weight: 20 kg
Max. Payload: 15 kg
Frame Material: Composite
Rotor Material: Wood
Motors: 2 × Zenoah G38 2-cylinder two-stroke engine
Available Colors: Grey
Max. Speed: 120 km/h (74,5 mp/h)
Max. Flying Altitude: 1,600 m (5,250 ft)
GPS Compatible: √
Altitude Hold: √
Position Hold: –
Flight Planning: √
Autonomous Flight: √
Target Tracking: –
Automatic Landing: √
Return To Home: √
Obstacle Avoidance: –
Headless Mode: –
Built-in Camera: √

Integrated Dynamics

Border Eagle Mk-II

© Integrated Dynamics

Release Date: November 2003
Price: Unknown
Number of Rotors: 1 (Fixed Wing)
Control System: Ground Control Station
Applications: Military Purposes, Border Patrol, Reconnaissance, Surveillance
Dimensions: 1.75 × 3.1 × 0.48 m
Wingspan: 3.1 m
Weight: 9 kg
Max. Payload: 4 kg
Frame Material: Composite
Motors: Two-cylinder piston engine
Available Colors: Grey
Max. Flight Time: 3 hours
Max. Speed: 160 km/h (99,5 mp/h)
Max. Operating Range: 30 km
GPS Compatible: √
Altitude Hold: √
Position Hold: –
Flight Planning: √
Autonomous Flight: √
Target Tracking: √
Automatic Landing: √
Return To Home: √
Obstacle Avoidance: √
Headless Mode: –
Built-in Camera: √

Guizhou Aviation Industry Group (CAIG)

WZ-2000

© popsci.com

Release Date: December 2003
Price: $35 million
Number of Rotors: – (Fixed Wing)
Control System: Ground Control Station
Applications: Military Purposes, Reconnaissance, Surveillance
Dimensions: 7.5 × 9.8 m
Wingspan: 9.8 m
Weight: 1,700 kg
Max. Payload: 80 kg
Motors: 1 × WS-11 turbofan
Available Colors: Grey
Max. Flight Time: 3 hours
Max. Speed: 800 km/h (497 mp/h)
Max. Flying Altitude: 18,000 m (59,055 ft)
Max. Operating Range: 2,400 km
GPS Compatible: √
Altitude Hold: √
Position Hold: –
Flight Planning: √
Autonomous Flight: √
Target Tracking: √
Automatic Landing: √
Return To Home: √
Obstacle Avoidance: √
Headless Mode: –
Built-in Camera: √

2004

EMT-Penzberg

UAV X-13

© EMT Penzberg

Release Date: April 2004
Price: Unknown
Number of Rotors: 1 (Fixed Wing)
Control System: Ground Control Station
Applications: Military Purposes, Reconnaissance, Surveillance
Wingspan: 5.1 m
Weight: 130 kg
Frame Material: Composite
Available Colors: White
Max. Flight Time: 6 hours
Max. Speed: 180 km/h (112 mp/h)
Max. Flying Altitude: 3,048 m (10,000 ft)
Max. Operating Range: 200 km
GPS Compatible: √
Altitude Hold: √
Position Hold: –
Flight Planning: √
Autonomous Flight: √
Target Tracking: –
Automatic Landing: √
Return To Home: √
Obstacle Avoidance: –
Headless Mode: –
Built-in Camera: √

Israel Aerospace Industries (IAI)

Bird-Eye 500

© IAI

Release Date: June 2004
Price: $200,000
Number of Rotors: 1 (Fixed Wing)
Control System: Ground Control Station
Applications: Military Purposes, Reconnaissance, Surveillance
Dimensions: 1.6 × 2 m
Wingspan: 2 m
Weight: 4.15 kg
Max. Payload: 850 gr
Frame Material: Composite
Available Colors: Grey
Max. Flight Time: 60 minutes
Max. Speed: 111 km/h (69 mp/h)
Max. Flying Altitude: 305 m (1,000 ft)
Max. Operating Range: 10 km
GPS Compatible: √
Altitude Hold: √
Position Hold: –
Flight Planning: √
Autonomous Flight: √
Target Tracking: √
Automatic Landing: –
Return To Home: √
Obstacle Avoidance: √
Headless Mode: –
Built-in Camera: √

Griffon Aerospace

MQM-170A Outlaw G1

© Griffon Aerospace

Release Date: July 2004
Price: Unknown
Number of Rotors: 1 (Fixed Wing)
Control System: Ground Control Station
Applications: Military Purposes, Aerial Target, Reconnaissance
Dimensions: 2.7 × 4.15 m
Wingspan: 4.15 m
Weight: 54 kg
Frame Material: Composite
Motors: Two-cylinder two-stroke piston engine
Available Colors: White
Max. Flight Time: 3 hours
Max. Speed: 193 km/h (120 mp/h)
Max. Flying Altitude: 4,900 m (16,076 ft)
GPS Compatible: √
Altitude Hold: √
Position Hold: –
Flight Planning: √
Autonomous Flight: √
Target Tracking: –
Automatic Landing: √
Return To Home: √
Obstacle Avoidance: –
Headless Mode: –
Built-in Camera: –

Textron Systems

RQ-7B Shadow

© US Marine Corps

Release Date: July 2004
Price: $750,000
Number of Rotors: 1 (Fixed Wing)
Control System: Ground Control Station
Applications: Military Purposes, Reconnaissance, Surveillance
Dimensions: 3.4 × 4.3 × 1 m
Wingspan: 4.3 m
Weight: 84 kg
Max. Payload: 45 kg
Frame Material: Composite
Motors: 1 × Wankel UAV engine 741-1101
Available Colors: Grey
Max. Flight Time: 6 hours
Max. Speed: 204 km/h (127 mp/h)
Max. Flying Altitude: 4,600 m (15,092 ft)
Max. Operating Range: 125 km
GPS Compatible: √
Altitude Hold: √
Position Hold: –
Flight Planning: √
Autonomous Flight: √
Target Tracking: √
Automatic Landing: √
Return To Home: √
Obstacle Avoidance: √
Headless Mode: –
Built-in Camera: √

SATUMA

Jasoos II/Bravo+

©Satuma

Release Date: August 2004
Price: $250,000
Number of Rotors: 1 (Fixed Wing)
Control System: Ground Control Station
Applications: Military Purposes, Reconnaissance, Surveillance
Dimensions: 4.27 × 4.92 m
Wingspan: 4.92 m
Weight: 145 kg
Max. Payload: 25 kg
Frame Material: Composite
Motors: Two-stroke gasoline engine
Available Colors: Blue
Max. Flight Time: 5 hours
Max. Speed: 130 km/h (81 mp/h)
Max. Flying Altitude: 3,048 m (10,000 ft)
Max. Operating Range: 100 km
GPS Compatible: √
Altitude Hold: √
Position Hold: –
Flight Planning: √
Autonomous Flight: √
Target Tracking: –
Automatic Landing: √
Return To Home: √
Obstacle Avoidance: √
Headless Mode: –
Built-in Camera: √

Innocon Ltd.

Mini Falcon

©jewishbusinessnews.com

Release Date: October 2004
Price: Unknown
Number of Rotors: 1 (Fixed Wing)
Control System: Ground Control Station
Applications: Military Purposes, Reconnaissance, Surveillance
Dimensions: 4 × 5 × 0.9 m
Wingspan: 5 m
Weight: 90 kg
Frame Material: Composite
Available Colors: Grey
Max. Speed: 210 km/h (130,5 mp/h)
Max. Flying Altitude: 6,096 m (20,000 ft)
Max. Operating Range: 200 km
GPS Compatible: √
Altitude Hold: √
Position Hold: –
Flight Planning: √
Autonomous Flight: √
Target Tracking: √
Automatic Landing: √
Return To Home: √
Obstacle Avoidance: √
Headless Mode: –
Built-in Camera: √

2005

Boeing Insitu

ScanEagle

© US Marine Corps/Shannon Arledge

Release Date: April 2005
Price: $800,000
Number of Rotors: 1 (Fixed Wing)
Control System: Ground Control Station
Applications: Military Purposes, Maritime Patrol, Reconnaissance, Surveillance
Dimensions: 1.6 × 3.11 m
Wingspan: 3.11 m
Weight: 16 kg
Max. Payload: 3.4 kg
Frame Material: Composite
Motors: Heavy fuel (JP-5/JP-8)
Available Colors: Grey
Max. Flight Time: 24 hours
Max. Speed: 148 km/h (92 mp/h)
Max. Flying Altitude: 5,791 m (19,000 ft)
Max. Operating Range: 550 km
GPS Compatible: √
Altitude Hold: √
Position Hold: –
Flight Planning: √
Autonomous Flight: √
Target Tracking: √
Automatic Landing: √
Return To Home: √
Obstacle Avoidance: √
Headless Mode: –
Built-in Camera: √

Aeronautics Ltd.

Orbiter 2

© aviationsmilitaires.net/Ansierra117

Release Date: March 2005
Price: Unknown
Number of Rotors: 1 (Fixed Wing)
Control System: Remote Control, Laptop
Applications: Reconnaissance, Surveillance
Wingspan: 3 m
Weight: 10.3 kg
Max. Payload: 1.5 kg
Available Colors: Grey
Max. Flight Time: 4 hours
Max. Speed: 112 km/h (70 mp/h)
Max. Flying Altitude: 5,486 m (18,000 ft)
Max. Operating Range: 80 km
GPS Compatible: √
Altitude Hold: √
Position Hold: –
Flight Planning: √
Autonomous Flight: √
Target Tracking: √
Automatic Landing: –
Return To Home: √
Obstacle Avoidance: √
Headless Mode: –
Built-in Camera: √

Leonardo S.p.A.

Sky-X

©Leonardo S.p.A.

Release Date: May 2005
Price: Unknown
Number of Rotors: – (Fixed Wing)
Control System: Ground Control Station
Applications: Demonstration, Research & Development, Reconnaissance, Surveillance
Dimensions: 7.8 × 5.95 × 1.86 m
Wingspan: 5.95 m
Weight: 1,000 kg
Max. Payload: 450 kg
Motors: Snecma Microturbo TRI60-268 turbofan
Available Colors: Grey
Max. Speed: 648 km/h (402 mp/h)
Max. Flying Altitude: 7,254 m (23,800 ft)
GPS Compatible: √
Altitude Hold: √
Position Hold: –
Flight Planning: √
Autonomous Flight: √
Target Tracking: –
Automatic Landing: √
Return To Home: √
Obstacle Avoidance: √
Headless Mode: –
Built-in Camera: √

Hellenic Aerospace Industry

E1-79 Pegasus II

©Konstantinos Panitsidis

Release Date: August 2005
Price: Unknown
Number of Rotors: 1 (Fixed Wing)
Control System: Ground Control Station
Applications: Military Purposes, Reconnaissance, Surveillance
Dimensions: 4.3 × 6.2 m
Wingspan: 6.2 m
Weight: 250 kg
Max. Payload: 50 kg
Frame Material: Composite
Available Colors: Grey
Max. Flight Time: 15 hours
GPS Compatible: √
Altitude Hold: √
Position Hold: –
Flight Planning: √
Autonomous Flight: √
Target Tracking: √
Automatic Landing: √
Return To Home: √
Obstacle Avoidance: √
Headless Mode: –
Built-in Camera: √

Israel Aerospace Industries (IAI)

Bird-Eye 400

© IAI

Release Date: September 2005
Price: $125,000
Number of Rotors: 1 (Fixed Wing)
Control System: Ground Control Station, Remote Control
Applications: Military Purposes, Reconnaissance, Surveillance
Dimensions: 0.8 × 2.2 m
Wingspan: 2.2 m
Weight: 5.6 kg
Available Colors: Red
Max. Flight Time: 2 hours
Max. Speed: 83 km/h (51,5 mp/h)
Max. Operating Range: 120 km
Battery Type: Lithium Polymer
GPS Compatible: √
Altitude Hold: √
Position Hold: –
Flight Planning: √
Autonomous Flight: √
Target Tracking: –
Automatic Landing: –
Return To Home: √
Obstacle Avoidance: √
Headless Mode: –
Built-in Camera: √

Israel Aerospace Industries (IAI)

Heron 1/Shovel

©US Air Force/Reynaldo Ramon

Release Date: September 2005
Price: $10 million
Number of Rotors: 1 (Fixed Wing)
Control System: Ground Control Station
Applications: Military Purposes, Reconnaissance, Surveillance
Dimensions: 8.5 × 16.6 m
Wingspan: 16.6 m
Weight: 1,150 kg
Frame Material: Composite
Motors: 1 × Rotax 914, 75kW
Available Colors: Grey
Max. Flight Time: 52 hours
Max. Speed: 209 km/h (130 mp/h)
Max. Flying Altitude: 10,000 m (32,808 ft)
Max. Operating Range: 350 km
GPS Compatible: √
Altitude Hold: √
Position Hold: –
Flight Planning: √
Autonomous Flight: √
Target Tracking: √
Automatic Landing: √
Return To Home: √
Obstacle Avoidance: √
Headless Mode: –
Built-in Camera: √

Aerodreams

UAV Strix

Release Date: September 2005
Price: Unknown
Number of Rotors: 1 (Fixed Wing)
Control System: Ground Control Station
Applications: Military Purposes, Inspection, Reconnaissance, Surveillance
Dimensions: 2.9 × 3.6 × 0.95 m
Wingspan: 3.6 m
Weight: 48 kg
Motors: 10 hp piston engine
Available Colors: White
Max. Flight Time: 15 hours
Max. Speed: 160 km/h (99,5 mp/h)
Max. Flying Altitude: 4,500 m (14,764 ft)
Max. Operating Range: 750 km
GPS Compatible: √
Altitude Hold: √
Position Hold: –
Flight Planning: √
Autonomous Flight: √
Target Tracking: –
Automatic Landing: √
Return To Home: √
Obstacle Avoidance: √
Headless Mode: –
Built-in Camera: √

Airbus

KZO

Release Date: November 2005
Price: $5 million
Number of Rotors: 1 (Fixed Wing)
Control System: Ground Control Station
Applications: Military Purposes, Reconnaissance, Surveillance, Target Acquisition
Dimensions: 2.28 × 3.42 × 0.95 m
Wingspan: 3.42 m
Weight: 161 kg
Frame Material: Composite
Motors: 1 × FZ Sachs 24 piston engine
Available Colors: Grey, White
Max. Flight Time: 4 hours
Max. Speed: 250 km/h (155 mp/h)
Max. Flying Altitude: 4,000 m (13,123 ft)
Max. Operating Range: 500 km
GPS Compatible: √
Altitude Hold: √
Position Hold: –
Flight Planning: √
Autonomous Flight: √
Target Tracking: √
Automatic Landing: –
Return To Home: √
Obstacle Avoidance: √
Headless Mode: –
Built-in Camera: √

Integrated Dynamics

Desert Hawk

Release Date: November 2005
Price: Unknown
Number of Rotors: 1 (Fixed Wing)
Control System: Ground Control Station
Applications: Military Purposes, Aerial Video, Surveillance
Dimensions: 1.8 × 2.43 m
Wingspan: 2.43 m
Weight: 6.8 kg
Frame Material: Carbon Fiber, Composite
Available Colors: Brown
Max. Flight Time: 3 hours
Max. Speed: 165 km/h (102,5 mp/h)
Max. Flying Altitude: 305 m (1,000 ft)
Max. Operating Range: 8 km
GPS Compatible: √
Altitude Hold: √
Position Hold: –
Flight Planning: √
Autonomous Flight: √
Target Tracking: √
Automatic Landing: √
Return To Home: √
Obstacle Avoidance: √
Headless Mode: –
Built-in Camera: √

2006

Boeing Insitu

Dominator

© Boeing Insitu

Release Date: April 2006
Price: Unknown
Number of Rotors: 1 (Fixed Wing)
Control System: Ground Control Station
Applications: Military Purposes, Target Acquisition
Wingspan: 3.7 m
Weight: 27 kg
Available Colors: Yellow
GPS Compatible: √
Altitude Hold: √
Position Hold: –
Flight Planning: √
Autonomous Flight: √
Target Tracking: √
Automatic Landing: √
Return To Home: √
Obstacle Avoidance: –
Headless Mode: –
Built-in Camera: √

Iran Aircraft Manufacturing Industrial Company

HESA Ababil 3

© IRNA

Release Date: July 2006
Price: $4 million
Number of Rotors: 1 (Fixed Wing)
Control System: Ground Control Station
Applications: Military Purposes, Reconnaissance, Surveillance
Dimensions: 2.88 × 3.25 × 0.91 m
Wingspan: 3.25 m
Weight: 30 kg
Max. Payload: 40 kg
Frame Material: Composite
Rotor Material: Wood
Motors: 1 × L-550E four-cylinder piston engine
Available Colors: Blue, Grey
Max. Flight Time: 10 hours
Max. Speed: 370 km/h (230 mp/h)
Max. Flying Altitude: 3,000 m (9,842 ft)
Max. Operating Range: 120 km
GPS Compatible: √
Altitude Hold: √
Position Hold: –
Flight Planning: √
Autonomous Flight: √
Target Tracking: √
Automatic Landing: √
Return To Home: √
Obstacle Avoidance: √
Headless Mode: –
Built-in Camera: √

Lockheed Martin

Desert Hawk III

©Lockheed Martin

Release Date: July 2006
Price: $100,000
Number of Rotors: 1 (Fixed Wing)
Control System: Ground Control Station
Applications: Military Purposes, Reconnaissance, Surveillance
Wingspan: 1.5 m
Weight: 3.7 kg
Max. Payload: 907 gr
Frame Material: Composite
Available Colors: Grey
Max. Flight Time: 90 minutes
Max. Speed: 92 km/h (57 mp/h)
Max. Flying Altitude: 3,400 m (11,155 ft)
Max. Operating Range: 15 km
Controller Frequency: 2.4 GHz
Battery Type: Lithium Polymer
GPS Compatible: √
Altitude Hold: √
Position Hold: –
Flight Planning: √
Autonomous Flight: √
Target Tracking: –
Automatic Landing: –
Return To Home: √
Obstacle Avoidance: –
Headless Mode: –
Built-in Camera: √

Nostromo Defensa

Yarara 01-B

©infotechnology.com

Release Date: August 2006
Price: Unknown
Number of Rotors: 1 (Fixed Wing)
Control System: Ground Control Station
Applications: Military Purposes, Reconnaissance, Surveillance
Dimensions: 2.46 × 3.98 × 0.77 m
Wingspan: 3.98 m
Weight: 29.5 kg
Max. Payload: 5.5 kg
Frame Material: Composite
Rotor Material: Wood
Motors: 1 × Wankel electric injection
Available Colors: Grey
Max. Flight Time: 6 hours
Max. Speed: 163 km/h (101 mp/h)
Max. Flying Altitude: 3,000 m (9,842 ft)
Max. Ascent Speed: 6 m/s
Max. Operating Range: 30 km
GPS Compatible: √
Altitude Hold: √
Position Hold: –
Flight Planning: √
Autonomous Flight: √
Target Tracking: –
Automatic Landing: √
Return To Home: √
Obstacle Avoidance: –
Headless Mode: –
Built-in Camera: √

Elbit Systems

Skylark II

©Elbit Systems

Release Date: October 2006
Price: Unknown
Number of Rotors: 1 (Fixed Wing)
Control System: Ground Control Station
Applications: Military Purposes, Reconnaissance, Surveillance
Wingspan: 4.2 m
Weight: 35 kg
Motors: 4 kilowatt electrical motor
Available Colors: Grey
Max. Flight Time: 6 hours
Max. Operating Range: 50 km
GPS Compatible: √
Altitude Hold: √
Position Hold: –
Flight Planning: √
Autonomous Flight: √
Target Tracking: –
Automatic Landing: √
Return To Home: √
Obstacle Avoidance: √
Headless Mode: –
Built-in Camera: √

Israel Aerospace Industries (IAI)

I-View MK 50

©bredow-web.de

Release Date: October 2006
Price: Unknown
Number of Rotors: 1 (Fixed Wing)
Control System: Ground Control Station
Applications: Military Purposes, Reconnaissance, Surveillance, Target Acquisition
Dimensions: 2.7 × 4 m
Wingspan: 4 m
Weight: 65 kg
Max. Payload: 10 kg
Frame Material: Composite
Available Colors: Green
Max. Flight Time: 6 hours
Max. Speed: 80 km/h (49,7 mp/h
Max. Flying Altitude: 4,572 m (15,000 ft)
Max. Operating Range: 50 km
GPS Compatible: √
Altitude Hold: √
Position Hold: –
Flight Planning: √
Autonomous Flight: √
Target Tracking: √
Automatic Landing: √
Return To Home: √
Obstacle Avoidance: √
Headless Mode: –
Built-in Camera: √

Kale-Baykar

Bayraktar Mini

© Kale-Baykar

Release Date: October 2006
Price: Unknown
Number of Rotors: 1 (Fixed Wing)
Control System: Ground Control Station
Applications: Military Purposes, Reconnaissance, Surveillance
Dimensions: 1.2 × 2 m
Wingspan: 2 m
Weight: 4.5 kg
Frame Material: Carbon Fiber, Composite
Available Colors: Grey
Max. Flight Time: 60 minutes
Max. Speed: 60 km/h (37 mp/h)
Max. Flying Altitude: 1,000 m (3,280 ft)
Min. Operating Temperature: -20°C
Max. Operating Temperature: 55°C
Max. Operating Range: 55 km
GPS Compatible: √
Altitude Hold: √
Position Hold: –
Flight Planning: √
Autonomous Flight: √
Target Tracking: √
Automatic Landing: –
Return To Home: √
Obstacle Avoidance: √
Headless Mode: –
Built-in Camera: √

Harbin Aviation Industrial Group (HAIG)

BZK-005 Giant Eagle

© sinodefenceforum.com

Release Date: November 2006
Price: $7.5 million
Number of Rotors: 1 (Fixed Wing)
Control System: Ground Control Station
Applications: Military Purposes, Reconnaissance, Surveillance
Weight: 1,250 kg
Max. Payload: 150 kg
Frame Material: Composite
Available Colors: Blue, Grey
Max. Flight Time: 40 hours
Max. Speed: 180 km/h (112 mp/h)
Max. Flying Altitude: 8,000 m (26,246 ft)
GPS Compatible: √
Altitude Hold: √
Position Hold: –
Flight Planning: √
Autonomous Flight: √
Target Tracking: √
Automatic Landing: √
Return To Home: √
Obstacle Avoidance: √
Headless Mode: –
Built-in Camera: √

Rafael

Skylite B

Release Date: November 2006
Price: $150,000
Number of Rotors: 1 (Fixed Wing)
Control System: Flight Computer, Remote Control
Applications: Military Purposes, Reconnaissance, Surveillance
Dimensions: 1.15 × 2.4 × 0.12 m
Wingspan: 2.4 m
Weight: 6 kg
Available Colors: Grey
Max. Flight Time: 90 minutes
Max. Speed: 100 km/h (62 mp/h)
Max. Flying Altitude: 600 m (1,968 ft)
Max. Operating Range: 10 km
GPS Compatible: √
Altitude Hold: √
Position Hold: –
Flight Planning: √
Autonomous Flight: √
Target Tracking: √
Automatic Landing: –
Return To Home: √
Obstacle Avoidance: √
Headless Mode: –
Built-in Camera: √

2007

Leonardo S.p.A.

Falco

Release Date: January 2007
Price: $10 million
Number of Rotors: 1 (Fixed Wing)
Control System: Ground Control Station
Applications: Military Purposes, Armed Reconnaissance, Surveillance, Target Acquisition
Dimensions: 5.25 × 7.2 × 1.8 m
Wingspan: 7.2 m
Weight: 420 kg
Max. Payload: 70 kg
Frame Material: Composite
Available Colors: Grey
Max. Speed: 216 km/h (134 mp/h)
Max. Flying Altitude: 5,000 m (16,404 ft)
Max. Operating Range: 200 km
GPS Compatible: √
Altitude Hold: √
Position Hold: √
Flight Planning: √
Autonomous Flight: √
Target Tracking: √
Automatic Landing: √
Return To Home: √
Obstacle Avoidance: √
Headless Mode: –
Built-in Camera: √

Composite Technology Research Malaysia

Aludra MK1

Release Date: March 2007
Price: Unknown
Number of Rotors: 1 (Fixed Wing)
Control System: Ground Control Station
Applications: Military Purposes, Research & Development, Surveillance
Dimensions: 4.57 × 6 m
Wingspan: 6 m
Weight: 200 kg
Max. Payload: 25 kg
Frame Material: Composite
Motors: 1 × two-cylinder gas engine
Available Colors: Grey
Max. Flight Time: 3 hours
Max. Speed: 220 km/h (137 mp/h)
Max. Flying Altitude: 3,657 m (12,000 ft)
Max. Operating Range: 100 km
GPS Compatible: √
Altitude Hold: √
Position Hold: –
Flight Planning: √
Autonomous Flight: √
Target Tracking: √
Automatic Landing: √
Return To Home: √
Obstacle Avoidance: √
Headless Mode: –
Built-in Camera: √

Composite Technology Research Malaysia

Aludra SR-08

©CTRM

Release Date: March 2007
Price: Unknown
Number of Rotors: 1 (Fixed Wing)
Control System: Ground Control Station
Applications: Military Purposes, Reconnaissance, Surveillance
Dimensions: 425 × 810 m
Wingspan: 810 mm
Weight: 2.1 kg
Available Colors: Grey
Max. Flight Time: 100 minutes
Max. Speed: 130 km/h (81 mp/h)
Max. Flying Altitude: 400 m (1,312 ft)
Max. Operating Range: 15 km
Controller Frequency: 2.4 GHz
Battery Type: Lithium Polymer
GPS Compatible: √
Altitude Hold: √
Position Hold: –
Flight Planning: √
Autonomous Flight: √
Target Tracking: –
Automatic Landing: –
Return To Home: √
Obstacle Avoidance: –
Headless Mode: –
Built-in Camera: √

Northrop Grumman

RQ-4B Global Hawk

©US Air Force/Christopher Okula

Release Date: March 2007
Price: $90 million
Number of Rotors: – (Fixed Wing)
Control System: Ground Control Station
Applications: Military Purposes, Reconnaissance, Surveillance
Dimensions: 14.5 × 39.9 × 4.7 m
Wingspan: 39.9 m
Weight: 14,628 kg
Max. Payload: 1,360 kg
Frame Material: Composite
Motors: 1 × Allison Rolls-Royce AE3007H turbofan engine
Available Colors: Grey
Max. Flight Time: 35 hours
Max. Speed: 575 km/h (357 mp/h)
Max. Flying Altitude: 19,800 m (64,960 ft)
Max. Operating Range: 22,780 km
GPS Compatible: √
Altitude Hold: √
Position Hold: –
Flight Planning: √
Autonomous Flight: √
Target Tracking: √
Automatic Landing: √
Return To Home: √
Obstacle Avoidance: √
Headless Mode: –
Built-in Camera: √

Turkish Aerospace Industries (TAI)

Gözcü

©Wikipedia/KIZILSUNGUR

Release Date: March 2007
Price: $3 million
Number of Rotors: 1 (Fixed Wing)
Control System: Ground Control Station
Applications: Military Purposes, Reconnaissance, Surveillance, Target Acquisition
Dimensions: 2.45 × 3.75 × 0.66 m
Wingspan: 3.75 m
Weight: 85 kg
Max. Payload: 8 kg
Frame Material: Composite
Motors: 1 × Rotary engine
Available Colors: Grey
Max. Flight Time: 2 hours
Max. Speed: 170 km/h (106 mp/h)
Max. Flying Altitude: 3,700 m (12,140 ft)
Max. Operating Range: 50 km
GPS Compatible: √
Altitude Hold: √
Position Hold: –
Flight Planning: √
Autonomous Flight: √
Target Tracking: –
Automatic Landing: –
Return To Home: √
Obstacle Avoidance: –
Headless Mode: –
Built-in Camera: √

Israel Aerospace Industries (IAI)

Heron TP/Eitan

©defpost.com

Release Date: April 2007
Price: $35 million
Number of Rotors: 1 (Fixed Wing)
Control System: Ground Control Station
Applications: Military Purposes, Armed Reconnaissance, Surveillance, Target Acquisition
Dimensions: 14 × 26 m
Wingspan: 26 m
Weight: 46,50 kg
Max. Payload: 2,700 kg
Frame Material: Composite
Motors: 1 × Pratt & Whitney Canada PT6-67A
Available Colors: Grey
Max. Flight Time: 36 hours
Max. Speed: 407 km/h (253 mp/h)
Max. Flying Altitude: 13,716 m (45,000 ft)
Max. Operating Range: 7,400 km
GPS Compatible: √
Altitude Hold: √
Position Hold: –
Flight Planning: √
Autonomous Flight: √
Target Tracking: √
Automatic Landing: √
Return To Home: √
Obstacle Avoidance: √
Headless Mode: –
Built-in Camera: √

General Atomics

MQ-9 Reaper (Predator B)

©US Air Force/Lt. Col. Leslie Pratt

Release Date: May 2007
Price: $27.5 million
Number of Rotors: 1 (Fixed Wing)
Control System: Ground Control Station
Applications: Military Purposes, Armed Reconnaissance, Surveillance, Target Acquisition
Dimensions: 11 × 20 × 3.81 m
Wingspan: 20 m
Weight: 2,223 kg
Max. Payload: 1,700 kg
Frame Material: Composite
Motors: 1 × Honeywell TPE331-10 turboprop
Available Colors: Grey
Max. Flight Time: 14 hours
Max. Speed: 482 km/h (300 mp/h)
Max. Flying Altitude: 15,240 m (50,000 ft)
Max. Operating Range: 1,852 km
GPS Compatible: √
Altitude Hold: √
Position Hold: –
Flight Planning: √
Autonomous Flight: √
Target Tracking: √
Automatic Landing: √
Return To Home: √
Obstacle Avoidance: √
Headless Mode: √
Built-in Camera: √

Integrated Dynamics

Vision MK-II

©Integrated Dynamics

Release Date: May 2007
Price: Unknown
Number of Rotors: 1 (Fixed Wing)
Control System: Ground Control Station
Applications: Military Purposes, Reconnaissance, Surveillance
Dimensions: 2.9 × 4.51 × 0.8 m
Wingspan: 4.51 m
Weight: 55 kg
Max. Payload: 15 kg
Frame Material: Composite
Motors: 22 BHP two-cylinder piston engine
Available Colors: Grey
Max. Flight Time: 4 hours
Max. Operating Range: 200 km
GPS Compatible: √
Altitude Hold: √
Position Hold: –
Flight Planning: √
Autonomous Flight: √
Target Tracking: –
Automatic Landing: √
Return To Home: √
Obstacle Avoidance: √
Headless Mode: –
Built-in Camera: √

Lehmann Aviation

LP960

©Lehmann Aviation

Release Date: June 2007
Price: $40,000
Number of Rotors: 1 (Fixed Wing)
Control System: Ground Control Station
Applications: Aerial Mapping, Agriculture, Inspection, Reconnaissance
Wingspan: 920 mm
Weight: 1.25 kg
Max. Payload: 350 gr
Frame Material: EPP Foam
Available Colors: White
Max. Flight Time: 25 minutes
Max. Speed: 45 km/h (28 mp/h)
Min. Operating Temperature: -25°C
Max. Operating Temperature: 60°C
Max. Operating Range: 5 km
GPS Compatible: √
Altitude Hold: √
Position Hold: –
Flight Planning: √
Autonomous Flight: √
Target Tracking: –
Automatic Landing: –
Return To Home: √
Obstacle Avoidance: –
Headless Mode: –
Built-in Camera: –

Leonardo S.p.A.

Sky-Y

©Alenia Aeronautica

Release Date: June 2007
Price: Unknown
Number of Rotors: 1 (Fixed Wing)
Control System: Ground Control Station
Applications: Demonstration, Research & Development, Reconnaissance, Surveillance
Dimensions: 9.72 × 9.94 × 1.86 m
Wingspan: 9.94 m
Weight: 1,050 kg
Max. Payload: 150 kg
Frame Material: Composite
Motors: 1 × DieselJet Fiat 1.9 JTD common rail FADEC Turbodiesel four-cylinder engine
Available Colors: White
Max. Flight Time: 8 hours
Max. Speed: 260 km/h (161,5 mp/h)
Max. Flying Altitude: 7,620 m (25,000 ft)
GPS Compatible: √
Altitude Hold: √
Position Hold: √
Flight Planning: √
Autonomous Flight: √
Target Tracking: √
Automatic Landing: √
Return To Home: √
Obstacle Avoidance: √
Headless Mode: –
Built-in Camera: √

Boeing Insitu

X48-B

© NASA/Tony Landis

Release Date: July 2007
Price: Unknown
Number of Rotors: – (Fixed Wing)
Control System: Ground Control Station
Applications: Military Purposes, Armed Reconnaissance, Surveillance, Target Acquisition
Dimensions: 4.6 × 6.4 × 0.9 m
Wingspan: 6.4 m
Weight: 178 kg
Max. Payload: 49 kg
Frame Material: Composite
Motors: 3 × JetCat P200 turbojet
Available Colors: White
Max. Flight Time: 40 minutes
Max. Speed: 219 km/h (136 mp/h)
Max. Flying Altitude: 3,000 m (9,842 ft)
Max. Operating Range: 145 km
GPS Compatible: √
Altitude Hold: √
Position Hold: –
Flight Planning: √
Autonomous Flight: √
Target Tracking: √
Automatic Landing: √
Return To Home: √
Obstacle Avoidance: √
Headless Mode: –
Built-in Camera: √

Boeing Insitu

Integrator

© Boeing Insitu

Release Date: August 2007
Price: Unknown
Number of Rotors: 1 (Fixed Wing)
Control System: Ground Control Station
Applications: Military Purposes, Maritime Patrol, Reconnaissance, Search & Rescue, Surveillance
Dimensions: 2.5 × 4.8 m
Wingspan: 4.8 m
Weight: 34 kg
Max. Payload: 18 kg
Frame Material: Composite
Available Colors: Grey
Max. Flight Time: 24 hours
Max. Flying Altitude: 5,950 m (19,521 ft)
GPS Compatible: √
Altitude Hold: √
Position Hold: –
Flight Planning: √
Autonomous Flight: √
Target Tracking: √
Automatic Landing: √
Return To Home: √
Obstacle Avoidance: √
Headless Mode: –
Built-in Camera: √

Avartek

AT-04

©Wikipedia/Markus Säynevirta

Release Date: September 2007
Price: $3,400
Number of Rotors: 1 (Fixed Wing)
Control System: Ground Control Station
Applications: Military Purposes, Reconnaissance, Surveillance, Aerial Target
Dimensions: 1.67 × 2 m
Wingspan: 2 m
Weight: 8.9 kg
Frame Material: Composite
Rotor Material: Wood
Motors: 1 × Zenoah G620PU two-stroke engine
Available Colors: Orange
Max. Flight Time: 2 hours
Max. Speed: 245 km/h (152 mp/h)
Max. Flying Altitude: 2,000 m (6,560 ft)
Max. Operating Range: 50 km
GPS Compatible: √
Altitude Hold: √
Position Hold: –
Flight Planning: √
Autonomous Flight: √
Target Tracking: –
Automatic Landing: –
Return To Home: √
Obstacle Avoidance: √
Headless Mode: –
Built-in Camera: √

Israel Aerospace Industries (IAI)

I-View MK 150

©israeli-weapons.com

Release Date: September 2007
Price: $4.5 million
Number of Rotors: 1 (Fixed Wing)
Control System: Ground Control Station
Applications: Military Purposes, Reconnaissance, Surveillance, Target Acquisition
Dimensions: 3.1 × 5.7 m
Wingspan: 5.7 m
Weight: 160 kg
Max. Payload: 20 kg
Frame Material: Composite
Available Colors: Brown
Max. Flight Time: 7 hours
Max. Speed: 80 km/h (49,7 mp/h
Max. Flying Altitude: 5,180 m (17,000 ft)
Max. Operating Range: 100 km
GPS Compatible: √
Altitude Hold: √
Position Hold: –
Flight Planning: √
Autonomous Flight: √
Target Tracking: √
Automatic Landing: √
Return To Home: √
Obstacle Avoidance: √
Headless Mode: –
Built-in Camera: √

Israel Aerospace Industries (IAI)

I-View MK 250

©Wikipedia/Jastrow

Release Date: September 2007
Price: Unknown
Number of Rotors: 1 (Fixed Wing)
Control System: Ground Control Station
Applications: Military Purposes, Reconnaissance, Surveillance, Target Acquisition
Dimensions: 4.1 × 7.1 m
Wingspan: 7.1 m
Weight: 250 kg
Max. Payload: 60 kg
Frame Material: Composite
Available Colors: Grey
Max. Flight Time: 8 hours
Max. Speed: 80 km/h (49,7 mp/h
Max. Flying Altitude: 6,706 m (22,000 ft)
Max. Operating Range: 150 km
GPS Compatible: √
Altitude Hold: √
Position Hold: –
Flight Planning: √
Autonomous Flight: √
Target Tracking: √
Automatic Landing: √
Return To Home: √
Obstacle Avoidance: √
Headless Mode: –
Built-in Camera: √

AeroVironment

Wasp III

©AeroVironment

Release Date: October 2007
Price: $49,000
Number of Rotors: 1 (Fixed Wing)
Control System: Ground Control Station
Applications: Military Purposes, Maritime Patrol, Reconnaissance, Surveillance
Dimensions: 380 × 723 mm
Wingspan: 723 mm
Weight: 430 gr
Available Colors: White
Max. Flight Time: 45 minutes
Max. Speed: 65 km/h (40 mp/h)
Max. Flying Altitude: 305 m (1,000 ft)
Max. Operating Range: 5 km
Battery Type: Lithium Ion
GPS Compatible: √
Altitude Hold: √
Position Hold: –
Flight Planning: √
Autonomous Flight: √
Target Tracking: √
Automatic Landing: √
Return To Home: √
Obstacle Avoidance: √
Headless Mode: –
Built-in Camera: √

Applied Research Associates

Nighthawk IV

©Applied Research Associates

Release Date: October 2007
Price: Unknown
Number of Rotors: 1 (Fixed Wing)
Control System: Flight Computer, Remote Control
Applications: Military Purposes, Reconnaissance, Surveillance
Wingspan: 660 mm
Weight: 860 gr
Frame Material: Carbon Fiber
Available Colors: Black
Max. Flight Time: 60 minutes
Max. Speed: 64 km/h (40 mp/h)
Max. Flying Altitude: 3,350 m (10,990 ft)
Max. Operating Range: 10 km
Battery Type: Lithium Polymer
GPS Compatible: √
Altitude Hold: √
Position Hold: √
Flight Planning: √
Autonomous Flight: √
Target Tracking: –
Automatic Landing: √
Return To Home: √
Obstacle Avoidance: –
Headless Mode: –
Built-in Camera: √

Lockheed Martin

RQ-170 Sentinel

©cdn.presstv.com

Release Date: October 2007
Price: $6 million
Number of Rotors: – (Fixed Wing)
Control System: Ground Control Station
Applications: Military Purposes, Reconnaissance, Research & Development, Surveillance
Dimensions: 4.5 × 20 × 1.8 m
Wingspan: 20 m
Weight: 3,900 kg
Frame Material: Composite
Motors: 1 × Garrett TFE731 engine
Available Colors: Grey
Max. Flying Altitude: 15,240 m (50,000 ft)
GPS Compatible: √
Altitude Hold: √
Position Hold: –
Flight Planning: √
Autonomous Flight: √
Target Tracking: √
Automatic Landing: √
Return To Home: √
Obstacle Avoidance: √
Headless Mode: –
Built-in Camera: √

National Chung-Shan Institute of Science and Technology (NCSIST)

Chung Shyang II

© ncsist.org.tw

Release Date: October 2007
Price: $5 million
Number of Rotors: 1 (Fixed Wing)
Control System: Ground Control Station
Applications: Military Purposes, Armed Reconnaissance, Surveillance
Available Colors: Green
GPS Compatible: √
Altitude Hold: √
Position Hold: –
Flight Planning: √
Autonomous Flight: √
Target Tracking: –
Automatic Landing: √
Return To Home: √
Obstacle Avoidance: √
Headless Mode: –
Built-in Camera: √

BAE Systems

HERTI

© BAE Systems

Release Date: November 2007
Price: $7.5 million
Number of Rotors: 1 (Fixed Wing)
Control System: Ground Control Station
Applications: Military Purposes, Armed Reconnaissance, Surveillance
Dimensions: 5 × 12 m
Wingspan: 12 m
Weight: 450 kg
Max. Payload: 150 kg
Frame Material: Composite
Motors: 1 × Rotax 914, 75kW
Available Colors: Grey
Max. Flight Time: 20 hours
Max. Speed: 232 km/h (145 mp/h)
Max. Flying Altitude: 6,096 m (20,000 ft)
GPS Compatible: √
Altitude Hold: √
Position Hold: –
Flight Planning: √
Autonomous Flight: √
Target Tracking: √
Automatic Landing: √
Return To Home: √
Obstacle Avoidance: √
Headless Mode: –
Built-in Camera: √

2008

Honeywell Aerospace

RQ-16 T-Hawk

© navy.mil/Kenneth G. Takada

Release Date: January 2008
Price: $450,000
Number of Rotors: –
Control System: Remote Control
Applications: Military Purposes, Reconnaissance, Surveillance
Weight: 8.4 kg
Frame Material: Carbon Fiber
Motors: 3W-56 56cc Boxer two-piston engine
Available Colors: Grey
Max. Flight Time: 40 minutes
Max. Speed: 130 km/h (81 mp/h)
Max. Flying Altitude: 3,000 m (9,842 ft)
Max. Ascent Speed: 3 m/s
Max. Descent Speed: 1.5 m/s
Min. Operating Temperature: -6°C
Max. Operating Temperature: 49°C
Max. Operating Range: 10 km
GPS Compatible: –
Altitude Hold: √
Position Hold: √
Flight Planning: √
Autonomous Flight: √
Target Tracking: –
Automatic Landing: –
Return To Home: –
Obstacle Avoidance: –
Headless Mode: –
Built-in Camera: √

Prioria Robotics

Maveric

© Prioria Robotics

Release Date: March 2008
Price: $240,000
Number of Rotors: 1 (Fixed Wing)
Control System: Ground Control Station
Applications: Military Purposes, Maritime Patrol, Reconnaissance, Search & Rescue, Surveillance
Dimensions: 914 × 750 mm
Wingspan: 750 mm
Weight: 1.16 kg
Available Colors: Black
Max. Flight Time: 60 minutes
Max. Speed: 100 km/h (62 mp/h)
Max. Flying Altitude: 7,620 m (25,000 ft)
Max. Operating Range: 15 km
Controller Frequency: 2.4 GHz
Battery Type: Lithium Polymer
GPS Compatible: √
Altitude Hold: √
Position Hold: –
Flight Planning: √
Autonomous Flight: √
Target Tracking: √
Automatic Landing: √
Return To Home: √
Obstacle Avoidance: –
Headless Mode: –
Built-in Camera: √

Enics

DSAS E25

© Enics

Release Date: April 2008
Price: Unknown
Number of Rotors: 1 (Fixed Wing)
Control System: Ground Control Station
Applications: Military Purposes, Reconnaissance, Surveillance
Available Colors: Black
Max. Flight Time: 60 minutes
Max. Speed: 110 km/h (68 mp/h)
Max. Operating Range: 1.2 km
Controller Frequency: 2.4 GHz
Battery Type: Lithium Polymer
GPS Compatible: √
Altitude Hold: √
Position Hold: –
Flight Planning: √
Autonomous Flight: √
Target Tracking: –
Automatic Landing: √
Return To Home: √
Obstacle Avoidance: –
Headless Mode: –
Built-in Camera: √

Composite Engineering Inc.

BQM-167 Skeeter

© US Air Force/Lisa Norman

Release Date: April 2008
Price: $570,000
Number of Rotors: – (Fixed Wing)
Control System: Ground Control Station
Applications: Military Purposes, Aerial Target
Dimensions: 6.1 × 10.5 × 1.2 m
Wingspan: 10.5 m
Weight: 313 kg
Max. Payload: 227 kg
Frame Material: Carbon Fiber, Epoxy
Motors: 1 × MicroTurbo TR 60-5 turbojet
Available Colors: Red
Max. Flight Time: 3 hours
Max. Speed: 1,115 km/h (693 mp/h)
Max. Flying Altitude: 15,240 m (50,000 ft)
Max. Operating Range: 2,600 km
GPS Compatible: √
Altitude Hold: √
Position Hold: –
Flight Planning: √
Autonomous Flight: √
Target Tracking: –
Automatic Landing: √
Return To Home: √
Obstacle Avoidance: –
Headless Mode: –
Built-in Camera: –

Turkish Aerospace Industries (TAI)

Turna-G

© defense-arab.com

Release Date: April 2008
Price: Unknown
Number of Rotors: 1 (Fixed Wing)
Control System: Remote Control
Applications: Military Purposes, Research & Development, Target Acquisition
Dimensions: 2.6 × 2.7 × 0.57 m
Wingspan: 2.7 m
Weight: 55 kg
Max. Payload: 15 kg
Frame Material: Composite
Motors: 1 × AR741-1100 engine
Available Colors: Red
Max. Flight Time: 90 minutes
Max. Speed: 330 km/h (205 mp/h)
Max. Flying Altitude: 3,650 m (11,975 ft)
GPS Compatible: √
Altitude Hold: √
Position Hold: –
Flight Planning: √
Autonomous Flight: √
Target Tracking: –
Automatic Landing: √
Return To Home: √
Obstacle Avoidance: –
Headless Mode: –
Built-in Camera: –

Utva Aviation Industry

Vrabac

© Srdan Popović

Release Date: May 2008
Price: $75,000
Number of Rotors: 1 (Fixed Wing)
Control System: Ground Control Station
Applications: Military Purposes, Reconnaissance, Surveillance, Target Acquisition
Dimensions: 1.94 × 2.8 m
Wingspan: 2.8 m
Weight: 5.5 kg
Max. Payload: 1.5 kg
Frame Material: Composite
Rotor Material: Graphite Composite
Available Colors: Green
Max. Flight Time: 60 minutes
Max. Speed: 120 km/h (74,5 mp/h)
Max. Flying Altitude: 500 m (1,640 ft)
Controller Frequency: 2.4 GHz
Battery Capacity: 8,000 mAh
Battery Type: Lithium Polymer
Battery Power: 28V
GPS Compatible: √
Altitude Hold: √
Position Hold: –
Flight Planning: √
Autonomous Flight: √
Target Tracking: –
Automatic Landing: √
Return To Home: √
Obstacle Avoidance: –
Headless Mode: –
Built-in Camera: √

AeroVironment

RQ-20 Puma

©US Army/Sgt. Mike MacLeod

Release Date: June 2008
Price: $250,000
Number of Rotors: 1 (Fixed Wing)
Control System: Ground Control Station
Applications: Military Purposes, Reconnaissance, Surveillance
Dimensions: 1.4 × 2.8 m
Wingspan: 2.8 m
Weight: 6.3 kg
Frame Material: Composite
Available Colors: Grey
Max. Flight Time: 3 hours
Max. Speed: 83 km/h (51,5 mp/h)
Max. Flying Altitude: 152 m (500 ft)
Min. Operating Temperature: -29°C
Max. Operating Temperature: 49°C
Max. Operating Range: 20 km
GPS Compatible: √
Altitude Hold: √
Position Hold: –
Flight Planning: √
Autonomous Flight: √
Target Tracking: √
Automatic Landing: –
Return To Home: √
Obstacle Avoidance: √
Headless Mode: –
Built-in Camera: √

Airbus

Harfang

©Airbus Defence and Space

Release Date: June 2008
Price: Unknown
Number of Rotors: 1 (Fixed Wing)
Control System: Ground Control Station
Applications: Military Purposes, Reconnaissance, Research & Development, Surveillance
Dimensions: 9.3 × 16.6 × 2.3 m
Wingspan: 16.6 m
Weight: 1,250 kg
Max. Payload: 250 kg
Frame Material: Composite
Motors: 1 × Rotax 914
Available Colors: Grey
Max. Flight Time: 24 hours
Max. Speed: 204 km/h (127 mp/h)
Max. Flying Altitude: 7,620 m (25,000 ft)
GPS Compatible: √
Altitude Hold: √
Position Hold: –
Flight Planning: √
Autonomous Flight: √
Target Tracking: √
Automatic Landing: √
Return To Home: √
Obstacle Avoidance: √
Headless Mode: –
Built-in Camera: √

Hydra Technologies

S4 Ehécatl

© Hydra Technologies

Release Date: June 2008
Price: $1 million
Number of Rotors: 1 (Fixed Wing)
Control System: Ground Control Station
Applications: Military Purposes, Reconnaissance, Surveillance
Wingspan: 3.7 m
Weight: 54 kg
Max. Payload: 9 kg
Frame Material: Composite
Available Colors: Grey
Max. Flight Time: 8 hours
Max. Speed: 167 km/h (103 mp/h)
Max. Flying Altitude: 4,572 m (15,000 ft)
GPS Compatible: √
Altitude Hold: √
Position Hold: –
Flight Planning: √
Autonomous Flight: √
Target Tracking: –
Automatic Landing: √
Return To Home: √
Obstacle Avoidance: √
Headless Mode: –
Built-in Camera: √

Hydra Technologies

E1 Gavilan

© Hydra Technologies

Release Date: June 2008
Price: $75,000
Number of Rotors: 1 (Fixed Wing)
Control System: Ground Control Station, Remote Control
Applications: Military Purposes, Reconnaissance, Surveillance
Dimensions: 1.61 × 2.34 m
Wingspan: 2.34 m
Weight: 5 kg
Frame Material: Carbon Fiber, Composite, Kevlar
Available Colors: Grey
Max. Flight Time: 90 minutes
Max. Speed: 65 km/h (40 mp/h)
Max. Flying Altitude: 2,440 m (8,000 ft)
Max. Operating Range: 10 km
Controller Frequency: 2.4 GHz
Battery Type: Lithium Polymer
GPS Compatible: √
Altitude Hold: √
Position Hold: –
Flight Planning: √
Autonomous Flight: √
Target Tracking: –
Automatic Landing: √
Return To Home: √
Obstacle Avoidance: √
Headless Mode: –
Built-in Camera: √

Griffon Aerospace

MQM-171 BroadSword

© Griffon Aerospace

Release Date: July 2008
Price: Unknown
Number of Rotors: 1 (Fixed Wing)
Control System: Ground Control Station
Applications: Military Purposes, Reconnaissance, Research & Development
Dimensions: 4.51 × 6.86 m
Wingspan: 6.86 m
Weight: 250 kg
Available Colors: Grey
Max. Flight Time: 4 hours
Max. Speed: 202 km/h (125,5 mp/h)
Max. Flying Altitude: 4,300 m (14,108 ft)
GPS Compatible: √
Altitude Hold: √
Position Hold: –
Flight Planning: √
Autonomous Flight: √
Target Tracking: –
Automatic Landing: –
Return To Home: √
Obstacle Avoidance: –
Headless Mode: –
Built-in Camera: √

SantosLab

Carcará I

© SantosLab

Release Date: July 2008
Price: $10,130
Number of Rotors: 1 (Fixed Wing)
Control System: Flight Computer
Applications: Aerial Mapping, Inspection
Dimensions: 0.5 × 1.5 × 0.2 m
Wingspan: 1.5 m
Weight: 2.3 kg
Frame Material: EPP Foam
Available Colors: Yellow
Max. Flight Time: 50 minutes
Max. Speed: 54 km/h (33,5 mp/h)
Max. Flying Altitude: 500 m (1,640 ft)
Max. Operating Range: 30 km
GPS Compatible: √
Altitude Hold: √
Position Hold: –
Flight Planning: √
Autonomous Flight: √
Target Tracking: –
Automatic Landing: √
Return To Home: √
Obstacle Avoidance: –
Headless Mode: –
Built-in Camera: √

ZALA Aero Group

Zala 421-8

©Wikipedia/Vitaly V. Kuzmin

Release Date: July 2008
Price: $80,000
Number of Rotors: 1 (Fixed Wing)
Control System: Ground Control Station
Applications: Military Purposes, Reconnaissance, Surveillance
Wingspan: 810 mm
Max. Payload: 300 gr
Available Colors: Grey
Max. Flight Time: 80 minutes
Max. Speed: 130 km/h (81 mp/h)
Max. Flying Altitude: 3,600 m (11,811 ft)
Min. Operating Temperature: -30°C
Max. Operating Temperature: 40°C
Max. Operating Range: 15 km
Controller Frequency: 2.4 GHz
Battery Type: Lithium Polymer
GPS Compatible: √
Altitude Hold: √
Position Hold: –
Flight Planning: √
Autonomous Flight: √
Target Tracking: √
Automatic Landing: –
Return To Home: √
Obstacle Avoidance: –
Headless Mode: –
Built-in Camera: √

SATUMA

Flamingo

©defence.pk

Release Date: September 2008
Price: Unknown
Number of Rotors: 1 (Fixed Wing)
Control System: Ground Control Station
Applications: Military Purposes, Reconnaissance, Surveillance
Dimensions: 5.2 × 6.6 m
Wingspan: 6.6 m
Weight: 245 kg
Max. Payload: 30 kg
Frame Material: Composite
Rotor Material: Wood
Motors: Two-stroke gasoline engine
Available Colors: Blue, Yellow
Max. Flight Time: 8 hours
Max. Speed: 130 km/h (81 mp/h)
Max. Flying Altitude: 4,572 m (15,000 ft)
Max. Operating Range: 200 km
GPS Compatible: √
Altitude Hold: √
Position Hold: –
Flight Planning: √
Autonomous Flight: √
Target Tracking: –
Automatic Landing: √
Return To Home: √
Obstacle Avoidance: √
Headless Mode: –
Built-in Camera: √

Vega Radio Engineering Corporation

Tipchak

© robotrends.ru

Release Date: September 2008
Price: Unknown
Number of Rotors: 1 (Fixed Wing)
Control System: Ground Control Station
Applications: Military Purposes, Reconnaissance, Surveillance, Target Acquisition
Weight: 50 kg
Max. Payload: 14.5 kg
Frame Material: Composite, Metal
Motors: 1 × Hirth piston engine
Available Colors: Grey
Max. Flight Time: 3 hours
Max. Speed: 55 km/h (34 mp/h)
Max. Flying Altitude: 3,000 m (9,842 ft)
Max. Operating Range: 40 km
GPS Compatible: √
Altitude Hold: √
Position Hold: –
Flight Planning: √
Autonomous Flight: √
Target Tracking: –
Automatic Landing: –
Return To Home: √
Obstacle Avoidance: –
Headless Mode: –
Built-in Camera: √

China Aerospace Science and Technology Corporation (CASC)

CH-802

© peopledaily.com.cn

Release Date: October 2008
Price: Unknown
Number of Rotors: 1 (Fixed Wing)
Control System: Ground Control Station
Applications: Military Purposes, Reconnaissance, Surveillance
Dimensions: 1.8 × 3 m
Wingspan: 3 m
Weight: 6.5 kg
Max. Payload: 1 kg
Frame Material: Composite
Available Colors: Grey
Max. Flight Time: 2.5 hours
Max. Speed: 60 km/h (37 mp/h)
Max. Flying Altitude: 4,000 m (13,123 ft)
Max. Operating Range: 30 km
GPS Compatible: √
Altitude Hold: √
Position Hold: –
Flight Planning: √
Autonomous Flight: √
Target Tracking: –
Automatic Landing: √
Return To Home: √
Obstacle Avoidance: √
Headless Mode: –
Built-in Camera: √

EMT-Penzberg

Fancopter

© EMT Penzberg

Release Date: October 2008
Price: Unknown
Number of Rotors: 2
Control System: Ground Control Station, Remote Control
Applications: Military Purposes, Reconnaissance, Surveillance
Dimensions: 730 × 440 mm
Weight: 1.5 kg
Available Colors: White
Max. Flight Time: 25 minutes
Max. Operating Range: 1 km
GPS Compatible: √
Altitude Hold: √
Position Hold: √
Flight Planning: √
Autonomous Flight: √
Target Tracking: √
Automatic Landing: √
Return To Home: √
Obstacle Avoidance: √
Headless Mode: –
Built-in Camera: √

Airbus

Barracuda

© Flickr/YearmanHJ

Release Date: November 2008
Price: Unknown
Number of Rotors: – (Fixed Wing)
Control System: Ground Control Station
Applications: Military Purposes, Reconnaissance, Surveillance
Dimensions: 8.25 × 7.22 m
Wingspan: 7.22 m
Weight: 2,300 kg
Max. Payload: 300 kg
Available Colors: Grey
Max. Speed: 1041 km/h (647 mp/h)
Max. Flying Altitude: 6,096 m (20,000 ft)
Max. Operating Range: 200 km
GPS Compatible: √
Altitude Hold: √
Position Hold: –
Flight Planning: √
Autonomous Flight: √
Target Tracking: √
Automatic Landing: √
Return To Home: √
Obstacle Avoidance: √
Headless Mode: –
Built-in Camera: √

China Aerospace Science and Industry Corporation (CASIC)

Sky Hawk 1

©forums.eagle.ru

Release Date: November 2008
Price: Unknown
Number of Rotors: 1 (Fixed Wing)
Control System: Ground Control Station
Applications: Military Purposes, Reconnaissance, Surveillance
Dimensions: 2.3 × 4.2 m
Wingspan: 4.2 m
Weight: 150 kg
Available Colors: White
Max. Flight Time: 6 hours
Max. Speed: 150 km/h (93 mp/h)
Max. Flying Altitude: 5,000 m (16,404 ft)
Max. Operating Range: 180 km
GPS Compatible: √
Altitude Hold: √
Position Hold: –
Flight Planning: √
Autonomous Flight: √
Target Tracking: –
Automatic Landing: –
Return To Home: √
Obstacle Avoidance: √
Headless Mode: –
Built-in Camera: √

2009

Integrated Dynamics

Shadow MK-I

©Integrated Dynamics

Release Date: January 2009
Price: Unknown
Number of Rotors: 1 (Fixed Wing)
Control System: Ground Control Station
Applications: Military Purposes, Reconnaissance, Surveillance
Dimensions: 2.95 × 5.2 × 0.89 m
Wingspan: 5.2 m
Weight: 55 kg
Max. Payload: 25 kg
Frame Material: Composite
Motors: 22 BHP two-cylinder piston engine
Available Colors: Grey
Max. Flight Time: 6 hours
Max. Speed: 208 km/h (130 mp/h)
Max. Operating Range: 160 km
GPS Compatible: √
Altitude Hold: √
Position Hold: –
Flight Planning: √
Autonomous Flight: √
Target Tracking: √
Automatic Landing: √
Return To Home: √
Obstacle Avoidance: √
Headless Mode: –
Built-in Camera: √

Israel Aerospace Industries (IAI)

Harop

©Wikipedia/Julian Herzog

Release Date: February 2009
Price: $3.5 million
Number of Rotors: 1 (Fixed Wing)
Control System: Ground Control Station
Applications: Loitering Munition
Dimensions: 2.5 × 3 m
Max. Payload: 23 kg
Frame Material: Composite
Available Colors: Grey
Max. Flight Time: 6 hours
Max. Operating Range: 1,000 km
GPS Compatible: √
Altitude Hold: √
Position Hold: –
Flight Planning: √
Autonomous Flight: √
Target Tracking: √
Automatic Landing: –
Return To Home: √
Obstacle Avoidance: √
Headless Mode: –
Built-in Camera: √

Chengdu Aircraft Industry Group (CAIG)

Sky Wing III

© litenews.hk

Release Date: February 2009
Price: Unknown
Number of Rotors: – (Fixed Wing)
Control System: Ground Control Station
Applications: Military Purposes, Reconnaissance, Surveillance
Dimensions: 8.9 × 18.8 × 3.44 m
Wingspan: 18.8 m
Weight: 9,100 kg
Max. Payload: 2,000 kg
Frame Material: Composite
Available Colors: Grey
Max. Flight Time: 6 hours
Max. Speed: 750 km/h (466 mp/h)
Max. Flying Altitude: 15,000 m (49,213 ft)
Max. Operating Range: 7,000 km
GPS Compatible: √
Altitude Hold: √
Position Hold: –
Flight Planning: √
Autonomous Flight: √
Target Tracking: √
Automatic Landing: √
Return To Home: √
Obstacle Avoidance: √
Headless Mode: –
Built-in Camera: √

Textron Systems

Aerosonde Mark 4.7

© si.wsj.net

Release Date: March 2009
Price: $1.5 million
Number of Rotors: 1 (Fixed Wing)
Control System: Ground Control Station
Applications: Military Purposes, Maritime Patrol, Reconnaissance, Surveillance
Dimensions: 1.7 × 3.6 × 0.6 m
Wingspan: 3.6 m
Weight: 36.4 kg
Max. Payload: 9.1 kg
Frame Material: Composite
Available Colors: Grey
Max. Flight Time: 14 hours
Max. Flying Altitude: 4,572 m (15,000 ft)
Max. Operating Range: 140 km
GPS Compatible: √
Altitude Hold: √
Position Hold: –
Flight Planning: √
Autonomous Flight: √
Target Tracking: √
Automatic Landing: –
Return To Home: √
Obstacle Avoidance: –
Headless Mode: –
Built-in Camera: √

SATUMA

Muhkbar

©Satuma

Release Date: March 2009
Price: Unknown
Number of Rotors: 1 (Fixed Wing)
Control System: Ground Control Station
Applications: Military Purposes, Reconnaissance, Surveillance
Dimensions: 2.86 × 3.56 m
Wingspan: 3.56 m
Weight: 40 kg
Max. Payload: 5 kg
Motors: 1 × 120cc, two-stroke gasoline engine
Available Colors: Black
Max. Flight Time: 90 minutes
Max. Speed: 120 km/h (74,5 mp/h)
Max. Flying Altitude: 2,133 m (7,000 ft)
Max. Operating Range: 50 km
GPS Compatible: √
Altitude Hold: √
Position Hold: –
Flight Planning: √
Autonomous Flight: √
Target Tracking: –
Automatic Landing: √
Return To Home: √
Obstacle Avoidance: –
Headless Mode: –
Built-in Camera: √

Sistemas de Control Remoto S.L.

SCRAB I

©temporal.rpaslife.es

Release Date: March 2009
Price: Unknown
Number of Rotors: – (Fixed Wing)
Control System: Ground Control Station
Applications: Military Purposes, Aerial Target
Dimensions: 1.95 × 1.62 m
Wingspan: 1.62 m
Weight: 32 kg
Max. Payload: 4 kg
Frame Material: Composite
Motors: Heavy fuel
Available Colors: Orange
Max. Flight Time: 45 minutes
Max. Speed: 360 km/h (223,6 mp/h)
Max. Operating Range: 80 km
GPS Compatible: √
Altitude Hold: √
Position Hold: –
Flight Planning: √
Autonomous Flight: –
Target Tracking: –
Automatic Landing: –
Return To Home: √
Obstacle Avoidance: –
Headless Mode: –
Built-in Camera: –

Sistemas de Control Remoto S.L.

SCRAB II

©Sistemas

Release Date: March 2009
Price: Unknown
Number of Rotors: – (Fixed Wing)
Control System: Ground Control Station
Applications: Military Purposes, Aerial Target
Dimensions: 2.94 × 2.52 m
Wingspan: 2.52 m
Weight: 90 kg
Max. Payload: 10 kg
Frame Material: Composite
Motors: Double twin turbojet
Available Colors: Orange
Max. Flight Time: 60 minutes
Max. Speed: 432 km/h (268 mp/h)
Max. Operating Range: 100 km
GPS Compatible: √
Altitude Hold: √
Position Hold: –
Flight Planning: √
Autonomous Flight: –
Target Tracking: –
Automatic Landing: –
Return To Home: √
Obstacle Avoidance: –
Headless Mode: –
Built-in Camera: –

General Atomics

Avenger (Predator C)

© General Atomics

Release Date: April 2009
Price: $30 million
Number of Rotors: – (Fixed Wing)
Control System: Ground Control Station
Applications: Military Purposes, Armed Reconnaissance, Surveillance, Target Acquisition
Dimensions: 13 × 20 m
Wingspan: 20 m
Weight: 8,255 kg
Frame Material: Composite
Motors: 1 × Pratt & Whitney Canada PW545B turbofan
Available Colors: Grey
Max. Flight Time: 18 hours
Max. Speed: 740 km/h (460 mp/h)
Max. Flying Altitude: 15,240 m (50,000 ft)
GPS Compatible: √
Altitude Hold: √
Position Hold: –
Flight Planning: √
Autonomous Flight: √
Target Tracking: √
Automatic Landing: √
Return To Home: √
Obstacle Avoidance: √
Headless Mode: –
Built-in Camera: √

Northern Research Institute (NORUT)

CryoWing

©NORUT

Release Date: May 2009
Price: Unknown
Number of Rotors: 1 (Fixed Wing)
Control System: Ground Control Station
Applications: Aerial Mapping, Inspection, Research & Development
Wingspan: 3.8 m
Max. Payload: 15 kg
Frame Material: Composite
Available Colors: White
Max. Speed: 120 km/h (74,5 mp/h)
Max. Flying Altitude: 5,000 m (16,404 ft)
Max. Operating Range: 500 km
GPS Compatible: √
Altitude Hold: √
Position Hold: –
Flight Planning: √
Autonomous Flight: √
Target Tracking: –
Automatic Landing: √
Return To Home: √
Obstacle Avoidance: √
Headless Mode: –
Built-in Camera: √

AscTec

Falcon 8

© AscTec

Release Date: June 2009
Price: On Request
Number of Rotors: 8
Control System: Ground Control Station, Remote Control
Applications: Film, Photography, Aerial Mapping, Inspection
Dimensions: 820 × 770 × 125 mm
Wingspan: 770 mm
Weight: 1.1 kg
Max. Payload: 800 gr
Frame Material: Carbon Fiber
Rotor Material: Carbon Fiber
Available Colors: White
Max. Flight Time: 22 minutes
Max. Speed: 54 km/h (33,5 mp/h)
Max. Operating Range: 1 km
Controller Frequency: 2.4 GHz
Battery Type: Lithium Polymer
GPS Compatible: √
Altitude Hold: √
Position Hold: √
Flight Planning: √
Autonomous Flight: √
Target Tracking: –
Automatic Landing: √
Return To Home: √
Obstacle Avoidance: √
Headless Mode: –
Built-in Camera: –

General Atomics

MQ-1C Gray Eagle

Release Date: June 2009
Price: $21.5 million
Number of Rotors: 1 (Fixed Wing)
Control System: Ground Control Station
Applications: Military Purposes, Armed Reconnaissance, Surveillance, Target Acquisition
Dimensions: 8.53 × 17 × 2.1 m
Wingspan: 17 m
Weight: 1,633 kg
Frame Material: Composite
Motors: 1 × Thielert Centurion 1.7 heavy-fuel engine
Available Colors: Grey
Max. Flight Time: 25 hours
Max. Speed: 309 km/h (192 mp/h)
Max. Flying Altitude: 8,840 m (29,000 ft)
GPS Compatible: √
Altitude Hold: √
Position Hold: –
Flight Planning: √
Autonomous Flight: √
Target Tracking: √
Automatic Landing: √
Return To Home: √
Obstacle Avoidance: √
Headless Mode: –
Built-in Camera: √

Lehmann Aviation

LV580

Release Date: June 2009
Price: $50,000
Number of Rotors: 1 (Fixed Wing)
Control System: Ground Control Station
Applications: Aerial Mapping, Agriculture, Inspection, Reconnaissance
Wingspan: 920 mm
Weight: 950 gr
Max. Payload: 350 gr
Frame Material: EPP Foam
Available Colors: White
Max. Flight Time: 25 minutes
Max. Speed: 45 km/h (28 mp/h)
Min. Operating Temperature: -25°C
Max. Operating Temperature: 60°C
Max. Operating Range: 5 km
GPS Compatible: √
Altitude Hold: √
Position Hold: –
Flight Planning: √
Autonomous Flight: √
Target Tracking: –
Automatic Landing: –
Return To Home: √
Obstacle Avoidance: –
Headless Mode: –
Built-in Camera: –

ZALA Aero Group

Zala 421-12

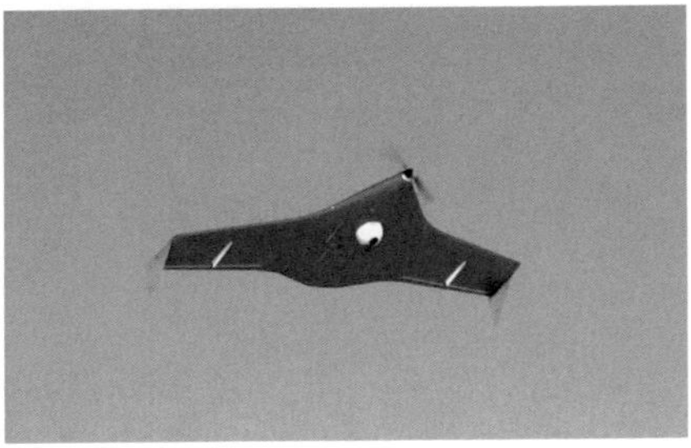

© Wikipedia/QwasERqwasER

Release Date: June 2009
Price: $340,000
Number of Rotors: 1 (Fixed Wing)
Control System: Ground Control Station
Applications: Military Purposes, Reconnaissance, Surveillance
Dimensions: 0.62 × 1.6 m
Wingspan: 1.6 m
Weight: 3.9 kg
Max. Payload: 1 kg
Available Colors: Red
Max. Flight Time: 2 hours
Max. Speed: 120 km/h (74,5 mp/h)
Max. Flying Altitude: 3,600 m (11,811 ft)
Controller Frequency: 2.4 GHz
Battery Type: Lithium Polymer
GPS Compatible: √
Altitude Hold: √
Position Hold: –
Flight Planning: √
Autonomous Flight: √
Target Tracking: √
Automatic Landing: –
Return To Home: √
Obstacle Avoidance: –
Headless Mode: –
Built-in Camera: √ (10 megapixels)

Raytheon

JLENS

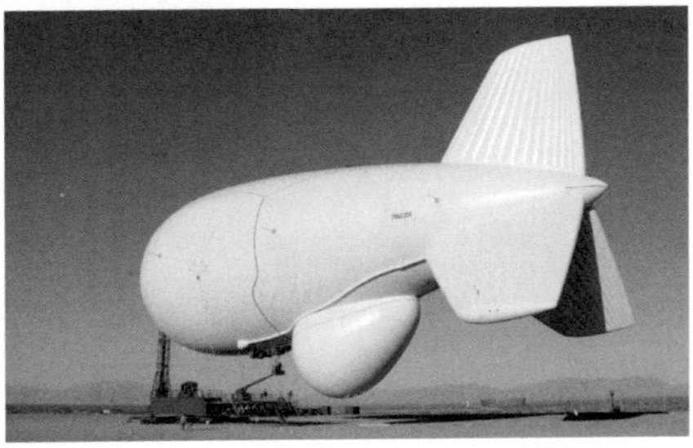

© US Air Force/Tiffany DeNault

Release Date: August 2009
Price: $175 million
Number of Rotors: –
Control System: Ground Control Station
Applications: Military Purposes, Research & Development, Surveillance
Dimensions: 70 m
Weight: 3,200 kg
Max. Payload: 1,600 kg
Available Colors: White
Max. Flight Time: 30 days
Max. Speed: – (tethered cable)
Max. Flying Altitude: 3,000 m (9,842 ft)
GPS Compatible: √
Altitude Hold: √
Position Hold: √
Flight Planning: –
Autonomous Flight: √
Target Tracking: –
Automatic Landing: –
Return To Home: –
Obstacle Avoidance: –
Headless Mode: –
Built-in Camera: –

Kronshtadt Group

Dozor-600

©Wikipedia/Pycckue

Release Date: August 2009
Price: $750,000
Number of Rotors: 1 (Fixed Wing)
Control System: Ground Control Station
Applications: Military Purposes, Reconnaissance, Surveillance
Dimensions: 7 × 12 × 2.5 m
Wingspan: 12 m
Weight: 500 kg
Max. Payload: 210 kg
Available Colors: Grey
Max. Flight Time: 24 hours
Max. Speed: 150 km/h (93 mp/h)
Max. Flying Altitude: 7,100 m (23,294 ft)
Max. Operating Range: 3,700 km
GPS Compatible: √
Altitude Hold: √
Position Hold: –
Flight Planning: √
Autonomous Flight: √
Target Tracking: √
Automatic Landing: √
Return To Home: √
Obstacle Avoidance: √
Headless Mode: –
Built-in Camera: √

Xmobots

Apoena 1000

©Wikipedia/Giovani Amianti

Release Date: August 2009
Price: Unknown
Number of Rotors: 1 (Fixed Wing)
Control System: Ground Control Station
Applications: Military Purposes, Reconnaissance, Surveillance
Dimensions: 2.66 × 2.51 × 1.13 m
Wingspan: 2.51 m
Weight: 35 kg
Frame Material: Composite
Rotor Material: Wood
Available Colors: Grey
Max. Flight Time: 8 hours
Max. Speed: 115 km/h (71 mp/h)
Max. Flying Altitude: 3,000 m (9,842 ft)
Max. Operating Range: 60 km
GPS Compatible: √
Altitude Hold: √
Position Hold: –
Flight Planning: √
Autonomous Flight: √
Target Tracking: –
Automatic Landing: √
Return To Home: √
Obstacle Avoidance: √
Headless Mode: –
Built-in Camera: √

Elbit Systems

Hermes 90

© Elbit Systems

Release Date: September 2009
Price: $1.5 million
Number of Rotors: 1 (Fixed Wing)
Control System: Ground Control Station
Applications: Military Purposes, Reconnaissance, Surveillance
Wingspan: 5 m
Weight: 125 kg
Max. Payload: 30 kg
Available Colors: Grey
Max. Flight Time: 15 hours
Max. Flying Altitude: 5,000 m (16,404 ft)
Max. Operating Range: 150 km
GPS Compatible: √
Altitude Hold: √
Position Hold: –
Flight Planning: √
Autonomous Flight: √
Target Tracking: –
Automatic Landing: √
Return To Home: √
Obstacle Avoidance: √
Headless Mode: –
Built-in Camera: √

Israel Aerospace Industries (IAI)

Mosquito

© IAI

Release Date: September 2009
Price: Unknown
Number of Rotors: 1 (Fixed Wing)
Control System: Remote Control, Laptop
Applications: Military Purposes, Reconnaissance, Surveillance
Dimensions: 300 × 400 mm
Wingspan: 400 mm
Weight: 500 gr
Available Colors: Yellow
Max. Flight Time: 40 minutes
Max. Flying Altitude: 152 m (500 ft)
Max. Operating Range: 1.5 km
GPS Compatible: √
Altitude Hold: √
Position Hold: –
Flight Planning: √
Autonomous Flight: √
Target Tracking: √
Automatic Landing: –
Return To Home: √
Obstacle Avoidance: –
Headless Mode: –
Built-in Camera: √

Aeryon Labs Inc.

Scout

© Reuters/Euan Rocha

Release Date: November 2009
Price: $100,000
Number of Rotors: 4
Control System: Flight Computer, Remote Control
Applications: Military Purposes, Reconnaissance, Surveillance
Dimensions: 800 × 800 × 300 mm
Wingspan: 800 mm
Weight: 1.4 kg
Max. Payload: 250 gr
Available Colors: Grey
Max. Flight Time: 25 minutes
Max. Speed: 50 km/h (31 mp/h)
Max. Flying Altitude: 5,000 m (16,404 ft)
Max. Ascent Speed: 2 m/s
Min. Operating Temperature: -30°C
Max. Operating Temperature: 50°C
Max. Operating Range: 3 km
Controller Frequency: 2.4 GHz
Battery Type: Lithium Polymer
GPS Compatible: √
Altitude Hold: √
Position Hold: √
Flight Planning: √
Autonomous Flight: √
Target Tracking: –
Automatic Landing: √
Return To Home: √
Obstacle Avoidance: √
Headless Mode: –
Built-in Camera: √

Chengdu Aircraft Industry Group (CAIG)

Pterodactyl I (Wing Loong)

© Wikipedia/Vitaly V. Kuzmin

Release Date: November 2009
Price: $15 million
Number of Rotors: 1 (Fixed Wing)
Control System: Ground Control Station
Applications: Military Purposes, Armed Reconnaissance, Surveillance, Target Acquisition
Dimensions: 9 × 14 × 2.8 m
Wingspan: 14 m
Weight: 1,150 kg
Max. Payload: 200 kg
Frame Material: Composite
Available Colors: Grey
Max. Flight Time: 20 hours
Max. Speed: 280 km/h (174 mp/h)
Max. Flying Altitude: 5,000 m (16,404 ft)
Max. Operating Range: 4,000 km
GPS Compatible: √
Altitude Hold: √
Position Hold: √
Flight Planning: √
Autonomous Flight: √
Target Tracking: √
Automatic Landing: √
Return To Home: √
Obstacle Avoidance: √
Headless Mode: √
Built-in Camera: √

Defence Research Development Organisation of India (DRDO)

Rustom 1

© aviationsmilitaires.net

Release Date: November 2009
Price: Unknown
Number of Rotors: 1 (Fixed Wing)
Control System: Ground Control Station
Applications: Military Purposes, Reconnaissance, Surveillance
Dimensions: 5.12 × 7.9 × 2.4 m
Wingspan: 7.9 m
Weight: 720 kg
Max. Payload: 95 kg
Frame Material: Composite
Motors: 1 × Lycoming O-320 engine
Available Colors: Red/White
Max. Flight Time: 12 hours
Max. Speed: 150 km/h (93 mp/h)
Max. Flying Altitude: 7,925 m (26,000 ft)
Max. Operating Range: 350 km
GPS Compatible: √
Altitude Hold: √
Position Hold: –
Flight Planning: √
Autonomous Flight: √
Target Tracking: √
Automatic Landing: √
Return To Home: √
Obstacle Avoidance: √
Headless Mode: –
Built-in Camera: √

EMT-Penzberg

Aladin

© tag-der-budeswehr.de

Release Date: November 2009
Price: Unknown
Number of Rotors: 1 (Fixed Wing)
Control System: Ground Control Station
Applications: Military Purposes, Reconnaissance, Surveillance
Dimensions: 1.57 × 1.46 × 0.38 m
Wingspan: 1.46 m
Weight: 4 kg
Frame Material: Composite
Available Colors: Grey
Max. Flight Time: 60 minutes
Max. Speed: 70 km/h (43 mp/h)
Max. Flying Altitude: 4,500 m (14,764 ft)
Max. Operating Range: 15 km
Controller Frequency: 2.4 GHz
Battery Type: Lithium Polymer
GPS Compatible: √
Altitude Hold: √
Position Hold: –
Flight Planning: √
Autonomous Flight: √
Target Tracking: –
Automatic Landing: √
Return To Home: √
Obstacle Avoidance: √
Headless Mode: –
Built-in Camera: √

Xi'an Aisheng Technology Group

ASN-207

©Chinese Military Review

Release Date: November 2009
Price: $1 million
Number of Rotors: 1 (Fixed Wing)
Control System: Ground Control Station
Applications: Military Purposes, Aerial Mapping, Armed Reconnaissance, Surveillance, Target Acquisition
Dimensions: 4.27 × 7.5 × 1.54 m
Wingspan: 7.5 m
Weight: 320 kg
Max. Payload: 50 kg
Available Colors: White
Max. Flight Time: 10 hours
Max. Speed: 180 km/h (112 mp/h)
Max. Flying Altitude: 5,000 m (16,404 ft)
Max. Operating Range: 200 km
GPS Compatible: √
Altitude Hold: √
Position Hold: –
Flight Planning: √
Autonomous Flight: √
Target Tracking: √
Automatic Landing: –
Return To Home: √
Obstacle Avoidance: √
Headless Mode: –
Built-in Camera: √

Xi'an Aisheng Technology Group

ASN-209

©Chinese Military Review

Release Date: November 2009
Price: $1.2 million
Number of Rotors: 1 (Fixed Wing)
Control System: Ground Control Station
Applications: Military Purposes, Aerial Mapping, Armed Reconnaissance, Surveillance, Target Acquisition
Dimensions: 6 × 9.3 × 2.1 m
Wingspan: 9.3 m
Weight: 480 kg
Max. Payload: 100 kg
Frame Material: Composite
Available Colors: White
Max. Flight Time: 16 hours
Max. Speed: 180 km/h (112 mp/h)
Max. Flying Altitude: 6,000 m (19,685 ft)
Max. Operating Range: 600 km
GPS Compatible: √
Altitude Hold: √
Position Hold: –
Flight Planning: √
Autonomous Flight: √
Target Tracking: √
Automatic Landing: –
Return To Home: √
Obstacle Avoidance: √
Headless Mode: –
Built-in Camera: √

Elbit Systems

Hermes 900

Release Date: December 2009
Price: $10 million
Number of Rotors: 1 (Fixed Wing)
Control System: Ground Control Station
Applications: Military Purposes, Maritime Patrol, Reconnaissance, Surveillance
Dimensions: 8.3 × 15 m
Wingspan: 15 m
Weight: 970 kg
Max. Payload: 350 kg
Motors: 1 × Rotax 914, 86kW
Available Colors: Grey
Max. Flight Time: 36 hours
Max. Speed: 220 km/h (137 mp/h)
Max. Flying Altitude: 9,100 m (29,855 ft)
GPS Compatible: √
Altitude Hold: √
Position Hold: –
Flight Planning: √
Autonomous Flight: √
Target Tracking: –
Automatic Landing: √
Return To Home: √
Obstacle Avoidance: √
Headless Mode: –
Built-in Camera: √

Microdrones.com

MD4-1000

Release Date: December 2009
Price: Unknown
Number of Rotors: 4
Control System: Remote Control
Applications: Film, Photography, Aerial Mapping
Wingspan: 1.03 m
Weight: 2.65 kg
Max. Payload: 1.2 kg
Frame Material: Carbon Fiber
Rotor Material: Carbon Fiber
Available Colors: White
Max. Flight Time: 45 minutes
Max. Speed: 43 km/h (26,8 mp/h)
Max. Flying Altitude: 4,500 m (14,764 ft)
Max. Ascent Speed: 7.5 m/s
Min. Operating Temperature: -10°C
Max. Operating Temperature: 50°C
Max. Operating Range: 20 km
Controller Frequency: 2.4 GHz
Battery Capacity: 13,000 mAh
Battery Type: Lithium Polymer 6S
Battery Power: 22.2V
GPS Compatible: √
Altitude Hold: √
Position Hold: √
Flight Planning: √
Autonomous Flight: √
Target Tracking: –
Automatic Landing: √
Return To Home: √
Obstacle Avoidance: –
Headless Mode: –
Built-in Camera: –

2010

Israel Aerospace Industries (IAI)

Bird-Eye 650

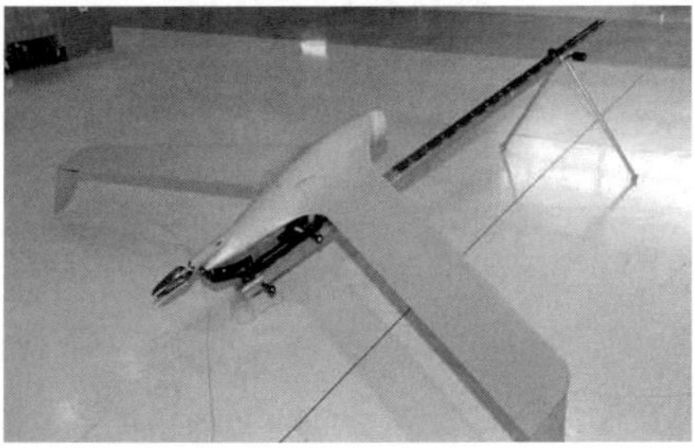

©IAI

Release Date: January 2010
Price: Unknown
Number of Rotors: 1 (Fixed Wing)
Control System: Ground Control Station
Applications: Military Purposes, Reconnaissance, Surveillance
Wingspan: 3 m
Weight: 9.8 kg
Max. Payload: 1.2 kg
Frame Material: Composite
Available Colors: Grey
Max. Flight Time: 6 hours
Max. Speed: 120 km/h (74,5 mp/h)
Max. Flying Altitude: 4,267 m (14,000 ft)
Max. Operating Range: 20 km
GPS Compatible: √
Altitude Hold: √
Position Hold: –
Flight Planning: √
Autonomous Flight: √
Target Tracking: –
Automatic Landing: –
Return To Home: √
Obstacle Avoidance: –
Headless Mode: –
Built-in Camera: √

Israel Aerospace Industries (IAI)

Searcher III

©abcblogs.abc.es

Release Date: January 2010
Price: Unknown
Number of Rotors: 1 (Fixed Wing)
Control System: Ground Control Station
Applications: Military Purposes, Reconnaissance, Surveillance, Target Acquisition
Dimensions: 5.85 × 8.55 m
Wingspan: 8.55 m
Weight: 436 kg
Max. Payload: 120 kg
Frame Material: Composite
Available Colors: Grey
Max. Flight Time: 18 hours
Max. Flying Altitude: 7,010 m (23,000 ft)
Max. Operating Range: 350 km
GPS Compatible: √
Altitude Hold: √
Position Hold: –
Flight Planning: √
Autonomous Flight: √
Target Tracking: √
Automatic Landing: √
Return To Home: √
Obstacle Avoidance: √
Headless Mode: –
Built-in Camera: √

Northrop Grumman

Bat

©US Navy/Lt. Jessica Crownover

Release Date: January 2010
Price: Unknown
Number of Rotors: 1 (Fixed Wing)
Control System: Ground Control Station
Applications: Military Purposes, Armed Reconnaissance, Surveillance
Dimensions: 2 × 4.3 m
Wingspan: 4.3 m
Weight: 158 kg
Max. Payload: 34 kg
Frame Material: Composite
Available Colors: Grey
Max. Flight Time: 18 hours
Max. Speed: 166 km/h (104 mp/h)
Max. Flying Altitude: 5,200 m (17,060 ft)
GPS Compatible: √
Altitude Hold: √
Position Hold: –
Flight Planning: √
Autonomous Flight: √
Target Tracking: √
Automatic Landing: √
Return To Home: √
Obstacle Avoidance: √
Headless Mode: –
Built-in Camera: √

China Aerospace Science and Industry Corporation (CASIC)

HW-300 Blade

©CNS Photo

Release Date: February 2010
Price: Unknown
Number of Rotors: 1 (Fixed Wing)
Control System: Ground Control Station, Remote Control
Applications: Military Purposes, Reconnaissance, Surveillance
Dimensions: 2.8 × 3.98 m
Wingspan: 3.98 m
Frame Material: Composite
Available Colors: Yellow
Max. Flight Time: 3 hours
Max. Speed: 120 km/h (74,5 mp/h)
Max. Flying Altitude: 4,000 m (13,123 ft)
GPS Compatible: √
Altitude Hold: √
Position Hold: –
Flight Planning: √
Autonomous Flight: √
Target Tracking: –
Automatic Landing: √
Return To Home: √
Obstacle Avoidance: √
Headless Mode: –
Built-in Camera: √

China Aerospace Science and Industry Corporation (CASIC)

HW-310

© mil.huanqiu.com

Release Date: February 2010
Price: Unknown
Number of Rotors: 1 (Fixed Wing)
Control System: Ground Control Station, Remote Control
Applications: Military Purposes, Reconnaissance, Surveillance
Dimensions: 3 × 4.6 m
Wingspan: 4.6 m
Weight: 85 kg
Max. Payload: 15 kg
Frame Material: Composite
Available Colors: Grey
Max. Flight Time: 6 hours
Max. Speed: 120 km/h (74,5 mp/h)
Max. Flying Altitude: 4,000 m (13,123 ft)
Max. Operating Range: 120 km
GPS Compatible: √
Altitude Hold: √
Position Hold: –
Flight Planning: √
Autonomous Flight: √
Target Tracking: –
Automatic Landing: √
Return To Home: √
Obstacle Avoidance: √
Headless Mode: –
Built-in Camera: √

Aerovel Corporation

Flexrotor

© Aerovel

Release Date: March 2010
Price: $200,000
Number of Rotors: 3 (Fixed Wing/VTOL)
Control System: Ground Control Station, Laptop
Applications: Inspection, Maritime Patrol, Reconnaissance, Surveillance
Dimensions: 2 × 3 m
Wingspan: 3 m
Weight: 22 kg
Motors: 28 cc two-stroke
Available Colors: Yellow
Max. Flight Time: 30 hours
Max. Speed: 160 km/h (99,5 mp/h)
Max. Flying Altitude: 5,486 m (18,000 ft)
Max. Operating Range: 100 km
GPS Compatible: √
Altitude Hold: √
Position Hold: √
Flight Planning: √
Autonomous Flight: √
Target Tracking: √
Automatic Landing: √
Return To Home: √
Obstacle Avoidance: √
Headless Mode: –
Built-in Camera: √

China Aerospace Science and Industry Corporation (CASIC)

WJ 600 Sky Hawk

©aviationsmilitaires.net

Release Date: March 2010
Price: Unknown
Number of Rotors: – (Fixed Wing)
Control System: Ground Control Station
Applications: Military Purposes, Armed Reconnaissance, Surveillance, Target Acquisition
Frame Material: Composite
Available Colors: White
Max. Flight Time: 5 hours
Max. Speed: 720 km/h (447 mp/h)
Max. Flying Altitude: 10,000 m (32,808 ft)
Max. Operating Range: 2,100 km
GPS Compatible: √
Altitude Hold: √
Position Hold: –
Flight Planning: √
Autonomous Flight: √
Target Tracking: √
Automatic Landing: √
Return To Home: √
Obstacle Avoidance: √
Headless Mode: –
Built-in Camera: √

Microdrones.com

MD4-200

©microdrones.com

Release Date: May 2010
Price: Unknown
Number of Rotors: 4
Control System: Remote Control
Applications: Film, Photography, Aerial Mapping
Wingspan: 540 mm
Weight: 800 gr
Max. Payload: 250 gr
Frame Material: Carbon Fiber
Rotor Material: Carbon Fiber
Available Colors: White
Max. Flight Time: 25 minutes
Max. Speed: 29 km/h (18 mp/h)
Max. Flying Altitude: 2,000 m (6,560 ft)
Max. Ascent Speed: 7 m/s
Min. Operating Temperature: -10°C
Max. Operating Temperature: 40°C
Max. Operating Range: 6 km
Controller Frequency: 2.4 GHz
Battery Capacity: 2,300 mAh
Battery Type: Lithium Polymer 4S
Battery Power: 14.8V
GPS Compatible: √
Altitude Hold: √
Position Hold: √
Flight Planning: √
Autonomous Flight: √
Target Tracking: –
Automatic Landing: √
Return To Home: √
Obstacle Avoidance: –
Headless Mode: –
Built-in Camera: –

Lehmann Aviation

LM450

©Lehmann Aviation

Release Date: June 2010
Price: $25,000
Number of Rotors: 1 (Fixed Wing)
Control System: Ground Control Station
Applications: Aerial Mapping, Agriculture, Inspection, Reconnaissance
Wingspan: 920 mm
Weight: 950 gr
Max. Payload: 200 gr
Frame Material: EPP Foam
Available Colors: White
Max. Flight Time: 45 minutes
Max. Speed: 45 km/h (28 mp/h)
Min. Operating Temperature: -25°C
Max. Operating Temperature: 60°C
Max. Operating Range: 5 km
GPS Compatible: √
Altitude Hold: √
Position Hold: –
Flight Planning: √
Autonomous Flight: √
Target Tracking: –
Automatic Landing: –
Return To Home: √
Obstacle Avoidance: –
Headless Mode: –
Built-in Camera: –

ZALA Aero Group

Zala 421-21

©innovudm.ru

Release Date: June 2010
Price: Unknown
Number of Rotors: 6
Control System: Ground Control Station
Applications: Military Purposes, Reconnaissance, Surveillance
Dimensions: 600 × 520 × 75 mm
Weight: 1.5 kg
Max. Payload: 300 gr
Frame Material: Carbon Fiber
Available Colors: Grey
Max. Flight Time: 40 minutes
Max. Speed: 40 km/h (25 mp/h)
Max. Flying Altitude: 1,000 m (3,280 ft)
Min. Operating Temperature: -30°C
Max. Operating Temperature: 40°C
Max. Operating Range: 2 km
Controller Frequency: 2.4 GHz
Battery Type: Lithium Polymer
GPS Compatible: √
Altitude Hold: √
Position Hold: √
Flight Planning: √
Autonomous Flight: √
Target Tracking: –
Automatic Landing: √
Return To Home: √
Obstacle Avoidance: –
Headless Mode: –
Built-in Camera: √

Irkut Corporation

Irkut-200

© suasnews.com

Release Date: July 2010
Price: Unknown
Number of Rotors: 1 (Fixed Wing)
Control System: Ground Control Station
Applications: Military Purposes, Inspection, Surveillance
Dimensions: 4.53 × 5.34 × 1.68 m
Wingspan: 5.34 m
Weight: 200 kg
Max. Payload: 50 kg
Frame Material: Composite
Available Colors: White
Max. Flight Time: 12 hours
Max. Speed: 210 km/h (130,5 mp/h)
Max. Flying Altitude: 5,000 m (16,404 ft)
Min. Operating Temperature: -40°C
Max. Operating Temperature: 50°C
Max. Operating Range: 200 km
GPS Compatible: √
Altitude Hold: √
Position Hold: –
Flight Planning: √
Autonomous Flight: √
Target Tracking: √
Automatic Landing: √
Return To Home: √
Obstacle Avoidance: √
Headless Mode: –
Built-in Camera: √

R.M.S.

Mantarraya

© R.M.S.

Release Date: July 2010
Price: Unknown
Number of Rotors: 1 (Fixed Wing)
Control System: Flight Computer
Applications: Military Purposes, Maritime Patrol, Reconnaissance, Surveillance
Dimensions: 1.4 × 2.52 m
Weight: 40 kg
Max. Payload: 15 kg
Frame Material: Composite
Rotor Material: Wood
Available Colors: White
Max. Flight Time: 4 hours
Max. Speed: 240 km/h (149 mp/h)
Max. Flying Altitude: 3,000 m (9,842 ft)
Max. Operating Range: 100 km
GPS Compatible: √
Altitude Hold: √
Position Hold: √
Flight Planning: √
Autonomous Flight: √
Target Tracking: √
Automatic Landing: √
Return To Home: √
Obstacle Avoidance: √
Headless Mode: –
Built-in Camera: √

SenseFly

Swinglet

© SenseFly

Release Date: July 2010
Price: Unknown
Number of Rotors: 1 (Fixed Wing)
Control System: Remote Control
Applications: Aerial Mapping, Inspection
Wingspan: 800 mm
Weight: 500 gr
Frame Material: EPP Foam, Carbon Fiber
Available Colors: Black
Max. Flight Time: 30 minutes
Max. Speed: 40 km/h (25 mp/h)
Max. Flying Altitude: 1,000 m (3,280 ft)
Max. Operating Range: 1 km
Controller Frequency: 2.4 GHz
Battery Type: Lithium Polymer 3S
GPS Compatible: √
Altitude Hold: √
Position Hold: –
Flight Planning: √
Autonomous Flight: √
Target Tracking: –
Automatic Landing: √
Return To Home: √
Obstacle Avoidance: –
Headless Mode: –
Built-in Camera: √

AscTec

Pelican

© AscTec

Release Date: August 2010
Price: On Request
Number of Rotors: 4
Control System: Remote Control
Applications: Film, Photography, Transport
Dimensions: 651 × 651 × 188 mm
Weight: 1.65 kg
Max. Payload: 650 gr
Frame Material: Carbon Fiber
Available Colors: Black
Max. Flight Time: 30 minutes
Max. Speed: 57 km/h (35,6 mp/h)
Max. Ascent Speed: 8 m/s
Max. Operating Range: 4.5 km
Controller Frequency: 2.4 GHz
Battery Capacity: 6,250 mAh
Battery Type: Lithium Polymer
GPS Compatible: √
Altitude Hold: √
Position Hold: √
Flight Planning: √
Autonomous Flight: √
Target Tracking: –
Automatic Landing: √
Return To Home: √
Obstacle Avoidance: √
Headless Mode: –
Built-in Camera: –

Bormatec

Maja

© Bormatec

Release Date: August 2010
Price: $275
Number of Rotors: 1 (Fixed Wing)
Control System: Flight Computer
Applications: Aerial Mapping, Agriculture, Inspection, Surveillance
Dimensions: 1.2 × 2.2 m
Wingspan: 2.2 m
Weight: 2.5 kg
Max. Payload: 500 gr
Frame Material: EPP, Coroplast
Available Colors: Black
Max. Flight Time: 60 minutes
Controller Frequency: 2.4 GHz
Battery Capacity: 8,000 mAh
Battery Type: Lithium Polymer 3S
GPS Compatible: –
Altitude Hold: √
Position Hold: –
Flight Planning: √
Autonomous Flight: √
Target Tracking: √
Automatic Landing: –
Return To Home: √
Obstacle Avoidance: –
Headless Mode: –
Built-in Camera: √

Iran Aircraft Manufacturing Industrial Company

HESA Karrar

© Fars News Agency/Mahdi Marizad

Release Date: August 2010
Price: $3 million
Number of Rotors: – (Fixed Wing)
Control System: Ground Control Station
Applications: Military Purposes, Armed Reconnaissance, Aerial Target
Dimensions: 4 × 2.5 m
Wingspan: 2.5 m
Weight: 700 kg
Max. Payload: 227 kg
Frame Material: Composite
Motors: 1 × MicroTurbo TR 60-5 turbojet
Available Colors: Green, Orange
Max. Speed: 900 km/h (560 mp/h)
Max. Operating Range: 500 km
GPS Compatible: √
Altitude Hold: √
Position Hold: –
Flight Planning: √
Autonomous Flight: √
Target Tracking: √
Automatic Landing: –
Return To Home: √
Obstacle Avoidance: –
Headless Mode: –
Built-in Camera: √

Parrot

AR.Drone

©Parrot

Release Date: August 2010
Price: $120
Number of Rotors: 4
Control System: Remote Control
Applications: Film, Photography, Recreation
Wingspan: 450 mm
Weight: 380 gr
Frame Material: EPP Foam, Plastic
Available Colors: Black
Max. Flight Time: 12 minutes
Max. Speed: 18 km/h (11,2 mp/h)
Max. Operating Range: 50 m
Controller Frequency: 2.4 GHz
Battery Capacity: 1,000 mAh
Battery Type: Lithium Polymer 3S
Battery Power: 11.1V
GPS Compatible: √
Altitude Hold: √
Position Hold: √
Flight Planning: –
Autonomous Flight: –
Target Tracking: –
Automatic Landing: –
Return To Home: –
Obstacle Avoidance: –
Headless Mode: –
Built-in Camera: √

China Aerospace Science and Industry Corporation (CASIC)

HW-210

©mil.cnr.cn

Release Date: September 2010
Price: Unknown
Number of Rotors: 1 (Fixed Wing)
Control System: Ground Control Station
Applications: Military Purposes, Reconnaissance, Surveillance
Dimensions: 2.1 × 3.5 m
Wingspan: 3.5 m
Weight: 22 kg
Max. Payload: 5 kg
Frame Material: Composite
Available Colors: White
Max. Flight Time: 16 hours
Max. Speed: 120 km/h (74,5 mp/h)
Max. Flying Altitude: 5,000 m (16,404 ft)
Max. Operating Range: 50 km
GPS Compatible: √
Altitude Hold: √
Position Hold: –
Flight Planning: √
Autonomous Flight: √
Target Tracking: –
Automatic Landing: –
Return To Home: √
Obstacle Avoidance: √
Headless Mode: –
Built-in Camera: √

Rzeszów University of Technology

PR-5 Wiewiór Plus

©Wikipedia/Marcin Marchewka

Release Date: September 2010
Price: $250,000
Number of Rotors: 2 (Fixed Wing)
Control System: Ground Control Station
Applications: Military Purposes, Reconnaissance, Surveillance, Target Acquisition
Dimensions: 1.4 × 2.14 × 0.3 m
Wingspan: 2.14 m
Weight: 2.5 kg
Max. Payload: 3 kg
Frame Material: Aluminum, Carbon Fiber, Kevlar
Motors: 2 × AXI 2820/14
Available Colors: Grey
Max. Flight Time: 50 minutes
GPS Compatible: √
Altitude Hold: √
Position Hold: –
Flight Planning: √
Autonomous Flight: √
Target Tracking: –
Automatic Landing: √
Return To Home: √
Obstacle Avoidance: –
Headless Mode: –
Built-in Camera: √

Aurora Integrated Systems

Altius MK-II

©Aurora Integrated Systems

Release Date: October 2010
Price: Unknown
Number of Rotors: 1 (Fixed Wing)
Control System: Ground Control Station
Applications: Military Purposes, Reconnaissance, Surveillance, Target Acquisition
Dimensions: 3.2 × 5.5 m
Wingspan: 5.5 m
Weight: 35 kg
Max. Payload: 20 kg
Frame Material: Composite
Motors: Two-stroke gasoline engine
Available Colors: White
Max. Flight Time: 10 hours
Max. Speed: 160 km/h (99,5 mp/h)
Max. Flying Altitude: 3,657 m (12,000 ft)
Max. Operating Range: 150 km
GPS Compatible: √
Altitude Hold: √
Position Hold: –
Flight Planning: √
Autonomous Flight: √
Target Tracking: –
Automatic Landing: √
Return To Home: √
Obstacle Avoidance: –
Headless Mode: –
Built-in Camera: √

Idetec

Stardust

©Idetec

Release Date: October 2010
Price: On Request
Number of Rotors: 1 (Fixed Wing)
Control System: Ground Control Station
Applications: Aerial Mapping, Agriculture, Inspection
Weight: 4.5 kg
Available Colors: Orange
Max. Flight Time: 2 hours
Max. Speed: 65 km/h (40 mp/h)
Max. Flying Altitude: 5,000 m (16,404 ft)
Max. Operating Range: 20 km
Controller Frequency: 2.4 GHz
Battery Type: Lithium Polymer
GPS Compatible: √
Altitude Hold: √
Position Hold: –
Flight Planning: √
Autonomous Flight: √
Target Tracking: –
Automatic Landing: –
Return To Home: √
Obstacle Avoidance: –
Headless Mode: –
Built-in Camera: √

Israel Aerospace Industries (IAI)

Panther

©Wikipedia/Vivawariors

Release Date: October 2010
Price: $4 million
Number of Rotors: 3 (Fixed Wing/VTOL)
Control System: Ground Control Station
Applications: Military Purposes, Inspection, Maritime Patrol, Reconnaissance, Surveillance
Wingspan: 8 m
Weight: 65 kg
Max. Payload: 8.5 kg
Frame Material: Composite
Available Colors: Grey
Max. Flight Time: 6 hours
Max. Speed: 92 km/h (57 mp/h)
Max. Flying Altitude: 914 m (3,000 ft)
Max. Operating Range: 130 km
GPS Compatible: √
Altitude Hold: √
Position Hold: √
Flight Planning: √
Autonomous Flight: √
Target Tracking: √
Automatic Landing: √
Return To Home: √
Obstacle Avoidance: √
Headless Mode: –
Built-in Camera: √

Israel Aerospace Industries (IAI)

Mini Panther

© Guillaume BELAN

Release Date: November 2010
Price: Unknown
Number of Rotors: 3 (Fixed Wing/VTOL)
Control System: Ground Control Station
Applications: Military Purposes, Inspection, Maritime Patrol, Reconnaissance, Surveillance
Wingspan: 3.5 m
Weight: 12 kg
Max. Payload: 2 kg
Frame Material: Composite
Available Colors: Grey
Max. Flight Time: 90 minutes
Max. Speed: 74 km/h (46 mp/h)
Max. Flying Altitude: 460 m (1,500 ft)
Max. Operating Range: 20 km
Battery Type: Lithium Polymer
GPS Compatible: √
Altitude Hold: √
Position Hold: √
Flight Planning: √
Autonomous Flight: √
Target Tracking: √
Automatic Landing: √
Return To Home: √
Obstacle Avoidance: √
Headless Mode: –
Built-in Camera: √

Xi'an Aisheng Technology Group

ASN-102

© worldwide-defence.blogspot.com

Release Date: November 2010
Price: Unknown
Number of Rotors: 1 (Fixed Wing)
Control System: Ground Control Station
Applications: Military Purposes, Reconnaissance, Surveillance
Dimensions: 2.65 × 2.68 × 0.65 m
Wingspan: 2.68 m
Weight: 92 kg
Max. Payload: 10 kg
Frame Material: Composite
Available Colors: White/Red
Max. Flight Time: 60 minutes
Max. Speed: 360 km/h (223,6 mp/h)
Max. Flying Altitude: 5,000 m (16,404 ft)
GPS Compatible: √
Altitude Hold: √
Position Hold: –
Flight Planning: √
Autonomous Flight: √
Target Tracking: –
Automatic Landing: –
Return To Home: √
Obstacle Avoidance: –
Headless Mode: –
Built-in Camera: √

Xi'an Aisheng Technology Group

ASN-216

Release Date: November 2010
Price: Unknown
Number of Rotors: 1 (Fixed Wing)
Control System: Ground Control Station
Applications: Military Purposes, Reconnaissance, Surveillance
Dimensions: 2.4 × 3.75 m
Wingspan: 3.75 m
Weight: 20 kg
Max. Payload: 6 kg
Frame Material: Composite
Rotor Material: Wood
Available Colors: Grey
Max. Flight Time: 6 hours
Max. Speed: 140 km/h (87 mp/h)
Max. Flying Altitude: 4,500 m (14,764 ft)
GPS Compatible: √
Altitude Hold: √
Position Hold: –
Flight Planning: √
Autonomous Flight: √
Target Tracking: –
Automatic Landing: –
Return To Home: √
Obstacle Avoidance: –
Headless Mode: –
Built-in Camera: √

Xi'an Aisheng Technology Group

ASN-229

Release Date: November 2010
Price: Unknown
Number of Rotors: 1 (Fixed Wing)
Control System: Ground Control Station
Applications: Military Purposes, Armed Reconnaissance, Surveillance, Target Acquisition
Dimensions: 5.5 × 11 m
Wingspan: 11 m
Weight: 800 kg
Max. Payload: 100 kg
Frame Material: Composite
Available Colors: Grey
Max. Flight Time: 20 hours
Max. Speed: 180 km/h (112 mp/h)
Max. Flying Altitude: 8,000 m (26,246 ft)
GPS Compatible: √
Altitude Hold: √
Position Hold: –
Flight Planning: √
Autonomous Flight: √
Target Tracking: –
Automatic Landing: –
Return To Home: √
Obstacle Avoidance: √
Headless Mode: –
Built-in Camera: √

Ejército Argentino

Lipán M3

©Wikipedia/Joaquín Alvarez Riera

Release Date: December 2010
Price: Unknown
Number of Rotors: 1 (Fixed Wing)
Control System: Ground Control Station
Applications: Military Purposes, Reconnaissance, Surveillance
Dimensions: 3.55 × 4.6 m
Wingspan: 4.6 m
Weight: 60 kg
Max. Payload: 20 kg
Frame Material: Composite
Available Colors: Grey
Max. Flight Time: 5 hours
Max. Speed: 170 km/h (106 mp/h)
Max. Flying Altitude: 2,000 m (6,560 ft)
Max. Operating Range: 40 km
GPS Compatible: √
Altitude Hold: √
Position Hold: –
Flight Planning: √
Autonomous Flight: √
Target Tracking: –
Automatic Landing: √
Return To Home: √
Obstacle Avoidance: –
Headless Mode: –
Built-in Camera: √

Elbit Systems

Skylark I-LE

©Elbit Systems

Release Date: December 2010
Price: Unknown
Number of Rotors: 1 (Fixed Wing)
Control System: Ground Control Station
Applications: Military Purposes, Reconnaissance, Surveillance
Wingspan: 3 m
Weight: 7 kg
Max. Payload: 1.2 kg
Available Colors: Grey
Max. Flight Time: 3 hours
Max. Flying Altitude: 5,000 m (16,404 ft)
Max. Operating Range: 40 km
GPS Compatible: √
Altitude Hold: √
Position Hold: –
Flight Planning: √
Autonomous Flight: √
Target Tracking: –
Automatic Landing: √
Return To Home: √
Obstacle Avoidance: √
Headless Mode: –
Built-in Camera: √

Nostromo Defensa

Cabure III

©Nostromo Defensa

Release Date: December 2010
Price: Unknown
Number of Rotors: 1 (Fixed Wing)
Control System: Ground Control Station
Applications: Military Purposes, Reconnaissance, Surveillance
Dimensions: 1.2 × 2.2 m
Wingspan: 2.2 m
Weight: 5 kg
Frame Material: Composite
Available Colors: Grey
Max. Flight Time: 90 minutes
Max. Speed: 120 km/h (74,5 mp/h)
Max. Flying Altitude: 4,000 m (13,123 ft)
Max. Operating Range: 15 km
GPS Compatible: √
Altitude Hold: √
Position Hold: –
Flight Planning: √
Autonomous Flight: √
Target Tracking: –
Automatic Landing: √
Return To Home: √
Obstacle Avoidance: –
Headless Mode: –
Built-in Camera: √

Izhmash Unmanned Systems

Granat-4

©robotrends.ru

Release Date: December 2010
Price: Unknown
Number of Rotors: 1 (Fixed Wing)
Control System: Ground Control Station
Applications: Military Purposes, Reconnaissance, Surveillance
Dimensions: 2.6 × 3.2 m
Wingspan: 3.2 m
Weight: 30 kg
Max. Payload: 3 kg
Frame Material: Composite
Available Colors: Grey
Max. Flight Time: 6 hours
Max. Speed: 145 km/h (90 mp/h)
Max. Flying Altitude: 4,000 m (13,123 ft)
Min. Operating Temperature: -30°C
Max. Operating Temperature: 40°C
Max. Operating Range: 100 km
GPS Compatible: √
Altitude Hold: √
Position Hold: –
Flight Planning: √
Autonomous Flight: √
Target Tracking: √
Automatic Landing: √
Return To Home: √
Obstacle Avoidance: √
Headless Mode: –
Built-in Camera: √

2011

Aurora Integrated Systems

Urban View

Release Date: February 2011
Price: Unknown
Number of Rotors: 1 (Fixed Wing)
Control System: Ground Control Station
Applications: Inspection, Surveillance, Reconnaissance
Dimensions: 0.8 × 1.5 m
Wingspan: 1.5 m
Weight: 2 kg
Frame Material: Composite
Available Colors: White
Max. Flight Time: 60 minutes
Max. Speed: 60 km/h (37 mp/h)
Max. Flying Altitude: 4,600 m (15,092 ft)
Max. Operating Range: 15 km
Controller Frequency: 2.4 GHz
Battery Type: Lithium Polymer
GPS Compatible: √
Altitude Hold: √
Position Hold: –
Flight Planning: √
Autonomous Flight: √
Target Tracking: –
Automatic Landing: –
Return To Home: √
Obstacle Avoidance: –
Headless Mode: –
Built-in Camera: √

Israel Aerospace Industries (IAI)

Ghost

Release Date: February 2011
Price: $100,000
Number of Rotors: 2
Control System: Ground Control Station
Applications: Military Purposes, Inspection, Surveillance
Dimensions: 1.55 × 0.75 m
Weight: 4 kg
Frame Material: Composite
Available Colors: Black
Max. Flight Time: 40 minutes
Max. Speed: 65 km/h (40 mp/h)
Max. Operating Range: 4 km
Controller Frequency: 2.4 GHz
Battery Type: Lithium Polymer
GPS Compatible: √
Altitude Hold: √
Position Hold: √
Flight Planning: √
Autonomous Flight: √
Target Tracking: √
Automatic Landing: √
Return To Home: √
Obstacle Avoidance: √
Headless Mode: –
Built-in Camera: √

Northrop Grumman

X-47B

©US Navy

Release Date: February 2011
Price: $90 million
Number of Rotors: – (Fixed Wing)
Control System: Ground Control Station
Applications: Military Purposes, Research & Development, Armed Reconnaissance, Surveillance
Dimensions: 11.6 × 18.9 × 3.1 m
Wingspan: 9.41 m
Weight: 19,000 kg
Max. Payload: 2,045 kg
Frame Material: Composite
Motors: 1 × Pratt & Whitney F100-PW-220E turbofan
Available Colors: Grey
Max. Speed: 980 km/h (609 mp/h)
Max. Flying Altitude: 12,800 m (42,000 ft)
Max. Operating Range: 3,900 km
GPS Compatible: √
Altitude Hold: √
Position Hold: –
Flight Planning: √
Autonomous Flight: √
Target Tracking: √
Automatic Landing: √
Return To Home: √
Obstacle Avoidance: √
Headless Mode: –
Built-in Camera: √

AltiGator

ALG-EOS

©AltiGator

Release Date: May 2011
Price: $2,383
Number of Rotors: 4
Control System: Remote Control
Applications: Film, Photography, Aerial Mapping
Wingspan: 500 mm
Weight: 1.5 kg
Frame Material: Carbon Fiber
Available Colors: Black
Max. Flight Time: 45 minutes
Controller Frequency: 2.4 GHz
Battery Capacity: 3,300 mAh
Battery Type: Lithium Polymer 4S
GPS Compatible: √
Altitude Hold: √
Position Hold: √
Flight Planning: √
Autonomous Flight: √
Target Tracking: –
Automatic Landing: √
Return To Home: √
Obstacle Avoidance: √
Headless Mode: –
Built-in Camera: –

ST Aerospace

Skyblade III

© ST Aerospace

Release Date: May 2011
Price: $500,000
Number of Rotors: 1 (Fixed Wing)
Control System: Ground Control Station
Applications: Military Purposes, Reconnaissance, Surveillance, Target Acquisition
Dimensions: 1.4 × 2.6 m
Wingspan: 2.6 m
Weight: 5 kg
Frame Material: Composite
Available Colors: Grey
Max. Flight Time: 60 minutes
Max. Speed: 65 km/h (40 mp/h)
Max. Flying Altitude: 460 m (1,500 ft)
Max. Operating Range: 8 km
GPS Compatible: √
Altitude Hold: √
Position Hold: –
Flight Planning: √
Autonomous Flight: √
Target Tracking: –
Automatic Landing: √
Return To Home: √
Obstacle Avoidance: –
Headless Mode: –
Built-in Camera: √

Aeronautics Ltd.

Orbiter 3

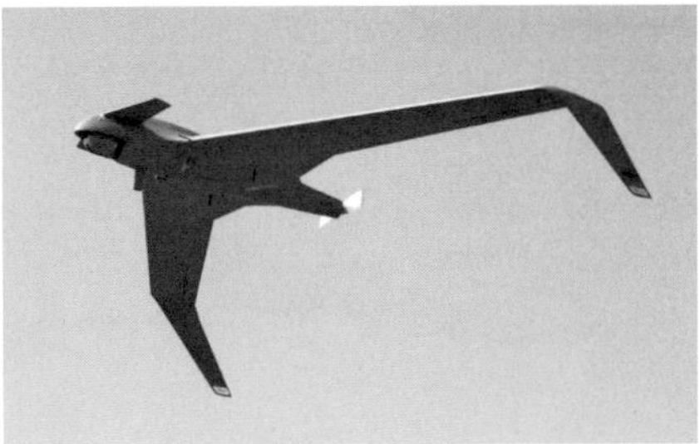

© Aeronautics

Release Date: June 2011
Price: Unknown
Number of Rotors: 1 (Fixed Wing)
Control System: Remote Control, Laptop
Applications: Reconnaissance, Surveillance
Wingspan: 4.4 m
Weight: 30 kg
Max. Payload: 5.5 kg
Available Colors: Grey
Max. Flight Time: 7 hours
Max. Speed: 112 km/h (70 mp/h)
Max. Operating Range: 150 km
GPS Compatible: √
Altitude Hold: √
Position Hold: –
Flight Planning: √
Autonomous Flight: √
Target Tracking: √
Automatic Landing: √
Return To Home: √
Obstacle Avoidance: √
Headless Mode: –
Built-in Camera: √

China Aerospace Science and Technology Corporation (CASC)

CH-803

©forcesdz.com

Release Date: October 2008
Price: Unknown
Number of Rotors: 1 (Fixed Wing)
Control System: Ground Control Station
Applications: Military Purposes, Reconnaissance, Surveillance
Dimensions: 1.8 × 3 m
Wingspan: 3 m
Weight: 18 kg
Max. Payload: 3.5 kg
Frame Material: Composite
Available Colors: Grey
Max. Flight Time: 5 hours
Max. Speed: 110 km/h (68 mp/h)
Max. Flying Altitude: 3,500 m (11,483 ft)
Max. Operating Range: 30 km
GPS Compatible: √
Altitude Hold: √
Position Hold: –
Flight Planning: √
Autonomous Flight: √
Target Tracking: –
Automatic Landing: –
Return To Home: √
Obstacle Avoidance: √
Headless Mode: –
Built-in Camera: √

BAE Systems

Telemos

©airport-data.com/Yves Q

Release Date: June 2011
Price: Unknown
Number of Rotors: 2 (Fixed Wing)
Control System: Ground Control Station
Dimensions: 19.8 × 22 m
Wingspan: 22 m
Weight: 9,000 kg
Frame Material: Composite
Motors: 2 × Rolls-Royce M250B-17 turboprop
Available Colors: Grey
Max. Flight Time: 30 hours
Max. Speed: 556 km/h (345 mp/h)
Max. Operating Range: 16,000 km
GPS Compatible: √
Altitude Hold: √
Position Hold: –
Flight Planning: √
Autonomous Flight: √
Target Tracking: √
Automatic Landing: √
Return To Home: √
Obstacle Avoidance: √
Headless Mode: –
Built-in Camera: √

Uvision

Sparrow

Release Date: June 2011
Price: Unknown
Number of Rotors: 1 (Fixed Wing)
Control System: Flight Computer
Applications: Loitering munition, Reconnaissance, Surveillance
Dimensions: 2.14 × 2.44 m
Wingspan: 2.44 m
Weight: 45 kg
Max. Payload: 12 kg
Available Colors: Grey
Max. Flight Time: 4 hours
Max. Speed: 129 km/h (80 mp/h)
Max. Operating Range: 120 km
GPS Compatible: √
Altitude Hold: √
Position Hold: –
Flight Planning: √
Autonomous Flight: √
Target Tracking: √
Automatic Landing: –
Return To Home: –
Obstacle Avoidance: √
Headless Mode: –
Built-in Camera: √

Global Industrial and Defefense Solutions (GIDS)

UQAB-II

© pakmilitary.blogspot.com

Release Date: July 2011
Price: $400,000
Number of Rotors: 1 (Fixed Wing)
Control System: Ground Control Station
Applications: Military Purposes, Inspection, Maritime Patrol, Reconnaissance, Surveillance
Dimensions: 4 × 5.5 × 1.2 m
Wingspan: 5.5 m
Weight: 200 kg
Max. Payload: 50 kg
Frame Material: Composite
Available Colors: Grey
Max. Flight Time: 6 hours
Max. Speed: 150 km/h (93 mp/h)
Max. Flying Altitude: 3,000 m (9,842 ft)
Max. Operating Range: 150 km
GPS Compatible: √
Altitude Hold: √
Position Hold: –
Flight Planning: √
Autonomous Flight: √
Target Tracking: –
Automatic Landing: √
Return To Home: √
Obstacle Avoidance: –
Headless Mode: –
Built-in Camera: √

SATUMA

Thunder SR

© Satuma

Release Date: July 2011
Price: Unknown
Number of Rotors: 1 (Fixed Wing)
Control System: Ground Control Station, Remote Control
Applications: Military Purposes, Research & Development, Target Acquisition
Dimensions: 2.83 × 3.05 × 0.5 m
Wingspan: 3.05 m
Weight: 68 kg
Frame Material: Composite
Rotor Material: Wood
Motors: Two-stroke gasoline engine
Available Colors: Red
Max. Speed: 300 km/h (186 mp/h)
Max. Flying Altitude: 5,000 m (16,404 ft)
Max. Operating Range: 10 km
GPS Compatible: √
Altitude Hold: √
Position Hold: –
Flight Planning: √
Autonomous Flight: √
Target Tracking: –
Automatic Landing: –
Return To Home: √
Obstacle Avoidance: –
Headless Mode: –
Built-in Camera: √

AeroVironment

Shrike VTOL

© AeroVironment

Release Date: October 2011
Price: Unknown
Number of Rotors: 4
Control System: Ground Control Station
Applications: Military Purposes, Inspection, Reconnaissance, Surveillance
Wingspan: 914 mm
Weight: 2.5 kg
Available Colors: White
Max. Speed: 55 km/h (34 mp/h)
Max. Flying Altitude: 152 m (500 ft)
Max. Operating Range: 5 km
GPS Compatible: √
Altitude Hold: √
Position Hold: √
Flight Planning: √
Autonomous Flight: √
Target Tracking: √
Automatic Landing: √
Return To Home: √
Obstacle Avoidance: √
Headless Mode: –
Built-in Camera: √

ZALA Aero Group

Zala 421-16E

© Vitaly V. Kuzmin

Release Date: August 2011
Price: Unknown
Number of Rotors: 1 (Fixed Wing)
Control System: Ground Control Station
Applications: Military Purposes, Inspection, Reconnaissance, Surveillance
Dimensions: 1 × 2.81 m
Wingspan: 2.81 m
Weight: 10 kg
Max. Payload: 1.5 kg
Frame Material: Composite
Available Colors: Grey
Max. Flight Time: 4 hours
Max. Speed: 110 km/h (68 mp/h)
Max. Flying Altitude: 3,600 m (11,811 ft)
Min. Operating Temperature: -30°C
Max. Operating Temperature: 40°C
Max. Operating Range: 50 km
Battery Capacity: 10,000 mAh
Battery Type: Lithium Polymer 10S
GPS Compatible: √
Altitude Hold: √
Position Hold: –
Flight Planning: √
Autonomous Flight: √
Target Tracking: √
Automatic Landing: –
Return To Home: √
Obstacle Avoidance: √
Headless Mode: –
Built-in Camera: √

Brock Technologies

Havoc UAS

© Brock Technologies

Release Date: October 2011
Price: Unknown
Number of Rotors: 1 (Fixed Wing)
Control System: Ground Control Station
Applications: Military Purposes, Reconnaissance, Surveillance
Max. Payload: 2.7 kg
Frame Material: Composite
Available Colors: Grey
Max. Flight Time: 2.5 hours
GPS Compatible: √
Altitude Hold: √
Position Hold: –
Flight Planning: √
Autonomous Flight: √
Target Tracking: √
Automatic Landing: √
Return To Home: √
Obstacle Avoidance: √
Headless Mode: –
Built-in Camera: √

Aeronautics Ltd.

Dominator XP

© Aeronautics Ltd.

Release Date: October 2011
Price: $17.5 million
Number of Rotors: 1 (Fixed Wing)
Control System: Ground Control Station
Applications: Military Purposes, Maritime Patrol, Reconnaissance, Surveillance
Dimensions: 8.6 × 13.5 × 2.5 m
Wingspan: 13.5 m
Weight: 1,910 kg
Max. Payload: 373 gr
Available Colors: White
Max. Flight Time: 20 hours
Max. Speed: 277 km/h (172 mp/h)
Max. Operating Range: 300 km
GPS Compatible: √
Altitude Hold: √
Position Hold: –
Flight Planning: √
Autonomous Flight: √
Target Tracking: –
Automatic Landing: √
Return To Home: √
Obstacle Avoidance: √
Headless Mode: –
Built-in Camera: √

Arcturus UAV

T-20

© Wikipedia/Efolkestad

Release Date: November 2011
Price: Unknown
Number of Rotors: 1 (Fixed Wing)
Control System: Ground Control Station
Applications: Aerial Mapping, Inspection
Dimensions: 2.87 × 5.26 m
Wingspan: 5.26 m
Weight: 36 kg
Max. Payload: 30 kg
Frame Material: Composite
Motors: 190cc 4 stroke fuel Injected
Available Colors: Grey
Max. Flight Time: 16 hours
Max. Speed: 167 km/h (103 mp/h)
Max. Flying Altitude: 4,572 m (15,000 ft)
GPS Compatible: √
Altitude Hold: √
Position Hold: –
Flight Planning: √
Autonomous Flight: √
Target Tracking: –
Automatic Landing: √
Return To Home: √
Obstacle Avoidance: –
Headless Mode: –
Built-in Camera: √

AirRobot

AR 100-B

©handelsblatt.com

Release Date: December 2011
Price: $110,000
Number of Rotors: 4
Control System: Flight Computer
Applications: Military Purposes, Inspection, Surveillance
Wingspan: 1 m
Weight: 1 kg
Max. Payload: 200 gr
Frame Material: Carbon Fiber
Available Colors: Black
Max. Flight Time: 30 minutes
Max. Flying Altitude: 1,000 m (3,280 ft)
Max. Operating Range: 1 km
GPS Compatible: √
Altitude Hold: √
Position Hold: √
Flight Planning: √
Autonomous Flight: √
Target Tracking: –
Automatic Landing: √
Return To Home: √
Obstacle Avoidance: √
Headless Mode: –
Built-in Camera: √ (10 megapixels)

AscTec

Hummingbird

©AscTec

Release Date: December 2011
Price: Unknown
Number of Rotors: 4
Control System: Remote Control
Applications: Research & Development
Dimensions: 540 × 540 × 55 mm
Weight: 710 gr
Max. Payload: 200 gr
Frame Material: Carbon Fiber
Available Colors: Black
Max. Flight Time: 20 minutes
Max. Speed: 54 km/h (33,5 mp/h)
Controller Frequency: 2.4 GHz
Battery Capacity: 2,100 mAh
Battery Type: Lithium Polymer
GPS Compatible: –
Altitude Hold: –
Position Hold: –
Flight Planning: –
Autonomous Flight: –
Target Tracking: –
Automatic Landing: –
Return To Home: –
Obstacle Avoidance: –
Headless Mode: –
Built-in Camera: –

Israel Aerospace Industries (IAI)

Electric Tethered Observation Platform (ETOP)

© diydrones.com

Release Date: December 2011
Price: Unknown
Number of Rotors: 4
Control System: Remote Control
Applications: Border Patrol, Reconnaissance, Surveillance
Dimensions: 1.6 × 1.6 × 0.2 m
Max. Payload: 20 kg
Available Colors: Grey
Max. Speed: – (tethered cable)
Max. Flying Altitude: 100 m (328 ft)
GPS Compatible: –
Altitude Hold: √
Position Hold: √
Flight Planning: –
Autonomous Flight: –
Target Tracking: √
Automatic Landing: √
Return To Home: –
Obstacle Avoidance: –
Headless Mode: –
Built-in Camera: √

National Aerospace Laboratories (NAL)

Slybird

© rpdefense.over-blog.com

Release Date: December 2011
Price: Unknown
Number of Rotors: 1 (Fixed Wing)
Control System: Ground Control Station
Applications: Military Purposes, Reconnaissance, Surveillance
Weight: 2 kg
Frame Material: Composite
Available Colors: Grey
Max. Flight Time: 60 minutes
Max. Flying Altitude: 4,267 m (14,000 ft)
Max. Operating Range: 10 km
GPS Compatible: √
Altitude Hold: √
Position Hold: √
Flight Planning: √
Autonomous Flight: √
Target Tracking: –
Automatic Landing: –
Return To Home: √
Obstacle Avoidance: –
Headless Mode: –
Built-in Camera: √

2012

Defence Research Development Organisation of India (DRDO)

Imperial Eagle

© aermech.in

Release Date: January 2012
Price: Unknown
Number of Rotors: 1 (Fixed Wing)
Control System: Ground Control Station
Applications: Military Purposes, Reconnaissance, Surveillance
Dimensions: 1.2 × 1.6 m
Wingspan: 1.6 m
Weight: 2.3 kg
Max. Payload: 250 gr
Frame Material: Composite
Motors: 1 × BLDC engine, wing-mounted turboprop
Available Colors: Grey
Max. Speed: 90 km/h (56 mp/h)
Max. Flying Altitude: 4,600 m (15,092 ft)
Max. Operating Range: 15 km
GPS Compatible: √
Altitude Hold: √
Position Hold: –
Flight Planning: √
Autonomous Flight: √
Target Tracking: √
Automatic Landing: √
Return To Home: √
Obstacle Avoidance: –
Headless Mode: –
Built-in Camera: √

Defence Research Development Organisation of India (DRDO)

Netra

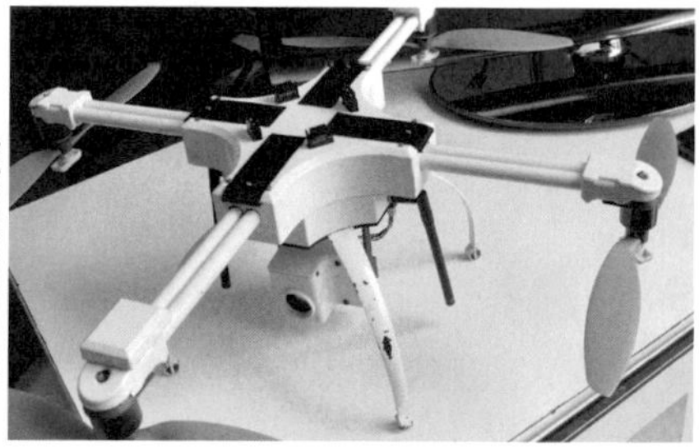

© Wikipedia/Biswarup Ganguly

Release Date: January 2012
Price: $27,850
Number of Rotors: 4
Control System: Flight Computer
Applications: Reconnaissance, Surveillance
Wingspan: 900 mm
Weight: 1.5 kg
Available Colors: White
Max. Flight Time: 30 minutes
Max. Speed: 30 km/h (18,6 mp/h)
Max. Flying Altitude: 300 m (985 ft)
Max. Operating Range: 2.5 km
Battery Type: Lithium Polymer
GPS Compatible: √
Altitude Hold: √
Position Hold: √
Flight Planning: √
Autonomous Flight: √
Target Tracking: √
Automatic Landing: √
Return To Home: √
Obstacle Avoidance: –
Headless Mode: –
Built-in Camera: √

Parrot

AR.Drone 2.0

© Parrot

Release Date: January 2012
Price: $299
Number of Rotors: 4
Control System: Remote Control, Phone/Tablet
Applications: Film, Photography, Recreation
Dimensions: 520 × 520 × 110 mm
Weight: 420 gr
Frame Material: EPP Foam, Plastic
Available Colors: Black, Brown
Max. Flight Time: 36 minutes
Max. Speed: 18 km/h (11,2 mp/h)
Controller Frequency: 2.4 GHz
Battery Capacity: 1,000 mAh
Battery Type: Lithium Polymer 3S
Battery Power: 11.1V
GPS Compatible: √
Altitude Hold: √
Position Hold: √
Flight Planning: –
Autonomous Flight: –
Target Tracking: –
Automatic Landing: –
Return To Home: –
Obstacle Avoidance: –
Headless Mode: –
Built-in Camera: √

Robota

Triton UAV

© Robota

Release Date: January 2012
Price: On Request
Number of Rotors: 1 (Fixed Wing)
Control System: Ground Control Station
Applications: Aerial Mapping, Agriculture, Inspection, Surveillance
Dimensions: 0.93 × 1.47 m
Wingspan: 1.47 m
Weight: 1.36 kg
Max. Payload: 250 gr
Frame Material: Composite
Available Colors: White
Max. Flight Time: 60 minutes
Max. Speed: 57 km/h (35,6 mp/h)
Max. Operating Range: 10 km
Battery Type: Lithium Polymer
GPS Compatible: √
Altitude Hold: √
Position Hold: –
Flight Planning: √
Autonomous Flight: √
Target Tracking: –
Automatic Landing: √
Return To Home: √
Obstacle Avoidance: –
Headless Mode: –
Built-in Camera: √

Bormatec

Cam Flyer Q

© Bormatec

Release Date: February 2012
Price: $180
Number of Rotors: 1 (Fixed Wing)
Control System: Flight Computer
Applications: Film, Photography, Aerial Mapping
Wingspan: 850 mm
Weight: 300 gr
Frame Material: EPP, Coroplast
Available Colors: Black
Controller Frequency: 2.4 GHz
Battery Capacity: 1,300 mAh
Battery Type: Lithium Polymer 3S
GPS Compatible: –
Altitude Hold: √
Position Hold: –
Flight Planning: √
Autonomous Flight: √
Target Tracking: √
Automatic Landing: –
Return To Home: √
Obstacle Avoidance: –
Headless Mode: –
Built-in Camera: √

ST Aerospace

Skyblade 360

© automacomp.blogspot.com

Release Date: February 2012
Price: Unknown
Number of Rotors: 1 (Fixed Wing)
Control System: Ground Control Station
Applications: Military Purposes, Reconnaissance, Surveillance, Target Acquisition
Dimensions: 1.8 × 3 m
Wingspan: 3 m
Weight: 9 kg
Frame Material: Composite
Available Colors: Black
Max. Flight Time: 6 hours
Max. Speed: 65 km/h (40 mp/h)
Max. Flying Altitude: 914 m (3,000 ft)
Max. Operating Range: 15 km
GPS Compatible: √
Altitude Hold: √
Position Hold: –
Flight Planning: √
Autonomous Flight: √
Target Tracking: –
Automatic Landing: –
Return To Home: √
Obstacle Avoidance: –
Headless Mode: –
Built-in Camera: √

MDBA

Fire Shadow

©MDBA

Release Date: March 2012
Price: Unknown
Number of Rotors: – (Fixed Wing)
Control System: Remote Control
Applications: Loitering Munition
Weight: 200 kg
Available Colors: Grey
Max. Flight Time: 6 hours
Max. Flying Altitude: 4,600 m (15,000 ft)
Max. Operating Range: 100 km
GPS Compatible: √
Altitude Hold: √
Position Hold: –
Flight Planning: √
Autonomous Flight: –
Target Tracking: √
Automatic Landing: –
Return To Home: –
Obstacle Avoidance: –
Headless Mode: –
Built-in Camera: –

Iran Aircraft Manufacturing Industrial Company

Shahed 129

©globalmilitaryreview.blogspot.com

Release Date: April 2012
Price: $12.5 million
Number of Rotors: 1 (Fixed Wing)
Control System: Ground Control Station
Applications: Military Purposes, Armed Reconnaissance, Surveillance, Target Acquisition
Dimensions: 6 × 10.5 m
Wingspan: 10.5 m
Weight: 990 kg
Max. Payload: 400 kg
Frame Material: Composite
Motors: 1 × Rotax 914 four-cylinder engine
Available Colors: Grey
Max. Flight Time: 24 hours
Max. Speed: 150 km/h (93 mp/h)
Max. Flying Altitude: 7,315 m (24,000 ft)
Max. Operating Range: 1,700 km
GPS Compatible: √
Altitude Hold: √
Position Hold: √
Flight Planning: –
Autonomous Flight: √
Target Tracking: √
Automatic Landing: √
Return To Home: √
Obstacle Avoidance: √
Headless Mode: –
Built-in Camera: √

AeroVironment

RQ-12A (Wasp AE)

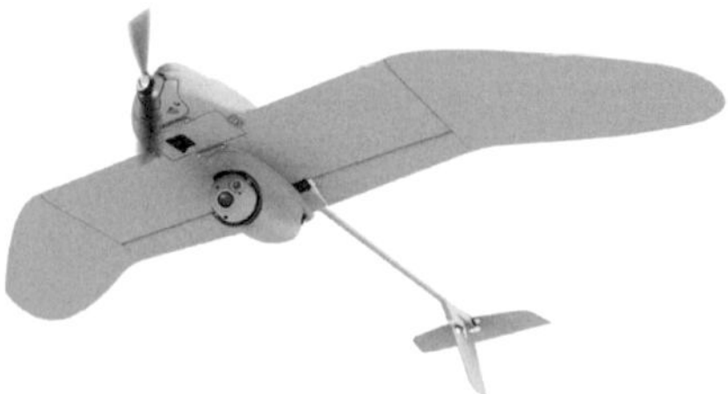

© AeroVironment

Release Date: May 2012
Price: Unknown
Number of Rotors: 1 (Fixed Wing)
Control System: Ground Control Station
Applications: Military Purposes, Maritime Patrol, Reconnaissance, Surveillance
Dimensions: 0.76 × 1.02 m
Wingspan: 1.02 m
Weight: 1.3 kg
Frame Material: Composite
Available Colors: White
Max. Flight Time: 50 minutes
Max. Speed: 83 km/h (51,5 mp/h)
Max. Flying Altitude: 304 m (1,000 ft)
Max. Operating Range: 5 km
GPS Compatible: √
Altitude Hold: √
Position Hold: –
Flight Planning: √
Autonomous Flight: √
Target Tracking: √
Automatic Landing: √
Return To Home: √
Obstacle Avoidance: √
Headless Mode: –
Built-in Camera: √

AeroVironment

Switchblade

© tehnomil.net

Release Date: May 2012
Price: Unknown
Number of Rotors: 1 (Fixed Wing)
Control System: Ground Control Station
Applications: Loitering Munition
Dimensions: 610 mm
Weight: 2.5 kg
Frame Material: Composite
Available Colors: Black, White
Max. Flight Time: 10 minutes
Max. Speed: 157 km/h (97,5 mp/h)
Max. Flying Altitude: 4,600 m (15,000 ft)
Max. Operating Range: 10 km
GPS Compatible: √
Altitude Hold: √
Position Hold: –
Flight Planning: √
Autonomous Flight: –
Target Tracking: √
Automatic Landing: –
Return To Home: –
Obstacle Avoidance: –
Headless Mode: –
Built-in Camera: √

Bormatec

Explorer

© Bormatec

Release Date: May 2012
Price: $375
Number of Rotors: 2 (Fixed Wing)
Control System: Flight Computer
Applications: Film, Photography, Aerial Mapping, Inspection, Surveillance
Dimensions: 1.4 × 2.2 m
Wingspan: 2.2 m
Weight: 2.5 kg
Max. Payload: 2.5 kg
Frame Material: EPP, Coroplast
Available Colors: Black
Max. Flight Time: 30 minutes
Battery Type: Lithium Polymer
GPS Compatible: –
Altitude Hold: √
Position Hold: –
Flight Planning: √
Autonomous Flight: √
Target Tracking: √
Automatic Landing: –
Return To Home: √
Obstacle Avoidance: –
Headless Mode: –
Built-in Camera: √

Draganfly

Draganflyer X4-P

© Draganfly

Release Date: May 2012
Price: $25,000
Number of Rotors: 4
Control System: Flight Computer
Applications: Film, Photography, Aerial Mapping
Dimensions: 870 × 870 × 300 mm
Wingspan: 1070 mm
Weight: 1.67 kg
Max. Payload: 800 gr
Frame Material: Carbon Fiber
Rotor Material: Carbon Fiber
Available Colors: Green, Red, White
Max. Flight Time: 30 minutes
Max. Speed: 50 km/h (31 mp/h)
Max. Flying Altitude: 2,438 m (8,000 ft)
Max. Ascent Speed: 2 m/s
Max. Descent Speed: 2 m/s
Controller Frequency: 2.4 GHz
Battery Capacity: 5,400 mAh
Battery Type: Lithium Polymer 4S
GPS Compatible: –
Altitude Hold: √
Position Hold: √
Flight Planning: √
Autonomous Flight: √
Target Tracking: –
Automatic Landing: √
Return To Home: √
Obstacle Avoidance: –
Headless Mode: –
Built-in Camera: –

Hobby Lord

ST550 Bumblebee RTF

© blog.synosystems.de

Release Date: May 2012
Price: $969
Number of Rotors: 4
Control System: Remote Control
Applications: Film, Photography, Aerial Mapping
Dimensions: 550 × 550 × 360 mm
Weight: 1.2 kg
Max. Payload: 500 gr
Frame Material: Carbon Fiber
Rotor Material: Plastic
Motors: 4 × HL3512
Available Colors: Black, Green, Red, White
Max. Flight Time: 10 minutes
Controller Frequency: 2.4 GHz
Battery Capacity: 5,500 mAh
Battery Type: Lithium Polymer 3S
Battery Power: 3.7V
GPS Compatible: –
Altitude Hold: √
Position Hold: √
Flight Planning: √
Autonomous Flight: √
Target Tracking: –
Automatic Landing: √
Return To Home: √
Obstacle Avoidance: –
Headless Mode: –
Built-in Camera: –

ZALA Aero Group

Zala 421-16EM

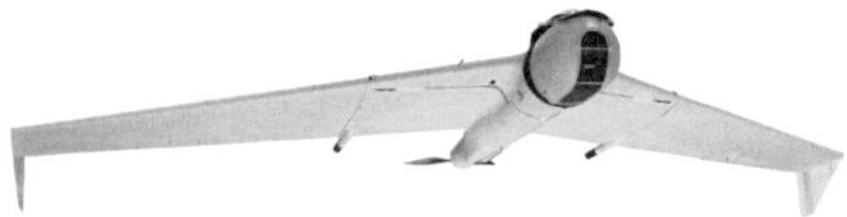

© innovudm.ru

Release Date: May 2012
Price: Unknown
Number of Rotors: 1 (Fixed Wing)
Control System: Ground Control Station
Applications: Military Purposes, Inspection, Reconnaissance, Surveillance
Wingspan: 1.81 m
Weight: 6.5 kg
Max. Payload: 1 kg
Frame Material: Composite
Available Colors: Grey
Max. Flight Time: 2.5 hours
Max. Speed: 110 km/h (68 mp/h)
Max. Flying Altitude: 3,600 m (11,811 ft)
Min. Operating Temperature: -30°C
Max. Operating Temperature: 40°C
Max. Operating Range: 25 km
Battery Capacity: 10,000 mAh
Battery Type: Lithium Polymer 10S
GPS Compatible: √
Altitude Hold: √
Position Hold: –
Flight Planning: √
Autonomous Flight: √
Target Tracking: √
Automatic Landing: –
Return To Home: √
Obstacle Avoidance: √
Headless Mode: –
Built-in Camera: √

Boeing Insitu

Phantom Eye

© Boeing Insitu

Release Date: June 2012
Price: $60 million
Number of Rotors: 2 (Fixed Wing)
Control System: Ground Control Station
Applications: Military Purposes, Reconnaissance, Surveillance
Wingspan: 46 m
Weight: 4,445 kg
Motors: 2 × Ford 2.3L, four-cylinder liquid-hydrogen fueled engine
Available Colors: White
Max. Flight Time: 96 hours
Max. Speed: 370 km/h (230 mp/h)
Max. Flying Altitude: 19,800 m (65,000 ft)
GPS Compatible: √
Altitude Hold: √
Position Hold: –
Flight Planning: √
Autonomous Flight: √
Target Tracking: √
Automatic Landing: √
Return To Home: √
Obstacle Avoidance: √
Headless Mode: –
Built-in Camera: √

Chengdu Aircraft Industry Group (CAIG)

VMA-01

© Chinese Military Review

Release Date: June 2012
Price: Unknown
Number of Rotors: 2 (Fixed Wing)
Control System: Ground Control Station
Applications: Military Purposes, Maritime Patrol, Reconnaissance, Surveillance
Wingspan: 4.8 m
Weight: 110 kg
Max. Payload: 15 kg
Frame Material: Composite
Available Colors: White
Max. Flight Time: 2 hours
Max. Speed: 220 km/h (137 mp/h)
Max. Flying Altitude: 3,000 m (9,842 ft)
Max. Operating Range: 100 km
GPS Compatible: √
Altitude Hold: √
Position Hold: √
Flight Planning: √
Autonomous Flight: √
Target Tracking: –
Automatic Landing: √
Return To Home: √
Obstacle Avoidance: –
Headless Mode: –
Built-in Camera: √

Innocon Ltd.

Micro Falcon

© asianmilitaryreview.com

Release Date: June 2012
Price: Unknown
Number of Rotors: 1 (Fixed Wing)
Control System: Ground Control Station
Applications: Military Purposes, Inspection, Reconnaissance, Surveillance
Dimensions: 1 × 2 × 0.3 m
Wingspan: 2 m
Weight: 6 kg
Max. Payload: 1 kg
Frame Material: Composite
Available Colors: White
Max. Flight Time: 2 hours
Max. Speed: 75 km/h (46,6 mp/h)
Max. Flying Altitude: 700 m (2,296 ft)
Max. Operating Range: 20 km
Controller Frequency: 2.4 GHz
Battery Type: Lithium Polymer
GPS Compatible: √
Altitude Hold: √
Position Hold: –
Flight Planning: √
Autonomous Flight: √
Target Tracking: –
Automatic Landing: –
Return To Home: √
Obstacle Avoidance: √
Headless Mode: –
Built-in Camera: √

Northrop Grumman

MQ-4C Triton

© Wikipedia/Alex Evers

Release Date: June 2012
Price: $120.689 million
Number of Rotors: – (Fixed Wing)
Control System: Ground Control Station
Applications: Military Purposes, Reconnaissance, Surveillance
Dimensions: 14.5 × 39.9 × 4.7 m
Weight: 14,628 kg
Frame Material: Composite
Motors: 1 × Rolls-Royce AE 3007 turbofan engine
Available Colors: Grey
Max. Flight Time: 30 hours
Max. Speed: 575 km/h (357 mp/h)
Max. Flying Altitude: 18,000 m (60,000 ft)
Max. Operating Range: 15,186 km
GPS Compatible: √
Altitude Hold: √
Position Hold: –
Flight Planning: √
Autonomous Flight: √
Target Tracking: √
Automatic Landing: √
Return To Home: √
Obstacle Avoidance: √
Headless Mode: –
Built-in Camera: √

Skycam UAV

Kahu

© uavataut.blogspot.com

Release Date: June 2012
Price: Unknown
Number of Rotors: 1 (Fixed Wing)
Control System: Ground Control Station
Applications: Military Purposes, Reconnaissance, Surveillance
Wingspan: 2.2 m
Weight: 3.9 kg
Frame Material: Composite
Available Colors: Grey
Max. Flight Time: 2 hours
Max. Speed: 100 km/h (62 mp/h)
Max. Flying Altitude: 5,030 m (16,500 ft)
Max. Operating Range: 25 km
Controller Frequency: 2.4 GHz
Battery Capacity: 8,400 mAh
Battery Type: Lithium Polymer 4S
Battery Power: 16.8V
GPS Compatible: √
Altitude Hold: √
Position Hold: –
Flight Planning: √
Autonomous Flight: √
Target Tracking: –
Automatic Landing: √
Return To Home: √
Obstacle Avoidance: –
Headless Mode: –
Built-in Camera: √

Team BlackSheep

Discovery

© Team BlackSheep

Release Date: June 2012
Price: $1,695
Number of Rotors: 4
Control System: Remote Control
Applications: FPV Racing
Weight: 2 kg
Frame Material: Fiberglass
Available Colors: Black
Max. Flight Time: 15 minutes
Max. Flying Altitude: 2,000 m (6,560 ft)
Max. Operating Range: 5 km
Controller Frequency: 5.8 GHz
Battery Capacity: 4,500 mAh
Battery Type: Lithium Polymer 4S
Battery Power: 14.8V
GPS Compatible: √
Altitude Hold: –
Position Hold: –
Flight Planning: –
Autonomous Flight: –
Target Tracking: –
Automatic Landing: –
Return To Home: √
Obstacle Avoidance: –
Headless Mode: –
Built-in Camera: √ (TBS59, TBS69)

Xi'an Aisheng Technology Group

ASN-218

© mil.ifeng.com

Release Date: June 2012
Price: Unknown
Number of Rotors: 1 (Fixed Wing)
Control System: Ground Control Station
Applications: Military Purposes, Reconnaissance, Surveillance
Weight: 40 kg
Frame Material: Composite
Available Colors: Grey
Max. Flight Time: 8 hours
Max. Speed: 140 km/h (87 mp/h)
Max. Flying Altitude: 4,600 m (15,092 ft)
Max. Operating Range: 42 km
GPS Compatible: √
Altitude Hold: √
Position Hold: –
Flight Planning: √
Autonomous Flight: √
Target Tracking: –
Automatic Landing: √
Return To Home: √
Obstacle Avoidance: –
Headless Mode: –
Built-in Camera: √

ADCOM Systems

Yabhon Smart Eye

©jangaavaran.ir

Release Date: July 2012
Price: Unknown
Number of Rotors: 1 (Fixed Wing)
Control System: Flight Computer
Applications: Military Purposes, Reconnaissance, Surveillance
Dimensions: 7 × 21 × 2 m
Wingspan: 21 m
Weight: 450 kg
Max. Payload: 550 kg
Available Colors: Grey
Max. Flight Time: 120 hours
Max. Speed: 220 km/h (137 mp/h)
Max. Flying Altitude: 7,315 m (24,000 ft)
GPS Compatible: √
Altitude Hold: √
Position Hold: √
Flight Planning: √
Autonomous Flight: √
Target Tracking: √
Automatic Landing: √
Return To Home: √
Obstacle Avoidance: √
Headless Mode: –
Built-in Camera: √

AirRobot

AR 200

© AirRobot

Release Date: July 2012
Price: On Request
Number of Rotors: 6
Control System: Flight Computer
Applications: Film, Photography, Aerial Mapping
Dimensions: 2.2 × 2.2 m
Weight: 10 kg
Max. Payload: 1.5 kg
Frame Material: Carbon Fiber
Available Colors: White
Max. Flight Time: 40 minutes
Max. Speed: 43 km/h (27 mp/h)
Min. Operating Temperature: -19°C
Max. Operating Temperature: 49°C
Max. Operating Range: 1 km
Controller Frequency: 2.4 GHz
Battery Type: Lithium Polymer
GPS Compatible: √
Altitude Hold: √
Position Hold: √
Flight Planning: √
Autonomous Flight: √
Target Tracking: –
Automatic Landing: √
Return To Home: √
Obstacle Avoidance: √
Headless Mode: –
Built-in Camera: √ (10 megapixels)

Lynxmotion

Hunter VTail 500

© robotshop.com

Release Date: July 2012
Price: $169
Number of Rotors: 4
Control System: Remote Control
Applications: Recreation, FPV Racing
Dimensions: 433 × 473 × 157 mm
Weight: 365 gr
Rotor Material: Fiberglass
Available Colors: Black
Controller Frequency: 2.4 GHz
Battery Capacity: 2,000 mAh
Battery Type: Lithium Polymer 3S
Battery Power: 11.1V
GPS Compatible: –
Altitude Hold: √
Position Hold: √
Flight Planning: –
Autonomous Flight: –
Target Tracking: –
Automatic Landing: –
Return To Home: –
Obstacle Avoidance: –
Headless Mode: –
Built-in Camera: –

Tekever

AR4 Light Ray Evolution

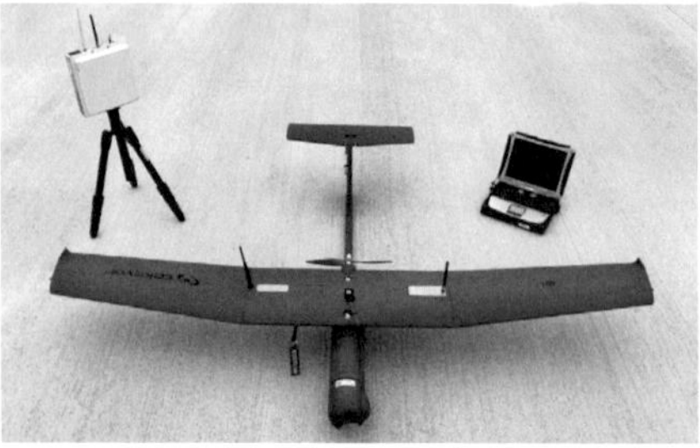

© Tekever

Release Date: July 2012
Price: On Request
Number of Rotors: 1 (Fixed Wing)
Control System: Ground Control Station
Applications: Military Purposes, Reconnaissance, Surveillance, Target Acquisition
Dimensions: 1.2 × 1.8 m
Wingspan: 1.8 m
Weight: 5 kg
Max. Payload: 1 kg
Frame Material: Composite
Available Colors: Grey
Max. Flight Time: 2 hours
Max. Speed: 80 km/h (49,7 mp/h
Max. Operating Range: 20 km
Controller Frequency: 2.4 GHz
GPS Compatible: √
Altitude Hold: √
Position Hold: √
Flight Planning: √
Autonomous Flight: √
Target Tracking: –
Automatic Landing: –
Return To Home: √
Obstacle Avoidance: –
Headless Mode: –
Built-in Camera: –

CATUAV

Argos

© euskaditecnologia.com

Release Date: August 2012
Price: Unknown
Number of Rotors: 1 (Fixed Wing)
Control System: Ground Control Station
Applications: Film, Photography, Aerial Mapping, Inspection, Surveillance
Weight: 12 kg
Max. Payload: 6.25 kg
Available Colors: Grey
Max. Flight Time: 14 hours
Max. Speed: 145 km/h (90 mp/h)
Max. Flying Altitude: 3,000 m (9,842 ft)
Max. Operating Range: 400 km
GPS Compatible: √
Altitude Hold: √
Position Hold: –
Flight Planning: √
Autonomous Flight: √
Target Tracking: –
Automatic Landing: √
Return To Home: √
Obstacle Avoidance: –
Headless Mode: –
Built-in Camera: √

Skycam UAV

Swamp Fox

© kahunet.co.nz

Release Date: August 2012
Price: Unknown
Number of Rotors: 1 (Fixed Wing)
Control System: Ground Control Station
Applications: Aerial Mapping, Inspection, Surveillance
Wingspan: 1.8 m
Weight: 4.5 kg
Frame Material: EPP Foam
Available Colors: Brown
Max. Flight Time: 50 minutes
Max. Speed: 100 km/h (62 mp/h)
Controller Frequency: 2.4 GHz
Battery Type: Lithium Polymer
Battery Power: 14.8V
GPS Compatible: √
Altitude Hold: √
Position Hold: –
Flight Planning: √
Autonomous Flight: √
Target Tracking: –
Automatic Landing: –
Return To Home: √
Obstacle Avoidance: –
Headless Mode: –
Built-in Camera: √

Turkish Aerospace Industries (TAI)

Şimşek

© defenceturk.net

Release Date: August 2012
Price: Unknown
Number of Rotors: – (Fixed Wing)
Control System: Ground Control Station
Applications: Military Purposes, Loitering Munition, Target Acquisition
Frame Material: Composite
Available Colors: Black
Max. Flight Time: 45 minutes
Max. Speed: 740 km/h (460 mp/h)
Max. Flying Altitude: 4,572 m (15,000 ft)
Max. Operating Range: 50 km
GPS Compatible: √
Altitude Hold: √
Position Hold: –
Flight Planning: √
Autonomous Flight: √
Target Tracking: √
Automatic Landing: –
Return To Home: √
Obstacle Avoidance: √
Headless Mode: –
Built-in Camera: –

Xmobots

Nauru 500A

© theatlantic.com

Release Date: August 2012
Price: Unknown
Number of Rotors: 1 (Fixed Wing)
Control System: Ground Control Station
Applications: Military Purposes, Inspection, Reconnaissance
Dimensions: 1.83 × 2.32 × 0.68 m
Wingspan: 2.32 m
Weight: 15 kg
Max. Payload: 6 kg
Frame Material: Composite
Available Colors: White
Max. Flight Time: 5.5 hours
Max. Speed: 108 km/h (67 mp/h)
Max. Flying Altitude: 3,000 m (9,842 ft)
GPS Compatible: √
Altitude Hold: √
Position Hold: √
Flight Planning: √
Autonomous Flight: √
Target Tracking: –
Automatic Landing: √
Return To Home: √
Obstacle Avoidance: √
Headless Mode: –
Built-in Camera: √

Integrated Dynamics

Pride

© Integrated Dynamics

Release Date: September 2012
Price: Unknown
Number of Rotors: 1 (Fixed Wing)
Control System: Ground Control Station
Applications: Aerial Mapping, Inspection, Surveillance
Weight: 5 kg
Frame Material: Carbon Fiber, Composite
Available Colors: White
Max. Flight Time: 60 minutes
Max. Flying Altitude: 1,000 m (3,280 ft)
Max. Operating Range: 15 km
GPS Compatible: √
Altitude Hold: √
Position Hold: –
Flight Planning: √
Autonomous Flight: √
Target Tracking: –
Automatic Landing: –
Return To Home: √
Obstacle Avoidance: –
Headless Mode: –
Built-in Camera: √

ST Aerospace

Skyblade IV

Release Date: September 2012
Price: $8.5 million
Number of Rotors: 1 (Fixed Wing)
Control System: Ground Control Station
Applications: Military Purposes, Reconnaissance, Surveillance, Target Acquisition
Dimensions: 2.4 × 3.7 m
Wingspan: 3.7 m
Weight: 70 kg
Max. Payload: 12 kg
Frame Material: Composite
Available Colors: Grey
Max. Flight Time: 12 hours
Max. Speed: 148 km/h (92 mp/h)
Max. Flying Altitude: 4,572 m (15,000 ft)
Max. Operating Range: 100 km
GPS Compatible: √
Altitude Hold: √
Position Hold: –
Flight Planning: √
Autonomous Flight: √
Target Tracking: –
Automatic Landing: √
Return To Home: √
Obstacle Avoidance: –
Headless Mode: –
Built-in Camera: √

SteadiDrone

EI8HT

Release Date: September 2012
Price: Unknown
Number of Rotors: 8
Control System: Remote Control
Applications: Film, Photography, Aerial Mapping, Transport
Dimensions: 1.1 × 1.1 × 0.45 m
Wingspan: 1.2 m
Weight: 4.8 kg
Max. Payload: 10 kg
Frame Material: Carbon Fiber
Rotor Material: Wood
Motors: 400kv 4014
Available Colors: Black
Max. Flight Time: 15 minutes
Max. Speed: 70 km/h (43 mp/h)
Max. Operating Range: 1.5 km
Controller Frequency: 2.4 GHz
Battery Capacity: 6,200 mAh
Battery Type: Lithium Polymer 6S
GPS Compatible: √
Altitude Hold: √
Position Hold: √
Flight Planning: √
Autonomous Flight: √
Target Tracking: –
Automatic Landing: √
Return To Home: √
Obstacle Avoidance: –
Headless Mode: –
Built-in Camera: –

Denel Dynamics

Hungwe

© Denel Dynamics

Release Date: October 2012
Price: Unknown
Number of Rotors: 1 (Fixed Wing)
Control System: Ground Control Station
Applications: Military Purposes, Inspection, Maritime Patrol, Reconnaissance, Surveillance
Dimensions: 2 × 3 × 0.45 m
Wingspan: 3 m
Weight: 35 kg
Available Colors: Grey
Max. Speed: 150 km/h (93 mp/h)
Max. Flying Altitude: 5,000 m (16,404 ft)
Max. Operating Range: 100 km
GPS Compatible: √
Altitude Hold: √
Position Hold: –
Flight Planning: √
Autonomous Flight: √
Target Tracking: –
Automatic Landing: √
Return To Home: √
Obstacle Avoidance: –
Headless Mode: –
Built-in Camera: √

Dronetools

DroneHexa

© Dronetools

Release Date: October 2012
Price: On Request
Number of Rotors: 4
Control System: Flight Computer, Laptop
Applications: Film, Photography, Aerial Mapping
Max. Payload: 3 kg
Frame Material: Carbon Fiber
Rotor Material: Carbon Fiber
Available Colors: Black
Max. Flight Time: 8 hours
Max. Operating Range: 2 km
Controller Frequency: 2.4 GHz
Battery Capacity: 7,000 mAh
Battery Type: Lithium Polymer
GPS Compatible: √
Altitude Hold: √
Position Hold: √
Flight Planning: √
Autonomous Flight: √
Target Tracking: –
Automatic Landing: √
Return To Home: √
Obstacle Avoidance: –
Headless Mode: –
Built-in Camera: –

Embention

Flamingo Firefighting

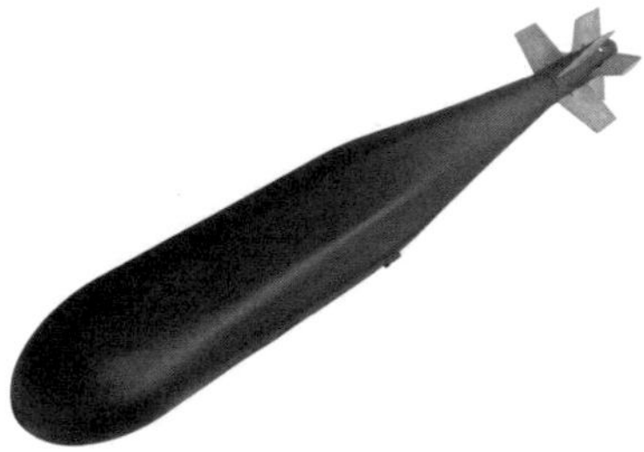

© Embention

Release Date: October 2012
Price: Unknown
Number of Rotors: –
Control System: Flight Computer
Applications: Firefighting
Dimensions: 2.65 m
Weight: 21.7 kg
Max. Payload: 200 l
Available Colors: Grey
Max. Flying Altitude: 3,962 m (13,00 ft)
Max. Operating Range: 3.5 km
GPS Compatible: √
Altitude Hold: √
Position Hold: –
Flight Planning: √
Autonomous Flight: √
Target Tracking: –
Automatic Landing: –
Return To Home: –
Obstacle Avoidance: –
Headless Mode: –
Built-in Camera: –

HeliMax

1SQ RTF

© amainhobbies.com

Release Date: October 2012
Price: $100
Number of Rotors: 4
Control System: Remote Control
Applications: Recreation
Dimensions: 143 × 145 × 41 mm
Weight: 33 gr
Frame Material: Plastic
Rotor Material: Plastic
Available Colors: White
Controller Frequency: 2.4 GHz
Battery Type: Lithium Polymer
GPS Compatible: –
Altitude Hold: √
Position Hold: √
Flight Planning: –
Autonomous Flight: –
Target Tracking: –
Automatic Landing: –
Return To Home: –
Obstacle Avoidance: –
Headless Mode: –
Built-in Camera: √

Irkut Corporation

Irkut-3

©Wikipedia/Vitaly V. Kuzmin

Release Date: October 2012
Price: Unknown
Number of Rotors: 1 (Fixed Wing)
Control System: Ground Control Station
Applications: Military Purposes, Inspection, Surveillance
Dimensions: 0.9 × 2 × 0.3 m
Wingspan: 2 m
Weight: 3 kg
Max. Payload: 0.5 kg
Frame Material: Composite
Available Colors: Grey
Max. Flight Time: 75 minutes
Max. Speed: 90 km/h (56 mp/h)
Max. Flying Altitude: 3,000 m (9,842 ft)
Min. Operating Temperature: -30°C
Max. Operating Temperature: 50°C
Max. Operating Range: 15 km
GPS Compatible: √
Altitude Hold: √
Position Hold: –
Flight Planning: √
Autonomous Flight: √
Target Tracking: √
Automatic Landing: √
Return To Home: √
Obstacle Avoidance: √
Headless Mode: –
Built-in Camera: √

Volt Robotics

WaveSight

©Volt Robotics

Release Date: October 2012
Price: Unknown
Number of Rotors: – (Fixed Wing)
Control System: Flight Computer
Applications: Military Purposes, Inspection, Surveillance
Dimensions: 1.36 × 2.3 × 0.38 m
Wingspan: 2.3 m
Weight: 9 kg
Frame Material: Composite
Available Colors: Grey
Max. Flight Time: 2 hours
Max. Speed: 72 km/h (45 mp/h)
Max. Operating Range: 10 km
GPS Compatible: √
Altitude Hold: √
Position Hold: –
Flight Planning: √
Autonomous Flight: √
Target Tracking: √
Automatic Landing: –
Return To Home: √
Obstacle Avoidance: √
Headless Mode: –
Built-in Camera: √ (20 megapixels)

China Aerospace Science and Industry Corporation (CASIC)

SF-460 Blade

©forums.eagle.ru

Release Date: November 2012
Price: Unknown
Number of Rotors: 1 (Fixed Wing)
Control System: Ground Control Station
Applications: Military Purposes, Inspection, Surveillance
Dimensions: 3.3 × 4.6 m
Wingspan: 4.6 m
Weight: 85 kg
Max. Payload: 15 kg
Frame Material: Composite
Available Colors: Yellow
Max. Flight Time: 6 hours
Max. Speed: 110 km/h (68 mp/h)
Max. Flying Altitude: 3,000 m (9,842 ft)
GPS Compatible: √
Altitude Hold: √
Position Hold: –
Flight Planning: √
Autonomous Flight: √
Target Tracking: –
Automatic Landing: √
Return To Home: √
Obstacle Avoidance: √
Headless Mode: –
Built-in Camera: √

China Aerospace Science and Technology Corporation (CASC)

CH-92

©Ken Duffey

Release Date: November 2012
Price: Unknown
Number of Rotors: 1 (Fixed Wing)
Control System: Ground Control Station
Applications: Military Purposes, Reconnaissance, Surveillance, Target Acquisition
Dimensions: 4.1 × 9 × 1.6 m
Wingspan: 9 m
Weight: 300 kg
Max. Payload: 60 kg
Frame Material: Composite
Available Colors: Grey
Max. Flight Time: 10 hours
Max. Speed: 190 km/h (118 mp/h)
Max. Flying Altitude: 6,000 m (19,685 ft)
Max. Operating Range: 250 km
GPS Compatible: √
Altitude Hold: √
Position Hold: –
Flight Planning: √
Autonomous Flight: √
Target Tracking: √
Automatic Landing: √
Return To Home: √
Obstacle Avoidance: √
Headless Mode: –
Built-in Camera: √

Global Industrial and Defefense Solutions (GIDS)

Shahpar

Release Date: November 2012
Price: $500,000
Number of Rotors: 1 (Fixed Wing)
Control System: Ground Control Station
Applications: Military Purposes, Reconnaissance, Surveillance
Dimensions: 4.2 × 6.6 m
Wingspan: 6.6 m
Weight: 480 kg
Max. Payload: 50 kg
Frame Material: Composite
Available Colors: Grey
Max. Flight Time: 7 hours
Max. Speed: 150 km/h (93 mp/h)
Max. Flying Altitude: 5,000 m (16,404 ft)
Max. Operating Range: 250 km
GPS Compatible: √
Altitude Hold: √
Position Hold: –
Flight Planning: √
Autonomous Flight: √
Target Tracking: √
Automatic Landing: √
Return To Home: √
Obstacle Avoidance: √
Headless Mode: –
Built-in Camera: √

Integrated Dynamics

Nishan MK-II

Release Date: November 2012
Price: Unknown
Number of Rotors: 1 (Fixed Wing)
Control System: Ground Control Station
Applications: Aerial Target
Dimensions: 3.06 × 2.87 × 0.89 m
Wingspan: 2.87 m
Weight: 30 kg
Max. Payload: 12 kg
Frame Material: Composite
Motors: Two-cylinder piston engine
Available Colors: Red
Max. Flight Time: 60 minutes
Max. Speed: 300 km/h (186 mp/h)
Max. Flying Altitude: 2,000 m (6,560 ft)
Max. Operating Range: 35 km
GPS Compatible: √
Altitude Hold: √
Position Hold: –
Flight Planning: √
Autonomous Flight: √
Target Tracking: –
Automatic Landing: √
Return To Home: √
Obstacle Avoidance: –
Headless Mode: –
Built-in Camera: –

Integrated Dynamics

Skycam UAV

Release Date: November 2012
Price: Unknown
Number of Rotors: 1 (Fixed Wing)
Control System: Ground Control Station
Applications: Aerial Mapping, Inspection, Surveillance
Weight: 2 kg
Frame Material: Composite, EPP Foam
Available Colors: Grey
Max. Flight Time: 30 minutes
Max. Flying Altitude: 304 m (1,000 ft)
Max. Operating Range: 5 km
Controller Frequency: 2.4 GHz
Battery Type: Lithium Polymer
GPS Compatible: √
Altitude Hold: √
Position Hold: –
Flight Planning: √
Autonomous Flight: √
Target Tracking: –
Automatic Landing: –
Return To Home: √
Obstacle Avoidance: –
Headless Mode: –
Built-in Camera: √ (1080P HD)

Lehmann Aviation

LA100

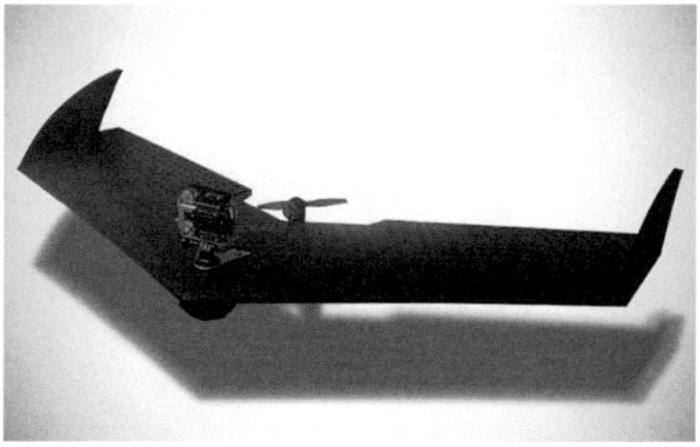

Release Date: November 2012
Price: Unknown
Number of Rotors: 1 (Fixed Wing)
Control System: Remote Control
Applications: Film, Photography
Wingspan: 920 mm
Weight: 850 gr
Max. Payload: 150 gr
Frame Material: EPP Foam
Rotor Material: Plastic
Available Colors: Black
Max. Flight Time: 5 minutes
Max. Speed: 35 km/h (21,7 mp/h)
Min. Operating Temperature: -25°C
Max. Operating Temperature: 60°C
Max. Operating Range: 500 m
Controller Frequency: 2.4 GHz
Battery Type: Lithium Polymer
GPS Compatible: √
Altitude Hold: –
Position Hold: –
Flight Planning: √
Autonomous Flight: √
Target Tracking: –
Automatic Landing: √
Return To Home: –
Obstacle Avoidance: –
Headless Mode: –
Built-in Camera: –

Lynxmotion

Hunter VTail 400

© robotshop.com

Release Date: November 2012
Price: $369
Number of Rotors: 4
Control System: Remote Control
Applications: Recreation, FPV Racing
Dimensions: 309 × 340 × 143 mm
Weight: 326 gr
Rotor Material: Fiberglass
Available Colors: Black
Controller Frequency: 2.4 GHz
Battery Capacity: 2,000 mAh
Battery Type: Lithium Polymer 3S
Battery Power: 11.1V
GPS Compatible: –
Altitude Hold: √
Position Hold: √
Flight Planning: –
Autonomous Flight: –
Target Tracking: –
Automatic Landing: –
Return To Home: –
Obstacle Avoidance: –
Headless Mode: –
Built-in Camera: –

Syma Toys

X1

© cdiscount.com

Release Date: November 2012
Price: $35
Number of Rotors: 4
Control System: Remote Control
Applications: Recreation
Dimensions: 360 × 360 × 60 mm
Weight: 600 gr
Frame Material: Plastic
Rotor Material: Plastic
Available Colors: Blue
Max. Flight Time: 6 minutes
Max. Operating Range: 40 m
Controller Frequency: 2.4 GHz
Battery Capacity: 350 mAh
Battery Type: Lithium Polymer
Battery Power: 3.7V
GPS Compatible: √
Altitude Hold: √
Position Hold: √
Flight Planning: √
Autonomous Flight: √
Target Tracking: √
Automatic Landing: √
Return To Home: √
Obstacle Avoidance: √
Headless Mode: √
Built-in Camera: √

Dassault Aviation

nEUROn

© Dassault Aviation

Release Date: December 2012
Price: $25 million
Number of Rotors: – (Fixed Wing)
Control System: Ground Control Station
Applications: Military Purposes, Armed Reconnaissance, Surveillance, Target Acquisition
Dimensions: 9.5 × 12.5 m
Wingspan: 12.5 m
Weight: 4,900 kg
Max. Payload: 2,100 kg
Frame Material: Composite
Motors: 1 × Rolls-Royce/Turbomeca Adour/Snecma M88
Available Colors: Grey
Max. Speed: 980 km/h (609 mp/h)
Max. Flying Altitude: 14,000 m (45,931 ft)
GPS Compatible: √
Altitude Hold: √
Position Hold: –
Flight Planning: √
Autonomous Flight: √
Target Tracking: √
Automatic Landing: √
Return To Home: √
Obstacle Avoidance: √
Headless Mode: –
Built-in Camera: √

Hongdu Aviation Industry Group (HAIG)

Sky Crossbow (Tian-Nu)

© HAIG

Release Date: December 2012
Price: Unknown
Number of Rotors: – (Fixed Wing)
Control System: Ground Control Station
Applications: Military Purposes, Reconnaissance, Surveillance
Dimensions: 2.15 × 2 × 0.6 m
Weight: 18.8 kg
Available Colors: White
Max. Speed: 150 km/h (93 mp/h)
GPS Compatible: √
Altitude Hold: √
Position Hold: –
Flight Planning: √
Autonomous Flight: √
Target Tracking: –
Automatic Landing: √
Return To Home: √
Obstacle Avoidance: –
Headless Mode: –
Built-in Camera: √

National Aerospace Laboratories (NAL)

Golden Hawk

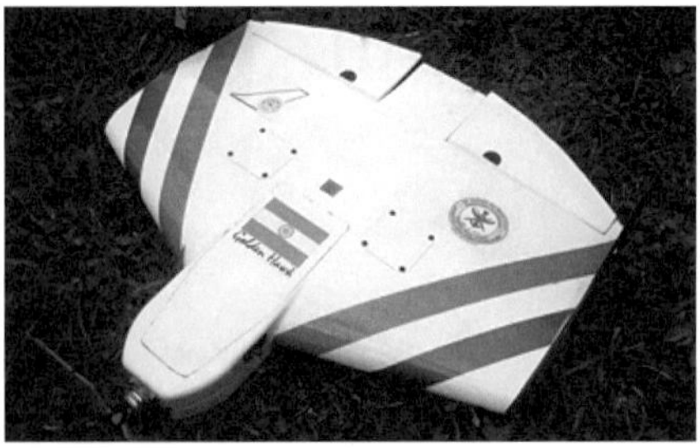
© sps-aviation.com

Release Date: December 2012
Price: Unknown
Number of Rotors: 1 (Fixed Wing)
Control System: Remote Control
Applications: Demonstration, Research & Development, Reconnaissance, Surveillance
Dimensions: 400 × 300 × 180 mm
Wingspan: 300 mm
Weight: 1.25 kg
Available Colors: White
Max. Speed: 55 km/h (34 mp/h)
Max. Flying Altitude: 300 m (985 ft)
Max. Operating Range: 2 km
Controller Frequency: 2.4 GHz
Battery Capacity: 1,350 mAh
Battery Type: Lithium Polymer 3S
Battery Power: 11.1V
GPS Compatible: –
Altitude Hold: √
Position Hold: –
Flight Planning: √
Autonomous Flight: √
Target Tracking: –
Automatic Landing: –
Return To Home: √
Obstacle Avoidance: –
Headless Mode: –
Built-in Camera: √

Popov Omsk Radio Factory

Iskatel

© Russian Ministry of Defense

Release Date: December 2012
Price: $100,000
Number of Rotors: 1 (Fixed Wing)
Control System: Ground Control Station
Applications: Military Purposes, Reconnaissance, Surveillance
Wingspan: 600 mm
Weight: 1.3 kg
Frame Material: Composite
Available Colors: Grey
Max. Flight Time: 40 minutes
Max. Flying Altitude: 4,000 m (13,123 ft)
GPS Compatible: √
Altitude Hold: √
Position Hold: –
Flight Planning: √
Autonomous Flight: √
Target Tracking: –
Automatic Landing: –
Return To Home: √
Obstacle Avoidance: –
Headless Mode: –
Built-in Camera: √

SATUMA

Mini UAV Salaar

©Satuma

Release Date: December 2012
Price: Unknown
Number of Rotors: 1 (Fixed Wing)
Control System: Ground Control Station
Applications: Military Purposes, Inspection, Reconnaissance, Surveillance
Dimensions: 1.35 × 2.16 m
Wingspan: 2.16 m
Weight: 4 kg
Frame Material: Composite
Available Colors: Grey
Max. Flight Time: 60 minutes
Max. Speed: 50 km/h (31 mp/h)
Max. Flying Altitude: 1,524 m (5,000 ft)
Max. Operating Range: 15 km
Controller Frequency: 2.4 GHz
Battery Type: Lithium Polymer
GPS Compatible: √
Altitude Hold: √
Position Hold: –
Flight Planning: √
Autonomous Flight: √
Target Tracking: –
Automatic Landing: √
Return To Home: √
Obstacle Avoidance: √
Headless Mode: –
Built-in Camera: √

2013

DJI

Phantom 1

Release Date: January 2013
Price: $679
Number of Rotors: 4
Control System: Remote Control
Applications: Film, Photography
Dimensions: 432 × 432 × 203 mm
Wingspan: 350 mm
Weight: 1.2 kg
Frame Material: Plastic
Rotor Material: Plastic
Available Colors: White
Max. Flight Time: 15 minutes
Max. Speed: 36 km/h (22,3 mp/h)
Max. Ascent Speed: 6 m/s
Max. Descent Speed: 6 m/s
Min. Operating Temperature: -10°C
Max. Operating Temperature: 50°C
Max. Operating Range: 100 m
Controller Frequency: 2.4 GHz
Battery Capacity: 2,200 mAh
Battery Type: Lithium Polymer 3S
Battery Weight: 170 gr
GPS Compatible: √
Altitude Hold: √
Position Hold: √
Flight Planning: √
Autonomous Flight: √
Target Tracking: –
Automatic Landing: √
Return To Home: √
Obstacle Avoidance: –
Headless Mode: –
Built-in Camera: –

Innocon Ltd.

Spider

© Innocon Ltd.

Release Date: January 2013
Price: Unknown
Number of Rotors: 1 (Fixed Wing)
Control System: Ground Control Station
Applications: Military Purposes, Reconnaissance, Surveillance
Dimensions: 0.5 × 1.5 × 0.16 m
Wingspan: 1.5 m
Weight: 2.5 kg
Max. Payload: 1.5 kg
Frame Material: Composite
Available Colors: Brown
Max. Speed: 100 km/h (62 mp/h)
Max. Flying Altitude: 4,100 m (13,450 ft)
Max. Operating Range: 5 km
Controller Frequency: 2.4 GHz
Battery Type: Lithium Polymer
GPS Compatible: √
Altitude Hold: √
Position Hold: –
Flight Planning: √
Autonomous Flight: √
Target Tracking: –
Automatic Landing: –
Return To Home: √
Obstacle Avoidance: –
Headless Mode: –
Built-in Camera: √

SteadiDrone

QU4D 2013

© SteadiDrone

Release Date: January 2013
Price: $2,350
Number of Rotors: 4
Control System: Remote Control
Applications: Film, Photography
Dimensions: 580 × 560 × 110 mm
Weight: 1.2 kg
Max. Payload: 800 gr
Frame Material: Carbon Fiber
Available Colors: Black
Max. Flight Time: 25 minutes
Max. Speed: 100 km/h (62 mp/h)
Max. Operating Range: 1.5 km
Controller Frequency: 2.4 GHz
Battery Capacity: 4,200 mAh
Battery Type: Lithium Polymer 4S
GPS Compatible: √
Altitude Hold: √
Position Hold: √
Flight Planning: √
Autonomous Flight: √
Target Tracking: –
Automatic Landing: √
Return To Home: √
Obstacle Avoidance: –
Headless Mode: –
Built-in Camera: –

Defence Research Development Organisation of India (DRDO)

Abhyas

© defence.pk

Release Date: February 2013
Price: Unknown
Number of Rotors: – (Fixed Wing)
Control System: Flight Computer
Applications: Anti-Aircraft, Target Acquisition
Weight: 25 kg
Frame Material: Aluminum, Composite
Available Colors: Orange
Max. Flight Time: 45 minutes
GPS Compatible: √
Altitude Hold: √
Position Hold: –
Flight Planning: √
Autonomous Flight: –
Target Tracking: –
Automatic Landing: –
Return To Home: –
Obstacle Avoidance: –
Headless Mode: –
Built-in Camera: –

Flytronic

Fly Eye

Release Date: February 2013
Price: Unknown
Number of Rotors: 1 (Fixed Wing)
Control System: Ground Control Station
Applications: Military Purposes, Inspection, Reconnaissance, Surveillance, Target Acquisition
Dimensions: 1.9 × 3.6 m
Wingspan: 3.6 m
Weight: 11 kg
Max. Payload: 4 kg
Frame Material: Composite
Available Colors: Grey
Max. Flight Time: 150 minutes
Max. Speed: 120 km/h (74,5 mp/h)
Max. Flying Altitude: 4,000 m (13,123 ft)
Max. Operating Range: 30 km
Controller Frequency: 2.4 GHz
Battery Type: Lithium Polymer
GPS Compatible: √
Altitude Hold: √
Position Hold: –
Flight Planning: √
Autonomous Flight: √
Target Tracking: –
Automatic Landing: √
Return To Home: √
Obstacle Avoidance: –
Headless Mode: –
Built-in Camera: √

Prox Dynamics

PD-100 Black hornet Nano

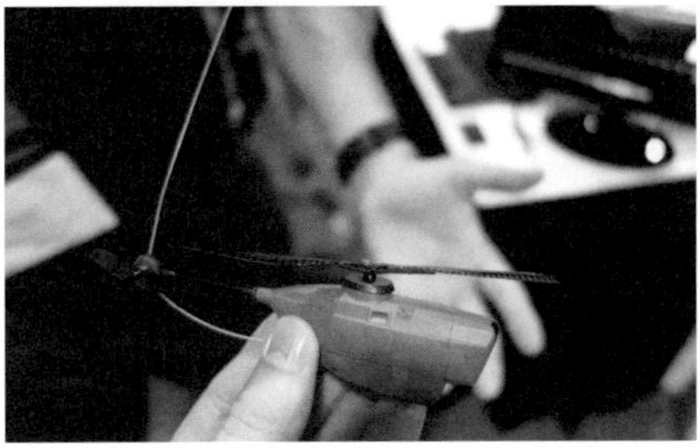

© defensereview.com/David Crane

Release Date: February 2013
Price: $40,000
Number of Rotors: 2
Control System: Ground Control Station, Remote Control
Applications: Military Purposes, Personal Reconnaissance, Target Acquisition
Dimensions: 100 × 30 mm
Weight: 18 gr
Frame Material: Composite
Rotor Material: Carbon Fiber
Available Colors: Black, White
Max. Flight Time: 25 minutes
Max. Speed: 18 km/h (11,2 mp/h)
Max. Operating Range: 1.6 km
Controller Frequency: 2.4 GHz
Battery Type: Lithium Polymer
GPS Compatible: √
Altitude Hold: √
Position Hold: √
Flight Planning: √
Autonomous Flight: √
Target Tracking: √
Automatic Landing: –
Return To Home: √
Obstacle Avoidance: –
Headless Mode: –
Built-in Camera: √

ADCOM Systems

Yabhon U-40

©Vitaly V. Kuzmin

Release Date: March 2013
Price: Unknown
Number of Rotors: 2 (Fixed Wing)
Control System: Flight Computer
Applications: Military Purposes, Armed Reconnaissance, Surveillance, Target Acquisition
Dimensions: 11.13 × 20 × 4.38 m
Wingspan: 20 m
Weight: 950 kg
Max. Payload: 1,000 kg
Available Colors: White
Max. Flight Time: 100 hours
Max. Speed: 220 km/h (137 mp/h)
Max. Flying Altitude: 7000 m (22,965 ft)
GPS Compatible: √
Altitude Hold: √
Position Hold: √
Flight Planning: √
Autonomous Flight: √
Target Tracking: √
Automatic Landing: √
Return To Home: √
Obstacle Avoidance: √
Headless Mode: –
Built-in Camera: √

Aermatica

Anteos

© Aermatica

Release Date: March 2013
Price: Unknown
Number of Rotors: 4
Control System: Remote Control, Laptop
Applications: Film, Photography, Aerial Mapping, Transport
Dimensions: 800 × 800 × 360 mm
Weight: 7 kg
Frame Material: Carbon Fiber
Rotor Material: Plastic
Available Colors: Black
Max. Flight Time: 40 minutes
Max. Speed: 40 km/h (25 mp/h)
Max. Flying Altitude: 3,500 m (11,483 ft)
Max. Operating Range: 5 km
Controller Frequency: 2.4 GHz
Battery Type: Lithium Polymer
GPS Compatible: √
Altitude Hold: √
Position Hold: √
Flight Planning: √
Autonomous Flight: √
Target Tracking: –
Automatic Landing: √
Return To Home: √
Obstacle Avoidance: –
Headless Mode: –
Built-in Camera: –

AeroSky

550

Release Date: March 2013
Price: $580
Number of Rotors: 4
Control System: Remote Control
Applications: Film, Photography, FPV Racing
Dimensions: 550 × 550 × 100 mm
Weight: 1.2 kg
Max. Payload: 1 kg
Frame Material: Plastic
Rotor Material: Plastic
Motors: 2212 Outrunner brushless 920kv
Available Colors: White
Controller Frequency: 2.4 GHz
Battery Capacity: 2,200 mAh
Battery Type: Lithium Polymer
Battery Power: 14.8V
GPS Compatible: –
Altitude Hold: –
Position Hold: –
Flight Planning: –
Autonomous Flight: –
Target Tracking: –
Automatic Landing: –
Return To Home: –
Obstacle Avoidance: –
Headless Mode: –
Built-in Camera: –

Guizhou Aviation Industry Group (CAIG)

Harrier Hawk I

© imgur.com

Release Date: March 2013
Price: $1.5 million
Number of Rotors: 1 (Fixed Wing)
Control System: Ground Control Station
Applications: Military Purposes, Reconnaissance, Surveillance
Dimensions: 6 × 10 × 2.2 m
Wingspan: 10 m
Weight: 700 kg
Max. Payload: 100 kg
Available Colors: White
Max. Flight Time: 16 hours
Max. Speed: 230 km/h (143 mp/h)
Max. Flying Altitude: 7,500 m (24,600 ft)
GPS Compatible: √
Altitude Hold: √
Position Hold: –
Flight Planning: √
Autonomous Flight: √
Target Tracking: √
Automatic Landing: √
Return To Home: √
Obstacle Avoidance: –
Headless Mode: –
Built-in Camera: √

Lynxmotion

Crazy2Fly

Release Date: March 2013
Price: $275
Number of Rotors: 4
Control System: Remote Control
Applications: Recreation, FPV Racing
Dimensions: 380 × 380 × 110 mm
Weight: 250 gr
Frame Material: Fiberglass
Available Colors: Black
Controller Frequency: 2.4 GHz
Battery Capacity: 2,000 mAh
Battery Type: Lithium Polymer 3S
Battery Power: 11.1V
GPS Compatible: –
Altitude Hold: √
Position Hold: √
Flight Planning: –
Autonomous Flight: –
Target Tracking: –
Automatic Landing: –
Return To Home: –
Obstacle Avoidance: –
Headless Mode: –
Built-in Camera: –

Iran Aircraft Manufacturing Industrial Company

HESA H110 Sarir

Release Date: April 2013
Price: $5.5 million
Number of Rotors: 1 (Fixed Wing)
Control System: Ground Control Station
Applications: Military Purposes, Armed Reconnaissance, Surveillance, Target Acquisition
Dimensions: 6 × 8 × 1.8 m
Wingspan: 8 m
Max. Payload: 100 kg
Frame Material: Composite
Available Colors: Grey
Max. Flight Time: 12 hours
Max. Flying Altitude: 5,486 m (18,000 ft)
GPS Compatible: √
Altitude Hold: √
Position Hold: –
Flight Planning: √
Autonomous Flight: √
Target Tracking: √
Automatic Landing: √
Return To Home: √
Obstacle Avoidance: √
Headless Mode: –
Built-in Camera: √

SantosLab

Carcará II

Release Date: April 2013
Price: $11,445
Number of Rotors: 1 (Fixed Wing)
Control System: Flight Computer
Applications: Aerial Mapping, Inspection
Dimensions: 1 × 2.1 × 0.31 m
Wingspan: 2.1 m
Weight: 4.5 kg
Frame Material: EPP Foam
Available Colors: Grey
Max. Flight Time: 2 hours
Max. Speed: 54 km/h (33,5 mp/h)
Max. Flying Altitude: 500 m (1,640 ft)
Max. Operating Range: 30 km
GPS Compatible: √
Altitude Hold: √
Position Hold: –
Flight Planning: √
Autonomous Flight: √
Target Tracking: –
Automatic Landing: √
Return To Home: √
Obstacle Avoidance: –
Headless Mode: –
Built-in Camera: √

ZALA Aero Group

Zala 421-22

©innovudm.ru

Release Date: April 2013
Price: Unknown
Number of Rotors: 8
Control System: Ground Control Station
Applications: Military Purposes, Reconnaissance, Surveillance
Dimensions: 1.06 × 1.06 × 0.24 m
Weight: 8 kg
Max. Payload: 2 kg
Frame Material: Composite
Available Colors: White
Max. Flight Time: 35 minutes
Max. Speed: 40 km/h (25 mp/h)
Max. Flying Altitude: 1,000 m (3,280 ft)
Min. Operating Temperature: -30°C
Max. Operating Temperature: 40°C
Max. Operating Range: 5 km
Controller Frequency: 2.4 GHz
Battery Type: Lithium Polymer
GPS Compatible: √
Altitude Hold: √
Position Hold: √
Flight Planning: √
Autonomous Flight: √
Target Tracking: √
Automatic Landing: √
Return To Home: √
Obstacle Avoidance: √
Headless Mode: √
Built-in Camera: √

Aeryon Labs Inc.

SkyRanger R60

© Aeryon Labs

Release Date: May 2013
Price: $150,000
Number of Rotors: 4
Control System: Flight Computer, Laptop, Tablet
Applications: Military Purposes, Inspection, Reconnaissance, Surveillance
Dimensions: 600 × 600 × 300 mm
Wingspan: 600 mm
Weight: 2.4 kg
Max. Payload: 900 gr
Available Colors: White
Max. Flight Time: 50 minutes
Max. Speed: 50 km/h (31 mp/h)
Max. Flying Altitude: 5,000 m (16,404 ft)
Max. Ascent Speed: 2 m/s
Min. Operating Temperature: -30°C
Max. Operating Temperature: 50°C
Max. Operating Range: 5 km
Controller Frequency: 2.4 GHz
Battery Type: Lithium Polymer
GPS Compatible: √
Altitude Hold: √
Position Hold: √
Flight Planning: √
Autonomous Flight: √
Target Tracking: –
Automatic Landing: √
Return To Home: √
Obstacle Avoidance: √
Headless Mode: –
Built-in Camera: √

Autonomous Systems Lab

Atlantiksolar

© atlantiksolar.ethz.ch

Release Date: May 2013
Price: Unknown
Number of Rotors: 1 (Fixed Wing)
Control System: Ground Control Station
Applications: Research & Development
Wingspan: 5.6 m
Weight: 6.9 kg
Frame Material: Carbon Fiber, Kevlar
Available Colors: White
Max. Flight Time: 10 days
Max. Speed: 75 km/h (46,6 mp/h)
Max. Operating Range: 4,500 km
Battery Type: Solar powered Lithium Ion
GPS Compatible: √
Altitude Hold: √
Position Hold: –
Flight Planning: √
Autonomous Flight: √
Target Tracking: –
Automatic Landing: –
Return To Home: √
Obstacle Avoidance: –
Headless Mode: –
Built-in Camera: √

HiSystems GmbH

Mikrokopter 6S12

©flydrotec.de

Release Date: May 2013
Price: $7,500
Number of Rotors: 8
Control System: Flight Computer
Applications: Aerial Mapping, Agriculture, Inspection, Surveillance
Dimensions: 1.08 × 1.16 × 0.45 m
Weight: 3.35 kg
Max. Payload: 4 kg
Frame Material: Carbon Fiber
Available Colors: Black
Max. Flight Time: 40 minutes
Max. Speed: 54 km/h (33,5 mp/h)
Max. Flying Altitude: 5,000 m (16,404 ft)
Min. Operating Temperature: -5°C
Max. Operating Temperature: 40°C
Max. Operating Range: 4 km
Controller Frequency: 2.4 GHz
Battery Capacity: 4,500 mAh
Battery Type: Lithium Polymer 6S
GPS Compatible: √
Altitude Hold: √
Position Hold: √
Flight Planning: √
Autonomous Flight: √
Target Tracking: –
Automatic Landing: √
Return To Home: √
Obstacle Avoidance: –
Headless Mode: –
Built-in Camera: √

Iran Aviation Industries Organisation

Hamaseh

© Fars News Agency/Hassan Mousavi

Release Date: May 2013
Price: Unknown
Number of Rotors: 1 (Fixed Wing)
Control System: Ground Control Station
Applications: Military Purposes, Armed Reconnaissance, Surveillance, Target Acquisition
Dimensions: 5 × 9 × 1.9 m
Wingspan: 9 m
Weight: 415 kg
Max. Payload: 185 kg
Frame Material: Composite
Available Colors: Red
Max. Flight Time: 11 hours
Max. Speed: 194 km/h (121 mp/h)
Max. Flying Altitude: 4,572 m (15,000 ft)
Max. Ascent Speed: 4 m/s
Max. Operating Range: 200 km
GPS Compatible: √
Altitude Hold: √
Position Hold: –
Flight Planning: √
Autonomous Flight: √
Target Tracking: √
Automatic Landing: √
Return To Home: √
Obstacle Avoidance: √
Headless Mode: –
Built-in Camera: √

WLtoys

V959 Pro Battleship

©amazon.com

Release Date: May 2013
Price: $90
Number of Rotors: 4
Control System: Remote Control
Applications: Recreation
Dimensions: 350 × 350 × 80 mm
Weight: 65 gr
Frame Material: Plastic
Rotor Material: Plastic
Available Colors: Grey
Max. Flight Time: 10 minutes
Max. Operating Range: 150 m
Controller Frequency: 2.4 GHz
Battery Capacity: 500 mAh
Battery Type: Lithium Polymer
Battery Power: 3.7V
GPS Compatible: –
Altitude Hold: √
Position Hold: √
Flight Planning: –
Autonomous Flight: –
Target Tracking: –
Automatic Landing: –
Return To Home: –
Obstacle Avoidance: –
Headless Mode: –
Built-in Camera: –

Jin Xing Da

385

©rc-today.ru

Release Date: June 2013
Price: $30
Number of Rotors: 4
Control System: Remote Control
Applications: Recreation
Dimensions: 123 × 123 × 30 mm
Weight: 645 gr
Frame Material: Plastic
Rotor Material: Plastic
Available Colors: Black
Max. Flight Time: 10 minutes
Max. Operating Range: 50 m
Controller Frequency: 2.4 GHz
Battery Capacity: 240 mAh
Battery Type: Lithium Polymer
Battery Power: 3.7V
GPS Compatible: –
Altitude Hold: –
Position Hold: –
Flight Planning: –
Autonomous Flight: –
Target Tracking: –
Automatic Landing: –
Return To Home: –
Obstacle Avoidance: –
Headless Mode: –
Built-in Camera: –

Syma Toys

X3 Pioneer

©aliexpress.com

Release Date: June 2013
Price: $25
Number of Rotors: 4
Control System: Remote Control
Applications: Recreation
Dimensions: 220 × 240 × 63 mm
Frame Material: Plastic
Rotor Material: Plastic
Available Colors: White
Max. Flight Time: 6 minutes
Controller Frequency: 2.4 GHz
Battery Capacity: 350 mAh
Battery Type: Lithium Polymer
Battery Power: 3.7V
GPS Compatible: –
Altitude Hold: –
Position Hold: –
Flight Planning: –
Autonomous Flight: –
Target Tracking: –
Automatic Landing: –
Return To Home: –
Obstacle Avoidance: –
Headless Mode: –
Built-in Camera: –

Physical Sciences Inc.

Instant Eye Mk-2 Gen3-A0

©epequip.com

Release Date: June 2013
Price: Unknown
Number of Rotors: 4
Control System: Ground Control Station, Remote Control
Applications: Military Purposes, Reconnaissance, Surveillance
Dimensions: 247 × 425 × 215 mm
Weight: 544 gr
Max. Payload: 303 gr
Frame Material: Carbon Fiber, Composite
Available Colors: Grey
Max. Flight Time: 30 minutes
Max. Speed: 56 km/h (35 mp/h)
Max. Flying Altitude: 3,657 m (12,000 ft)
Max. Operating Range: 2 km
Controller Frequency: 2.4 GHz
Battery Type: Lithium Polymer
GPS Compatible: √
Altitude Hold: √
Position Hold: √
Flight Planning: √
Autonomous Flight: √
Target Tracking: –
Automatic Landing: √
Return To Home: √
Obstacle Avoidance: –
Headless Mode: –
Built-in Camera: √

Team BlackSheep

Caipirinha

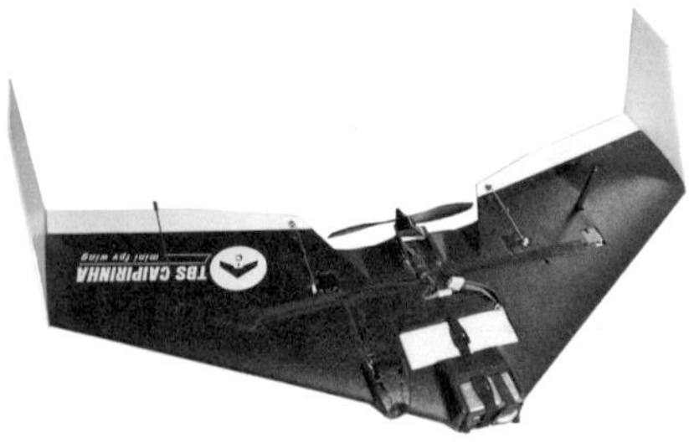

© Team BlackSheep

Release Date: June 2013
Price: $99
Number of Rotors: 1 (Fixed Wing)
Control System: Remote Control
Applications: FPV Racing, Recreation
Wingspan: 850 mm
Weight: 620 gr
Frame Material: EPP Foam
Rotor Material: Plastic
Available Colors: Black
Max. Flight Time: 35 minutes
Max. Operating Range: 10 km
Controller Frequency: 2.4 GHz
Battery Capacity: 3,300 mAh
Battery Type: Lithium Polymer 2S
GPS Compatible: √
Altitude Hold: √
Position Hold: –
Flight Planning: √
Autonomous Flight: –
Target Tracking: –
Automatic Landing: –
Return To Home: √
Obstacle Avoidance: –
Headless Mode: –
Built-in Camera: √ (TBS59, TBS69)

Versadrones

Versa X6

© hexcam.files.wordpress.com

Release Date: June 2013
Price: $14,675
Number of Rotors: 6
Control System: Remote Control
Applications: Film, Photography, Aerial Mapping
Dimensions: 850 × 850 × 340 mm
Weight: 5 kg
Max. Payload: 2.5 kg
Frame Material: Carbon Fiber
Rotor Material: Carbon Fiber
Available Colors: Black
Max. Speed: 70 km/h (43 mp/h)
Max. Flying Altitude: 2,000 m (6,560 ft)
Max. Ascent Speed: 6 m/s
Battery Type: Lithium Polymer
GPS Compatible: √
Altitude Hold: √
Position Hold: √
Flight Planning: √
Autonomous Flight: √
Target Tracking: –
Automatic Landing: √
Return To Home: √
Obstacle Avoidance: –
Headless Mode: –
Built-in Camera: –

3D Robotics

Y6 RTF

© theuavguy.files.wordpress.com

Release Date: July 2013
Price: Unknown
Number of Rotors: 6
Propellers: 10 × 4.7 slow-fly APC
Control System: Remote Control
Applications: Film, Photography
Dimensions: 508 × 355 × 203 mm
Wingspan: 500 mm
Max. Payload: 600 gr
Frame Material: Aluminium
Rotor Material: Plastic
Motors: 850kv brushless
Available Colors: Blue
Max. Flight Time: 15 minutes
Controller Frequency: 2.4 GHz
Battery Capacity: 4,000 mAh
Battery Type: Lithium Polymer 4S
GPS Compatible: √
Altitude Hold: √
Position Hold: √
Flight Planning: √
Autonomous Flight: √
Target Tracking: –
Automatic Landing: √
Return To Home: √
Obstacle Avoidance: –
Headless Mode: –
Built-in Camera: –

Airbus

Atlante

© Airbus

Release Date: July 2013
Price: Unknown
Number of Rotors: 1 (Fixed Wing)
Control System: Ground Control Station
Applications: Military Purposes, Inspection, Reconnaissance, Surveillance, Target Acquisition
Dimensions: 4.6 × 8 × 1.8 m
Wingspan: 8 m
Weight: 520 kg
Max. Payload: 100 kg
Frame Material: Composite
Available Colors: Grey
Max. Flight Time: 20 hours
Max. Flying Altitude: 6,096 m (20,000 ft)
Max. Operating Range: 400 km
GPS Compatible: √
Altitude Hold: √
Position Hold: –
Flight Planning: √
Autonomous Flight: √
Target Tracking: √
Automatic Landing: √
Return To Home: √
Obstacle Avoidance: √
Headless Mode: –
Built-in Camera: √

Airhogs

Helix X4 Stunt

© Airhogs

Release Date: July 2013
Price: $65
Number of Rotors: 4
Control System: Remote Control
Applications: Recreation
Dimensions: 150 × 200 × 50 mm
Weight: 65 gr
Frame Material: EPP Foam
Rotor Material: Plastic
Available Colors: Blue, Green
Max. Flight Time: 7 minutes
Max. Operating Range: 70 m
Controller Frequency: 2.4 GHz
Battery Capacity: 140 mAh
Battery Type: Lithium Polymer
GPS Compatible: –
Altitude Hold: –
Position Hold: –
Flight Planning: –
Autonomous Flight: –
Target Tracking: –
Automatic Landing: –
Return To Home: –
Obstacle Avoidance: –
Headless Mode: –
Built-in Camera: –

Blade

Nano QX

© hobbytown.com

Release Date: July 2013
Price: $50
Number of Rotors: 4
Control System: Remote Control
Applications: Recreation
Dimensions: 140 × 140 × 30 mm
Wingspan: 154 mm
Weight: 18 gr
Frame Material: Plastic
Rotor Material: Plastic
Available Colors: Green, Yellow
Controller Frequency: 2.4 GHz
Battery Capacity: 150 mAh
Battery Type: Lithium Polymer
Battery Power: 3.7V
GPS Compatible: –
Altitude Hold: √
Position Hold: √
Flight Planning: –
Autonomous Flight: –
Target Tracking: –
Automatic Landing: –
Return To Home: –
Obstacle Avoidance: –
Headless Mode: –
Built-in Camera: –

DJI

Spreading Wings S800

© empire-broadcast.com

Release Date: July 2013
Price: $2,080
Number of Rotors: 6
Control System: Remote Control
Applications: Film, Photography, Aerial Mapping
Dimensions: 1.18 × 1 × 0.5 m
Wingspan: 800 mm
Weight: 2.6 kg
Max. Payload: 2.5 kg
Frame Material: Carbon Fiber
Rotor Material: Carbon Fiber
Available Colors: Black
Max. Flight Time: 16 minutes
Max. Speed: 57,6 km/h (35,8 mp/h)
Max. Ascent Speed: 6 m/s
Max. Descent Speed: 6 m/s
Min. Operating Temperature: -10°C
Max. Operating Temperature: 40°C
Max. Operating Range: 2 km
Controller Frequency: 2.4 GHz
Battery Capacity: 15,000 mAh
Battery Type: Lithium Polymer 6S
Battery Power: 22.2V
GPS Compatible: √
Altitude Hold: √
Position Hold: √
Flight Planning: √
Autonomous Flight: √
Target Tracking: –
Automatic Landing: √
Return To Home: √
Obstacle Avoidance: –
Headless Mode: –
Built-in Camera: –

Hubsan

Sky Cruiser H302F

© banggood.com

Release Date: July 2013
Price: $350
Number of Rotors: 1 (Fixed Wing)
Control System: Remote Control
Applications: Film, Photography, Recreation
Dimensions: 0.62 × 1.4 m
Wingspan: 1.4 m
Weight: 400 gr
Frame Material: EPO Foam (carbon fiber reinforced)
Rotor Material: Plastic
Motors: 2210kv 1400 brushless
Available Colors: White
Max. Flight Time: 30 minutes
Max. Operating Range: 1 km
Controller Frequency: 2.4 GHz
Battery Capacity: 1,000 mAh
Battery Type: Lithium Polymer
Battery Power: 11.1V
GPS Compatible: √
Altitude Hold: √
Position Hold: –
Flight Planning: √
Autonomous Flight: –
Target Tracking: –
Automatic Landing: –
Return To Home: √
Obstacle Avoidance: –
Headless Mode: –
Built-in Camera: √ (5 megapixels)

Lehmann Aviation

LA200

© Lehmann Aviation

Release Date: July 2013
Price: $3,300
Number of Rotors: 1 (Fixed Wing)
Control System: Remote Control
Applications: Film, Photography
Wingspan: 920 mm
Weight: 850 gr
Max. Payload: 200 gr
Frame Material: EPP Foam
Rotor Material: Plastic
Available Colors: Black
Max. Flight Time: 30 minutes
Max. Speed: 35 km/h (21,7 mp/h)
Min. Operating Temperature: -25°C
Max. Operating Temperature: 60°C
Max. Operating Range: 3 km
Controller Frequency: 2.4 GHz
Battery Type: Lithium Polymer
GPS Compatible: √
Altitude Hold: –
Position Hold: –
Flight Planning: √
Autonomous Flight: √
Target Tracking: –
Automatic Landing: √
Return To Home: –
Obstacle Avoidance: –
Headless Mode: –
Built-in Camera: –

Lehmann Aviation

LA300

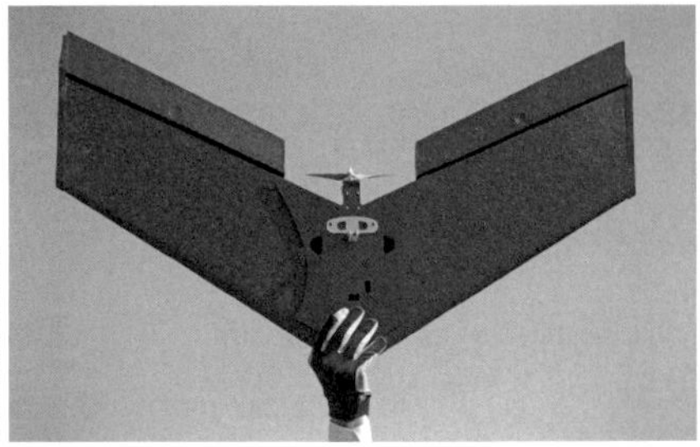

© Lehmann Aviation

Release Date: July 2013
Price: $6,640
Number of Rotors: 1 (Fixed Wing)
Control System: Remote Control
Applications: Aerial Mapping, Agriculture, Inspection, Mining
Wingspan: 920 mm
Weight: 850 gr
Max. Payload: 200 gr
Frame Material: EPP Foam
Rotor Material: Plastic
Available Colors: Black
Max. Flight Time: 45 minutes
Max. Speed: 35 km/h (21,7 mp/h)
Min. Operating Temperature: -25°C
Max. Operating Temperature: 60°C
Max. Operating Range: 30 km
Controller Frequency: 2.4 GHz
Battery Type: Lithium Polymer
GPS Compatible: √
Altitude Hold: –
Position Hold: –
Flight Planning: √
Autonomous Flight: √
Target Tracking: –
Automatic Landing: √
Return To Home: –
Obstacle Avoidance: –
Headless Mode: –
Built-in Camera: –

QuadH2o

QuadH2o

© QuadH2o

Release Date: July 2013
Price: $2,895
Number of Rotors: 4
Control System: Remote Control
Applications: Inspection, Search & Rescue
Dimensions: 370 × 540 × 260 mm
Wingspan: 610 mm
Weight: 1.6 kg
Max. Payload: 600 gr
Frame Material: Carbon Fiber, Perspex
Available Colors: Yellow
Max. Flight Time: 12 minutes
Max. Speed: 72 km/h (45 mp/h)
Max. Operating Range: 1 km
Controller Frequency: 2.4 GHz
Battery Capacity: 5,000 mAh
Battery Type: Lithium Polymer 4S
GPS Compatible: √
Altitude Hold: √
Position Hold: √
Flight Planning: √
Autonomous Flight: √
Target Tracking: –
Automatic Landing: √
Return To Home: √
Obstacle Avoidance: –
Headless Mode: –
Built-in Camera: –

Revell Control

Atomium

© Revell Control

Release Date: July 2013
Price: $69
Number of Rotors: 4
Control System: Remote Control
Applications: Recreation
Dimensions: 190 × 190 × 120 mm
Weight: 748 gr
Frame Material: Plastic
Rotor Material: Plastic
Available Colors: Black
Controller Frequency: 2.4 GHz
Battery Type: Lithium Polymer
GPS Compatible: –
Altitude Hold: –
Position Hold: –
Flight Planning: –
Autonomous Flight: –
Target Tracking: –
Automatic Landing: –
Return To Home: –
Obstacle Avoidance: –
Headless Mode: –
Built-in Camera: –

Tekever

AR1 Blue Ray

© pplware.sapo.pt

Release Date: July 2013
Price: On Request
Number of Rotors: 1 (Fixed Wing)
Control System: Ground Control Station
Applications: Military Purposes, Inspection, Reconnaissance, Surveillance, Target Acquisition
Dimensions: 1.4 × 1.8 m
Wingspan: 1.8 m
Weight: 5 kg
Max. Payload: 1.5 kg
Frame Material: Composite
Available Colors: White
Max. Flight Time: 3 hours
Max. Speed: 80 km/h (49 mp/h)
Max. Operating Range: 20 km
Controller Frequency: 2.4 GHz
Battery Type: Lithium Polymer
GPS Compatible: √
Altitude Hold: √
Position Hold: –
Flight Planning: √
Autonomous Flight: √
Target Tracking: –
Automatic Landing: –
Return To Home: √
Obstacle Avoidance: –
Headless Mode: –
Built-in Camera: √

Art-Tech

Mini X6

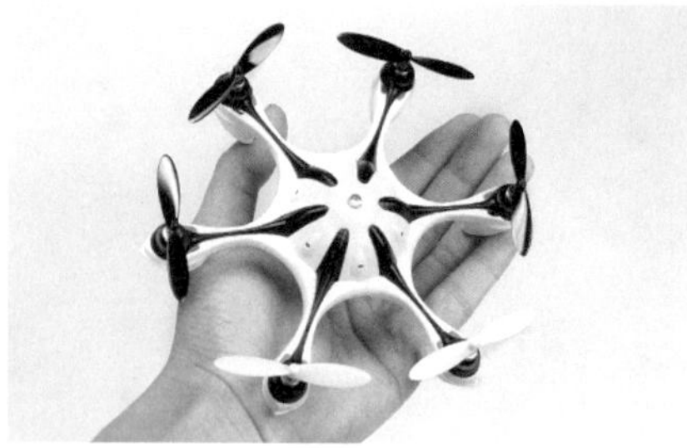

© Hobbyking

Release Date: August 2013
Price: $89
Number of Rotors: 6
Control System: Remote Control
Applications: Recreation
Wingspan: 130 mm
Weight: 52 gr
Frame Material: Plastic
Rotor Material: Plastic
Available Colors: Black, White
Max. Flight Time: 8 minutes
Max. Operating Range: 100 m
Controller Frequency: 2.4 GHz
Battery Capacity: 500 mAh
Battery Type: Lithium Polymer
Battery Power: 3.7V
GPS Compatible: –
Altitude Hold: –
Position Hold: √
Flight Planning: –
Autonomous Flight: –
Target Tracking: –
Automatic Landing: –
Return To Home: –
Obstacle Avoidance: –
Headless Mode: –
Built-in Camera: –

Aurora Flight Sciences

Orion

© Aurora Flight Sciences

Release Date: August 2013
Price: Unknown
Number of Rotors: 2 (Fixed Wing)
Control System: Ground Control Station
Applications: Military Purposes, Research & Development, Armed Reconnaissance, Surveillance
Weight: 4,650 kg
Max. Payload: 1,100 kg
Frame Material: Composite
Motors: 2 × Austro engine AE300 turbo-diesel
Available Colors: Grey
Max. Flight Time: 120 hours
Max. Speed: 170 km/h (106 mp/h)
Max. Flying Altitude: 6,096 m (20,000 ft)
Max. Operating Range: 17,600 km
GPS Compatible: √
Altitude Hold: √
Position Hold: –
Flight Planning: √
Autonomous Flight: √
Target Tracking: √
Automatic Landing: √
Return To Home: √
Obstacle Avoidance: √
Headless Mode: –
Built-in Camera: √

Blade

350 QX

© idrones.fr

Release Date: August 2013
Price: $479
Number of Rotors: 4
Control System: Remote Control
Applications: Film, Photography
Dimensions: 465 × 465 × 138 mm
Wingspan: 580 mm
Weight: 680 gr
Frame Material: Plastic
Rotor Material: Plastic
Motors: 1100kv brushless
Available Colors: White
Max. Flight Time: 15 minutes
Controller Frequency: 2.4 GHz
Battery Capacity: 2,200 mAh
Battery Type: Lithium Polymer 3S
Battery Power: 11.1V
GPS Compatible: –
Altitude Hold: √
Position Hold: √
Flight Planning: √
Autonomous Flight: √
Target Tracking: –
Automatic Landing: √
Return To Home: √
Obstacle Avoidance: –
Headless Mode: –
Built-in Camera: –

C-Astral

Bramor C4EYE

© C-Astral

Release Date: August 2013
Price: On Request
Number of Rotors: 1 (Fixed Wing)
Control System: Ground Control Station
Applications: Military Purposes, Aerial Mapping, Inspection, Search & Rescue, Surveillance, Target Acquisition
Dimensions: 0.96 × 2.3 m
Wingspan: 2.3 m
Weight: 4.7 kg
Max. Payload: 1 kg
Frame Material: Carbon Fiber, Kevlar
Available Colors: Grey
Max. Flight Time: 3.5 hours
Max. Speed: 79 km/h (49 mp/h)
Max. Flying Altitude: 5,000 m (16,404 ft)
Max. Operating Range: 150 km
Controller Frequency: 2.4 GHz
Battery Type: Lithium Polymer
GPS Compatible: √
Altitude Hold: √
Position Hold: –
Flight Planning: √
Autonomous Flight: √
Target Tracking: √
Automatic Landing: –
Return To Home: √
Obstacle Avoidance: √
Headless Mode: –
Built-in Camera: √

Ewatt Aerospace

EWZ-S8

© Ewatt Aerospace

Release Date: August 2013
Price: $13,999
Number of Rotors: 8
Control System: Remote Control, Phone/Tablet
Applications: Film, Photography, Aerial Mapping
Dimensions: 1.13 × 1.13 × 0.52 m
Weight: 9.8 kg
Max. Payload: 5 kg
Frame Material: Carbon Fiber
Rotor Material: Carbon Fiber
Available Colors: Red
Max. Flight Time: 42 minutes
Max. Speed: 72 km/h (45 mp/h)
Max. Flying Altitude: 5,200 m (17,060 ft)
Min. Operating Temperature: -30°C
Max. Operating Temperature: 50°C
Max. Operating Range: 20 km
Controller Frequency: 2.4 GHz
Battery Capacity: 32,000 mAh
Battery Type: Lithium Polymer
GPS Compatible: √
Altitude Hold: √
Position Hold: √
Flight Planning: √
Autonomous Flight: √
Target Tracking: –
Automatic Landing: √
Return To Home: √
Obstacle Avoidance: √
Headless Mode: –
Built-in Camera: √

Hubsan
X4 H107C

© Hubsan

Release Date: August 2013
Price: $35
Number of Rotors: 4
Control System: Remote Control
Applications: Recreation
Dimensions: 80 × 80 × 25 mm
Wingspan: 70 mm
Frame Material: Plastic
Rotor Material: Plastic
Available Colors: Black, Red
Max. Flight Time: 7 minutes
Max. Operating Range: 100 m
Controller Frequency: 2.4 GHz
Battery Capacity: 380 mAh
Battery Type: Lithium Polymer
Battery Power: 3.7V
GPS Compatible: –
Altitude Hold: –
Position Hold: –
Flight Planning: –
Autonomous Flight: –
Target Tracking: –
Automatic Landing: –
Return To Home: –
Obstacle Avoidance: –
Headless Mode: –
Built-in Camera: √

Lockheed Martin
Stalker XE

© Lockheed Martin

Release Date: August 2013
Price: Unknown
Number of Rotors: 1 (Fixed Wing)
Control System: Ground Control Station
Applications: Military Purposes, Reconnaissance, Surveillance, Target Acquisition
Wingspan: 3.65 m
Weight: 10.8 kg
Max. Payload: 2.5 kg
Frame Material: Composite
Available Colors: Grey
Max. Flight Time: 8 hours
Max. Speed: 80 km/h (49,7 mp/h
Max. Flying Altitude: 3,657 m (12,000 ft)
GPS Compatible: √
Altitude Hold: √
Position Hold: –
Flight Planning: √
Autonomous Flight: √
Target Tracking: √
Automatic Landing: –
Return To Home: √
Obstacle Avoidance: √
Headless Mode: –
Built-in Camera: √

Skyartec

Butterfly

©amazon.com

Release Date: August 2013
Price: $150
Number of Rotors: 4
Control System: Remote Control
Applications: Film, Photography
Dimensions: 267 × 267 × 100 mm
Weight: 300 gr
Max. Payload: 80 gr
Frame Material: Plastic
Rotor Material: Plastic
Motors: 1804-2000kv
Available Colors: Black
Max. Operating Range: 500 m
Controller Frequency: 2.4 GHz
Battery Capacity: 1,300 mAh
Battery Type: Lithium Polymer 2S
Battery Power: 7.4V
GPS Compatible: –
Altitude Hold: –
Position Hold: –
Flight Planning: –
Autonomous Flight: –
Target Tracking: –
Automatic Landing: –
Return To Home: –
Obstacle Avoidance: –
Headless Mode: –
Built-in Camera: –

UAV Factory

Penguin B

©unmannedsystemstechnology.com

Release Date: August 2013
Price: $25,000
Number of Rotors: 1 (Fixed Wing)
Control System: Ground Control Station
Applications: Military Purposes, Aerial Mapping, Inspection, Maritime Patrol, Surveillance, Reconnaissance
Dimensions: 2.27 × 3.3 × 0.9 m
Wingspan: 3.3 m
Weight: 10 kg
Max. Payload: 10 kg
Frame Material: Composite
Motors: UAV28-EFI Turnkey fuel injected engine
Available Colors: Grey
Max. Flight Time: 20 hours
Max. Speed: 129 km/h (80 mp/h)
GPS Compatible: √
Altitude Hold: √
Position Hold: –
Flight Planning: √
Autonomous Flight: √
Target Tracking: –
Automatic Landing: √
Return To Home: √
Obstacle Avoidance: –
Headless Mode: –
Built-in Camera: √

Hobby Lord

ST800

©emaxmodel.com

Release Date: September 2013
Price: $445
Number of Rotors: 6
Control System: Remote Control
Applications: Film, Photography, Aerial Mapping
Dimensions: 820 × 820 × 440 mm
Weight: 1.09 kg
Max. Payload: 2.5 kg
Frame Material: Carbon Fiber
Rotor Material: Carbon Fiber
Motors: ST4708 360kv brushless
Available Colors: Black
Max. Flight Time: 10 minutes
Controller Frequency: 2.4 GHz
Battery Capacity: 10,000 mAh
Battery Type: Lithium Polymer 6S
GPS Compatible: √
Altitude Hold: √
Position Hold: √
Flight Planning: √
Autonomous Flight: √
Target Tracking: –
Automatic Landing: √
Return To Home: √
Obstacle Avoidance: –
Headless Mode: –
Built-in Camera: –

Iran Aviation Industries Organisation

Yasir

©Amir Pourmand

Release Date: September 2013
Price: $4.9 million
Number of Rotors: 1 (Fixed Wing)
Control System: Ground Control Station
Applications: Military Purposes, Inspection, Reconnaissance
Dimensions: 1.7 × 3.05 m
Wingspan: 3.05 m
Weight: 19 kg
Max. Payload: 6 kg
Frame Material: Composite
Rotor Material: Wood
Available Colors: Grey
Max. Flight Time: 8 hours
Max. Speed: 120 km/h (74,5 mp/h)
Max. Flying Altitude: 4,600 m (15,092 ft)
GPS Compatible: √
Altitude Hold: √
Position Hold: –
Flight Planning: √
Autonomous Flight: √
Target Tracking: –
Automatic Landing: –
Return To Home: √
Obstacle Avoidance: √
Headless Mode: –
Built-in Camera: √

PowerUp

Power Up 2.0

© PowerUp

Release Date: September 2013
Price: $17
Number of Rotors: 1 (Fixed Wing)
Control System: Remote Control
Applications: Recreation
Frame Material: Paper, Plastic
Rotor Material: Plastic
Available Colors: Black
Max. Flight Time: 30 seconds
Controller Frequency: 2.4 GHz
Battery Type: Lithium Polymer
GPS Compatible: –
Altitude Hold: –
Position Hold: –
Flight Planning: –
Autonomous Flight: –
Target Tracking: –
Automatic Landing: –
Return To Home: –
Obstacle Avoidance: –
Headless Mode: –
Built-in Camera: –

Sky-watch

Huginn X1D

© aeroexpo.online

Release Date: September 2013
Price: $50,000
Number of Rotors: 4
Control System: Ground Control Station
Applications: Military Purposes, Inspection, Reconnaissance, Surveillance
Dimensions: 500 × 500 × 220 mm
Weight: 940 gr
Frame Material: Composite
Rotor Material: Carbon Fiber
Available Colors: Grey
Max. Flight Time: 25 minutes
Max. Speed: 21,6 km/h (13,4 mp/h)
Max. Flying Altitude: 3,048 m (10,000 ft)
Max. Ascent Speed: 6 m/s
Min. Operating Temperature: -20°C
Max. Operating Temperature: 45°C
Max. Operating Range: 10 km
Controller Frequency: 5.8 GHz
Battery Capacity: 5,000 mAh
Battery Type: Lithium Polymer
Battery Power: 14.8V
Battery Weight: 450 gr
GPS Compatible: √
Altitude Hold: √
Position Hold: √
Flight Planning: √
Autonomous Flight: √
Target Tracking: √
Automatic Landing: √
Return To Home: √
Obstacle Avoidance: –
Headless Mode: –
Built-in Camera: –

Izhmash Unmanned Systems

Granat-2

©i-korotchenko.livejournal.com

Release Date: September 2013
Price: Unknown
Number of Rotors: 1 (Fixed Wing)
Control System: Ground Control Station
Applications: Military Purposes, Inspection, Reconnaissance
Wingspan: 2 m
Weight: 3.5 kg
Frame Material: Composite
Available Colors: Grey
Max. Flight Time: 60 minutes
Max. Speed: 120 km/h (74,5 mp/h)
Max. Flying Altitude: 3,500 m (11,483 ft)
Min. Operating Temperature: -30°C
Max. Operating Temperature: 40°C
Max. Operating Range: 15 km
GPS Compatible: √
Altitude Hold: √
Position Hold: –
Flight Planning: √
Autonomous Flight: √
Target Tracking: –
Automatic Landing: –
Return To Home: √
Obstacle Avoidance: –
Headless Mode: –
Built-in Camera: √

Blade

180 QX

©Horizonhobby

Release Date: October 2013
Price: $190
Number of Rotors: 4
Control System: Remote Control
Applications: Film, Photography
Dimensions: 292 × 292 × 76 mm
Weight: 95 gr
Frame Material: Plastic
Rotor Material: Plastic
Available Colors: Black
Max. Flight Time: 10 minutes
Controller Frequency: 2.4 GHz
Battery Capacity: 500 mAh
Battery Type: Lithium Polymer 1S
Battery Power: 3.7V
GPS Compatible: –
Altitude Hold: √
Position Hold: √
Flight Planning: –
Autonomous Flight: –
Target Tracking: –
Automatic Landing: –
Return To Home: –
Obstacle Avoidance: –
Headless Mode: –
Built-in Camera: √ (1.3 megapixels)

DJI

Phantom 2 Vision

© DJI

Release Date: October 2013
Price: $999
Number of Rotors: 4
Control System: Remote Control
Applications: Film, Photography
Dimensions: 432 × 432 × 203 mm
Wingspan: 350 mm
Weight: 1.16 kg
Frame Material: Plastic
Rotor Material: Plastic
Available Colors: White
Max. Flight Time: 25 minutes
Max. Speed: 54 km/h (33,5 mp/h)
Max. Ascent Speed: 6 m/s
Max. Descent Speed: 2 m/s
Min. Operating Temperature: -10°C
Max. Operating Temperature: 50°C
Max. Operating Range: 500 m
Controller Frequency: 2.4 GHz
Battery Capacity: 5,200 mAh
Battery Type: Lithium Polymer 3S
Battery Power: 11.1V
Battery Weight: 170 gr
GPS Compatible: √
Altitude Hold: √
Position Hold: √
Flight Planning: √
Autonomous Flight: √
Target Tracking: –
Automatic Landing: √
Return To Home: √
Obstacle Avoidance: –
Headless Mode: –
Built-in Camera: √ (14 megapixels)

IDS Corporation

IA-3 Colibri

© IDS Corporation

Release Date: October 2013
Price: On Request
Number of Rotors: 4
Control System: Ground Control Station, Remote Control
Applications: Aerial Mapping, Inspection, Reconnaissance
Wingspan: 810 mm
Weight: 2.2 kg
Max. Payload: 1 kg
Frame Material: Carbon Fiber
Available Colors: White
Max. Flight Time: 40 minutes
Max. Speed: 60 km/h (37 mp/h)
Max. Flying Altitude: 500 m (1,640 ft)
Max. Ascent Speed: 6 m/s
Min. Operating Temperature: -10°C
Max. Operating Temperature: 45°C
Max. Operating Range: 1 km
Controller Frequency: 2.4 GHz
Battery Capacity: 10,000 mAh
Battery Type: Lithium Polymer 6S
GPS Compatible: √
Altitude Hold: √
Position Hold: √
Flight Planning: √
Autonomous Flight: √
Target Tracking: –
Automatic Landing: √
Return To Home: √
Obstacle Avoidance: –
Headless Mode: –
Built-in Camera: √ (16 megapixels)

IDS Corporation

IA-17 manta

© IDS Corporation

Release Date: October 2013
Price: On Request
Number of Rotors: 1 (Fixed Wing)
Control System: Ground Control Station, Remote Control
Applications: Aerial Mapping, Inspection, Maritime Patrol
Dimensions: 1.27 × 2.8 × 0.21 m
Wingspan: 2.8 m
Weight: 25 kg
Max. Payload: 2.5 kg
Frame Material: Composite
Available Colors: Black
Max. Flight Time: 5 hours
Max. Speed: 160 km/h (99,4 mp/h)
Max. Flying Altitude: 4,500 m (14,764 ft)
Max. Operating Range: 60 km
GPS Compatible: √
Altitude Hold: √
Position Hold: –
Flight Planning: √
Autonomous Flight: √
Target Tracking: –
Automatic Landing: –
Return To Home: √
Obstacle Avoidance: √
Headless Mode: –
Built-in Camera: √ (100 megapixels)

ING Robotic Aviation

Serenity

© ING Robotic Aviation

Release Date: October 2013
Price: On Request
Number of Rotors: 1 (Fixed Wing)
Control System: Ground Control Station
Applications: Aerial Mapping, Inspection, Search & Rescue, Surveillance
Max. Payload: 5 kg
Frame Material: Composite
Available Colors: Grey
Max. Flight Time: 8 hours
Max. Speed: 80 km/h (49,7 mp/h
Max. Operating Range: 100 km
GPS Compatible: √
Altitude Hold: √
Position Hold: –
Flight Planning: √
Autonomous Flight: √
Target Tracking: –
Automatic Landing: √
Return To Home: √
Obstacle Avoidance: √
Headless Mode: –
Built-in Camera: √

L3 Unnamed Systems

Cutlass

© armyrecognition.com

Release Date: October 2013
Price: Unknown
Number of Rotors: 1 (Fixed Wing)
Control System: Ground Control Station, Remote Control
Applications: Military Purposes, Reconnaissance, Surveillance
Dimensions: 0.83 × 1.4 × 1.14 m
Wingspan: 1.4 m
Weight: 6.8 kg
Frame Material: Composite
Motors: AXI 2826
Available Colors: Grey, Orange
Max. Flight Time: 60 minutes
Max. Speed: 157 km/h (97,5 mp/h)
Battery Type: Lithium Polymer
GPS Compatible: √
Altitude Hold: √
Position Hold: –
Flight Planning: √
Autonomous Flight: √
Target Tracking: √
Automatic Landing: –
Return To Home: √
Obstacle Avoidance: –
Headless Mode: –
Built-in Camera: √

Utva Aviation Industry

Pegaz 011

© Wikipedia/Matricasrpska

Release Date: October 2013
Price: $600,000
Number of Rotors: 1 (Fixed Wing)
Control System: Ground Control Station
Applications: Military Purposes, Armed Reconnaissance, Surveillance
Dimensions: 5.4 × 6.34 × 1.6 m
Wingspan: 6.34 m
Weight: 190 kg
Max. Payload: 40 kg
Frame Material: Composite
Motors: 1 × two-cylinders boxer, 32 kW
Available Colors: Green
Max. Flight Time: 12 hours
Max. Speed: 200 km/h (124 mp/h)
Max. Flying Altitude: 3,000 m (9,842 ft)
Max. Operating Range: 100 km
GPS Compatible: √
Altitude Hold: √
Position Hold: –
Flight Planning: √
Autonomous Flight: √
Target Tracking: –
Automatic Landing: √
Return To Home: √
Obstacle Avoidance: –
Headless Mode: –
Built-in Camera: √

Ares RC

Ethos QX 130

© blog.miniplanes.fr

Release Date: November 2013
Price: $100
Number of Rotors: 4
Control System: Remote Control
Applications: Recreation
Wingspan: 145 mm
Weight: 90 gr
Frame Material: Plastic
Rotor Material: Plastic
Available Colors: White
Controller Frequency: 2.4 GHz
Battery Capacity: 500 mAh
Battery Type: Lithium Polymer 1S
Battery Power: 3.7V
GPS Compatible: –
Altitude Hold: –
Position Hold: –
Flight Planning: –
Autonomous Flight: –
Target Tracking: –
Automatic Landing: –
Return To Home: –
Obstacle Avoidance: –
Headless Mode: –
Built-in Camera: √

Chengdu Aircraft Industry Group (CAIG)

VD200

© forcesdz.com

Release Date: November 2013
Price: Unknown
Number of Rotors: 2 (Fixed Wing)
Control System: Flight Computer
Applications: Military Purposes, Inspection, Maritime Patrol, Reconnaissance, Surveillance
Dimensions: 1.8 × 4.6 m
Wingspan: 4.6 m
Weight: 200 kg
Max. Payload: 20 kg
Available Colors: Grey
Max. Flight Time: 3 hours
Max. Speed: 260 km/h (161,5 mp/h)
Max. Operating Range: 150 km
GPS Compatible: √
Altitude Hold: √
Position Hold: √
Flight Planning: √
Autonomous Flight: √
Target Tracking: –
Automatic Landing: √
Return To Home: √
Obstacle Avoidance: –
Headless Mode: –
Built-in Camera: √

Estes Rockets

Proto-X

© Hobbico

Release Date: November 2013
Price: $30
Number of Rotors: 4
Control System: Remote Control
Applications: Recreation
Dimensions: 140 × 100 × 63 mm
Weight: 113 gr
Frame Material: Plastic
Rotor Material: Plastic
Available Colors: Black, Red
Controller Frequency: 2.4 GHz
Battery Capacity: 100 mAh
Battery Type: Lithium Polymer
Battery Power: 3.7V
GPS Compatible: –
Altitude Hold: –
Position Hold: –
Flight Planning: –
Autonomous Flight: –
Target Tracking: –
Automatic Landing: –
Return To Home: –
Obstacle Avoidance: –
Headless Mode: –
Built-in Camera: –

HeliMax

1Si

© bhphotovideo.com

Release Date: November 2013
Price: $120
Number of Rotors: 4
Control System: Remote Control
Applications: Recreation
Dimensions: 138 × 138 × 45 mm
Wingspan: 125 mm
Weight: 41 gr
Frame Material: Plastic
Rotor Material: Plastic
Available Colors: Grey
Controller Frequency: 2.4 GHz
Battery Capacity: 350 mAh
Battery Type: Lithium Polymer
GPS Compatible: –
Altitude Hold: –
Position Hold: √
Flight Planning: –
Autonomous Flight: –
Target Tracking: √
Automatic Landing: –
Return To Home: √
Obstacle Avoidance: –
Headless Mode: –
Built-in Camera: √

Hongdu Avation Industry Group (HAIG)

Sharp Sword (Li-Jian)

© it.cjboy.net

Release Date: November 2013
Price: $50 million
Number of Rotors: – (Fixed Wing)
Control System: Ground Control Station
Applications: Military Purposes, Armed Reconnaissance, Surveillance, Target Acquisition
Dimensions: 10 × 14 × 2.5 m
Wingspan: 14 m
Max. Payload: 2,000 kg
Motors: 1 × Shenyang WP7 turbofan engine
Available Colors: Grey
Max. Operating Range: 4,000 km
GPS Compatible: √
Altitude Hold: √
Position Hold: –
Flight Planning: √
Autonomous Flight: √
Target Tracking: √
Automatic Landing: √
Return To Home: √
Obstacle Avoidance: √
Headless Mode: –
Built-in Camera: √

Iran Aviation Industries Organisation

Fotros

© theaviationist.com

Release Date: November 2013
Price: $7.5 million
Number of Rotors: 1 (Fixed Wing)
Control System: Ground Control Station
Applications: Military Purposes, Armed Reconnaissance, Surveillance, Target Acquisition
Dimensions: 8.25 × 9 × 1.86 m
Wingspan: 9 m
Weight: 1,025 kg
Max. Payload: 520 kg
Frame Material: Composite
Available Colors: Grey
Max. Flight Time: 30 hours
Max. Speed: 400 km/h (250 mp/h)
Max. Flying Altitude: 7,620 m (25,000 ft)
Max. Operating Range: 2,000 km
GPS Compatible: √
Altitude Hold: √
Position Hold: –
Flight Planning: √
Autonomous Flight: √
Target Tracking: √
Automatic Landing: √
Return To Home: √
Obstacle Avoidance: √
Headless Mode: –
Built-in Camera: √

National Engineering and Scientific Commission (NESCOM)

Burraq

© nation.com.pk

Release Date: November 2013
Price: $2 million
Number of Rotors: 1 (Fixed Wing)
Control System: Ground Control Station
Applications: Military Purposes, Armed Reconnaissance, Surveillance, Target Acquisition
Dimensions: 7 × 9 × 2 m
Wingspan: 9 m
Weight: 500 kg
Max. Payload: 500 kg
Frame Material: Composite
Available Colors: Grey
Max. Speed: 215 km/h (134 mp/h)
Max. Flying Altitude: 7,500 m (24,600 ft)
Max. Operating Range: 1,000 km
GPS Compatible: √
Altitude Hold: √
Position Hold: –
Flight Planning: √
Autonomous Flight: √
Target Tracking: √
Automatic Landing: √
Return To Home: √
Obstacle Avoidance: √
Headless Mode: –
Built-in Camera: √

Prioria Robotics

Hex

© Prioria Robotics

Release Date: November 2013
Price: $17,500
Number of Rotors: 6
Control System: Ground Control Station, Tablet
Applications: Film, Photography, Aerial Mapping, Agriculture, Inspection, Surveillance
Wingspan: 800 mm
Weight: 6 kg
Frame Material: Carbon Fiber
Available Colors: Black
Max. Flight Time: 20 minutes
Max. Speed: 22.5 km/h (14 mp/h)
Max. Operating Range: 2 km
Controller Frequency: 2.4 GHz
Battery Type: Lithium Polymer
GPS Compatible: √
Altitude Hold: √
Position Hold: √
Flight Planning: √
Autonomous Flight: √
Target Tracking: –
Automatic Landing: √
Return To Home: √
Obstacle Avoidance: –
Headless Mode: –
Built-in Camera: –

Prioria Robotics

Hex Mini

© Prioria Robotics

Release Date: November 2013
Price: Unknown
Number of Rotors: 6
Control System: Ground Control Station, Tablet
Applications: Film, Photography, Aerial Mapping, Agriculture, Inspection, Surveillance
Wingspan: 550 mm
Weight: 2.2 kg
Frame Material: Carbon Fiber
Available Colors: Black
Max. Flight Time: 15 minutes
Max. Speed: 35.4 km/h (22 mp/h)
Max. Operating Range: 2 km
Controller Frequency: 2.4 GHz
Battery Type: Lithium Polymer
GPS Compatible: √
Altitude Hold: √
Position Hold: √
Flight Planning: √
Autonomous Flight: √
Target Tracking: –
Automatic Landing: √
Return To Home: √
Obstacle Avoidance: –
Headless Mode: –
Built-in Camera: –

ScaraBot Technologies

Scarabot X8

© Scarabot Technologies

Release Date: November 2013
Price: $35.000
Number of Rotors: 8
Control System: Remote Control
Applications: Film, Photography, Aerial Mapping, Agriculture, Inspection, Surveillance
Wingspan: 500 mm
Weight: 5 kg
Frame Material: Carbon Fiber
Available Colors: Red
Controller Frequency: 2.4 GHz
Battery Capacity: 20,300 mAh
Battery Type: Lithium Polymer
Battery Power: 18V
GPS Compatible: √
Altitude Hold: √
Position Hold: √
Flight Planning: √
Autonomous Flight: √
Target Tracking: –
Automatic Landing: √
Return To Home: √
Obstacle Avoidance: –
Headless Mode: –
Built-in Camera: √

Action Drone

AD1

© Action Drone

Release Date: December 2013
Price: Unknown
Number of Rotors: 4
Propellers: 11-12 Inch
Control System: Remote Control, Phone/Tablet
Applications: Film, Photography, Aerial Mapping
Wingspan: 597 mm
Weight: 2.18 kg
Frame Material: Carbon Fiber
Rotor Material: Carbon Fiber
Available Colors: Black
Max. Flight Time: 20 minutes
Max. Speed: 29 km/h (18 mp/h)
Max. Operating Range: 1.6 km
Battery Capacity: 6,200 mAh
Battery Type: Lithium Polymer
GPS Compatible: √
Altitude Hold: √
Position Hold: √
Flight Planning: √
Autonomous Flight: √
Target Tracking: –
Automatic Landing: √
Return To Home: √
Obstacle Avoidance: –
Headless Mode: –
Built-in Camera: –

Align

M424

© amainhobbies.com

Release Date: December 2013
Price: $119
Number of Rotors: 4
Control System: Remote Control
Applications: Recreation
Wingspan: 240 mm
Weight: 96 gr
Frame Material: Plastic
Rotor Material: Plastic
Available Colors: White
Battery Capacity: 530 mAh
Battery Type: Lithium Polymer
Battery Power: 3.7V
GPS Compatible: –
Altitude Hold: –
Position Hold: –
Flight Planning: –
Autonomous Flight: –
Target Tracking: –
Automatic Landing: –
Return To Home: –
Obstacle Avoidance: –
Headless Mode: –
Built-in Camera: –

DJI

Phantom 2

© DJI

Release Date: December 2013
Price: $679
Number of Rotors: 4
Control System: Remote Control
Applications: Film, Photography
Dimensions: 432 × 432 × 203 mm
Weight: 1 kg
Max. Payload: 300 gr
Frame Material: Plastic
Rotor Material: Plastic
Available Colors: White
Max. Flight Time: 25 minutes
Max. Speed: 54 km/h (33,5 mp/h)
Max. Ascent Speed: 6 m/s
Max. Descent Speed: 2 m/s
Min. Operating Temperature: -10°C
Max. Operating Temperature: 50°C
Max. Operating Range: 300 m
Controller Frequency: 2.4 GHz
Battery Capacity: 5,200 mAh
Battery Type: Lithium Polymer 3S
Battery Power: 11.1V
Battery Weight: 170 gr
GPS Compatible: √
Altitude Hold: √
Position Hold: √
Flight Planning: √
Autonomous Flight: √
Target Tracking: –
Automatic Landing: √
Return To Home: √
Obstacle Avoidance: –
Headless Mode: –
Built-in Camera: –

Drone America

Phoebus

© Drone America

Release Date: December 2013
Price: $400,000
Number of Rotors: 2 (Fixed Wing)
Control System: Ground Control Station, Laptop, Tablet
Applications: Aerial Mapping, Inspection, Maritime Patrol, Surveillance
Dimensions: 3.81 × 7.92 m
Wingspan: 7.92 m
Weight: 81 kg
Max. Payload: 68 kg
Frame Material: Carbon Fiber
Motors: 2 × 342cc two-stroke engine
Available Colors: White
Max. Flight Time: 7 hours
Max. Speed: 161 km/h (100 mp/h)
Max. Flying Altitude: 5,486 m (18,000 ft)
Max. Operating Range: 563 km
GPS Compatible: √
Altitude Hold: √
Position Hold: √
Flight Planning: √
Autonomous Flight: √
Target Tracking: √
Automatic Landing: √
Return To Home: √
Obstacle Avoidance: √
Headless Mode: –
Built-in Camera: –

Enics

Eleron 3SV

© Enics

Release Date: December 2013
Price: $200,000
Number of Rotors: 1 (Fixed Wing)
Control System: Ground Control Station
Applications: Military Purposes, Aerial Mapping, Inspection, Maritime Patrol, Reconnaissance, Surveillance
Dimensions: 0.63 × 1.47 × 0.15 m
Wingspan: 1.47 m
Weight: 5.5 kg
Frame Material: Composite
Available Colors: Green
Max. Flight Time: 100 minutes
Max. Speed: 130 km/h (81 mp/h)
Max. Flying Altitude: 5,000 m (16,404 ft)
Max. Ascent Speed: 3 m/s
Min. Operating Temperature: -20°C
Max. Operating Temperature: 40°C
Max. Operating Range: 50 km
GPS Compatible: √
Altitude Hold: √
Position Hold: –
Flight Planning: √
Autonomous Flight: √
Target Tracking: –
Automatic Landing: –
Return To Home: √
Obstacle Avoidance: –
Headless Mode: –
Built-in Camera: √ (10 megapixels)

Extreme Fliers

Microdrone 2.0

© Extreme Fliers

Release Date: December 2013
Price: $145
Number of Rotors: 4
Control System: Remote Control
Applications: Recreation
Dimensions: 150 × 150 × 40 mm
Weight: 34 gr
Frame Material: Plastic
Rotor Material: Plastic
Available Colors: Black
Controller Frequency: 2.4 GHz
Battery Type: Lithium Polymer
Battery Power: 3.7V
GPS Compatible: –
Altitude Hold: √
Position Hold: √
Flight Planning: –
Autonomous Flight: –
Target Tracking: –
Automatic Landing: –
Return To Home: –
Obstacle Avoidance: –
Headless Mode: –
Built-in Camera: √

2014

AeroSky

C-17

© Aerosky

Release Date: January 2014
Price: $1,599
Number of Rotors: 6
Propellers: 304 × 114 mm
Control System: Remote Control
Applications: Film, Photography
Wingspan: 680 mm
Weight: 2 kg
Max. Payload: 2.5 kg
Frame Material: Carbon Fiber
Rotor Material: Plastic
Motors: Aerosky 840kv
Available Colors: Black
Max. Flight Time: 30 minutes
Min. Operating Temperature: -40°C
Max. Operating Temperature: 85°C
Controller Frequency: 2.4 GHz
Battery Capacity: 2,800 mAh
Battery Type: Lithium Polymer
Battery Power: 14.8V
GPS Compatible: √
Altitude Hold: √
Position Hold: √
Flight Planning: √
Autonomous Flight: √
Target Tracking: –
Automatic Landing: –
Return To Home: –
Obstacle Avoidance: –
Headless Mode: –
Built-in Camera: –

Blade

200 QX

© hubschrauber-ferngesteuert.net

Release Date: January 2014
Price: $229
Number of Rotors: 4
Control System: Remote Control, Phone/Tablet
Applications: Film, Photography
Dimensions: 142 × 142 × 90 mm
Wingspan: 200 mm
Weight: 194 gr
Max. Payload: 60 gr
Frame Material: Plastic
Rotor Material: Plastic
Available Colors: White
Max. Flight Time: 15 minutes
Controller Frequency: 2.4 GHz
Battery Capacity: 800 mAh
Battery Type: Lithium Polymer
Battery Power: 7V
Battery Weight: 45 gr
GPS Compatible: –
Altitude Hold: √
Position Hold: √
Flight Planning: –
Autonomous Flight: –
Target Tracking: –
Automatic Landing: –
Return To Home: –
Obstacle Avoidance: –
Headless Mode: –
Built-in Camera: –

Blade

350 QX2

Release Date: January 2014
Price: $900
Number of Rotors: 4
Control System: Remote Control, Phone/Tablet
Applications: Film, Photography
Dimensions: 465 × 465 × 203 mm
Wingspan: 580 mm
Weight: 1 kg
Max. Payload: 250 gr
Frame Material: Plastic, Carbon Fiber
Rotor Material: Plastic, Carbon Fiber
Available Colors: White
Max. Flight Time: 15 minutes
Max. Speed: 72 km/h (45 mp/h)
Max. Operating Range: 5 km
Controller Frequency: 2.4 GHz
Battery Capacity: 3,000 mAh
Battery Type: Lithium Polymer 3S
Battery Power: 11V
GPS Compatible: √
Altitude Hold: √
Position Hold: √
Flight Planning: –
Autonomous Flight: –
Target Tracking: –
Automatic Landing: –
Return To Home: √
Obstacle Avoidance: –
Headless Mode: –
Built-in Camera: –

DJI

Spreading Wings S1000

© droneshop.com

Release Date: January 2014
Price: $3,999
Number of Rotors: 8
Propellers: 1552 High Performance
Control System: Remote Control
Applications: Film, Photography, Aerial Mapping
Wingspan: 1.05 m
Weight: 4.2 kg
Max. Payload: 1.8 kg
Frame Material: Carbon Fiber
Rotor Material: Plastic
Motors: DJI 4114-400kv
Available Colors: Black
Max. Flight Time: 15 minutes
Max. Ascent Speed: 4 m/s
Min. Operating Temperature: -10°C
Max. Operating Temperature: 40°C
Max. Operating Range: 2 km
Controller Frequency: 5.8 GHz
Battery Capacity: 20,000 mAh
Battery Type: Lithium Polymer 6S
Battery Power: 22.2V
Battery Weight: 670 gr
GPS Compatible: √
Altitude Hold: √
Position Hold: √
Flight Planning: √
Autonomous Flight: √
Target Tracking: √
Automatic Landing: √
Return To Home: √
Obstacle Avoidance: –
Headless Mode: –
Built-in Camera: –

Enics

Eleron 10SV

© Enics

Release Date: January 2014
Price: Unknown
Number of Rotors: 1 (Fixed Wing)
Control System: Ground Control Station
Applications: Military Purposes, Inspection, Reconnaissance, Surveillance
Dimensions: 0.88 × 2.2 × 0.38 m
Wingspan: 2.2 m
Weight: 15.5 kg
Frame Material: Composite
Available Colors: Grey
Max. Flight Time: 2.5 hours
Max. Speed: 135 km/h (84 mp/h)
Max. Flying Altitude: 4,000 m (13,123 ft)
Max. Ascent Speed: 2 m/s
Min. Operating Temperature: -20°C
Max. Operating Temperature: 40°C
Max. Operating Range: 60 km
GPS Compatible: √
Altitude Hold: √
Position Hold: –
Flight Planning: √
Autonomous Flight: √
Target Tracking: –
Automatic Landing: –
Return To Home: √
Obstacle Avoidance: –
Headless Mode: –
Built-in Camera: √ (10 megapixels)

Hubsan

X4 H107L

© droneshop.com

Release Date: January 2014
Price: $40
Number of Rotors: 4
Control System: Remote Control
Applications: Recreation
Dimensions: 80 × 80 × 25 mm
Wingspan: 105 mm
Weight: 500 gr
Frame Material: Plastic
Rotor Material: Plastic
Available Colors: Black
Max. Flight Time: 9 minutes
Max. Operating Range: 100 m
Controller Frequency: 2.4 GHz
Battery Capacity: 240 mAh
Battery Type: Lithium Polymer
Battery Power: 3.7V
GPS Compatible: –
Altitude Hold: –
Position Hold: –
Flight Planning: –
Autonomous Flight: –
Target Tracking: –
Automatic Landing: –
Return To Home: –
Obstacle Avoidance: –
Headless Mode: –
Built-in Camera: –

Izhmash Unmanned Systems

Granat-3

© i-korotchenko.livejournal.com

Release Date: January 2014
Price: Unknown
Number of Rotors: 1 (Fixed Wing)
Control System: Ground Control Station
Applications: Military Purposes, Inspection, Reconnaissance
Wingspan: 2 m
Weight: 7 kg
Max. Payload: 1 kg
Frame Material: Composite
Available Colors: Grey
Max. Flight Time: 2 hours
Max. Speed: 120 km/h (74,5 mp/h)
Max. Flying Altitude: 2,000 m (6,560 ft)
Min. Operating Temperature: -30°C
Max. Operating Temperature: 40°C
Max. Operating Range: 25 km
GPS Compatible: √
Altitude Hold: √
Position Hold: –
Flight Planning: √
Autonomous Flight: √
Target Tracking: √
Automatic Landing: √
Return To Home: √
Obstacle Avoidance: √
Headless Mode: –
Built-in Camera: √

Monstertronic

Sky Agent V2

© nextmodelrc.com

Release Date: January 2014
Price: $150
Number of Rotors: 4
Control System: Remote Control
Applications: Recreation
Dimensions: 630 × 630 × 100 mm
Weight: 220 gr
Frame Material: Foam, Plastic
Rotor Material: Plastic
Available Colors: Black
Max. Flight Time: 7 minutes
Controller Frequency: 2.4 GHz
Battery Capacity: 850 mAh
Battery Type: Lithium Polymer 1S
Battery Power: 7.4V
GPS Compatible: –
Altitude Hold: √
Position Hold: √
Flight Planning: –
Autonomous Flight: –
Target Tracking: –
Automatic Landing: –
Return To Home: √
Obstacle Avoidance: –
Headless Mode: –
Built-in Camera: √

PMG Multi-Rotors

Recon

©PMG

Release Date: January 2014
Price: $5,049
Number of Rotors: 8
Control System: Remote Control
Applications: Film, Photography, Aerial Mapping, Inspection, Surveillance
Frame Material: Carbon Fiber
Rotor Material: Carbon Fiber
Available Colors: Black
Max. Flight Time: 50 minutes
Controller Frequency: 2.4 GHz
Battery Capacity: 10,000 mAh
Battery Type: Lithium Polymer 6S
Battery Power: 22V
GPS Compatible: √
Altitude Hold: √
Position Hold: √
Flight Planning: √
Autonomous Flight: √
Target Tracking: √
Automatic Landing: √
Return To Home: √
Obstacle Avoidance: –
Headless Mode: –
Built-in Camera: –

RC Logger

RC Eye One Xtreme

©conrad.fr

Release Date: January 2014
Price: $269
Number of Rotors: 4
Control System: Remote Control
Applications: Film, Photography
Dimensions: 162 × 160 × 80 mm
Wingspan: 225 mm
Weight: 157 gr
Max. Payload: 100 gr
Frame Material: Plastic
Rotor Material: Plastic
Available Colors: Black
Max. Flight Time: 7 minutes
Max. Operating Range: 122 m
Controller Frequency: 2.4 GHz
Battery Capacity: 2,200 mAh
Battery Type: Lithium Polymer
Battery Power: 7V
GPS Compatible: –
Altitude Hold: √
Position Hold: √
Flight Planning: –
Autonomous Flight: –
Target Tracking: –
Automatic Landing: –
Return To Home: –
Obstacle Avoidance: –
Headless Mode: –
Built-in Camera: √ (5 megapixels)

Syma Toys

X4 Assault

© prisguide.no

Release Date: January 2014
Price: $35
Number of Rotors: 4
Control System: Remote Control
Applications: Recreation
Dimensions: 140 × 150 × 30 mm
Wingspan: 209 mm
Weight: 200 gr
Frame Material: Plastic, Foam
Rotor Material: Plastic
Available Colors: Silver
Max. Flight Time: 8 minutes
Max. Operating Range: 60 m
Controller Frequency: 2.4 GHz
Battery Capacity: 350 mAh
Battery Type: Lithium Polymer
Battery Power: 3.7V
GPS Compatible: –
Altitude Hold: –
Position Hold: –
Flight Planning: –
Autonomous Flight: –
Target Tracking: –
Automatic Landing: –
Return To Home: –
Obstacle Avoidance: –
Headless Mode: –
Built-in Camera: –

Vulcan

Black Widow

© Vulcan

Release Date: January 2014
Price: $3,795
Number of Rotors: 8
Control System: Remote Control
Applications: Film, Photography
Frame Material: Carbon Fiber
Rotor Material: Carbon Fiber
Available Colors: Black
Controller Frequency: 2.4 GHz
Battery Type: Lithium Polymer
GPS Compatible: –
Altitude Hold: √
Position Hold: √
Flight Planning: √
Autonomous Flight: –
Target Tracking: √
Automatic Landing: √
Return To Home: √
Obstacle Avoidance: –
Headless Mode: –
Built-in Camera: –

AeroSky

P12

Release Date: February 2014
Price: $599
Number of Rotors: 4
Propellers: 10 inch
Control System: Remote Control
Applications: Film, Photography, FPV Racing
Dimensions: 470 × 470 × 120 mm
Weight: 750 gr
Max. Payload: 1.3 kg
Frame Material: Plastic, Glass Fiber
Rotor Material: Plastic
Motors: 2212 Outrunner brushless 920kv
Available Colors: Black
Min. Operating Temperature: -40°C
Max. Operating Temperature: 85°C
Controller Frequency: 2.4 GHz
Battery Capacity: 2,200 mAh
Battery Type: Lithium Polymer 3S
Battery Power: 11.1V
GPS Compatible: √
Altitude Hold: √
Position Hold: √
Flight Planning: √
Autonomous Flight: √
Target Tracking: –
Automatic Landing: √
Return To Home: √
Obstacle Avoidance: –
Headless Mode: –
Built-in Camera: –

Aibotix

Aibot X6 V2

Release Date: February 2014
Price: $30,000
Number of Rotors: 6
Control System: Remote Control
Applications: Film, Photography, Aerial Mapping, Surveillance
Dimensions: 1.05 × 1.05 × 0.45 m
Wingspan: 1.05 m
Weight: 3.4 kg
Max. Payload: 2 kg
Frame Material: Carbon Fiber
Rotor Material: Carbon Fiber
Available Colors: Orange
Max. Flight Time: 20 minutes
Max. Speed: 40 km/h (25 mp/h)
Max. Flying Altitude: 500 m (1,640 ft)
Max. Ascent Speed: 8 m/s
Min. Operating Temperature: -20°C
Max. Operating Temperature: 40°C
Max. Operating Range: 3 km
Controller Frequency: 2.4 GHz
Battery Capacity: 10,000 mAh
Battery Type: Lithium Polymer
Battery Weight: 1.2 kg
GPS Compatible: √
Altitude Hold: √
Position Hold: √
Flight Planning: √
Autonomous Flight: √
Target Tracking: –
Automatic Landing: √
Return To Home: √
Obstacle Avoidance: √
Headless Mode: –
Built-in Camera: –

Cheerson

CX-20

© Cheerson Hobby

Release Date: February 2014
Price: $199
Number of Rotors: 4
Control System: Remote Control, Phone/Tablet
Applications: Film, Photography
Dimensions: 360 × 360 × 200 mm
Wingspan: 509 mm
Weight: 980 gr
Frame Material: Plastic
Rotor Material: Carbon Fiber
Available Colors: White
Max. Flight Time: 15 minutes
Max. Speed: 29 km/h (18 mp/h)
Max. Flying Altitude: 300 m (985 ft)
Max. Ascent Speed: 6 m/s
Max. Operating Range: 1.5 km
Controller Frequency: 2.4 GHz
Battery Capacity: 2,700 mAh
Battery Type: Lithium Polymer 3S
Battery Power: 11.1V
Battery Weight: 180 gr
GPS Compatible: √
Altitude Hold: √
Position Hold: √
Flight Planning: –
Autonomous Flight: –
Target Tracking: –
Automatic Landing: √
Return To Home: √
Obstacle Avoidance: –
Headless Mode: –
Built-in Camera: –

Israel Aerospace Industries (IAI)

Super Heron

© IAI

Release Date: February 2014
Price: $15 million
Number of Rotors: 1 (Fixed Wing)
Control System: Flight Computer
Applications: Reconnaissance, Surveillance, Maritime Patrol
Wingspan: 17 m
Weight: 1,000 kg
Max. Payload: 450 kg
Motors: 200 Hp heavy-fuel engine
Available Colors: White/Blue
Max. Flight Time: 45 hours
Max. Speed: 278 km/h (172,6 mp/h)
Max. Flying Altitude: 9,144 m (30,000 ft)
Max. Operating Range: 1,000 km
GPS Compatible: √
Altitude Hold: √
Position Hold: √
Flight Planning: √
Autonomous Flight: √
Target Tracking: –
Automatic Landing: √
Return To Home: √
Obstacle Avoidance: –
Headless Mode: –
Built-in Camera: √

Korea Aerospace Industries (KAI)

Devil Killer

©suasnews.com

Release Date: February 2014
Price: Unknown
Number of Rotors: – (Fixed Wing)
Control System: Flight Computer
Applications: Loitering munition
Dimensions: 1.5 × 1.3 m
Wingspan: 1.3 m
Weight: 25 kg
Available Colors: Grey
Max. Speed: 400 km/h (250 mp/h)
Max. Operating Range: 40 km
GPS Compatible: √
Altitude Hold: √
Position Hold: –
Flight Planning: √
Autonomous Flight: √
Target Tracking: √
Automatic Landing: –
Return To Home: √
Obstacle Avoidance: √
Headless Mode: –
Built-in Camera: √

PowerUp

Power Up 3.0

©gadetrecap.com

Release Date: February 2014
Price: $50
Number of Rotors: 1 (Fixed Wing)
Control System: Phone/Tablet
Applications: Recreation
Frame Material: Paper, Plastic
Rotor Material: Plastic
Available Colors: Red
Max. Flight Time: 10 minutes
Max. Speed: 24 km/h (15 mp/h)
GPS Compatible: –
Altitude Hold: –
Position Hold: –
Flight Planning: –
Autonomous Flight: –
Target Tracking: –
Automatic Landing: –
Return To Home: –
Obstacle Avoidance: –
Headless Mode: –
Built-in Camera: –

PT Dirgantara Indonesia

Puna Wulung

©kaskus.co.id

Release Date: February 2014
Price: Unknown
Number of Rotors: 1 (Fixed Wing)
Control System: Ground Control Station
Applications: Military Purposes, Inspection, Reconnaissance, Surveillance
Dimensions: 4.42 × 6.34 × 1.48 m
Wingspan: 6.34 m
Weight: 65 kg
Max. Payload: 35 kg
Frame Material: Composite
Available Colors: Grey, Green
Max. Flight Time: 4 hours
Max. Speed: 110 km/h (68 mp/h)
Max. Flying Altitude: 4,500 m (14,764 ft)
Max. Operating Range: 120 km
GPS Compatible: √
Altitude Hold: √
Position Hold: –
Flight Planning: √
Autonomous Flight: √
Target Tracking: –
Automatic Landing: √
Return To Home: √
Obstacle Avoidance: √
Headless Mode: –
Built-in Camera: √

SenseFly

Ebee

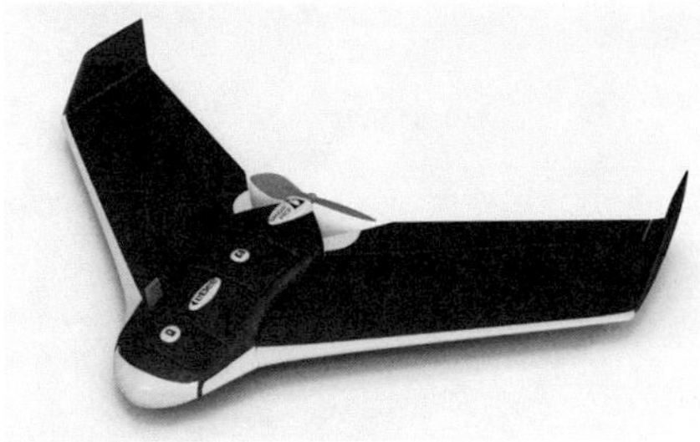

©SenseFly

Release Date: February 2014
Price: $25,000
Number of Rotors: 1 (Fixed Wing)
Control System: Remote Control
Applications: Aerial Mapping, Surveillance
Wingspan: 960 mm
Weight: 690 gr
Frame Material: EPP Foam, Carbon Fiber
Rotor Material: Carbon Fiber
Available Colors: Black
Max. Flight Time: 45 minutes
Max. Speed: 90 km/h (56 mp/h)
Min. Operating Temperature: -10°C
Max. Operating Temperature: 40°C
Max. Operating Range: 3 km
Controller Frequency: 2.4 GHz
Battery Capacity: 2,150 mAh
Battery Type: Lithium Polymer
Battery Power: 11.1V
GPS Compatible: √
Altitude Hold: √
Position Hold: –
Flight Planning: √
Autonomous Flight: √
Target Tracking: √
Automatic Landing: –
Return To Home: √
Obstacle Avoidance: –
Headless Mode: –
Built-in Camera: √

Sky Hero

Spy

© Sky Hero

Release Date: February 2014
Price: $399
Number of Rotors: 6
Control System: Remote Control
Applications: Film, Photography
Dimensions: 505 × 565 × 260 mm
Wingspan: 600 mm
Weight: 890 gr
Frame Material: Plastic, Carbon Fiber
Rotor Material: Plastic
Available Colors: White
Controller Frequency: 2.4 GHz
Battery Capacity: 3,700 mAh
Battery Type: Lithium Polymer
GPS Compatible: √
Altitude Hold: √
Position Hold: √
Flight Planning: √
Autonomous Flight: –
Target Tracking: √
Automatic Landing: –
Return To Home: √
Obstacle Avoidance: –
Headless Mode: –
Built-in Camera: –

ST Aerospace

Ustar-Y

© ST Aerospace

Release Date: February 2014
Price: Unknown
Number of Rotors: 8
Control System: Flight Computer
Applications: Aerial Mapping, Inspection, Search & Rescue, Surveillance
Wingspan: 560 mm
Weight: 5 kg
Max. Payload: 1 kg
Frame Material: Carbon Fiber
Available Colors: Black
Max. Flight Time: 25 minutes
Max. Flying Altitude: 914 m (3,000 ft)
Max. Operating Range: 900 m
Controller Frequency: 2.4 GHz
Battery Type: Lithium Polymer
GPS Compatible: √
Altitude Hold: √
Position Hold: √
Flight Planning: √
Autonomous Flight: √
Target Tracking: √
Automatic Landing: √
Return To Home: √
Obstacle Avoidance: √
Headless Mode: –
Built-in Camera: √

SteadiDrone

QU4D 2014

©suasnews.com

Release Date: February 2014
Price: $15,000
Number of Rotors: 4
Control System: Remote Control
Applications: Film, Photography
Wingspan: 645 mm
Weight: 1.5 kg
Max. Payload: 800 gr
Frame Material: Carbon Fiber
Available Colors: Black
Max. Flight Time: 18 minutes
Max. Operating Range: 2 km
Controller Frequency: 2.4 GHz
Battery Type: Lithium Polymer
GPS Compatible: √
Altitude Hold: √
Position Hold: √
Flight Planning: √
Autonomous Flight: √
Target Tracking: –
Automatic Landing: √
Return To Home: √
Obstacle Avoidance: –
Headless Mode: –
Built-in Camera: –

UDIRC Toys

U830 Demon

©UDIRC Toys

Release Date: February 2014
Price: $29
Number of Rotors: 4
Control System: Remote Control
Applications: Recreation
Dimensions: 110 × 110 × 43 mm
Weight: 36 gr
Frame Material: Plastic
Rotor Material: Plastic
Available Colors: Black
Max. Flight Time: 10 minutes
Max. Operating Range: 80 m
Controller Frequency: 2.4 GHz
Battery Capacity: 250 mAh
Battery Type: Lithium Polymer
Battery Power: 3.7V
GPS Compatible: –
Altitude Hold: –
Position Hold: –
Flight Planning: –
Autonomous Flight: –
Target Tracking: –
Automatic Landing: –
Return To Home: –
Obstacle Avoidance: –
Headless Mode: –
Built-in Camera: –

UMS AERO Group

F-720 (Discoverer II)

©UMS Skeldar

Release Date: February 2014
Price: Unknown
Number of Rotors: 1 (Fixed Wing)
Control System: Ground Control Station
Applications: Military Purposes, Inspection, Reconnaissance, Surveillance, Target Acquisition
Dimensions: 4.8 × 7.2 × 1.35 m
Wingspan: 7.2 m
Weight: 250 kg
Max. Payload: 70 kg
Frame Material: Composite
Available Colors: Grey
Max. Flight Time: 12 hours
Max. Speed: 200 km/h (124 mp/h)
Max. Flying Altitude: 4,900 m (16,076 ft)
Max. Operating Range: 200 km
GPS Compatible: √
Altitude Hold: √
Position Hold: –
Flight Planning: √
Autonomous Flight: √
Target Tracking: –
Automatic Landing: √
Return To Home: √
Obstacle Avoidance: √
Headless Mode: –
Built-in Camera: √

Walkera

QR X350 Pro

©Walkera

Release Date: February 2014
Price: $399
Number of Rotors: 4
Control System: Remote Control, Phone/Tablet
Applications: Film, Photography
Dimensions: 289 × 289 × 200 mm
Wingspan: 408 mm
Weight: 1.34 kg
Max. Payload: 200 gr
Frame Material: Plastic, Carbon Fiber
Rotor Material: Plastic
Available Colors: White
Max. Flight Time: 25 minutes
Max. Speed: 72 km/h (45 mp/h)
Min. Operating Temperature: -10°C
Max. Operating Temperature: 40°C
Max. Operating Range: 3.2 km
Controller Frequency: 2.4 GHz
Battery Capacity: 5,200 mAh
Battery Type: Lithium Polymer
Battery Power: 11V
GPS Compatible: √
Altitude Hold: √
Position Hold: √
Flight Planning: √
Autonomous Flight: –
Target Tracking: –
Automatic Landing: √
Return To Home: √
Obstacle Avoidance: –
Headless Mode: –
Built-in Camera: –

Yuneec

Flying Eyes HX3

© dronemakers.com

Release Date: February 2014
Price: $3,000
Number of Rotors: 6
Control System: Remote Control, Phone/Tablet
Applications: Film, Photography
Dimensions: 720 × 840 × 500 mm
Wingspan: 1100 mm
Weight: 1.4 kg
Max. Payload: 2 kg
Frame Material: Carbon Fiber
Rotor Material: Plastic
Available Colors: Black
Max. Flight Time: 45 minutes
Max. Speed: 54 km/h (33,5 mp/h)
Max. Flying Altitude: 4,000 m (13,123 ft)
Max. Ascent Speed: 8 m/s
Max. Descent Speed: 8 m/s
Max. Operating Range: 2 km
Controller Frequency: 2.4 GHz
Battery Capacity: 8,800 mAh
Battery Type: Lithium Polymer 4S
Battery Power: 14.7V
GPS Compatible: √
Altitude Hold: √
Position Hold: √
Flight Planning: √
Autonomous Flight: √
Target Tracking: √
Automatic Landing: –
Return To Home: √
Obstacle Avoidance: –
Headless Mode: –
Built-in Camera: –

AEE Technology

F50

© AEE Technologies

Release Date: March 2014
Price: $37,000
Number of Rotors: 4
Control System: Remote Control
Applications: Film, Photography, Aerial Mapping, Military Purposes, Surveillance
Weight: 1.8 kg
Max. Payload: 1 kg
Frame Material: Carbon Fiber
Rotor Material: Carbon Fiber
Available Colors: Black, Blue, White
Max. Flight Time: 40 minutes
Max. Speed: 80 km/h (49 mp/h)
Max. Flying Altitude: 1,500 m (4,921 ft)
Max. Ascent Speed: 5 m/s
Min. Operating Temperature: -20°C
Max. Operating Temperature: 60°C
Max. Operating Range: 2 km
Controller Frequency: 2.4 GHz
Battery Capacity: 8,000 mAh
Battery Type: Lithium Polymer
GPS Compatible: √
Altitude Hold: √
Position Hold: √
Flight Planning: √
Autonomous Flight: √
Target Tracking: –
Automatic Landing: √
Return To Home: √
Obstacle Avoidance: –
Headless Mode: –
Built-in Camera: √ (16 megapixels)

Aeronavics

Skyjib X4

© Aeronavics

Release Date: March 2014
Price: $15,600
Number of Rotors: 8
Control System: Remote Control
Applications: Film, Photography, Aerial Mapping, Surveillance
Wingspan: 912 mm
Weight: 5 kg
Max. Payload: 3.5 kg
Frame Material: Carbon Fiber
Rotor Material: Carbon Fiber
Available Colors: Black
Max. Flight Time: 20 minutes
Max. Speed: 80 km/h (49 mp/h)
Max. Operating Range: 3 km
Controller Frequency: 2.4 GHz
Battery Capacity: 10,000 mAh
Battery Type: Lithium Polymer 6S
GPS Compatible: √
Altitude Hold: √
Position Hold: √
Flight Planning: √
Autonomous Flight: √
Target Tracking: –
Automatic Landing: √
Return To Home: √
Obstacle Avoidance: –
Headless Mode: –
Built-in Camera: –

AirRobot

AR 180

© spektreworks.com

Release Date: March 2014
Price: On Request
Number of Rotors: 4
Control System: Flight Computer
Applications: Film, Photography, Aerial Mapping
Dimensions: 1.9 × 1.9 m
Weight: 5 kg
Max. Payload: 1.5 kg
Frame Material: Carbon Fiber
Available Colors: Silver
Max. Flight Time: 40 minutes
Max. Speed: 40 km/h (25 mp/h)
Min. Operating Temperature: -19°C
Max. Operating Temperature: 49°C
Max. Operating Range: 1 km
Controller Frequency: 2.4 GHz
Battery Type: Lithium Polymer
GPS Compatible: √
Altitude Hold: √
Position Hold: √
Flight Planning: √
Autonomous Flight: √
Target Tracking: –
Automatic Landing: √
Return To Home: √
Obstacle Avoidance: √
Headless Mode: –
Built-in Camera: √ (14 megapixels)

Altavian

Nova F6500

©toledoblade.com

Release Date: March 2014
Price: Unknown
Number of Rotors: 1 (Fixed Wing)
Control System: Flight Computer
Applications: Aerial Mapping, Inspection
Dimensions: 1.7 × 2.75 m
Wingspan: 2.75 m
Weight: 6.8 kg
Available Colors: Grey
Max. Flight Time: 90 minutes
Max. Speed: 112 km/h (70 mp/h)
Max. Flying Altitude: 3,048 m (10,000 ft)
Max. Operating Range: 80 km
Controller Frequency: 2.4 GHz
Battery Type: Lithium Polymer
GPS Compatible: √
Altitude Hold: √
Position Hold: –
Flight Planning: √
Autonomous Flight: √
Target Tracking: –
Automatic Landing: √
Return To Home: √
Obstacle Avoidance: –
Headless Mode: –
Built-in Camera: √ (29 megapixels)

Huajun Toys

W608-1

©Huajun Toys

Release Date: March 2014
Price: $50
Number of Rotors: 4
Control System: Remote Control
Applications: Recreation
Dimensions: 175 × 173 × 48 mm
Weight: 47 gr
Frame Material: Plastic
Rotor Material: Plastic
Available Colors: Black, Blue
Max. Flight Time: 8 minutes
Max. Operating Range: 30 m
Controller Frequency: 2.4 GHz
Battery Capacity: 300 mAh
Battery Type: Lithium Polymer
GPS Compatible: –
Altitude Hold: √
Position Hold: √
Flight Planning: –
Autonomous Flight: –
Target Tracking: –
Automatic Landing: –
Return To Home: –
Obstacle Avoidance: –
Headless Mode: –
Built-in Camera: –

Jin Xing Da

388

© Jin Xing Da

Release Date: March 2014
Price: $35
Number of Rotors: 4
Control System: Remote Control
Applications: Recreation
Dimensions: 133 × 133 × 3.2 mm
Weight: 247 gr
Frame Material: Plastic
Rotor Material: Plastic
Available Colors: Black, Red, Yellow
Max. Flight Time: 12 minutes
Max. Operating Range: 100 m
Controller Frequency: 2.4 GHz
Battery Capacity: 300 mAh
Battery Type: Lithium Polymer
Battery Power: 3.7V
GPS Compatible: –
Altitude Hold: √
Position Hold: √
Flight Planning: –
Autonomous Flight: –
Target Tracking: –
Automatic Landing: –
Return To Home: –
Obstacle Avoidance: –
Headless Mode: –
Built-in Camera: –

Jin Xing Da

392

© electricdragonfly.com

Release Date: March 2014
Price: $45
Number of Rotors: 4
Control System: Remote Control
Applications: Recreation
Dimensions: 100 × 95 × 25 mm
Weight: 319 gr
Frame Material: Plastic
Rotor Material: Plastic
Available Colors: Black
Max. Flight Time: 5 minutes
Max. Operating Range: 50 m
Controller Frequency: 2.4 GHz
Battery Capacity: 150 mAh
Battery Type: Lithium Polymer
Battery Power: 3.7V
GPS Compatible: –
Altitude Hold: –
Position Hold: –
Flight Planning: –
Autonomous Flight: –
Target Tracking: –
Automatic Landing: –
Return To Home: –
Obstacle Avoidance: –
Headless Mode: –
Built-in Camera: –

Monarch Inc.

AG

©Monarch Inc.

Release Date: March 2014
Price: On Request
Number of Rotors: 4
Control System: Remote Control
Applications: Aerial Mapping, Agriculture, Inspection, Surveillance
Dimensions: 1.04 × 1.04 × 0.46 m
Weight: 6.8 kg
Frame Material: Carbon Fiber
Rotor Material: Carbon Fiber
Available Colors: White
Max. Flight Time: 30 minutes
Min. Operating Temperature: -40°C
Max. Operating Temperature: 85°C
Controller Frequency: 2.4 GHz
Battery Capacity: 22,000 mAh
Battery Type: Lithium Polymer
GPS Compatible: √
Altitude Hold: √
Position Hold: √
Flight Planning: √
Autonomous Flight: √
Target Tracking: √
Automatic Landing: √
Return To Home: √
Obstacle Avoidance: –
Headless Mode: –
Built-in Camera: –

OM UAV

Guardian Fixedwing

©Novadem

Release Date: March 2014
Price: Unknown
Number of Rotors: 1 (Fixed Wing)
Control System: Remote Control
Applications: Aerial Mapping, Inspection, Surveillance
Dimensions: 1.3 × 1.76 × 0.2 m
Wingspan: 1.76 m
Weight: 2.4 kg
Max. Payload: 300 gr
Frame Material: Carbon Fiber
Rotor Material: Carbon Fiber
Available Colors: White
Max. Flight Time: 40 minutes
Max. Speed: 60 km/h (37 mp/h)
Max. Operating Range: 1 km
Controller Frequency: 2.4 GHz
Battery Capacity: 6,800 mAh
Battery Type: Lithium Polymer
Battery Power: 11.1V
GPS Compatible: √
Altitude Hold: √
Position Hold: √
Flight Planning: √
Autonomous Flight: √
Target Tracking: √
Automatic Landing: √
Return To Home: √
Obstacle Avoidance: √
Headless Mode: –
Built-in Camera: √

SeaHex

SeaHex

© Onne van der Wal

Release Date: March 2014
Price: $7,700
Number of Rotors: 6
Control System: Remote Control
Applications: Aerial Mapping, Surveillance
Dimensions: 750 × 688 × 191 mm
Wingspan: 1020 mm
Weight: 2.6 kg
Frame Material: Carbon Fiber
Rotor Material: Carbon Fiber
Available Colors: Yellow
Max. Flight Time: 20 minutes
Max. Operating Range: 1 km
Controller Frequency: 2.4 GHz
Battery Capacity: 10,000 mAh
Battery Type: Lithium Polymer 4S
GPS Compatible: √
Altitude Hold: –
Position Hold: √
Flight Planning: √
Autonomous Flight: –
Target Tracking: –
Automatic Landing: –
Return To Home: √
Obstacle Avoidance: –
Headless Mode: –
Built-in Camera: –

WLtoys

V303 Seeker

© eforsmart.com

Release Date: March 2014
Price: $330
Number of Rotors: 4
Control System: Remote Control
Applications: Film, Photography
Dimensions: 305 × 305 × 185 mm
Wingspan: 420 mm
Weight: 200 gr
Frame Material: Plastic
Rotor Material: Plastic
Available Colors: White
Max. Flight Time: 15 minutes
Max. Operating Range: 500 m
Controller Frequency: 2.4 GHz
Battery Capacity: 2,700 mAh
Battery Type: Lithium Polymer
Battery Power: 7.3V
GPS Compatible: –
Altitude Hold: √
Position Hold: √
Flight Planning: –
Autonomous Flight: –
Target Tracking: –
Automatic Landing: –
Return To Home: √
Obstacle Avoidance: –
Headless Mode: –
Built-in Camera: –

WLtoys

V323 Skywalker

© maisondudrone.com

Release Date: March 2014
Price: $100
Number of Rotors: 6
Control System: Remote Control
Applications: Recreation
Dimensions: 800 × 800 × 110 mm
Weight: 1.9 kg
Frame Material: Plastic, Foam
Rotor Material: Plastic
Available Colors: Black
Max. Flight Time: 9 minutes
Max. Operating Range: 150 m
Controller Frequency: 2.4 GHz
Battery Capacity: 2,700 mAh
Battery Type: Lithium Polymer
Battery Power: 7.3V
GPS Compatible: –
Altitude Hold: √
Position Hold: √
Flight Planning: –
Autonomous Flight: –
Target Tracking: –
Automatic Landing: –
Return To Home: √
Obstacle Avoidance: –
Headless Mode: –
Built-in Camera: –

Arcturus UAV

Jump 15

© Arcturus

Release Date: April 2014
Price: Unknown
Number of Rotors: 5 (Fixed Wing)
Control System: Ground Control Station
Applications: Aerial Mapping, Inspection
Dimensions: 1.8 × 3.3 m
Wingspan: 3.3 m
Weight: 24 kg
Frame Material: Composite
Motors: 40cc
Available Colors: Grey
Max. Flight Time: 6 hours
Max. Speed: 100 km/h (62 mp/h)
Max. Flying Altitude: 4,572 m (15,000 ft)
Max. Operating Range: 125 km
Controller Frequency: 2.4 GHz
GPS Compatible: √
Altitude Hold: √
Position Hold: √
Flight Planning: √
Autonomous Flight: √
Target Tracking: √
Automatic Landing: √
Return To Home: √
Obstacle Avoidance: –
Headless Mode: –
Built-in Camera: √

Boeing Insitu

RQ-21A Blackjack

© Boeing Insitu

Release Date: April 2014
Price: $5 million
Number of Rotors: 1 (Fixed Wing)
Control System: Flight Computer
Applications: Search & Rescue, Surveillance, Target Acquisition
Dimensions: 2.5 × 4.8 m
Wingspan: 4.8 m
Weight: 34 kg
Max. Payload: 18 kg
Motors: 100-cc heavy-fuel
Available Colors: Grey
Max. Flight Time: 24 hours
Max. Flight Time with Payload: 13 hours
Max. Speed: 166 km/h (103 mp/h)
Max. Flying Altitude: 5,950 m (19,521 ft)
GPS Compatible: √
Altitude Hold: √
Position Hold: √
Flight Planning: √
Autonomous Flight: √
Target Tracking: √
Automatic Landing: √
Return To Home: √
Obstacle Avoidance: √
Headless Mode: –
Built-in Camera: √

DJI

Phantom 2 Vision+

© quadcopters.co.uk

Release Date: April 2014
Price: $1,299
Number of Rotors: 4
Control System: Remote Control, Phone/Tablet
Applications: Film, Photography
Wingspan: 350 mm
Weight: 1.24 kg
Max. Payload: 300 gr
Frame Material: Plastic
Rotor Material: Plastic, Carbon Fiber
Available Colors: White
Max. Flight Time: 25 minutes
Max. Speed: 54 km/h (33,5 mp/h)
Max. Flying Altitude: 760 m (2,493 ft)
Max. Ascent Speed: 6 m/s
Max. Descent Speed: 2 m/s
Min. Operating Temperature: 0°C
Max. Operating Temperature: 40°C
Max. Operating Range: 800 m
Controller Frequency: 5.8 GHz
Battery Capacity: 5,200 mAh
Battery Type: Lithium Polymer 4S
Battery Power: 11V
Battery Weight: 300 gr
GPS Compatible: √
Altitude Hold: √
Position Hold: √
Flight Planning: √
Autonomous Flight: √
Target Tracking: √
Automatic Landing: √
Return To Home: √
Obstacle Avoidance: –
Headless Mode: √
Built-in Camera: √ (12.4 megapixels)

Flytronic

Flysar tactical UAV

©WB Group

Release Date: April 2014
Price: Unknown
Number of Rotors: 1 (Fixed Wing)
Control System: Ground Control Station
Applications: Inspection, Research & Development, Surveillance
Dimensions: 3.2 × 5.7 m
Wingspan: 5.7 m
Weight: 45 kg
Max. Payload: 45 kg
Frame Material: Composite
Available Colors: Grey
Max. Flight Time: 15 hours
Max. Speed: 160 km/h (99,4 mp/h)
GPS Compatible: √
Altitude Hold: √
Position Hold: –
Flight Planning: √
Autonomous Flight: √
Target Tracking: –
Automatic Landing: √
Return To Home: √
Obstacle Avoidance: –
Headless Mode: –
Built-in Camera: √

Jin Xing Da

389

©Jin Xing Da

Release Date: April 2014
Price: $50
Number of Rotors: 4
Control System: Remote Control
Applications: Recreation
Dimensions: 120 × 120 × 40 mm
Weight: 230 gr
Frame Material: Plastic
Rotor Material: Plastic
Available Colors: Blue, Red
Max. Flight Time: 10 minutes
Max. Operating Range: 100 m
Controller Frequency: 2.4 GHz
Battery Capacity: 500 mAh
Battery Type: Lithium Polymer
Battery Power: 3.7V
GPS Compatible: –
Altitude Hold: √
Position Hold: √
Flight Planning: –
Autonomous Flight: –
Target Tracking: –
Automatic Landing: –
Return To Home: –
Obstacle Avoidance: –
Headless Mode: –
Built-in Camera: –

Kale-Baykar

Bayraktar TB2

©Wikipedia/Bayhaluk

Release Date: April 2014
Price: $75,000
Number of Rotors: 1 (Fixed Wing)
Control System: Ground Control Station
Applications: Military Purposes, Armed Reconnaissance, Surveillance, Target Acquisition
Dimensions: 6.5 × 12 m
Wingspan: 12 m
Weight: 600 kg
Max. Payload: 55 kg
Frame Material: Composite
Motors: 1 × Rotax 912 internal combustion engine
Available Colors: Grey
Max. Flight Time: 24 hours
Max. Speed: 203 km/h (126 mp/h)
Max. Flying Altitude: 7,620 m (25,000 ft)
Max. Operating Range: 150 km
GPS Compatible: √
Altitude Hold: √
Position Hold: –
Flight Planning: √
Autonomous Flight: √
Target Tracking: √
Automatic Landing: √
Return To Home: √
Obstacle Avoidance: √
Headless Mode: –
Built-in Camera: √

Lockheed Martin

Indago

©Lockheed Martin

Release Date: April 2014
Price: $25,000
Number of Rotors: 4
Propellers: PP11A
Control System: Remote Control, Phone/Tablet
Applications: Agriculture, Inspection, Surveillance
Dimensions: 813 × 813 × 178 mm
Wingspan: 813 mm
Weight: 2.2 kg
Max. Payload: 200 gr
Frame Material: Carbon Fiber
Rotor Material: Carbon Fiber
Available Colors: Black
Max. Flight Time: 50 minutes
Max. Speed: 72 km/h (45 mp/h)
Max. Flying Altitude: 152 m (500 ft)
Max. Operating Range: 2 km
GPS Compatible: √
Altitude Hold: √
Position Hold: √
Flight Planning: √
Autonomous Flight: –
Target Tracking: √
Automatic Landing: √
Return To Home: √
Obstacle Avoidance: –
Headless Mode: –
Built-in Camera: √ (10 megapixels)

Onyxstar

Fox-C8 HD

© Onyxstar

Release Date: April 2014
Price: $8,425
Number of Rotors: 6
Control System: Remote Control
Applications: Film, Photography, Aerial Mapping, Agriculture, Inspection, Surveillance
Weight: 6 kg
Max. Payload: 5 kg
Frame Material: Carbon Fiber
Rotor Material: Carbon Fiber
Available Colors: Black
Max. Flight Time: 37 minutes
GPS Compatible: √
Altitude Hold: √
Position Hold: √
Flight Planning: √
Autonomous Flight: √
Target Tracking: √
Automatic Landing: √
Return To Home: √
Obstacle Avoidance: –
Headless Mode: –
Built-in Camera: –

QuestUAV

Q-200 Surveyor Pro

© QuestUAV

Release Date: April 2014
Price: $23,500
Number of Rotors: 1 (Fixed Wing)
Control System: Flight Computer
Applications: Aerial Mapping, Inspection
Dimensions: 1.25 × 0.75 × 0.13 m
Wingspan: 1.2 m
Weight: 3.8 kg
Frame Material: EPP Foam
Rotor Material: Carbon Fiber
Available Colors: Yellow
Max. Flight Time: 60 minutes
Max. Speed: 86 km/h (53,4 mp/h)
Max. Flying Altitude: 2,500 m (8,200 ft)
Max. Operating Range: 20 km
Controller Frequency: 2.4 GHz
Battery Capacity: 5,000 mAh
Battery Type: Lithium Polymer
Battery Power: 11.1V
GPS Compatible: √
Altitude Hold: √
Position Hold: √
Flight Planning: √
Autonomous Flight: –
Target Tracking: √
Automatic Landing: √
Return To Home: √
Obstacle Avoidance: –
Headless Mode: –
Built-in Camera: √ (24.3 megapixels)

Sky Hero

Spyder 6

© Sky Hero

Release Date: April 2014
Price: $529
Number of Rotors: 6, 12
Control System: Remote Control
Applications: Film, Photography
Dimensions: 0.93 × 1.07 × 0.3 mm
Wingspan: 1 m
Weight: 2.29 kg
Frame Material: Plastic, Carbon Fiber
Rotor Material: Plastic
Available Colors: White
Controller Frequency: 2.4 GHz
Battery Capacity: 3,700 mAh
Battery Type: Lithium Polymer
GPS Compatible: √
Altitude Hold: √
Position Hold: √
Flight Planning: √
Autonomous Flight: –
Target Tracking: √
Automatic Landing: –
Return To Home: √
Obstacle Avoidance: –
Headless Mode: –
Built-in Camera: –

Survey-Copter

DVF2000

© Survey-Copter

Release Date: April 2014
Price: Unknown
Number of Rotors: 1 (Fixed Wing)
Control System: Ground Control Station
Applications: Military Purposes, Maritime Patrol, Reconnaissance, Surveillance
Dimensions: 1.2 × 3 m
Wingspan: 3 m
Weight: 11 kg
Max. Payload: 1.1 kg
Frame Material: Composite
Available Colors: Grey
Max. Flight Time: 2 hours
Max. Speed: 97 km/h (60 mp/h)
Max. Flying Altitude: 2,500 m (8,200 ft)
Min. Operating Temperature: 0°C
Max. Operating Temperature: 50°C
Max. Operating Range: 50 km
GPS Compatible: √
Altitude Hold: √
Position Hold: √
Flight Planning: √
Autonomous Flight: √
Target Tracking: –
Automatic Landing: –
Return To Home: √
Obstacle Avoidance: √
Headless Mode: –
Built-in Camera: √

Virtualrobotix

VR Tiger Shark X8

© Virtualrobotix

Release Date: April 2014
Price: Unknown
Number of Rotors: 4
Control System: Remote Control
Applications: Film, Photography
Wingspan: 510 mm
Weight: 750 gr
Max. Payload: 350 gr
Frame Material: Plastic, Carbon Fiber
Rotor Material: Plastic
Available Colors: Black
Max. Flight Time: 30 minutes
Max. Operating Range: 1.5 km
Controller Frequency: 2.4 GHz
Battery Capacity: 10,000 mAh
Battery Type: Lithium Polymer 3S
Battery Power: 14.8V
GPS Compatible: √
Altitude Hold: √
Position Hold: √
Flight Planning: √
Autonomous Flight: –
Target Tracking: √
Automatic Landing: √
Return To Home: √
Obstacle Avoidance: –
Headless Mode: –
Built-in Camera: –

WLtoys

V252 Skylark

© WLtoys

Release Date: April 2014
Price: $25
Number of Rotors: 4
Control System: Remote Control
Applications: Recreation
Dimensions: 130 × 130 × 60 mm
Weight: 37 gr
Frame Material: Plastic
Rotor Material: Plastic
Available Colors: Purple, Red
Max. Flight Time: 8 minutes
Max. Operating Range: 150 m
Controller Frequency: 2.4 GHz
Battery Capacity: 250 mAh
Battery Type: Lithium Polymer
Battery Power: 3.7V
GPS Compatible: –
Altitude Hold: √
Position Hold: √
Flight Planning: –
Autonomous Flight: –
Target Tracking: –
Automatic Landing: –
Return To Home: –
Obstacle Avoidance: –
Headless Mode: –
Built-in Camera: –

Aeromao

Aeromapper Talon

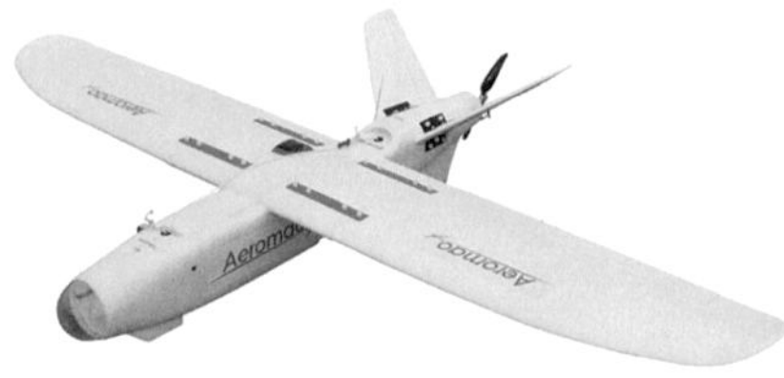

© Aeromao

Release Date: May 2014
Price: $9,600
Number of Rotors: 1 (Fixed Wing)
Control System: Remote Control
Applications: Aerial Mapping, Agriculture
Wingspan: 2 m
Weight: 3.5 kg
Max. Payload: 650 gr
Frame Material: EPO Foam (carbon fiber reinforced)
Available Colors: White
Max. Flight Time: 80 minutes
Max. Speed: 85 km/h (50 mp/h)
Max. Operating Range: 50 km
Controller Frequency: 2.4 GHz
Battery Type: Lithium Polymer
GPS Compatible: √
Altitude Hold: √
Position Hold: √
Flight Planning: √
Autonomous Flight: √
Target Tracking: √
Automatic Landing: √
Return To Home: √
Obstacle Avoidance: –
Headless Mode: –
Built-in Camera: √ (24 megapixels)

Aeronavics

Bot

© Aeronavics

Release Date: May 2014
Price: $1,000
Number of Rotors: 4
Propellers: T-Motors 15x5
Control System: Remote Control
Applications: Film, Photography, Aerial Mapping, Surveillance
Dimensions: 608 × 608 × 160 mm
Wingspan: 631 mm
Weight: 5 kg
Max. Payload: 1.2 kg
Frame Material: Carbon fiber
Rotor Material: Carbon Fiber
Motors: T-Motors MN3515-400kv brushless
Available Colors: Black, Orange
Max. Flight Time: 40 minutes
Controller Frequency: 2.4 GHz
Battery Capacity: 12,000 mAh
Battery Type: Lithium Polymer 6S
GPS Compatible: √
Altitude Hold: √
Position Hold: √
Flight Planning: √
Autonomous Flight: √
Target Tracking: –
Automatic Landing: √
Return To Home: √
Obstacle Avoidance: –
Headless Mode: –
Built-in Camera: –

Aeronavics

XM6

Release Date: May 2014
Price: $6,500
Number of Rotors: 6
Control System: Remote Control
Applications: Film, Photography, Aerial Mapping, Surveillance
Wingspan: 770 mm
Weight: 3 kg
Max. Payload: 1 kg
Frame Material: Carbon Fiber, Titanium
Rotor Material: Carbon Fiber
Available Colors: Black, Orange
Max. Flight Time: 25 minutes
Max. Speed: 80 km/h (48 mp/h)
Max. Operating Range: 3 km
Controller Frequency: 2.4 GHz
Battery Capacity: 6,000 mAh
Battery Type: Lithium Polymer 6S
Battery Power: 12V
GPS Compatible: √
Altitude Hold: √
Position Hold: √
Flight Planning: √
Autonomous Flight: √
Target Tracking: –
Automatic Landing: √
Return To Home: √
Obstacle Avoidance: –
Headless Mode: –
Built-in Camera: –

BHE Robotics

BXAP15

Release Date: May 2014
Price: Unknown
Number of Rotors: 1 (Fixed Wing)
Control System: Ground Control Station
Applications: Military Purposes, Armed Reconnaissance, Inspection, Surveillance
Dimensions: 1.7 × 3.7 m
Wingspan: 3.7 m
Weight: 17 kg
Max. Payload: 3 kg
Frame Material: Composite
Available Colors: Red
Max. Flight Time: 100 minutes
Max. Speed: 100 km/h (62 mp/h)
Max. Flying Altitude: 4,000 m (13,123 ft)
Max. Operating Range: 160 km
Battery Type: Lithium Polymer
GPS Compatible: √
Altitude Hold: √
Position Hold: √
Flight Planning: √
Autonomous Flight: √
Target Tracking: √
Automatic Landing: –
Return To Home: √
Obstacle Avoidance: √
Headless Mode: –
Built-in Camera: √

BirdsEyeView Aerobotics

FireFly6

© gregcovey.com

Release Date: May 2014
Price: $499
Number of Rotors: 6 (Fixed Wing)
Control System: Ground Control Station
Applications: Film, Photography, Aerial Mapping, Inspection
Dimensions: 0.95 × 1.52 m
Wingspan: 1.52 m
Weight: 3.7 kg
Max. Payload: 700 gr
Frame Material: EPO Foam (carbon fiber reinforced)
Available Colors: White/Red
Max. Flight Time: 15 minutes
Max. Speed: 80 km/h (48 mp/h)
Controller Frequency: 2.4 GHz
Battery Capacity: 3,900 mAh
Battery Type: Lithium Polymer
GPS Compatible: √
Altitude Hold: √
Position Hold: √
Flight Planning: –
Autonomous Flight: –
Target Tracking: –
Automatic Landing: √
Return To Home: √
Obstacle Avoidance: –
Headless Mode: –
Built-in Camera: –

Cheerson

CX-10

© dronetrest.com

Release Date: May 2014
Price: $15
Number of Rotors: 4
Control System: Remote Control
Applications: Recreation
Dimensions: 40 × 40 × 22 mm
Wingspan: 56,5 mm
Weight: 12 gr
Frame Material: Plastic
Rotor Material: Plastic
Available Colors: Blue, Green, Orange, Pink
Max. Flight Time: 4 minutes
Max. Operating Range: 200 m
Controller Frequency: 2.4 GHz
Battery Capacity: 100 mAh
Battery Type: Lithium Polymer
Battery Power: 3.7V
GPS Compatible: –
Altitude Hold: –
Position Hold: –
Flight Planning: –
Autonomous Flight: –
Target Tracking: –
Automatic Landing: –
Return To Home: –
Obstacle Avoidance: –
Headless Mode: –
Built-in Camera: –

Cheerson
CX-11

Release Date: May 2014
Price: $20
Number of Rotors: 4
Control System: Remote Control
Applications: Recreation
Dimensions: 45 × 45 × 23 mm
Wingspan: 63 mm
Weight: 29 gr
Frame Material: Plastic
Rotor Material: Plastic
Available Colors: White
Max. Flight Time: 5 minutes
Max. Operating Range: 20 m
Controller Frequency: 2.4 GHz
Battery Capacity: 100 mAh
Battery Type: Lithium Polymer
Battery Power: 3.7V
GPS Compatible: –
Altitude Hold: –
Position Hold: –
Flight Planning: –
Autonomous Flight: –
Target Tracking: –
Automatic Landing: –
Return To Home: –
Obstacle Avoidance: –
Headless Mode: –
Built-in Camera: –

Dronetech UAV
AV-1 Albatross

© Dronetech

Release Date: May 2014
Price: Unknown
Number of Rotors: 1 (Fixed Wing)
Control System: Ground Control Station
Applications: Military Purposes, Reconnaissance, Surveillance
Dimensions: 3 × 5.4 m
Wingspan: 5.4 m
Weight: 70 kg
Max. Payload: 55 kg
Frame Material: Composite
Available Colors: White
Max. Flight Time: 2 hours
Max. Speed: 148 km/h (92 mp/h)
Max. Operating Range: 100 km
GPS Compatible: √
Altitude Hold: √
Position Hold: –
Flight Planning: √
Autonomous Flight: √
Target Tracking: –
Automatic Landing: √
Return To Home: √
Obstacle Avoidance: –
Headless Mode: –
Built-in Camera: √

Event38

E384

© Event38

Release Date: May 2014
Price: $2,399
Number of Rotors: 1 (Fixed Wing)
Control System: Flight Computer, Laptop
Applications: Aerial Mapping, Agriculture, Surveillance
Dimensions: 1.3 × 1.9 m
Wingspan: 1.9 m
Weight: 2.5 kg
Max. Payload: 1 kg
Frame Material: Foam, Carbon Fiber
Rotor Material: Plastic
Available Colors: White
Max. Flight Time: 100 minutes
Max. Speed: 52 km/h (29 mp/h)
Max. Flying Altitude: 3,960 m (12,992 ft)
Min. Operating Temperature: -20°C
Max. Operating Temperature: 50°C
Max. Operating Range: 10 km
Controller Frequency: 2.4 GHz
GPS Compatible: √
Altitude Hold: √
Position Hold: √
Flight Planning: √
Autonomous Flight: √
Target Tracking: √
Automatic Landing: √
Return To Home: √
Obstacle Avoidance: √
Headless Mode: –
Built-in Camera: √

Fineco

FX1 Nano

© aliexpress.com

Release Date: May 2014
Price: $15
Number of Rotors: 4
Control System: Remote Control
Applications: Recreation
Dimensions: 45 × 45 × 20 mm
Frame Material: Plastic
Rotor Material: Plastic
Motors: 4 CH Gyro
Available Colors: White
Max. Flight Time: 7 minutes
Max. Operating Range: 30 m
Controller Frequency: 2.4 GHz
Battery Capacity: 350 mAh
Battery Type: Lithium Polymer
Battery Power: 3.7V
GPS Compatible: –
Altitude Hold: –
Position Hold: –
Flight Planning: –
Autonomous Flight: –
Target Tracking: –
Automatic Landing: –
Return To Home: –
Obstacle Avoidance: –
Headless Mode: –
Built-in Camera: –

Flypro
X400

Release Date: May 2014
Price: $339
Number of Rotors: 4
Control System: Remote Control
Applications: Film, Photography, Surveillance
Dimensions: 500 × 370 × 40 mm
Wingspan: 450 mm
Weight: 500 gr
Frame Material: Carbon Fiber
Rotor Material: Plastic
Motors: 2208 Brushless
Available Colors: Black
Max. Flight Time: 12 minutes
Controller Frequency: 2.4 GHz
Battery Capacity: 2,200 mAh
Battery Type: Lithium Polymer 3S
Battery Power: 11.1V
GPS Compatible: –
Altitude Hold: √
Position Hold: √
Flight Planning: –
Autonomous Flight: –
Target Tracking: –
Automatic Landing: √
Return To Home: √
Obstacle Avoidance: –
Headless Mode: –
Built-in Camera: –

HiSky Hobby
HMX120

©tmart.com

Release Date: May 2014
Price: $75
Number of Rotors: 4
Control System: Remote Control
Applications: Recreation
Dimensions: 128 × 128 × 36 mm
Weight: 309 gr
Frame Material: Plastic
Rotor Material: Plastic
Available Colors: Orange
Max. Flight Time: 7 minutes
Controller Frequency: 2.4 GHz
Battery Capacity: 250 mAh
Battery Type: Lithium Polymer
Battery Power: 3.7V
GPS Compatible: –
Altitude Hold: –
Position Hold: –
Flight Planning: –
Autonomous Flight: –
Target Tracking: –
Automatic Landing: –
Return To Home: –
Obstacle Avoidance: –
Headless Mode: –
Built-in Camera: –

Jin Xing Da

391 Space Trek Galaxy

©Jin Xing Da

Release Date: May 2014
Price: $110
Number of Rotors: 4
Control System: Remote Control, Phone/Tablet
Applications: Recreation
Dimensions: 530 × 530 × 120 mm
Frame Material: Plastic
Rotor Material: Plastic
Available Colors: Black
Max. Flight Time: 8 minutes
Max. Operating Range: 80 m
Controller Frequency: 2.4 GHz
Battery Capacity: 100 mAh
Battery Type: Lithium Polymer
Battery Power: 3.7V
GPS Compatible: –
Altitude Hold: √
Position Hold: √
Flight Planning: –
Autonomous Flight: –
Target Tracking: –
Automatic Landing: –
Return To Home: √
Obstacle Avoidance: –
Headless Mode: –
Built-in Camera: √ (0.3 megapixels)

Jin Xing Da

393 Space Trek Defiant

©Jin Xing Da

Release Date: May 2014
Price: $44
Number of Rotors: 4
Control System: Remote Control
Applications: Recreation
Dimensions: 390 × 390 × 80 mm
Frame Material: Plastic
Rotor Material: Plastic
Available Colors: Black
Max. Flight Time: 15 minutes
Max. Operating Range: 80 m
Controller Frequency: 2.4 GHz
Battery Capacity: 550 mAh
Battery Type: Lithium Polymer
Battery Power: 3.7V
GPS Compatible: –
Altitude Hold: √
Position Hold: √
Flight Planning: –
Autonomous Flight: –
Target Tracking: –
Automatic Landing: –
Return To Home: –
Obstacle Avoidance: –
Headless Mode: –
Built-in Camera: –

Jin Xing Da

395 Airbus

Release Date: May 2014
Price: $20
Number of Rotors: 4
Control System: Remote Control
Applications: Recreation
Dimensions: 65 × 65 × 25 mm
Weight: 13 gr
Frame Material: Plastic
Rotor Material: Plastic
Available Colors: Green, Yellow
Max. Flight Time: 5 minutes
Max. Operating Range: 20 m
Controller Frequency: 2.4 GHz
Battery Capacity: 100 mAh
Battery Type: Lithium Polymer
Battery Power: 3.7V
GPS Compatible: –
Altitude Hold: –
Position Hold: –
Flight Planning: –
Autonomous Flight: –
Target Tracking: –
Automatic Landing: –
Return To Home: –
Obstacle Avoidance: –
Headless Mode: –
Built-in Camera: –

OM UAV

Baaz Mini

©OM UAV

Release Date: May 2014
Price: Unknown
Number of Rotors: 1 (Fixed Wing)
Control System: Remote Control
Applications: Inspection, Surveillance
Dimensions: 450 × 1200 × 50 mm
Wingspan: 1200 mm
Weight: 800 gr
Max. Payload: 100 gr
Frame Material: Carbon Fiber
Rotor Material: Carbon Fiber
Available Colors: White
Max. Flight Time: 40 minutes
Max. Speed: 65 km/h (40 mp/h)
Max. Flying Altitude: 1,000 m (3,280 ft)
Max. Operating Range: 7 km
Controller Frequency: 2.4 GHz
Battery Capacity: 6,800 mAh
Battery Type: Lithium Polymer
Battery Power: 11.1V
GPS Compatible: √
Altitude Hold: √
Position Hold: √
Flight Planning: √
Autonomous Flight: √
Target Tracking: √
Automatic Landing: √
Return To Home: √
Obstacle Avoidance: √
Headless Mode: –
Built-in Camera: √

Syma Toys

X5C Explorer

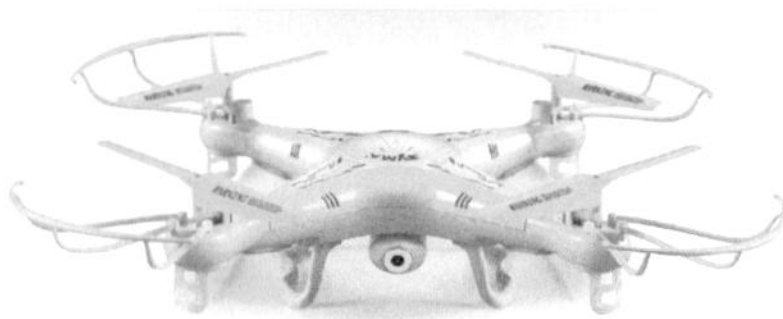

©hitoys.org

Release Date: May 2014
Price: $50
Number of Rotors: 4
Control System: Remote Control
Applications: Film, Photography
Dimensions: 310 × 310 × 80 mm
Wingspan: 438 mm
Weight: 916 gr
Frame Material: Plastic
Rotor Material: Plastic
Available Colors: White
Max. Flight Time: 7 minutes
Max. Speed: 54 km/h (33,5 mp/h)
Max. Flying Altitude: 50 m (164 ft)
Max. Operating Range: 50 m
Controller Frequency: 2.4 GHz
Battery Capacity: 500 mAh
Battery Type: Lithium Polymer
Battery Power: 4V
GPS Compatible: –
Altitude Hold: –
Position Hold: –
Flight Planning: –
Autonomous Flight: –
Target Tracking: –
Automatic Landing: –
Return To Home: –
Obstacle Avoidance: –
Headless Mode: –
Built-in Camera: √ (2 megapixels)

Syma Toys

X6 Super Ship

©prisguide.no

Release Date: May 2014
Price: $69
Number of Rotors: 4
Control System: Remote Control
Applications: Recreation
Dimensions: 580 × 560 × 100 mm
Wingspan: 806 mm
Weight: 320 gr
Frame Material: Plastic, Foam
Rotor Material: Plastic
Available Colors: Red
Max. Flight Time: 8 minutes
Max. Operating Range: 30 m
Controller Frequency: 2.4 GHz
Battery Capacity: 850 mAh
Battery Type: Lithium Polymer
Battery Power: 3.7V
GPS Compatible: –
Altitude Hold: –
Position Hold: –
Flight Planning: –
Autonomous Flight: –
Target Tracking: –
Automatic Landing: –
Return To Home: –
Obstacle Avoidance: –
Headless Mode: –
Built-in Camera: –

UMS AERO Group

F-330

©nfz.ch

Release Date: May 2014
Price: Unknown
Number of Rotors: 1 (Fixed Wing)
Control System: Ground Control Station
Applications: Military Purposes, Inspection, Reconnaissance, Surveillance, Target Acquisition
Dimensions: 2.27 × 3.3 × 0.9 m
Wingspan: 3.3 m
Weight: 23.7 kg
Max. Payload: 8 kg
Frame Material: Composite
Available Colors: Grey
Max. Flight Time: 8 hours
Max. Speed: 120 km/h (74 mp/h)
Max. Flying Altitude: 4,900 m (16,076 ft)
GPS Compatible: √
Altitude Hold: √
Position Hold: –
Flight Planning: √
Autonomous Flight: √
Target Tracking: –
Automatic Landing: √
Return To Home: √
Obstacle Avoidance: √
Headless Mode: –
Built-in Camera: √

XactSense

Titan1

©XactSense

Release Date: May 2014
Price: $120,000
Number of Rotors: 4
Control System: Remote Control, Phone/Tablet
Applications: Aerial Mapping, Inspection, Surveillance
Wingspan: 1.3 m
Weight: 8.2 kg
Max. Payload: 22.6 kg
Frame Material: Carbon Fiber
Rotor Material: Carbon Fiber
Available Colors: White
Max. Flight Time: 30 minutes
Max. Flying Altitude: 55 m (180 ft)
Max. Operating Range: 16 km
Controller Frequency: 2.4 GHz
Battery Type: Lithium Polymer
GPS Compatible: √
Altitude Hold: –
Position Hold: √
Flight Planning: √
Autonomous Flight: √
Target Tracking: √
Automatic Landing: √
Return To Home: √
Obstacle Avoidance: –
Headless Mode: –
Built-in Camera: –

Acme

Zoopa Q165

Release Date: June 2014
Price: $69
Number of Rotors: 4
Control System: Remote Control
Applications: Recreation
Dimensions: 160 × 160 × 35 mm
Wingspan: 200 mm
Weight: 40 gr
Frame Material: Plastic
Rotor Material: Plastic
Available Colors: Black
Max. Flight Time: 9 minutes
Max. Operating Range: 120 m
Controller Frequency: 2.4 GHz
Battery Capacity: 240 mAh
Battery Type: Lithium Polymer 1S
Battery Power: 3.7V
GPS Compatible: –
Altitude Hold: –
Position Hold: –
Flight Planning: –
Autonomous Flight: –
Target Tracking: –
Automatic Landing: –
Return To Home: –
Obstacle Avoidance: –
Headless Mode: –
Built-in Camera: –

Acme

Zoopa Q410 Movie

Release Date: June 2014
Price: $110
Number of Rotors: 4
Control System: Remote Control
Applications: Film, Photography
Dimensions: 325 × 325 × 80 mm
Wingspan: 410 mm
Weight: 170 gr
Frame Material: Plastic
Rotor Material: Plastic
Available Colors: Black
Max. Flight Time: 9 minutes
Max. Operating Range: 120 m
Controller Frequency: 2.4 GHz
Battery Capacity: 500 mAh
Battery Type: Lithium Polymer 1S
Battery Power: 3.7V
GPS Compatible: –
Altitude Hold: –
Position Hold: –
Flight Planning: –
Autonomous Flight: –
Target Tracking: –
Automatic Landing: –
Return To Home: –
Obstacle Avoidance: –
Headless Mode: –
Built-in Camera: √

All Tech

KWT-Z4M

© All Tech

Release Date: June 2014
Price: On Request
Number of Rotors: 4
Control System: Ground Control Station, Remote Control
Applications: Police, Inspection, Surveillance
Wingspan: 940 mm
Weight: 7.5 kg
Max. Payload: 2.5 kg
Frame Material: Carbon Fiber
Rotor Material: Carbon Fiber
Available Colors: White
Max. Flight Time: 40 minutes
Max. Speed: 54 km/h (33,5 mp/h)
Max. Flying Altitude: 4,000 m (13,123 ft)
Max. Operating Range: 10 km
Controller Frequency: 2.4 GHz
Battery Type: Lithium Polymer
GPS Compatible: √
Altitude Hold: √
Position Hold: √
Flight Planning: √
Autonomous Flight: √
Target Tracking: √
Automatic Landing: √
Return To Home: √
Obstacle Avoidance: –
Headless Mode: –
Built-in Camera: √

Delair-Tech

DT18

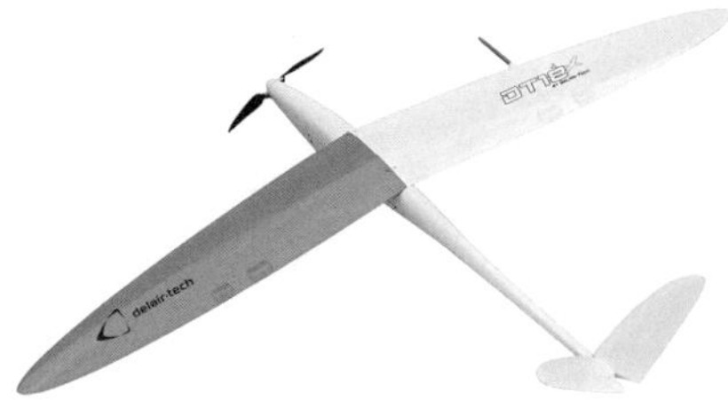

© Delair-Tech

Release Date: June 2014
Price: $18.000
Number of Rotors: 1 (Fixed Wing)
Control System: Flight Computer
Applications: Aerial Mapping, Agriculture
Dimensions: 1.2 m
Wingspan: 1.8 m
Weight: 2 kg
Frame Material: Carbon Fiber
Available Colors: Black
Max. Flight Time: 2 hours
Max. Speed: 50 km/h (31 mp/h)
Max. Flying Altitude: 3,000 m (9,842 ft)
Min. Operating Temperature: -20°C
Max. Operating Temperature: 50°C
Max. Operating Range: 100 km
Controller Frequency: 2.4 GHz
GPS Compatible: √
Altitude Hold: √
Position Hold: √
Flight Planning: √
Autonomous Flight: √
Target Tracking: √
Automatic Landing: √
Return To Home: √
Obstacle Avoidance: –
Headless Mode: –
Built-in Camera: √

Flypro

X600

© helicomicro.com

Release Date: June 2014
Price: $850
Number of Rotors: 6
Control System: Remote Control
Applications: Film, Photography, Aerial Mapping, Surveillance
Weight: 2.5 kg
Frame Material: Carbon Fiber
Rotor Material: Carbon Fiber
Available Colors: Black
Max. Flight Time: 10 minutes
Max. Speed: 20 km/h (12,4 mp/h)
Controller Frequency: 2.4 GHz
Battery Capacity: 5,200 mAh
Battery Type: Lithium Polymer
GPS Compatible: √
Altitude Hold: √
Position Hold: √
Flight Planning: √
Autonomous Flight: √
Target Tracking: –
Automatic Landing: –
Return To Home: √
Obstacle Avoidance: √
Headless Mode: –
Built-in Camera: –

General Atomics

Predator XP

© General Atomics

Release Date: June 2014
Price: $3 to $4 million
Number of Rotors: 1 (Fixed Wing)
Control System: Flight Computer
Applications: Armed Reconnaissance, Surveillance, Target Acquisition
Dimensions: 8.22 × 2.1 m
Wingspan: 17 m
Weight: 1,020 kg
Max. Payload: 204 kg
Motors: Rotax 914F
Available Colors: Grey
Max. Flight Time: 35 hours
Max. Speed: 217 km/h (135 mp/h)
Max. Flying Altitude: 7,620 m (25,000 ft)
Min. Operating Temperature: -60°C
Max. Operating Temperature: 60°C
Max. Operating Range: 1,100 km
GPS Compatible: √
Altitude Hold: √
Position Hold: √
Flight Planning: √
Autonomous Flight: √
Target Tracking: √
Automatic Landing: √
Return To Home: √
Obstacle Avoidance: –
Headless Mode: –
Built-in Camera: –

HeliMax

230Si

© bhphotovideo.com

Release Date: June 2014
Price: $160
Number of Rotors: 4
Control System: Remote Control, Phone/Tablet
Applications: Recreation
Dimensions: 306 × 306 × 75 mm
Wingspan: 230 mm
Weight: 132 gr
Frame Material: Plastic
Rotor Material: Plastic
Available Colors: White
Max. Flight Time: 15 minutes
Controller Frequency: 2.4 GHz
Battery Capacity: 1,200 mAh
Battery Type: Lithium Polymer
Battery Power: 3.7V
GPS Compatible: –
Altitude Hold: √
Position Hold: √
Flight Planning: –
Autonomous Flight: –
Target Tracking: –
Automatic Landing: √
Return To Home: √
Obstacle Avoidance: –
Headless Mode: –
Built-in Camera: √

HuaXiang

8953 Camera

© HuaXiang

Release Date: June 2014
Price: $59
Number of Rotors: 4
Control System: Remote Control
Applications: Recreation
Dimensions: 85 × 85 × 35 mm
Frame Material: Plastic
Rotor Material: Plastic
Available Colors: Blue
Max. Flight Time: 10 minutes
Max. Operating Range: 100 m
Controller Frequency: 2.4 GHz
Battery Capacity: 350 mAh
Battery Type: Lithium Polymer
Battery Power: 3.7V
GPS Compatible: –
Altitude Hold: √
Position Hold: √
Flight Planning: –
Autonomous Flight: –
Target Tracking: –
Automatic Landing: –
Return To Home: –
Obstacle Avoidance: –
Headless Mode: –
Built-in Camera: √

HuaXiang

8957 Raider

© HuaXiang

Release Date: June 2014
Price: $50
Number of Rotors: 4
Control System: Remote Control
Applications: Recreation
Dimensions: 130 × 130 × 40 mm
Frame Material: Plastic
Rotor Material: Plastic
Available Colors: Green, White
Max. Flight Time: 6 minutes
Max. Operating Range: 100 m
Controller Frequency: 2.4 GHz
Battery Capacity: 350 mAh
Battery Type: Lithium Polymer
Battery Power: 3.7V
GPS Compatible: –
Altitude Hold: √
Position Hold: √
Flight Planning: –
Autonomous Flight: –
Target Tracking: –
Automatic Landing: –
Return To Home: –
Obstacle Avoidance: –
Headless Mode: –
Built-in Camera: √ (0.3 megapixels)

Idea-Fly

Storm-800

© Idea-Fly

Release Date: June 2014
Price: $950
Number of Rotors: 6
Control System: Remote Control
Applications: Film, Photography
Wingspan: 350 mm
Weight: 1.1 kg
Max. Payload: 700 gr
Frame Material: Carbon Fiber
Rotor Material: Carbon Fiber
Available Colors: White
Max. Flight Time: 8 minutes
Max. Speed: 36 km/h (22,3 mp/h)
Max. Ascent Speed: 4 m/s
Max. Descent Speed: 5 m/s
Min. Operating Temperature: -20°C
Max. Operating Temperature: 50°C
Max. Operating Range: 500 m
Controller Frequency: 2.4 GHz
Battery Capacity: 2,200 mAh
Battery Type: Lithium Polymer 3S
Battery Power: 11.1V
GPS Compatible: √
Altitude Hold: √
Position Hold: √
Flight Planning: √
Autonomous Flight: –
Target Tracking: –
Automatic Landing: –
Return To Home: √
Obstacle Avoidance: –
Headless Mode: –
Built-in Camera: –

JJRC

F182

Release Date: June 2014
Price: $75
Number of Rotors: 4
Control System: Remote Control
Applications: Recreation
Dimensions: 420 × 420 × 90 mm
Wingspan: 594 mm
Frame Material: Plastic
Rotor Material: Plastic
Available Colors: Red
Max. Flight Time: 8 minutes
Max. Operating Range: 100 m
Controller Frequency: 2.4 GHz
Battery Capacity: 500 mAh
Battery Type: Lithium Polymer
Battery Power: 7.4V
GPS Compatible: –
Altitude Hold: –
Position Hold: –
Flight Planning: –
Autonomous Flight: –
Target Tracking: –
Automatic Landing: –
Return To Home: √
Obstacle Avoidance: –
Headless Mode: –
Built-in Camera: √ (2 megapixels)

Quanum

Nova

Release Date: June 2014
Price: $319
Number of Rotors: 4
Control System: Remote Control
Applications: Film, Photography
Dimensions: 300 × 300 × 200 mm
Wingspan: 365 mm
Weight: 875 gr
Frame Material: Plastic
Rotor Material: Plastic
Available Colors: White
Max. Flight Time: 15 minutes
Max. Speed: 36 km/h (22,3 mp/h)
Max. Ascent Speed: 6 m/s
Max. Descent Speed: 6 m/s
Max. Operating Range: 300 m
Controller Frequency: 2.4 GHz
Battery Capacity: 2,700 mAh
Battery Type: Lithium Polymer 3S
GPS Compatible: √
Altitude Hold: √
Position Hold: √
Flight Planning: √
Autonomous Flight: –
Target Tracking: –
Automatic Landing: –
Return To Home: √
Obstacle Avoidance: –
Headless Mode: –
Built-in Camera: –

UAV Systems Group

Supercam S350

© unmanned.ru

Release Date: June 2014
Price: Unknown
Number of Rotors: 1 (Fixed Wing)
Control System: Ground Control Station
Applications: Aerial Mapping, Inspection
Wingspan: 3.2 m
Weight: 4.5 kg
Available Colors: Orange
Max. Flight Time: 4 hours
Max. Speed: 120 km/h (74 mp/h)
Max. Flying Altitude: 5,500 m (18,045 ft)
Min. Operating Temperature: -30°C
Max. Operating Temperature: 30°C
Max. Operating Range: 360 km
GPS Compatible: √
Altitude Hold: √
Position Hold: √
Flight Planning: √
Autonomous Flight: √
Target Tracking: –
Automatic Landing: √
Return To Home: √
Obstacle Avoidance: –
Headless Mode: –
Built-in Camera: √

Walkera

QR Y100

© Walkera

Release Date: June 2014
Price: $200
Number of Rotors: 6
Control System: Remote Control, Phone/Tablet
Applications: Film, Photography
Dimensions: 222 × 252 × 90 mm
Wingspan: 335 mm
Weight: 146 gr
Max. Payload: 320 gr
Frame Material: Aluminum
Rotor Material: Plastic
Available Colors: Silver
Max. Flight Time: 20 minutes
Max. Speed: 79 km/h (49 mp/h)
Min. Operating Temperature: -10°C
Max. Operating Temperature: 40°C
Max. Operating Range: 1 km
Controller Frequency: 2.4 GHz
Battery Capacity: 1,600 mAh
Battery Type: Lithium Polymer
Battery Power: 4V
GPS Compatible: –
Altitude Hold: √
Position Hold: √
Flight Planning: –
Autonomous Flight: –
Target Tracking: –
Automatic Landing: √
Return To Home: √
Obstacle Avoidance: –
Headless Mode: –
Built-in Camera: √

WLtoys

V333N Ufo

Release Date: June 2014
Price: $110
Number of Rotors: 4
Control System: Remote Control
Applications: Recreation
Dimensions: 520 × 520 × 33 mm
Weight: 219 gr
Frame Material: Plastic, Foam
Rotor Material: Plastic
Available Colors: Black, Purple
Max. Flight Time: 10 minutes
Max. Operating Range: 150 m
Controller Frequency: 2.4 GHz
Battery Capacity: 730 mAh
Battery Type: Lithium Polymer
Battery Power: 3.7V
GPS Compatible: –
Altitude Hold: √
Position Hold: √
Flight Planning: –
Autonomous Flight: –
Target Tracking: –
Automatic Landing: –
Return To Home: √
Obstacle Avoidance: –
Headless Mode: –
Built-in Camera: –

Agribotix

Hornet

Release Date: July 2014
Price: $11,900
Number of Rotors: 1 (Fixed Wing)
Control System: Remote Control
Applications: Aerial Mapping, Agriculture
Wingspan: 1.95 m
Weight: 2.27 kg
Available Colors: White
Max. Flight Time: 80 minutes
Max. Speed: 53 km/h (33 mp/h)
Max. Operating Range: 1.6 km
Controller Frequency: 2.4 GHz
Battery Capacity: 7,000 mAh
Battery Type: Lithium Polymer
Battery Power: 11.1V
GPS Compatible: √
Altitude Hold: √
Position Hold: √
Flight Planning: √
Autonomous Flight: √
Target Tracking: √
Automatic Landing: √
Return To Home: √
Obstacle Avoidance: –
Headless Mode: –
Built-in Camera: √

Bluebird Aero Systems Ltd.

Micro B

© directindustry.fr

Release Date: July 2014
Price: $75,000
Number of Rotors: 1 (Fixed Wing)
Control System: Flight Computer
Applications: Military Purposes, Aerial Mapping, Inspection, Reconnaissance, Surveillance
Dimensions: 1 × 1.7 m
Wingspan: 1.7 m
Weight: 2.2 kg
Max. Payload: 300 gr
Frame Material: Composite
Available Colors: Grey
Max. Flight Time: 2 hours
Max. Speed: 75 km/h (46,6 mp/h)
Max. Flying Altitude: 1,000 m (3,280 ft)
Max. Operating Range: 10 km
Battery Type: Lithium Polymer
GPS Compatible: √
Altitude Hold: √
Position Hold: –
Flight Planning: √
Autonomous Flight: √
Target Tracking: √
Automatic Landing: √
Return To Home: √
Obstacle Avoidance: √
Headless Mode: –
Built-in Camera: √

Drone America

DAx8

© Drone America

Release Date: July 2014
Price: On Request
Number of Rotors: 8
Control System: Laptop, Tablet
Applications: Aerial Mapping, Agriculture, Surveillance
Wingspan: 1.23 m
Weight: 8 kg
Frame Material: Carbon Fiber
Rotor Material: Carbon Fiber
Available Colors: White
Max. Flight Time: 30 minutes
Max. Speed: 18 km/h (11 mp/h)
Max. Flying Altitude: 122 m (400 ft)
GPS Compatible: √
Altitude Hold: √
Position Hold: √
Flight Planning: √
Autonomous Flight: √
Target Tracking: √
Automatic Landing: √
Return To Home: √
Obstacle Avoidance: –
Headless Mode: –
Built-in Camera: –

Drone America

HuginnAir

© Dromida

Release Date: July 2014
Price: On Request
Number of Rotors: 1 (Fixed Wing)
Control System: Laptop, Tablet
Applications: Aerial Mapping, Agriculture, Search & Rescue, Surveillance
Dimensions: 2.67 m
Wingspan: 4.7 m
Weight: 15 kg
Max. Payload: 10 kg
Frame Material: Carbon Fiber
Rotor Material: Carbon Fiber
Available Colors: White
Max. Flight Time: 8.5 hours
Max. Speed: 193 km/h (120 mp/h)
Max. Flying Altitude: 3,650 m (11,975 ft)
GPS Compatible: √
Altitude Hold: √
Position Hold: √
Flight Planning: √
Autonomous Flight: √
Target Tracking: √
Automatic Landing: √
Return To Home: √
Obstacle Avoidance: –
Headless Mode: –
Built-in Camera: –

Eachine

H1 Sky Walker

© helicomicro.com

Release Date: July 2014
Price: $30
Number of Rotors: 4
Control System: Remote Control
Applications: Recreation
Dimensions: 45 × 45 × 24 mm
Weight: 80 gr
Frame Material: Plastic
Rotor Material: Plastic
Available Colors: Black
Max. Flight Time: 6 minutes
Max. Operating Range: 20 m
Controller Frequency: 2.4 GHz
Battery Capacity: 100 mAh
Battery Type: Lithium Polymer 1S
Battery Power: 3V
GPS Compatible: –
Altitude Hold: √
Position Hold: √
Flight Planning: –
Autonomous Flight: –
Target Tracking: –
Automatic Landing: –
Return To Home: –
Obstacle Avoidance: –
Headless Mode: –
Built-in Camera: –

Hubsan

Nano Q4 H111

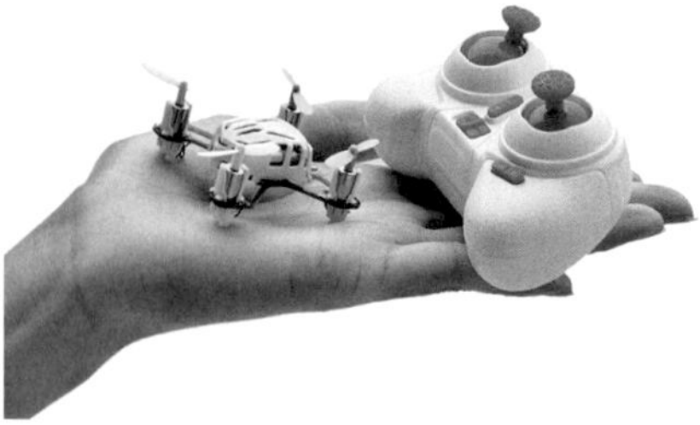

© Hubsan

Release Date: July 2014
Price: $24
Number of Rotors: 4
Control System: Remote Control
Applications: Recreation
Dimensions: 50 × 50 × 25 mm
Wingspan: 45 mm
Weight: 12 gr
Frame Material: Plastic
Rotor Material: Plastic
Available Colors: Black, Green, Purple, Red, White, Yellow
Max. Flight Time: 5 minutes
Max. Operating Range: 50 m
Controller Frequency: 2.4 GHz
Battery Capacity: 100 mAh
Battery Type: Lithium Polymer 1S
Battery Power: 3.7V
GPS Compatible: –
Altitude Hold: –
Position Hold: –
Flight Planning: –
Autonomous Flight: –
Target Tracking: –
Automatic Landing: –
Return To Home: –
Obstacle Avoidance: –
Headless Mode: –
Built-in Camera: –

Israel Aerospace Industries (IAI)

Bird-Eye 650D

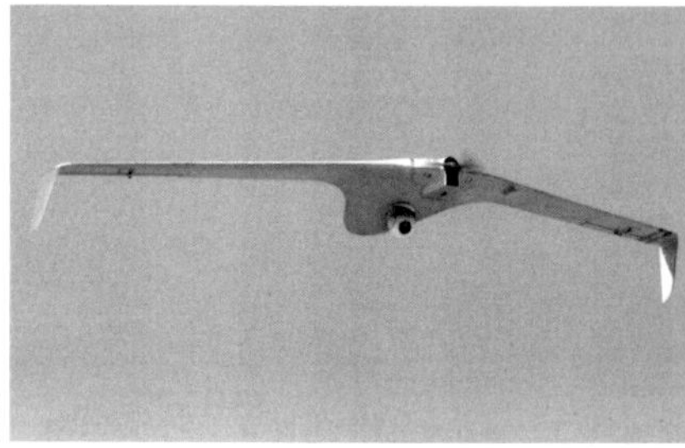

© IAI

Release Date: July 2014
Price: $450,000
Number of Rotors: 1 (Fixed Wing)
Control System: Ground Control Station
Applications: Military Purposes, Reconnaissance, Surveillance
Weight: 30 kg
Max. Payload: 5.5 kg
Frame Material: Composite
Available Colors: Grey
Max. Flight Time: 24 hours
Max. Operating Range: 50 km
GPS Compatible: √
Altitude Hold: √
Position Hold: –
Flight Planning: √
Autonomous Flight: √
Target Tracking: –
Automatic Landing: –
Return To Home: √
Obstacle Avoidance: –
Headless Mode: –
Built-in Camera: √

Kongcopter

FH-800

©rc-terminal.de

Release Date: July 2014
Price: $800
Number of Rotors: 6
Control System: Remote Control
Applications: Film, Photography
Wingspan: 810 mm
Weight: 3.5 kg
Max. Payload: 3.7 kg
Frame Material: Plastic
Rotor Material: Plastic
Motors: 4114-380kv
Available Colors: Black
Max. Flight Time: 17 minutes
Controller Frequency: 2.4 GHz
Battery Capacity: 10,000 mAh
Battery Type: Lithium Polymer 6S
GPS Compatible: √
Altitude Hold: √
Position Hold: √
Flight Planning: √
Autonomous Flight: √
Target Tracking: –
Automatic Landing: √
Return To Home: √
Obstacle Avoidance: –
Headless Mode: –
Built-in Camera: –

Navmar Applied Sciences Corporation

TigerShark XP

©logostech.net

Release Date: July 2014
Price: Unknown
Number of Rotors: 1 (Fixed Wing)
Control System: Ground Control Station
Applications: Military Purposes, Aerial Mapping, Research & Development, Reconnaissance, Surveillance
Dimensions: 4.57 × 6.7 m
Wingspan: 6.7 m
Weight: 226 kg
Max. Payload: 45 kg
Frame Material: Composite
Available Colors: Grey
Max. Flight Time: 12 hours
Max. Speed: 148 km/h (92 mp/h)
Max. Flying Altitude: 4,267 m (14,000 ft)
GPS Compatible: √
Altitude Hold: √
Position Hold: –
Flight Planning: √
Autonomous Flight: √
Target Tracking: √
Automatic Landing: √
Return To Home: √
Obstacle Avoidance: √
Headless Mode: –
Built-in Camera: √

Nine Eagles

Galaxy Visitor 3

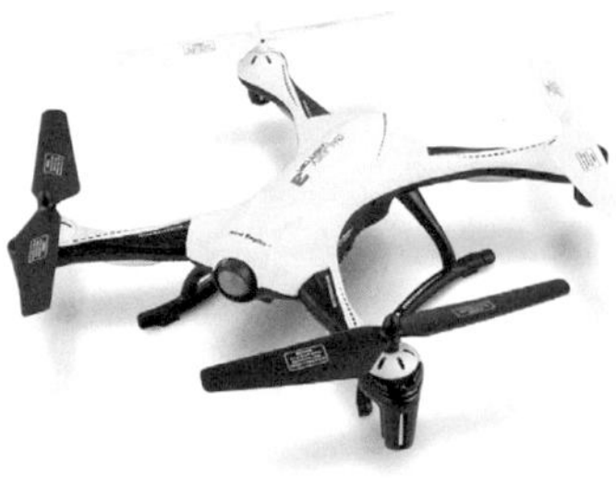

©Nine Eagles

Release Date: July 2014
Price: $325
Number of Rotors: 4
Control System: Remote Control, Phone/Tablet
Applications: Agriculture, Inspection, Surveillance
Dimensions: 163 × 163 × 78 mm
Wingspan: 230 mm
Weight: 135 gr
Frame Material: Plastic
Rotor Material: Plastic
Available Colors: White
Max. Flight Time: 12 minutes
Max. Flying Altitude: 100 m (328 ft)
Controller Frequency: 2.4 GHz
Battery Capacity: 1,200 mAh
Battery Type: Lithium Polymer
Battery Power: 3.7V
GPS Compatible: –
Altitude Hold: √
Position Hold: √
Flight Planning: –
Autonomous Flight: –
Target Tracking: –
Automatic Landing: –
Return To Home: √
Obstacle Avoidance: –
Headless Mode: –
Built-in Camera: –

Parrot

Rolling Spider

©Parrot

Release Date: July 2014
Price: $99
Number of Rotors: 4
Control System: Phone/Tablet
Applications: Recreation
Dimensions: 140 × 140 × 38 mm
Wingspan: 180 mm
Weight: 65 gr
Frame Material: Plastic
Rotor Material: Plastic
Available Colors: Blue, Red, White
Max. Flight Time: 8 minutes
Max. Speed: 18 km/h (11,2 mp/h)
Max. Operating Range: 20 m
Controller Frequency: 2.4 GHz
Battery Capacity: 550 mAh
Battery Type: Lithium Polymer 3S
GPS Compatible: –
Altitude Hold: √
Position Hold: √
Flight Planning: √
Autonomous Flight: –
Target Tracking: –
Automatic Landing: –
Return To Home: –
Obstacle Avoidance: –
Headless Mode: –
Built-in Camera: √

Syma Toys

X2

© Syma Toys

Release Date: July 2014
Price: $39
Number of Rotors: 4
Control System: Remote Control
Applications: Recreation
Frame Material: Plastic
Rotor Material: Plastic
Available Colors: Black, Red
Max. Flight Time: 5 minutes
Controller Frequency: 2.4 GHz
Battery Capacity: 150 mAh
Battery Type: Lithium Polymer
Battery Power: 3.7V
GPS Compatible: –
Altitude Hold: –
Position Hold: –
Flight Planning: –
Autonomous Flight: –
Target Tracking: –
Automatic Landing: –
Return To Home: –
Obstacle Avoidance: –
Headless Mode: –
Built-in Camera: –

Tekever

AR2 Carcará

© Tekever

Release Date: July 2014
Price: On Request
Number of Rotors: 1 (Fixed Wing)
Control System: Ground Control Station, Laptop, Tablet
Applications: Military Purposes, Maritime Patrol, Reconnaissance, Surveillance
Dimensions: 1 × 2.2 m
Wingspan: 2.2 m
Weight: 4 kg
Frame Material: Composite
Available Colors: Grey
Max. Flight Time: 2.5 hours
Max. Speed: 40 km/h (25 mp/h)
Max. Operating Range: 20 km
Controller Frequency: 2.4 GHz
Battery Type: Lithium Polymer
GPS Compatible: √
Altitude Hold: √
Position Hold: –
Flight Planning: √
Autonomous Flight: √
Target Tracking: –
Automatic Landing: √
Return To Home: √
Obstacle Avoidance: –
Headless Mode: –
Built-in Camera: √

Thunder Tiger Robotix

Super Hornet X650

©TTRobotix

Release Date: July 2014
Price: $699
Number of Rotors: 4
Control System: Remote Control
Applications: Film, Photography
Wingspan: 650 mm
Weight: 2.47 kg
Max. Payload: 730 gr
Frame Material: Carbon fiber
Rotor Material: Plastic
Available Colors: Black
Max. Flight Time: 18 minutes
Controller Frequency: 2.4 GHz
Battery Capacity: 6,000 mAh
Battery Type: Lithium Polymer 6S
Battery Power: 22.2V
GPS Compatible: –
Altitude Hold: √
Position Hold: √
Flight Planning: –
Autonomous Flight: –
Target Tracking: –
Automatic Landing: √
Return To Home: √
Obstacle Avoidance: –
Headless Mode: –
Built-in Camera: –

WLtoys

V282/V292 mini Skylark Egg

©WLtoys

Release Date: July 2014
Price: $23
Number of Rotors: 4
Control System: Remote Control
Applications: Recreation
Dimensions: 45 × 45 × 25 mm
Weight: 14 gr
Frame Material: Plastic
Rotor Material: Plastic
Available Colors: Black, Green, Yellow
Max. Flight Time: 5 minutes
Max. Operating Range: 30 m
Controller Frequency: 2.4 GHz
Battery Capacity: 150 mAh
Battery Type: Lithium Polymer
Battery Power: 3.7V
GPS Compatible: –
Altitude Hold: –
Position Hold: –
Flight Planning: –
Autonomous Flight: –
Target Tracking: –
Automatic Landing: –
Return To Home: –
Obstacle Avoidance: –
Headless Mode: –
Built-in Camera: –

WLtoys

V333 Ufo

©WLtoys

Release Date: July 2014
Price: $90
Number of Rotors: 4
Control System: Remote Control
Applications: Recreation
Dimensions: 520 × 520 × 33 mm
Weight: 219 gr
Frame Material: Plastic, Foam
Rotor Material: Plastic
Available Colors: Black, Purple
Max. Flight Time: 10 minutes
Max. Operating Range: 150 m
Controller Frequency: 2.4 GHz
Battery Capacity: 730 mAh
Battery Type: Lithium Polymer
Battery Power: 3.7V
GPS Compatible: –
Altitude Hold: √
Position Hold: √
Flight Planning: –
Autonomous Flight: –
Target Tracking: –
Automatic Landing: –
Return To Home: √
Obstacle Avoidance: –
Headless Mode: –
Built-in Camera: –

WLtoys

V353 Galaxy

©WLtoys

Release Date: July 2014
Price: $50
Number of Rotors: 4
Control System: Remote Control
Applications: Recreation
Dimensions: 300 × 300 × 65 mm
Weight: 205 gr
Frame Material: Plastic
Rotor Material: Plastic
Available Colors: White
Max. Flight Time: 10 minutes
Max. Operating Range: 150 m
Controller Frequency: 2.4 GHz
Battery Capacity: 730 mAh
Battery Type: Lithium Polymer
Battery Power: 3.7V
GPS Compatible: –
Altitude Hold: √
Position Hold: √
Flight Planning: –
Autonomous Flight: –
Target Tracking: –
Automatic Landing: –
Return To Home: √
Obstacle Avoidance: –
Headless Mode: –
Built-in Camera: –

Xamen Technologies
BE 4000

©directindustry.fr

Release Date: July 2014
Price: On Request
Number of Rotors: 4
Control System: Ground Control Stration, Remote Control
Applications: Film, Photography, Aerial Mapping
Dimensions: 315 × 190 × 90 mm
Weight: 1.2 kg
Max. Payload: 800 gr
Frame Material: Carbon Fiber
Rotor Material: Wood
Available Colors: Black
Max. Flight Time: 15 minutes
Max. Speed: 54 km/h (33,5 mp/h)
Max. Flying Altitude: 1,500 m (4,921 ft)
Min. Operating Temperature: -15°C
Max. Operating Temperature: 55°C
Max. Operating Range: 1 km
Controller Frequency: 2.4 GHz
Battery Type: Lithium Polymer
GPS Compatible: √
Altitude Hold: √
Position Hold: √
Flight Planning: √
Autonomous Flight: √
Target Tracking: –
Automatic Landing: √
Return To Home: √
Obstacle Avoidance: –
Headless Mode: –
Built-in Camera: –

Zerotech
HighOne

©uavonic.co.uk

Release Date: July 2014
Price: $6,250
Number of Rotors: 4
Control System: Remote Control
Applications: Film, Photography, Aerial Mapping, Agriculture, Inspection, Surveillance
Dimensions: 450 × 410 × 610 mm
Wingspan: 720 mm
Weight: 2.42 kg
Max. Payload: 4.4 kg
Frame Material: Carbon Fiber
Rotor Material: Carbon Fiber
Motors: 6210-380kv
Available Colors: Black
Max. Flight Time: 16 minutes
Controller Frequency: 2.4 GHz
Battery Capacity: 16,000 mAh
Battery Type: Lithium Polymer 6S
Battery Power: 22.2V
GPS Compatible: √
Altitude Hold: √
Position Hold: √
Flight Planning: √
Autonomous Flight: –
Target Tracking: √
Automatic Landing: √
Return To Home: √
Obstacle Avoidance: –
Headless Mode: –
Built-in Camera: –

Aeromao

Aeromapper 300

Release Date: August 2014
Price: $16,000
Number of Rotors: 1 (Fixed Wing)
Control System: Remote Control
Applications: Aerial Mapping, Agriculture, Surveillance
Wingspan: 3 m
Weight: 5.35 kg
Max. Payload: 650 gr
Frame Material: EPO Foam (carbon fiber reinforced)
Available Colors: White
Max. Flight Time: 100 minutes
Max. Speed: 90 km/h (56 mp/h)
Max. Flying Altitude: 4,500 m (14,764 ft)
Max. Operating Range: 50 km
Controller Frequency: 2.4 GHz
Battery Type: Lithium Polymer
GPS Compatible: √
Altitude Hold: √
Position Hold: √
Flight Planning: √
Autonomous Flight: √
Target Tracking: √
Automatic Landing: √
Return To Home: √
Obstacle Avoidance: –
Headless Mode: –
Built-in Camera: √ (24 megapixels)

Aosenma

CG022

Release Date: August 2014
Price: $25
Number of Rotors: 4
Control System: Remote Control
Applications: Recreation
Dimensions: 110 × 110 × 40 mm
Frame Material: Plastic
Rotor Material: Plastic
Available Colors: Gold, Red, White
Max. Flight Time: 10 minutes
Max. Operating Range: 60 m
Controller Frequency: 2.4 GHz
Battery Type: Lithium Polymer
GPS Compatible: –
Altitude Hold: –
Position Hold: –
Flight Planning: –
Autonomous Flight: –
Target Tracking: –
Automatic Landing: –
Return To Home: √
Obstacle Avoidance: –
Headless Mode: √
Built-in Camera: –

Cheerson

CX-30

Release Date: August 2014
Price: $60
Number of Rotors: 4
Control System: Remote Control
Applications: Recreation
Dimensions: 150 × 150 × 45 mm
Wingspan: 212 m
Weight: 400 gr
Max. Payload: 500 gr
Frame Material: Plastic
Rotor Material: Plastic
Available Colors: Blue, Green, Orange, Yellow
Max. Flight Time: 6 minutes
Max. Flying Altitude: 30 m (98 ft)
Controller Frequency: 5.8 GHz
Battery Capacity: 5,400 mAh
Battery Type: Lithium Polymer 3S
Battery Power: 11.1V
GPS Compatible: –
Altitude Hold: √
Position Hold: √
Flight Planning: –
Autonomous Flight: –
Target Tracking: –
Automatic Landing: –
Return To Home: –
Obstacle Avoidance: –
Headless Mode: –
Built-in Camera: √

DJI

Spreading Wings S900

Release Date: August 2014
Price: $1,400
Number of Rotors: 6
Propellers: DJI 1345T
Control System: Remote Control
Applications: Film, Photography, Aerial Mapping
Wingspan: 900 mm
Weight: 3.3 kg
Max. Payload: 4.9 kg
Frame Material: Carbon Fiber
Rotor Material: Carbon Fiber
Available Colors: Black
Max. Flight Time: 18 minutes
Max. Speed: 57,6 km/h (35,8 mp/h)
Max. Flying Altitude: 2,000 m (6,560 ft)
Max. Ascent Speed: 6 m/s
Min. Operating Temperature: -10°C
Max. Operating Temperature: 40°C
Max. Operating Range: 2 km
Controller Frequency: 2.4 GHz
Battery Capacity: 15,000 mAh
Battery Type: Lithium Polymer 6S
Battery Power: 22.2V
Battery Weight: 35 gr
GPS Compatible: √
Altitude Hold: √
Position Hold: √
Flight Planning: –
Autonomous Flight: –
Target Tracking: –
Automatic Landing: √
Return To Home: √
Obstacle Avoidance: –
Headless Mode: –
Built-in Camera: –

Eachine

X6

©banggood.com

Release Date: August 2014
Price: $60
Number of Rotors: 6
Control System: Remote Control
Applications: Recreation
Dimensions: 194 × 194 × 45.7 mm
Wingspan: 130 mm
Frame Material: Plastic
Rotor Material: Plastic
Available Colors: Blue, Gold, Red
Max. Flight Time: 12 minutes
Max. Operating Range: 100 m
Controller Frequency: 2.4 GHz
Battery Capacity: 550 mAh
Battery Type: Lithium Polymer 1S
Battery Power: 4V
GPS Compatible: –
Altitude Hold: –
Position Hold: –
Flight Planning: –
Autonomous Flight: –
Target Tracking: –
Automatic Landing: –
Return To Home: –
Obstacle Avoidance: –
Headless Mode: –
Built-in Camera: √ (2 megapixels)

Flairics

Journalist

©Flairics

Release Date: August 2014
Price: $17,400
Number of Rotors: 4
Control System: Remote Control
Applications: Film, Photography, Aerial Mapping, Surveillance
Dimensions: 845 × 845 × 175 mm
Wingspan: 1.2 m
Weight: 2.2 kg
Max. Payload: 1.5 kg
Frame Material: Plastic, Carbon Fiber
Rotor Material: Carbon Fiber
Available Colors: Black
Max. Flight Time: 15 minutes
Max. Speed: 60 km/h (37 mp/h)
Max. Flying Altitude: 2,000 m (6,560 ft)
Max. Ascent Speed: 8 m/s
Min. Operating Temperature: -20°C
Max. Operating Temperature: 40°C
Max. Operating Range: 800 m
Controller Frequency: 2.4 GHz
Battery Capacity: 8,000 mAh
Battery Type: Lithium Polymer 5S
GPS Compatible: √
Altitude Hold: √
Position Hold: √
Flight Planning: √
Autonomous Flight: √
Target Tracking: √
Automatic Landing: –
Return To Home: √
Obstacle Avoidance: –
Headless Mode: –
Built-in Camera: –

Hobby King

Super-X

© Hobby King

Release Date: August 2014
Price: $125
Number of Rotors: 4
Control System: Remote Control
Applications: Recreation
Dimensions: 125 × 125 mm
Weight: 38 gr
Frame Material: Carbon fiber
Rotor Material: Plastic
Motors: MR-03-H 6400kv
Available Colors: Black
Controller Frequency: 2.4 GHz
Battery Capacity: 500 mAh
Battery Type: Lithium Polymer 1S
Battery Power: 5V
GPS Compatible: √
Altitude Hold: √
Position Hold: √
Flight Planning: –
Autonomous Flight: –
Target Tracking: –
Automatic Landing: –
Return To Home: √
Obstacle Avoidance: –
Headless Mode: –
Built-in Camera: –

Nihui Model Technology

U207 Intruder

© ekladata.com

Release Date: August 2014
Price: $15
Number of Rotors: 4
Control System: Remote Control
Applications: Recreation
Dimensions: 98 × 98 × 13 mm
Weight: 22 gr
Frame Material: Plastic
Rotor Material: Plastic
Available Colors: Black
Max. Operating Range: 45 m
Controller Frequency: 2.4 GHz
Battery Capacity: 150 mAh
Battery Type: Lithium Polymer 1S
Battery Weight: 5 gr
GPS Compatible: –
Altitude Hold: –
Position Hold: –
Flight Planning: –
Autonomous Flight: –
Target Tracking: –
Automatic Landing: –
Return To Home: –
Obstacle Avoidance: –
Headless Mode: –
Built-in Camera: –

Qods Aviation Industries

Mohajer 4B

Release Date: August 2014
Price: Unknown
Number of Rotors: 1 (Fixed Wing)
Control System: Ground Control Station
Applications: Military Purposes, Inspection, Reconnaissance, Surveillance
Wingspan: 7 m
Weight: 242 kg
Frame Material: Composite
Available Colors: Blue, Grey
Max. Flight Time: 6 hours
Max. Speed: 200 km/h (124 mp/h)
Max. Flying Altitude: 4,572 m (15,000 ft)
Max. Operating Range: 200 km
GPS Compatible: √
Altitude Hold: √
Position Hold: –
Flight Planning: √
Autonomous Flight: √
Target Tracking: –
Automatic Landing: √
Return To Home: √
Obstacle Avoidance: –
Headless Mode: –
Built-in Camera: √

Revell Control

Nano Quad Pro

© Youtube/Robert de Wit

Release Date: August 2014
Price: $40
Number of Rotors: 4
Control System: Remote Control
Applications: Recreation
Dimensions: 45 × 45 × 10 mm
Weight: 12 gr
Frame Material: Plastic
Rotor Material: Plastic
Available Colors: Red
Max. Flight Time: 5 minutes
Max. Operating Range: 50 m
Controller Frequency: 2.4 GHz
Battery Capacity: 100 mAh
Battery Type: Lithium Polymer
Battery Power: 3.7V
GPS Compatible: –
Altitude Hold: –
Position Hold: –
Flight Planning: –
Autonomous Flight: –
Target Tracking: –
Automatic Landing: –
Return To Home: –
Obstacle Avoidance: –
Headless Mode: –
Built-in Camera: –

Thales

Watchkeeper WK450

© Thales Group

Release Date: August 2014
Price: $19.5 million
Number of Rotors: 1 (Fixed Wing)
Control System: Ground Control Station
Applications: Military Purposes, Reconnaissance, Surveillance, Target Acquisition
Dimensions: 6 × 10.5 m
Weight: 450 kg
Max. Payload: 150 kg
Frame Material: Composite
Motors: 1 × Wankel rotary engine
Available Colors: Grey
Max. Flight Time: 17 hours
Max. Speed: 175 km/h (109 mp/h)
Max. Flying Altitude: 5,500 m (18,045 ft)
Max. Ascent Speed: 4.5 m/s
Max. Operating Range: 150 km
GPS Compatible: √
Altitude Hold: √
Position Hold: –
Flight Planning: √
Autonomous Flight: √
Target Tracking: √
Automatic Landing: √
Return To Home: √
Obstacle Avoidance: √
Headless Mode: –
Built-in Camera: √

Trigger Composites

Pteryx

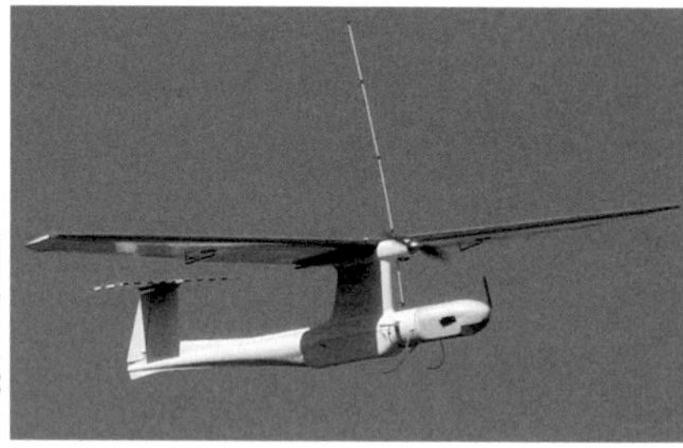

© Trigger Composites

Release Date: August 2014
Price: Unknown
Number of Rotors: 1 (Fixed Wing)
Control System: Ground Control Station
Applications: Film, Photography, Aerial Mapping
Dimensions: 1.4 × 2.8 × 0.33 m
Wingspan: 2.8 m
Frame Material: Carbon Fiber, Fiberglass Composite
Available Colors: Red
Max. Flight Time: 2 hours
Max. Speed: 160 km/h (99,4 mp/h)
Max. Flying Altitude: 3,000 m (9,842 ft)
GPS Compatible: √
Altitude Hold: √
Position Hold: –
Flight Planning: √
Autonomous Flight: √
Target Tracking: –
Automatic Landing: √
Return To Home: √
Obstacle Avoidance: –
Headless Mode: –
Built-in Camera: √

Vulcan
Raven

© Vulcan

Release Date: August 2014
Price: $13,000
Number of Rotors: 8
Control System: Remote Control
Applications: Film, Photography, Aerial Mapping, Agriculture, Inspection, Surveillance
Dimensions: 800 × 800 × 600 mm
Wingspan: 1.13 m
Weight: 15 kg
Frame Material: Carbon fiber
Rotor Material: Carbon Fiber
Available Colors: Black
GPS Compatible: √
Altitude Hold: √
Position Hold: √
Flight Planning: √
Autonomous Flight: –
Target Tracking: √
Automatic Landing: √
Return To Home: √
Obstacle Avoidance: –
Headless Mode: –
Built-in Camera: –

Walkera
Tali H500

© Walkera

Release Date: August 2014
Price: $999
Number of Rotors: 4
Control System: Remote Control, Phone/Tablet
Applications: Film, Photography
Dimensions: 471 × 536 × 270 mm
Wingspan: 713 mm
Weight: 2.5 kg
Max. Payload: 480 gr
Frame Material: Plastic, Carbon Fiber
Rotor Material: Plastic
Available Colors: Black, White
Max. Flight Time: 25 minutes
Max. Speed: 79 km/h (49 mp/h)
Min. Operating Temperature: -10°C
Max. Operating Temperature: 40°C
Max. Operating Range: 1.4 km
Controller Frequency: 2.4 GHz
Battery Capacity: 5,400 mAh
Battery Type: Lithium Polymer 6S
Battery Power: 22V
GPS Compatible: √
Altitude Hold: √
Position Hold: √
Flight Planning: √
Autonomous Flight: √
Target Tracking: √
Automatic Landing: √
Return To Home: √
Obstacle Avoidance: –
Headless Mode: –
Built-in Camera: –

Yi Zhan Toys

X4

© beginnerflyer.com

Release Date: August 2014
Price: $30
Number of Rotors: 4
Control System: Remote Control
Applications: Recreation
Dimensions: 145 × 135 × 25 mm
Wingspan: 198 mm
Weight: 33 gr
Frame Material: Plastic
Rotor Material: Plastic
Available Colors: Black, Gold, White
Max. Flight Time: 5 minutes
Max. Operating Range: 50 m
Controller Frequency: 2.4 GHz
Battery Capacity: 350 mAh
Battery Type: Lithium Ion
Battery Power: 3.7V
GPS Compatible: –
Altitude Hold: –
Position Hold: –
Flight Planning: –
Autonomous Flight: –
Target Tracking: –
Automatic Landing: –
Return To Home: √
Obstacle Avoidance: –
Headless Mode: –
Built-in Camera: –

3D Robotics

Iris+

© 3DRobotics

Release Date: September 2014
Price: $600
Number of Rotors: 4
Propellers: PRC0009 9.5 × 4.5
Control System: Remote Control, Phone/Tablet
Applications: Film, Photography
Wingspan: 550 mm
Weight: 1.28 kg
Max. Payload: 400 gr
Frame Material: Plastic
Rotor Material: Carbon Fiber
Motors: MT2213-950kv
Available Colors: Black, Blue
Max. Flight Time: 25 minutes
Max. Flight Time with Payload: 22 minutes
Max. Speed: 82 km/h (51 mp/h)
Max. Operating Range: 1 km
Controller Frequency: 2.4 GHz
Battery Capacity: 5,100 mAh
Battery Type: Lithium Polymer
Battery Power: 13V
Battery Weight: 320 gr
GPS Compatible: √
Altitude Hold: √
Position Hold: √
Flight Planning: √
Autonomous Flight: √
Target Tracking: √
Automatic Landing: √
Return To Home: √
Obstacle Avoidance: –
Headless Mode: –
Built-in Camera: –

AEE Technology
Toruk AP10

© droneshop.com

Release Date: September 2014
Price: $399
Number of Rotors: 4
Control System: Remote Control, Phone/Tablet
Applications: Film, Photography
Dimensions: 370 × 370 × 140 mm
Wingspan: 450 mm
Weight: 1.3 kg
Max. Payload: 300 gr
Frame Material: Plastic
Rotor Material: Plastic
Available Colors: White
Max. Flight Time: 22 minutes
Max. Speed: 90 km/h (56 mp/h)
Max. Flying Altitude: 500 m (1,640 ft)
Max. Ascent Speed: 6 m/s
Max. Descent Speed: 4 m/s
Min. Operating Temperature: 0°C
Max. Operating Temperature: 50°C
Max. Operating Range: 500 m
Controller Frequency: 2.4 GHz
Battery Capacity: 5,300 mAh
Battery Type: Lithium Polymer
Battery Power: 11V
Battery Weight: 300 gr
GPS Compatible: √
Altitude Hold: √
Position Hold: √
Flight Planning: √
Autonomous Flight: –
Target Tracking: –
Automatic Landing: √
Return To Home: √
Obstacle Avoidance: –
Headless Mode: –
Built-in Camera: √ (16 megapixels)

Align
M480L

© Align

Release Date: September 2014
Price: $1,275
Number of Rotors: 4
Control System: Remote Control, Phone/Tablet
Applications: Film, Photography
Dimensions: 590 × 484 × 430 mm
Wingspan: 740 mm
Weight: 2.7 kg
Frame Material: Carbon Fiber
Rotor Material: Carbon Fiber
Motors: BL4213 brushless (370KV)
Available Colors: Red
Max. Speed: 54 km/h (33,5 mp/h)
Max. Flying Altitude: 700 m (2,296 ft)
Max. Ascent Speed: 6 m/s
Min. Operating Temperature: -20°C
Max. Operating Temperature: 65°C
Controller Frequency: 2.4 GHz
Battery Capacity: 12,000 mAh
Battery Type: Lithium Polymer 6S
Battery Power: 22.2V
GPS Compatible: √
Altitude Hold: √
Position Hold: √
Flight Planning: √
Autonomous Flight: √
Target Tracking: √
Automatic Landing: √
Return To Home: √
Obstacle Avoidance: –
Headless Mode: –
Built-in Camera: –

China Aerospace Science and Technology Corporation (CASC)

CH-4 Rainbow

© images.china.cn

Release Date: September 2014
Price: Unknown
Number of Rotors: 1 (Fixed Wing)
Control System: Ground Control Station
Applications: Military Purposes, Armed Reconnaissance, Surveillance, Target Acquisition
Dimensions: 11 × 20 × 3.8 m
Wingspan: 20 m
Weight: 1,600 kg
Frame Material: Composite
Available Colors: Grey
Max. Flight Time: 30 hours
Max. Speed: 350 km/h (217 mp/h)
Max. Flying Altitude: 14,400 m (47,244 ft)
Max. Operating Range: 2,750 km
GPS Compatible: √
Altitude Hold: √
Position Hold: √
Flight Planning: √
Autonomous Flight: √
Target Tracking: √
Automatic Landing: √
Return To Home: √
Obstacle Avoidance: √
Headless Mode: √
Built-in Camera: √

CyPhy Works

Pocket Flyer

© CyPhy Works

Release Date: September 2014
Price: Unknown
Number of Rotors: 6
Control System: Phone/Tablet
Applications: Recreation
Wingspan: 177 mm
Weight: 80 gr
Frame Material: Plastic
Rotor Material: Plastic
Available Colors: Black
Max. Flight Time: 2 hours
Max. Operating Range: 75 m
Controller Frequency: 2.4 GHz
GPS Compatible: –
Altitude Hold: –
Position Hold: –
Flight Planning: –
Autonomous Flight: –
Target Tracking: –
Automatic Landing: –
Return To Home: √
Obstacle Avoidance: –
Headless Mode: –
Built-in Camera: –

Dromida

Omnius

© Dromida

Release Date: September 2014
Price: $40
Number of Rotors: 4
Control System: Remote Control
Applications: FPV Racing
Wingspan: 238 mm
Weight: 101 gr
Frame Material: Plastic
Rotor Material: Plastic
Available Colors: Blue, Green, Red, Yellow
Max. Flight Time: 15 minutes
Controller Frequency: 2.4 GHz
Battery Capacity: 700 mAh
Battery Type: Lithium Polymer 1S
Battery Power: 3.7V
GPS Compatible: –
Altitude Hold: –
Position Hold: –
Flight Planning: –
Autonomous Flight: –
Target Tracking: –
Automatic Landing: –
Return To Home: –
Obstacle Avoidance: –
Headless Mode: –
Built-in Camera: √

DYS

D800 X4

©DYS

Release Date: September 2014
Price: $1,330
Number of Rotors: 4
Propellers: 1555 CW/CCW
Control System: Remote Control
Applications: Film, Photography, Aerial Mapping, Agriculture, Inspection, Surveillance
Wingspan: 800 mm
Weight: 2.8 kg
Max. Payload: 6.5 kg
Frame Material: Carbon Fiber
Rotor Material: Carbon Fiber
Motors: BE4114-400kv
Available Colors: Black
Controller Frequency: 2.4 GHz
Battery Capacity: 10,000 mAh
Battery Type: Lithium Polymer 6S
Battery Power: 11.1V
Battery Weight: 1.4 kg
GPS Compatible: √
Altitude Hold: √
Position Hold: √
Flight Planning: √
Autonomous Flight: √
Target Tracking: √
Automatic Landing: √
Return To Home: √
Obstacle Avoidance: –
Headless Mode: –
Built-in Camera: –

Huajun Toys

W608-2

©Huajun Toys

Release Date: September 2014
Price: $40
Number of Rotors: 4
Control System: Remote Control
Applications: Recreation
Dimensions: 510 × 510 × 120 mm
Weight: 47 gr
Available Colors: Black, Blue, Red
Max. Flight Time: 8 minutes
Max. Operating Range: 100 m
Controller Frequency: 2.4 GHz
Battery Capacity: 650 mAh
Battery Type: Lithium Polymer
Battery Power: 7.4V
GPS Compatible: –
Altitude Hold: √
Position Hold: √
Flight Planning: –
Autonomous Flight: –
Target Tracking: –
Automatic Landing: –
Return To Home: –
Obstacle Avoidance: –
Headless Mode: –
Built-in Camera: –

Raytheon

Coyote

©NOAA

Release Date: September 2014
Price: Unknown
Number of Rotors: 1 (Fixed Wing)
Control System: Ground Control Station
Applications: Military Purposes, Armed Reconnaissance, Surveillance, Target Acquisition
Dimensions: 0.91 × 1.5 m
Wingspan: 1.5 m
Weight: 5.9 kg
Frame Material: Carbon Fiber, Composite
Available Colors: Black
Max. Flight Time: 60 minutes
Max. Speed: 130 km/h (81 mp/h)
Max. Flying Altitude: 9,100 m (29,855 ft)
Max. Operating Range: 130 km
GPS Compatible: √
Altitude Hold: √
Position Hold: –
Flight Planning: √
Autonomous Flight: √
Target Tracking: √
Automatic Landing: –
Return To Home: √
Obstacle Avoidance: √
Headless Mode: –
Built-in Camera: √

SteadiDrone

Mavrik

© SteadiDrone

Release Date: September 2014
Price: $5,995
Number of Rotors: 8
Control System: Remote Control
Applications: Film, Photography
Dimensions: 500 × 775 × 205 mm
Wingspan: 814 mm
Weight: 2.2 kg
Max. Payload: 1.5 kg
Frame Material: Aluminum
Rotor Material: Carbon Fiber
Available Colors: Red
Max. Flight Time: 25 minutes
Max. Speed: 65 km/h (40 mp/h)
Max. Flying Altitude: 4,000 m (13,123 ft)
Max. Ascent Speed: 9 m/s
Max. Descent Speed: 6 m/s
Min. Operating Temperature: -5°C
Max. Operating Temperature: 50°C
Max. Operating Range: 1.5 km
Controller Frequency: 2.4 GHz
Battery Capacity: 8,000 mAh
Battery Type: Lithium Polymer 4S
Battery Power: 15V
Battery Weight: 750 gr
GPS Compatible: √
Altitude Hold: √
Position Hold: √
Flight Planning: √
Autonomous Flight: √
Target Tracking: –
Automatic Landing: √
Return To Home: √
Obstacle Avoidance: –
Headless Mode: –
Built-in Camera: –

Team BlackSheep

Gemini FPV Racer

©Team BlackSheep

Release Date: September 2014
Price: $599
Number of Rotors: 6
Control System: Remote Control
Applications: FPV Racing
Wingspan: 315 mm
Weight: 370 gr
Frame Material: Plastic
Rotor Material: Plastic
Available Colors: Black
Max. Flight Time: 10 minutes
Max. Speed: 79 km/h (49 mp/h)
Max. Operating Range: 2 km
Controller Frequency: 5.8 GHz
Battery Capacity: 1,700 mAh
Battery Type: Lithium Polymer 3S
GPS Compatible: –
Altitude Hold: –
Position Hold: –
Flight Planning: –
Autonomous Flight: –
Target Tracking: –
Automatic Landing: –
Return To Home: –
Obstacle Avoidance: –
Headless Mode: –
Built-in Camera: √

TurboAce

Infinity 6

© TurboAce

Release Date: September 2014
Price: $1,897
Number of Rotors: 6
Propellers: PP11A
Control System: Remote Control, Phone/Tablet
Applications: Film, Photography, Inspection, Surveillance
Wingspan: 933 mm
Weight: 9.7 kg
Frame Material: Carbon Fiber
Rotor Material: Carbon Fiber
Available Colors: Black
Max. Flight Time with Payload: 35 minutes
Max. Speed: 40 km/h (25 mp/h)
Controller Frequency: 2.4 GHz
Battery Capacity: 22,000 mAh
Battery Type: Lithium Polymer 6S
Battery Power: 22.2V
GPS Compatible: √
Altitude Hold: √
Position Hold: √
Flight Planning: √
Autonomous Flight: –
Target Tracking: √
Automatic Landing: √
Return To Home: √
Obstacle Avoidance: –
Headless Mode: –
Built-in Camera: –

Walkera

Scout X4

© Walkera

Release Date: September 2014
Price: $739
Number of Rotors: 4
Control System: Remote Control
Applications: Film, Photography
Dimensions: 335 × 335 × 275 mm
Wingspan: 473 mm
Weight: 1.7 kg
Max. Payload: 500 gr
Frame Material: Plastic, Carbon Fiber
Rotor Material: Plastic
Available Colors: Black, White
Max. Flight Time: 25 minutes
Max. Speed: 61 km/h (38 mp/h)
Min. Operating Temperature: -10°C
Max. Operating Temperature: 40°C
Max. Operating Range: 1.5 km
Controller Frequency: 2.4 GHz
Battery Capacity: 5,400 mAh
Battery Type: Lithium Polymer
Battery Power: 22V
GPS Compatible: √
Altitude Hold: √
Position Hold: √
Flight Planning: √
Autonomous Flight: √
Target Tracking: √
Automatic Landing: √
Return To Home: √
Obstacle Avoidance: –
Headless Mode: –
Built-in Camera: –

Zerotech

E-Epic

Release Date: September 2014
Price: On Request
Number of Rotors: 8
Control System: Remote Control
Applications: Film, Photography, Aerial Mapping, Agriculture, Inspection, Surveillance
Wingspan: 1.32 m
Weight: 12 kg
Max. Payload: 6 kg
Frame Material: Carbon Fiber
Rotor Material: Carbon Fiber
Motors: 7010-390kv
Available Colors: Black
Max. Flight Time: 10 minutes
Controller Frequency: 2.4 GHz
Battery Type: Lithium Polymer 6S
Battery Power: 22.2V
GPS Compatible: √
Altitude Hold: √
Position Hold: √
Flight Planning: √
Autonomous Flight: –
Target Tracking: √
Automatic Landing: √
Return To Home: √
Obstacle Avoidance: –
Headless Mode: –
Built-in Camera: –

3D Robotics

Aero-M Fixed Wing

Release Date: October 2014
Price: $5,400
Number of Rotors: 1 (Fixed Wing)
Propellers: Gemfan 11 × 7
Control System: Remote Control
Applications: Aerial Mapping
Dimensions: 129 cm
Wingspan: 1880 mm
Weight: 3 kg
Max. Payload: 500 gr
Frame Material: Styrofoam
Rotor Material: Styrofoam
Motors: Tiger AT2820-7 830kv
Available Colors: White
Max. Flight Time: 40 minutes
Max. Speed: 53 km/h (33 mp/h)
Max. Operating Range: 1 km
Controller Frequency: 915 mHz
Battery Capacity: 6,000 mAh
Battery Type: Lithium Polymer
GPS Compatible: √
Altitude Hold: √
Position Hold: √
Flight Planning: √
Autonomous Flight: √
Target Tracking: –
Automatic Landing: √
Return To Home: √
Obstacle Avoidance: –
Headless Mode: –
Built-in Camera: √ (12.1 megapixels)

3D Robotics

X8-M

©3DRobotics

Release Date: October 2014
Price: $5,400
Number of Rotors: 8
Propellers: APC 11x4.7 SFP
Control System: Remote Control
Applications: Aerial Mapping, Surveillance
Dimensions: 35 × 51 × 30 cm
Wingspan: 618 mm
Weight: 3.5 kg
Max. Payload: 200 gr
Frame Material: Aluminum
Rotor Material: Carbon Fiber
Motors: SunnySky V2216-12 800kv II
Available Colors: Black
Max. Flight Time: 14 minutes
Max. Speed: 40 km/h (25 mp/h)
Max. Flying Altitude: 120 m (394 ft)
Max. Ascent Speed: 10 m/s
Min. Operating Temperature: 0°C
Max. Operating Temperature: 40°C
Max. Operating Range: 1 km
Controller Frequency: 2.4 GHz
Battery Capacity: 10,000 mAh
Battery Type: Lithium Polymer 4S
Battery Power: 15V
Battery Weight: 800 gr
GPS Compatible: √
Altitude Hold: √
Position Hold: √
Flight Planning: √
Autonomous Flight: √
Target Tracking: √
Automatic Landing: √
Return To Home: √
Obstacle Avoidance: –
Headless Mode: –
Built-in Camera: –

4 Front Robotics

Navig8 Electric

©4 Front Robotics

Release Date: October 2014
Price: Unknown
Number of Rotors: 2
Control System: Laptop, Tablet
Applications: Film, Photography, Aerial Mapping, Inspection
Wingspan: 650 mm
Weight: 6 kg
Frame Material: Carbon Fiber
Available Colors: Grey, Yellow
Max. Flight Time: 45 minutes
Max. Speed: 45 km/h (28 mp/h)
Max. Flying Altitude: 300 m (984 ft)
Min. Operating Temperature: -30°C
Max. Operating Temperature: 50°C
Controller Frequency: 2.4 GHz
Battery Type: Lithium Polymer
Battery Weight: 1 kg
GPS Compatible: √
Altitude Hold: √
Position Hold: √
Flight Planning: √
Autonomous Flight: √
Target Tracking: √
Automatic Landing: √
Return To Home: √
Obstacle Avoidance: √
Headless Mode: –
Built-in Camera: √

Acme

Zoopa Q650

© Acme

Release Date: October 2014
Price: $135
Number of Rotors: 4
Control System: Remote Control
Applications: Film, Photography
Dimensions: 520 × 520 × 85 mm
Wingspan: 650 mm
Weight: 310 gr
Frame Material: Carbon Fiber, Foam
Rotor Material: Plastic
Available Colors: Black
Max. Flight Time: 10 minutes
Max. Operating Range: 500 m
Controller Frequency: 2.4 GHz
Battery Capacity: 850 mAh
Battery Type: Lithium Polymer 2S
Battery Power: 7.4V
GPS Compatible: –
Altitude Hold: –
Position Hold: –
Flight Planning: –
Autonomous Flight: –
Target Tracking: –
Automatic Landing: –
Return To Home: –
Obstacle Avoidance: –
Headless Mode: –
Built-in Camera: –

Align

M690L

© Align

Release Date: October 2014
Price: $1,530
Number of Rotors: 6
Control System: Remote Control, Phone/Tablet
Applications: Film, Photography
Dimensions: 875 × 768 × 446 mm
Wingspan: 900 mm
Weight: 3.4 kg
Frame Material: Carbon Fiber
Rotor Material: Carbon Fiber
Motors: BL4213 brushless
Available Colors: Red
Max. Speed: 54 km/h (33,5 mp/h)
Max. Flying Altitude: 700 m (2,296 ft)
Max. Ascent Speed: 6 m/s
Min. Operating Temperature: -20°C
Max. Operating Temperature: 65°C
Controller Frequency: 2.4 GHz
Battery Capacity: 12,000 mAh
Battery Type: Lithium Polymer 6S
Battery Power: 22.2V
GPS Compatible: √
Altitude Hold: √
Position Hold: √
Flight Planning: √
Autonomous Flight: √
Target Tracking: √
Automatic Landing: √
Return To Home: √
Obstacle Avoidance: –
Headless Mode: –
Built-in Camera: –

Ares

Ethos HD

Release Date: October 2014
Price: $170
Number of Rotors: 4
Control System: Remote Control
Applications: Film, Photography
Dimensions: 292 × 200 × 127 mm
Wingspan: 320 mm
Weight: 312 gr
Frame Material: Plastic
Rotor Material: Plastic
Motors: 350 brushed
Available Colors: White
Controller Frequency: 2.4 GHz
Battery Capacity: 1,200 mAh
Battery Type: Lithium Polymer 2S
Battery Power: 7.4V
GPS Compatible: –
Altitude Hold: –
Position Hold: √
Flight Planning: –
Autonomous Flight: –
Target Tracking: –
Automatic Landing: –
Return To Home: –
Obstacle Avoidance: –
Headless Mode: –
Built-in Camera: √

Ares

Ethos FPV

Release Date: October 2014
Price: $240
Number of Rotors: 4
Control System: Remote Control
Applications: FPV Racing
Dimensions: 292 × 200 × 127 mm
Wingspan: 320 mm
Weight: 312 gr
Frame Material: Plastic
Rotor Material: Plastic
Motors: 350 brushed
Available Colors: White
Controller Frequency: 2.4 GHz
Battery Capacity: 1,200 mAh
Battery Type: Lithium Polymer 2S
Battery Power: 7.4V
GPS Compatible: –
Altitude Hold: –
Position Hold: √
Flight Planning: –
Autonomous Flight: –
Target Tracking: –
Automatic Landing: –
Return To Home: –
Obstacle Avoidance: –
Headless Mode: –
Built-in Camera: √

AscTec

Firefly

Release Date: October 2014
Price: Unknown
Number of Rotors: 4
Control System: Remote Control
Applications: Film, Photography, Aerial Mapping, Inspection
Dimensions: 605 × 665 × 165 mm
Weight: 1.6 kg
Max. Payload: 600 gr
Frame Material: Carbon fiber
Rotor Material: Plastic
Available Colors: White
Max. Flight Time: 22 minutes
Max. Flight Time with Payload: 14 minutes
Max. Speed: 54 km/h (33,5 mp/h)
Max. Ascent Speed: 8 m/s
Max. Operating Range: 4.5 km
Controller Frequency: 2.4 GHz
Battery Capacity: 5,500 mAh
Battery Type: Lithium Polymer
GPS Compatible: √
Altitude Hold: √
Position Hold: √
Flight Planning: √
Autonomous Flight: √
Target Tracking: √
Automatic Landing: √
Return To Home: √
Obstacle Avoidance: √
Headless Mode: –
Built-in Camera: –

Blade

Nano QX FPV

Release Date: October 2014
Price: $90
Number of Rotors: 4
Control System: Remote Control
Applications: FPV Racing
Dimensions: 182 × 160 × 63,5 mm
Frame Material: Plastic
Rotor Material: Plastic
Available Colors: Black, Green, Yellow
Max. Flight Time: 8 minutes
Controller Frequency: 2.4 GHz
Battery Capacity: 150 mAh
Battery Type: Lithium Polymer 1S
Battery Power: 3.7V
GPS Compatible: –
Altitude Hold: –
Position Hold: –
Flight Planning: –
Autonomous Flight: –
Target Tracking: –
Automatic Landing: –
Return To Home: –
Obstacle Avoidance: –
Headless Mode: –
Built-in Camera: √

Blade

Pico QX

©team-legit.com

Release Date: October 2014
Price: $50
Number of Rotors: 4
Control System: Remote Control, Phone/Tablet
Applications: Recreation
Wingspan: 92 mm
Weight: 8 gr
Frame Material: Plastic
Rotor Material: Plastic
Available Colors: Black, White
Max. Flight Time: 8 minutes
Controller Frequency: 2.4 GHz
Battery Type: Lithium Polymer
GPS Compatible: –
Altitude Hold: –
Position Hold: –
Flight Planning: –
Autonomous Flight: –
Target Tracking: –
Automatic Landing: –
Return To Home: –
Obstacle Avoidance: –
Headless Mode: –
Built-in Camera: –

Boeing Insitu

ScanEagle 2

©flickr.com/Alan Wilson

Release Date: October 2014
Price: $5 million
Number of Rotors: 1 (Fixed Wing)
Control System: Ground Control Station
Applications: Military Purposes, Maritime Patrol, Reconnaissance, Surveillance
Dimensions: 1.71 × 3.11 m
Wingspan: 3.11 m
Weight: 17 kg
Max. Payload: 5 kg
Frame Material: Composite
Motors: 1 × two-stroke 3W piston engine
Available Colors: Grey
Max. Flight Time: 18 hours
Max. Speed: 148 km/h (92 mp/h)
Max. Flying Altitude: 6,000 m (19,685 ft)
Max. Operating Range: 550 km
GPS Compatible: √
Altitude Hold: √
Position Hold: –
Flight Planning: √
Autonomous Flight: √
Target Tracking: √
Automatic Landing: √
Return To Home: √
Obstacle Avoidance: √
Headless Mode: –
Built-in Camera: √

C-Astral

Bramor rTK

©C-Astral

Release Date: October 2014
Price: Unknown
Number of Rotors: 1 (Fixed Wing)
Control System: Flight Computer
Applications: Aerial Mapping, Surveillance
Wingspan: 2.3 m
Weight: 4.5 kg
Frame Material: Carbon Fiber, Kevlar
Available Colors: White
Max. Flight Time: 2.5 hours
Max. Speed: 83 km/h (51,5 mp/h)
Max. Flying Altitude: 5,000 m (16,404 ft)
Min. Operating Temperature: -25°C
Max. Operating Temperature: 45°C
Max. Operating Range: 40 km
Battery Type: Lithium Polymer
GPS Compatible: √
Altitude Hold: √
Position Hold: √
Flight Planning: √
Autonomous Flight: √
Target Tracking: √
Automatic Landing: √
Return To Home: √
Obstacle Avoidance: –
Headless Mode: –
Built-in Camera: √ (24.3 megapixels)

DJI

Spreading Wings S1000+

©goodlux-studio.be

Release Date: October 2014
Price: $1,999
Number of Rotors: 8
Propellers: DJI 1345T
Control System: Remote Control
Applications: Film, Photography, Aerial Mapping
Wingspan: 1.05 m
Weight: 4.4 kg
Max. Payload: 1.8 kg
Frame Material: Carbon Fiber
Rotor Material: Plastic
Motors: 400rpm/V kv
Available Colors: Black
Max. Flight Time: 15 minutes
Max. Ascent Speed: 4 m/s
Min. Operating Temperature: -10°C
Max. Operating Temperature: 40°C
Max. Operating Range: 2 km
Controller Frequency: 5.8 GHz
Battery Capacity: 20,000 mAh
Battery Type: Lithium Polymer 6S
Battery Power: 22.8V
Battery Weight: 670 gr
GPS Compatible: √
Altitude Hold: √
Position Hold: √
Flight Planning: –
Autonomous Flight: –
Target Tracking: –
Automatic Landing: √
Return To Home: √
Obstacle Avoidance: –
Headless Mode: –
Built-in Camera: –

Dromida

Kodo

© bhphotovideo.com

Release Date: October 2014
Price: $45
Number of Rotors: 4
Control System: Remote Control
Applications: Recreation
Wingspan: 105 mm
Weight: 52 gr
Frame Material: Plastic
Rotor Material: Plastic
Available Colors: Black
Max. Flight Time: 6 minutes
Controller Frequency: 2.4 GHz
Battery Capacity: 390 mAh
Battery Type: Lithium Polymer 1S
Battery Power: 3.7V
GPS Compatible: –
Altitude Hold: –
Position Hold: –
Flight Planning: –
Autonomous Flight: –
Target Tracking: –
Automatic Landing: –
Return To Home: –
Obstacle Avoidance: –
Headless Mode: –
Built-in Camera: √

DYS

D800 V6

© banggood.com

Release Date: October 2014
Price: $1,990
Number of Rotors: 6
Propellers: 1555 CW/CCW
Control System: Remote Control
Applications: Film, Photography, Aerial Mapping, Agriculture, Surveillance
Dimensions: 310 × 430 × 320 mm
Wingspan: 800 mm
Weight: 3.4 kg
Max. Payload: 6.6 kg
Frame Material: Carbon Fiber
Rotor Material: Carbon Fiber
Motors: BE4114-400kv
Available Colors: Black
Max. Flight Time: 20 minutes
Min. Operating Temperature: -10°C
Max. Operating Temperature: 40°C
Controller Frequency: 2.4 GHz
Battery Capacity: 20,000 mAh
Battery Type: Lithium Polymer 6S
Battery Power: 22.2V
GPS Compatible: √
Altitude Hold: √
Position Hold: √
Flight Planning: √
Autonomous Flight: √
Target Tracking: √
Automatic Landing: √
Return To Home: √
Obstacle Avoidance: –
Headless Mode: –
Built-in Camera: –

DYS

D800 X8

© DYS

Release Date: October 2014
Price: $1,390
Number of Rotors: 8
Propellers: 1555 CW/CCW
Control System: Remote Control
Applications: Film, Photography, Aerial Mapping, Agriculture, Surveillance
Dimensions: 420 × 430 × 400 mm
Wingspan: 800 mm
Weight: 3.7 kg
Max. Payload: 6.7 kg
Frame Material: Carbon Fiber
Rotor Material: Carbon Fiber
Motors: BE4114-400kv
Available Colors: Black
Controller Frequency: 2.4 GHz
Battery Capacity: 20,000 mAh
Battery Type: Lithium Polymer 6S
Battery Power: 22.2V
GPS Compatible: √
Altitude Hold: √
Position Hold: √
Flight Planning: √
Autonomous Flight: √
Target Tracking: √
Automatic Landing: √
Return To Home: √
Obstacle Avoidance: –
Headless Mode: –
Built-in Camera: –

Hobbyzone

Faze

© bhphotovideo.com

Release Date: October 2014
Price: $20
Number of Rotors: 4
Control System: Remote Control
Applications: Recreation
Dimensions: 80 × 66 × 50 mm
Weight: 12 gr
Frame Material: Plastic
Rotor Material: Plastic
Available Colors: Red
Max. Flight Time: 7 minutes
Controller Frequency: 2.4 GHz
Battery Capacity: 100 mAh
Battery Type: Lithium Polymer 1S
Battery Power: 3.7V
GPS Compatible: –
Altitude Hold: –
Position Hold: –
Flight Planning: –
Autonomous Flight: –
Target Tracking: –
Automatic Landing: –
Return To Home: –
Obstacle Avoidance: –
Headless Mode: –
Built-in Camera: –

Idea-Fly

Hero-550

© Idea-Fly

Release Date: October 2014
Price: $1,250
Number of Rotors: 4
Control System: Remote Control
Applications: Film, Photography
Wingspan: 550 mm
Weight: 2.5 kg
Max. Payload: 800 gr
Frame Material: Carbon Fiber, Plastic
Rotor Material: Carbon Fiber
Available Colors: Black
Max. Flight Time: 23 minutes
Max. Speed: 29 km/h (18 mp/h)
Max. Flying Altitude: 500 m (1,640 ft)
Max. Ascent Speed: 10 m/s
Max. Descent Speed: 4 m/s
Min. Operating Temperature: -10°C
Max. Operating Temperature: 50°C
Max. Operating Range: 2 km
Controller Frequency: 2.4 GHz
Battery Capacity: 10,000 mAh
Battery Type: Lithium Polymer 4S
Battery Power: 14.8V
Battery Weight: 420 gr
GPS Compatible: √
Altitude Hold: √
Position Hold: √
Flight Planning: –
Autonomous Flight: –
Target Tracking: –
Automatic Landing: –
Return To Home: √
Obstacle Avoidance: –
Headless Mode: –
Built-in Camera: –

JJRC

JJ600A

© banggood.com

Release Date: October 2014
Price: $45
Number of Rotors: 4
Control System: Remote Control
Applications: Recreation
Dimensions: 250 × 250 × 60
Wingspan: 354 mm
Weight: 750 gr
Frame Material: Plastic
Rotor Material: Plastic
Available Colors: Black
Max. Flight Time: 7 minutes
Max. Operating Range: 100 m
Controller Frequency: 2.4 GHz
Battery Capacity: 380 mAh
Battery Type: Lithium Polymer
Battery Power: 3.7V
GPS Compatible: –
Altitude Hold: –
Position Hold: –
Flight Planning: –
Autonomous Flight: –
Target Tracking: –
Automatic Landing: –
Return To Home: √
Obstacle Avoidance: –
Headless Mode: –
Built-in Camera: √ (0.3 megapixels)

Kongcopter

FQ-700

©x-camtech.com

Release Date: October 2014
Price: $800
Number of Rotors: 4
Control System: Remote Control
Applications: Film, Photography
Wingspan: 700 mm
Weight: 1.1 kg
Frame Material: Plastic
Rotor Material: Plastic
Motors: 4615-420kv
Available Colors: Black
Max. Flight Time: 17 minutes
Controller Frequency: 2.4 GHz
Battery Capacity: 10,000 mAh
Battery Type: Lithium Polymer 6S
GPS Compatible: √
Altitude Hold: √
Position Hold: √
Flight Planning: √
Autonomous Flight: √
Target Tracking: –
Automatic Landing: √
Return To Home: √
Obstacle Avoidance: –
Headless Mode: –
Built-in Camera: –

Nine Eagles

Galaxy Visitor 6

©Nine Eagles

Release Date: October 2014
Price: $179
Number of Rotors: 4
Control System: Remote Control, Phone/Tablet
Applications: Agriculture, Inspection, Surveillance
Dimensions: 199 × 199 × 54 mm
Wingspan: 240 mm
Weight: 115 gr
Frame Material: Plastic
Rotor Material: Plastic
Available Colors: Black
Max. Flight Time: 10 minutes
Max. Flying Altitude: 100 m (328 ft)
Controller Frequency: 2.4 GHz
Battery Capacity: 700 mAh
Battery Type: Lithium Polymer
Battery Power: 3.7V
GPS Compatible: –
Altitude Hold: √
Position Hold: √
Flight Planning: –
Autonomous Flight: –
Target Tracking: –
Automatic Landing: –
Return To Home: √
Obstacle Avoidance: –
Headless Mode: –
Built-in Camera: √

Precision Drone

Pacesetter

©newhollandrochester.com

Release Date: October 2014
Price: Unknown
Number of Rotors: 6
Control System: Remote Control
Applications: Film, Photography, Aerial Mapping, Agriculture
Frame Material: Carbon Fiber
Rotor Material: Carbon Fiber
Available Colors: Black
Controller Frequency: 2.4 GHz
GPS Compatible: √
Altitude Hold: √
Position Hold: √
Flight Planning: √
Autonomous Flight: √
Target Tracking: –
Automatic Landing: –
Return To Home: √
Obstacle Avoidance: –
Headless Mode: –
Built-in Camera: –

Quanum

Venture

©helicomicro.com

Release Date: October 2014
Price: $240
Number of Rotors: 4
Control System: Remote Control
Applications: FPV Racing
Wingspan: 430 mm
Weight: 440 gr
Frame Material: Plastic
Rotor Material: Plastic
Available Colors: Black
Controller Frequency: 5.8 GHz
Battery Capacity: 2,200 mAh
Battery Type: Lithium Polymer 3S
GPS Compatible: –
Altitude Hold: –
Position Hold: –
Flight Planning: –
Autonomous Flight: –
Target Tracking: –
Automatic Landing: –
Return To Home: –
Obstacle Avoidance: –
Headless Mode: –
Built-in Camera: √ (600TVL)

Service-drone.com

G4 Surveying Robot

Release Date: October 2014
Price: $25,500
Number of Rotors: 6
Control System: Remote Control
Applications: Film, Photography, Aerial Mapping, Surveillance
Wingspan: 850 mm
Weight: 2.7 kg
Max. Payload: 2.3 kg
Frame Material: Carbon Fiber
Rotor Material: Carbon Fiber
Available Colors: Silver, Yellow
Max. Flight Time: 20 minutes
Max. Speed: 30 km/h (18,5 mp/h)
Min. Operating Temperature: -10°C
Max. Operating Temperature: 40°C
Max. Operating Range: 3 km
Controller Frequency: 2.4 GHz
Battery Capacity: 3,200 mAh
Battery Type: Lithium Polymer
GPS Compatible: √
Altitude Hold: √
Position Hold: √
Flight Planning: √
Autonomous Flight: √
Target Tracking: –
Automatic Landing: √
Return To Home: √
Obstacle Avoidance: –
Headless Mode: –
Built-in Camera: –

UDIRC Toys

U841-1 HD

Release Date: October 2014
Price: $60
Number of Rotors: 4
Control System: Remote Control
Applications: Recreation
Dimensions: 151 × 151 × 36 mm
Wingspan: 214 mm
Weight: 64 gr
Frame Material: Plastic
Rotor Material: Plastic
Available Colors: Black
Max. Flight Time: 7 minutes
Max. Operating Range: 25 m
Controller Frequency: 2.4 GHz
Battery Capacity: 450 mAh
Battery Type: Lithium Polymer
Battery Power: 3.7V
GPS Compatible: –
Altitude Hold: √
Position Hold: √
Flight Planning: –
Autonomous Flight: –
Target Tracking: –
Automatic Landing: –
Return To Home: –
Obstacle Avoidance: –
Headless Mode: –
Built-in Camera: –

WLtoys

V393 Brushless Explorer

© aliexpress.com

Release Date: October 2014
Price: $270
Number of Rotors: 4
Control System: Remote Control
Applications: Recreation
Dimensions: 61 × 61 × 20 mm
Weight: 600 gr
Frame Material: Plastic, Foam
Rotor Material: Plastic
Motors: 2212-950kv
Available Colors: Black
Max. Flight Time: 13 minutes
Max. Operating Range: 300 m
Controller Frequency: 2.4 GHz
Battery Capacity: 2,700 mAh
Battery Type: Lithium Polymer
Battery Power: 11.1V
GPS Compatible: –
Altitude Hold: √
Position Hold: √
Flight Planning: –
Autonomous Flight: –
Target Tracking: –
Automatic Landing: –
Return To Home: √
Obstacle Avoidance: –
Headless Mode: –
Built-in Camera: √

WLtoys

V636 Skylark

© Wltoys

Release Date: October 2014
Price: $50
Number of Rotors: 4
Control System: Remote Control
Applications: Recreation
Dimensions: 195 × 195 × 10 mm
Weight: 80 gr
Frame Material: Plastic
Rotor Material: Plastic
Available Colors: Black
Max. Flight Time: 10 minutes
Max. Operating Range: 150 m
Controller Frequency: 2.4 GHz
Battery Capacity: 730 mAh
Battery Type: Lithium Polymer
Battery Power: 7.3V
GPS Compatible: –
Altitude Hold: √
Position Hold: √
Flight Planning: –
Autonomous Flight: –
Target Tracking: –
Automatic Landing: –
Return To Home: √
Obstacle Avoidance: –
Headless Mode: –
Built-in Camera: √

Xmobots

Echar 20A

© abce.com.br

Release Date: October 2014
Price: Unknown
Number of Rotors: 1 (Fixed Wing)
Control System: Ground Control Station
Applications: Aerial Mapping, Agriculture, Inspection
Dimensions: 0.85 × 2.17 × 0.39 m
Wingspan: 2.17 m
Weight: 5.6 kg
Max. Payload: 1 kg
Frame Material: Composite
Rotor Material: Plastic
Available Colors: Grey
Max. Flight Time: 45 minutes
Max. Speed: 75 km/h (46,6 mp/h)
Max. Flying Altitude: 3,000 m (9,842 ft)
Controller Frequency: 2.4 GHz
Battery Type: Lithium Polymer
GPS Compatible: √
Altitude Hold: √
Position Hold: –
Flight Planning: √
Autonomous Flight: √
Target Tracking: –
Automatic Landing: –
Return To Home: √
Obstacle Avoidance: –
Headless Mode: –
Built-in Camera: √

Xtreem

Atom Quadcopter

© Swann

Release Date: October 2014
Price: $40
Number of Rotors: 4
Control System: Remote Control
Applications: Recreation
Dimensions: 70 × 70 × 27 mm
Weight: 14 gr
Frame Material: Plastic
Rotor Material: Plastic
Available Colors: Blue
Max. Flight Time: 7 minutes
Max. Operating Range: 30 m
Controller Frequency: 2.4 GHz
Battery Capacity: 150 mAh
Battery Type: Lithium Polymer
Battery Power: 4V
GPS Compatible: –
Altitude Hold: –
Position Hold: –
Flight Planning: –
Autonomous Flight: –
Target Tracking: –
Automatic Landing: –
Return To Home: –
Obstacle Avoidance: –
Headless Mode: –
Built-in Camera: –

3D Robotics

X8+

© gigaom.com

Release Date: November 2014
Price: $1,350
Number of Rotors: 8
Propellers: APC 10 × 4.7 SFP
Control System: Remote Control
Applications: Film, Photography
Wingspan: 800 mm
Weight: 2.6 kg
Max. Payload: 800 gr
Frame Material: Aluminum
Rotor Material: Carbon Fiber
Motors: SunnySky V2216-12 800 kv II
Available Colors: Black
Max. Flight Time: 15 minutes
Max. Speed: 108 km/h (67 mp/h)
Max. Flying Altitude: 100 m (328 ft)
Max. Operating Range: 300 m
Controller Frequency: 2.4 GHz
Battery Capacity: 10,000 mAh
Battery Type: Lithium Polymer 4S
Battery Power: 15V
Battery Weight: 800 gr
GPS Compatible: √
Altitude Hold: √
Position Hold: √
Flight Planning: √
Autonomous Flight: √
Target Tracking: √
Automatic Landing: √
Return To Home: √
Obstacle Avoidance: –
Headless Mode: –
Built-in Camera: –

Arca Space Corporation

AirStrato Accelerator

© Arca Space Corporation

Release Date: November 2014
Price: $20,000
Number of Rotors: 6 (Fixed Wing)
Control System: Flight Computer
Applications: Film, Photography, Aerial Mapping, Surveillance
Dimensions: 3 × 9 × 1.8 m
Wingspan: 9 m
Available Colors: Blue, Gray, Red, Yellow
Max. Speed: 55 km/h (34 mp/h)
Min. Operating Temperature: -20°C
Max. Operating Temperature: 50°C
GPS Compatible: √
Altitude Hold: √
Position Hold: √
Flight Planning: √
Autonomous Flight: √
Target Tracking: √
Automatic Landing: √
Return To Home: √
Obstacle Avoidance: –
Headless Mode: –
Built-in Camera: √

Arca Space Corporation

AirStrato Explorer

© Arca Space Corporation

Release Date: November 2014
Price: $140,000
Number of Rotors: 6 (Fixed Wing)
Control System: Flight Computer
Applications: Film, Photography, Aerial Mapping, Surveillance
Dimensions: 7 × 16 × 1.8 m
Wingspan: 16 m
Weight: 185 kg
Max. Payload: 45 kg
Motors: Robbe 8085/10 45V
Available Colors: Blue, Gray, Red, Yellow
Max. Flight Time: 20 hours
Max. Speed: 170 km/h (106 mp/h)
Max. Flying Altitude: 18,000 m (59,055 ft)
Max. Ascent Speed: 3 m/s
Min. Operating Temperature: -20°C
Max. Operating Temperature: 50°C
Max. Operating Range: 1,800 km
GPS Compatible: √
Altitude Hold: √
Position Hold: √
Flight Planning: √
Autonomous Flight: √
Target Tracking: √
Automatic Landing: √
Return To Home: √
Obstacle Avoidance: –
Headless Mode: –
Built-in Camera: √

Arca Space Corporation

AirStrato Pioneer

© Arca Space Corporation

Release Date: November 2014
Price: $80,000
Number of Rotors: 4 (Fixed Wing)
Control System: Flight Computer
Applications: Film, Photography, Aerial Mapping, Surveillance
Dimensions: 7 × 12 × 1.6 m
Wingspan: 12 m
Weight: 175 kg
Max. Payload: 45 kg
Motors: Robbe 6362/10 38V
Available Colors: Blue, Gray, Red, Yellow
Max. Flight Time: 12 hours
Max. Speed: 120 km/h (74 mp/h)
Max. Flying Altitude: 8,000 m (26,246 ft)
Max. Ascent Speed: 3 m/s
Min. Operating Temperature: -20°C
Max. Operating Temperature: 50°C
Max. Operating Range: 1,000 km
GPS Compatible: √
Altitude Hold: √
Position Hold: √
Flight Planning: √
Autonomous Flight: √
Target Tracking: √
Automatic Landing: √
Return To Home: √
Obstacle Avoidance: –
Headless Mode: –
Built-in Camera: √

Ares

Spectre X

© Ares-RC

Release Date: November 2014
Price: $90
Number of Rotors: 4
Control System: Remote Control, Phone/Tablet
Applications: Film, Recreation
Dimensions: 100 × 100 × 45 mm
Wingspan: 140 mm
Weight: 61 gr
Frame Material: Plastic
Rotor Material: Plastic
Motors: Micro coreless
Available Colors: Green, Orange
Controller Frequency: 2.4 GHz
Battery Capacity: 700 mAh
Battery Type: Lithium Polymer 1S
Battery Power: 3.7V
GPS Compatible: –
Altitude Hold: –
Position Hold: √
Flight Planning: –
Autonomous Flight: –
Target Tracking: –
Automatic Landing: –
Return To Home: –
Obstacle Avoidance: –
Headless Mode: –
Built-in Camera: √

Blade

350 QX3

© bhphotovideo.com

Release Date: November 2014
Price: $500
Number of Rotors: 4
Control System: Remote Control, Phone/Tablet
Applications: Film, Photography
Dimensions: 465 × 465 × 190 mm
Wingspan: 580 mm
Weight: 955 gr
Frame Material: Plastic
Rotor Material: Plastic
Available Colors: White
Max. Flight Time: 15 minutes
Max. Speed: 72 km/h (45 mp/h)
Max. Operating Range: 1 km
Controller Frequency: 2.4 GHz
Battery Capacity: 3,000 mAh
Battery Type: Lithium Polymer 3S
Battery Power: 11.1V
GPS Compatible: √
Altitude Hold: √
Position Hold: √
Flight Planning: √
Autonomous Flight: –
Target Tracking: –
Automatic Landing: √
Return To Home: √
Obstacle Avoidance: –
Headless Mode: –
Built-in Camera: –

Carrera RC

Quadrocopter X1

Release Date: November 2014
Price: $60
Number of Rotors: 4
Control System: Remote Control
Applications: Recreation
Dimensions: 174 × 174 mm
Weight: 900 gr
Frame Material: EPP Foam
Rotor Material: Plastic
Available Colors: Black
Max. Operating Range: 60 m
Controller Frequency: 2.4 GHz
Battery Type: Lithium Polymer
GPS Compatible: –
Altitude Hold: –
Position Hold: √
Flight Planning: –
Autonomous Flight: –
Target Tracking: –
Automatic Landing: –
Return To Home: –
Obstacle Avoidance: –
Headless Mode: –
Built-in Camera: –

China Aerospace Science and Industry Corporation (CASIC)

HW-510

Release Date: November 2014
Price: Unknown
Number of Rotors: 1 (Fixed Wing)
Control System: Ground Control Station
Applications: Military Purposes, Maritime Patrol, Reconnaissance, Surveillance
Dimensions: 5.6 × 8 m
Wingspan: 8 m
Weight: 350 kg
Max. Payload: 70 kg
Frame Material: Composite
Rotor Material: Wood
Motors: piston engine
Available Colors: White
Max. Flight Time: 12 hours
Max. Speed: 180 km/h (112 mp/h)
Max. Flying Altitude: 5,000 m (16,404 ft)
Max. Operating Range: 150 km
GPS Compatible: √
Altitude Hold: √
Position Hold: –
Flight Planning: √
Autonomous Flight: √
Target Tracking: –
Automatic Landing: √
Return To Home: √
Obstacle Avoidance: √
Headless Mode: –
Built-in Camera: √

DJI

Inspire 1 V2.0

© DJI

Release Date: November 2014
Price: $1,999
Number of Rotors: 4
Propellers: DJI 1345T
Control System: Remote Control, Phone/Tablet
Applications: Film, Photography
Dimensions: 438 × 451 × 301 mm
Wingspan: 560 mm
Weight: 2.94 kg
Max. Payload: 1.7 kg
Frame Material: Carbon Fiber
Rotor Material: Carbon Fiber
Motors: DJI 3510H
Available Colors: Black, White
Max. Flight Time: 18 minutes
Max. Speed: 79 km/h (49 mp/h)
Max. Flying Altitude: 4,500 m (14,764 ft)
Max. Ascent Speed: 5 m/s
Min. Operating Temperature: -10°C
Max. Operating Temperature: 40°C
Max. Operating Range: 3.5 km
Controller Frequency: 5.8 GHz
Battery Capacity: 5,700 mAh
Battery Type: Lithium Polymer 6S
Battery Power: 22.8V
Battery Weight: 670 gr
GPS Compatible: √
Altitude Hold: √
Position Hold: √
Flight Planning: √
Autonomous Flight: √
Target Tracking: √
Automatic Landing: √
Return To Home: √
Obstacle Avoidance: –
Headless Mode: –
Built-in Camera: √ (12.4 megapixels)

Draganfly

Draganflyer Guardian

© Draganfly

Release Date: November 2014
Price: $7,000
Number of Rotors: 4
Control System: Remote Control
Applications: Film, Photography, Aerial Mapping, Surveillance
Dimensions: 595 × 595 × 255 mm
Wingspan: 725 mm
Weight: 1.05 kg
Max. Payload: 420 gr
Frame Material: Carbon Fiber
Rotor Material: Carbon Fiber
Available Colors: Black
Max. Flight Time: 31 minutes
Max. Speed: 50 km/h (31 mp/h)
Max. Flying Altitude: 2,400 m (7,874 ft)
Max. Ascent Speed: 2 m/s
Max. Descent Speed: 2 m/s
Max. Operating Range: 200 m
Controller Frequency: 2.4 GHz
Battery Capacity: 2,700 mAh
Battery Type: Lithium Polymer
Battery Power: 15V
GPS Compatible: √
Altitude Hold: √
Position Hold: √
Flight Planning: –
Autonomous Flight: –
Target Tracking: –
Automatic Landing: √
Return To Home: √
Obstacle Avoidance: –
Headless Mode: –
Built-in Camera: –

DYS

D1100 V8

Release Date: November 2014
Price: $1,690
Number of Rotors: 8
Propellers: 1555 CW/CCW
Control System: Remote Control
Applications: Film, Photography, Aerial Mapping, Agriculture, Surveillance
Wingspan: 1.1 m
Weight: 4.2 kg
Max. Payload: 8.8 kg
Frame Material: Carbon Fiber
Rotor Material: Carbon Fiber
Motors: BE4114-400kv
Available Colors: Black
Max. Flight Time: 16 minutes
Min. Operating Temperature: -10°C
Max. Operating Temperature: 40°C
Controller Frequency: 2.4 GHz
Battery Capacity: 15,000 mAh
Battery Type: Lithium Polymer 6S
Battery Power: 22.2V
GPS Compatible: √
Altitude Hold: √
Position Hold: √
Flight Planning: √
Autonomous Flight: √
Target Tracking: √
Automatic Landing: √
Return To Home: √
Obstacle Avoidance: –
Headless Mode: –
Built-in Camera: –

Eachine

CG023

Release Date: November 2014
Price: $17
Number of Rotors: 4
Control System: Remote Control
Applications: Recreation
Dimensions: 173 × 173 × 36 mm
Frame Material: Plastic
Rotor Material: Plastic
Available Colors: Black, White
Max. Flight Time: 7 minutes
Max. Operating Range: 60 m
Controller Frequency: 2.4 GHz
Battery Capacity: 150 mAh
Battery Type: Lithium Polymer 1S
Battery Power: 3V
GPS Compatible: –
Altitude Hold: √
Position Hold: √
Flight Planning: –
Autonomous Flight: –
Target Tracking: –
Automatic Landing: –
Return To Home: –
Obstacle Avoidance: –
Headless Mode: –
Built-in Camera: –

Huajun Toys

W608-8

©Huajun Toys

Release Date: November 2014
Price: Unknown
Number of Rotors: 4
Control System: Remote Control
Applications: Film, Photography, Recreation
Dimensions: 560 × 450 × 170 mm
Weight: 450 gr
Frame Material: Plastic, Foam
Rotor Material: Plastic
Available Colors: White
Max. Flight Time: 7 minutes
Max. Operating Range: 150 m
Controller Frequency: 2.4 GHz
Battery Capacity: 650 mAh
Battery Type: Lithium Polymer
Battery Power: 7.4V
GPS Compatible: –
Altitude Hold: √
Position Hold: √
Flight Planning: –
Autonomous Flight: –
Target Tracking: –
Automatic Landing: √
Return To Home: –
Obstacle Avoidance: –
Headless Mode: –
Built-in Camera: √ (2 megapixels)

Idetec

AG-Wing

©Idetec

Release Date: November 2014
Price: On Request
Number of Rotors: 1 (Fixed Wing)
Control System: Remote Control
Applications: Aerial Mapping, Agriculture, Inspection
Frame Material: EPP Foam
Available Colors: White
Max. Flight Time: 45 minutes
Controller Frequency: 2.4 GHz
Battery Type: Lithium Polymer
GPS Compatible: √
Altitude Hold: √
Position Hold: –
Flight Planning: √
Autonomous Flight: √
Target Tracking: –
Automatic Landing: –
Return To Home: √
Obstacle Avoidance: –
Headless Mode: –
Built-in Camera: √ (20 megapixels)

JJRC

H8C

© planetdrone.com

Release Date: November 2014
Price: $45
Number of Rotors: 4
Control System: Remote Control
Applications: Recreation
Dimensions: 400 × 400 × 115 mm
Wingspan: 566 mm
Weight: 733 gr
Frame Material: Plastic
Rotor Material: Plastic
Available Colors: Black, White
Max. Flight Time: 8 minutes
Max. Operating Range: 300 m
Controller Frequency: 2.4 GHz
Battery Capacity: 500 mAh
Battery Type: Lithium Polymer
Battery Power: 7.4V
GPS Compatible: –
Altitude Hold: –
Position Hold: –
Flight Planning: –
Autonomous Flight: –
Target Tracking: –
Automatic Landing: –
Return To Home: √
Obstacle Avoidance: –
Headless Mode: √
Built-in Camera: √ (2 megapixels)

JJRC

H3

© toysmini.com

Release Date: November 2014
Price: $65
Number of Rotors: 4
Control System: Remote Control
Applications: Recreation
Dimensions: 285 × 175 × 135 mm
Wingspan: 334 mm
Frame Material: Plastic
Rotor Material: Plastic
Available Colors: Blue, Gold, White
Max. Flight Time: 6 minutes
Max. Operating Range: 300 m
Controller Frequency: 2.4 GHz
Battery Capacity: 380 mAh
Battery Type: Lithium Polymer
Battery Power: 7.4V
GPS Compatible: –
Altitude Hold: –
Position Hold: –
Flight Planning: –
Autonomous Flight: –
Target Tracking: –
Automatic Landing: –
Return To Home: –
Obstacle Avoidance: –
Headless Mode: –
Built-in Camera: –

Parrot

Bebop

© Parrot

Release Date: November 2014
Price: $500
Number of Rotors: 4
Propellers: PF070105
Control System: Remote Control, Phone/Tablet
Applications: Film, Photography
Dimensions: 280 × 320 × 36 mm
Wingspan: 248 mm
Weight: 420 gr
Max. Payload: 20 gr
Frame Material: Plastic
Rotor Material: Plastic
Motors: PF070079
Available Colors: Blue, Red, Yellow
Max. Flight Time: 22 minutes
Max. Speed: 47 km/h (29 mp/h)
Max. Flying Altitude: 150 m (492 ft)
Max. Ascent Speed: 6 m/s
Max. Operating Range: 250 m
Controller Frequency: 2.4 GHz
Battery Capacity: 1,200 mAh
Battery Type: Lithium Polymer 3S
GPS Compatible: √
Altitude Hold: √
Position Hold: √
Flight Planning: √
Autonomous Flight: √
Target Tracking: √
Automatic Landing: √
Return To Home: √
Obstacle Avoidance: –
Headless Mode: –
Built-in Camera: √ (14 megapixels)

Qods Aviation Industries

Mohajer 2N

© Tasnim News Agency

Release Date: November 2014
Price: Unknown
Number of Rotors: 1 (Fixed Wing)
Control System: Ground Control Station
Applications: Military Purposes, Inspection, Reconnaissance, Surveillance
Dimensions: 2.92 × 4 m
Wingspan: 4 m
Weight: 75 kg
Max. Payload: 15 kg
Frame Material: Composite
Motors: 1 × WAE-342 two-cylinder piston engine
Available Colors: Grey
Max. Flight Time: 6 hours
Max. Speed: 200 km/h (124 mp/h)
Max. Flying Altitude: 3350 m (10,990 ft)
Max. Operating Range: 150 km
GPS Compatible: √
Altitude Hold: √
Position Hold: –
Flight Planning: √
Autonomous Flight: √
Target Tracking: –
Automatic Landing: √
Return To Home: √
Obstacle Avoidance: –
Headless Mode: –
Built-in Camera: √

Revell Control

Hexatron

© Revell Control

Release Date: November 2014
Price: $155
Number of Rotors: 6
Control System: Remote Control
Applications: Recreation
Dimensions: 800 × 800 × 120 mm
Wingspan: 650 mm
Weight: 410 gr
Frame Material: Plastic
Rotor Material: Plastic
Available Colors: Black
Max. Flight Time: 7 minutes
Max. Operating Range: 50 m
Controller Frequency: 2.4 GHz
Battery Capacity: 1,500 mAh
Battery Type: Lithium Polymer
Battery Power: 7.4V
GPS Compatible: –
Altitude Hold: –
Position Hold: –
Flight Planning: –
Autonomous Flight: –
Target Tracking: –
Automatic Landing: –
Return To Home: –
Obstacle Avoidance: –
Headless Mode: –
Built-in Camera: –

Sky Hero

Little Spyder

© Sky Hero

Release Date: November 2014
Price: $699
Number of Rotors: 4
Control System: Remote Control
Applications: Film, Photography
Frame Material: Plastic, Carbon Fiber
Rotor Material: Plastic
Motors: Little 2806-950kv
Available Colors: White
Controller Frequency: 2.4 GHz
Battery Capacity: 3,700 mAh
Battery Type: Lithium Polymer 4S
GPS Compatible: √
Altitude Hold: √
Position Hold: √
Flight Planning: √
Autonomous Flight: –
Target Tracking: √
Automatic Landing: –
Return To Home: √
Obstacle Avoidance: –
Headless Mode: –
Built-in Camera: –

Speedwolf

Vajra 80

© Speedwolf

Release Date: November 2014
Price: $599
Number of Rotors: 4
Control System: Remote Control
Applications: Film, Photography
Dimensions: 370 × 370 × 92 mm
Weight: 1.1 kg
Max. Payload: 1.2 kg
Frame Material: Carbon Fiber
Rotor Material: Carbon Fiber
Motors: 2216-910kv
Available Colors: Black
Max. Flight Time: 25 minutes
Max. Speed: 126 km/h (78 mp/h)
Min. Operating Temperature: -10°C
Max. Operating Temperature: 50°C
Max. Operating Range: 800 m
Controller Frequency: 2.4 GHz
Battery Capacity: 10,000 mAh
Battery Type: Lithium Polymer
Battery Power: 11.1V
Battery Weight: 320 gr
GPS Compatible: √
Altitude Hold: √
Position Hold: √
Flight Planning: √
Autonomous Flight: √
Target Tracking: √
Automatic Landing: √
Return To Home: √
Obstacle Avoidance: –
Headless Mode: –
Built-in Camera: –

Special Technological Centre

Orlan-10

© military.ru

Release Date: November 2014
Price: $120,000
Number of Rotors: 1 (Fixed Wing)
Control System: Ground Control Station
Applications: Military Purposes, Reconnaissance, Surveillance
Weight: 9 kg
Max. Payload: 6 kg
Frame Material: Composite
Motors: 1 × A-95 gasoline engine
Available Colors: Grey
Max. Flight Time: 16 hours
Max. Speed: 150 km/h (93,2 mp/h)
Max. Flying Altitude: 5,000 m (16,404 ft)
Min. Operating Temperature: -30°C
Max. Operating Temperature: 40°C
Max. Operating Range: 150 km
GPS Compatible: √
Altitude Hold: √
Position Hold: √
Flight Planning: √
Autonomous Flight: √
Target Tracking: –
Automatic Landing: –
Return To Home: √
Obstacle Avoidance: √
Headless Mode: –
Built-in Camera: √

Syma Toys

X11 Hornet

©prisguide.no

Release Date: November 2014
Price: $30
Number of Rotors: 4
Control System: Remote Control
Applications: Recreation
Dimensions: 152 × 152 × 37 mm
Wingspan: 600 mm
Weight: 215 gr
Frame Material: Plastic, Foam
Rotor Material: Plastic
Available Colors: Black, Red, White
Max. Flight Time: 8 minutes
Max. Operating Range: 40 m
Controller Frequency: 2.4 GHz
Battery Capacity: 200 mAh
Battery Type: Lithium Polymer
Battery Power: 3.7V
GPS Compatible: –
Altitude Hold: –
Position Hold: –
Flight Planning: –
Autonomous Flight: –
Target Tracking: –
Automatic Landing: –
Return To Home: –
Obstacle Avoidance: –
Headless Mode: –
Built-in Camera: –

Syma Toys

X12S Nano

©prisguide.no

Release Date: November 2014
Price: $25
Number of Rotors: 4
Control System: Remote Control
Applications: Recreation
Dimensions: 77 × 77 × 27 mm
Wingspan: 109 mm
Weight: 320 gr
Frame Material: Plastic, Foam
Rotor Material: Plastic
Available Colors: Blue, Green, Red, White
Max. Flight Time: 8 minutes
Max. Operating Range: 30 m
Controller Frequency: 2.4 GHz
Battery Capacity: 100 mAh
Battery Type: Lithium Polymer
Battery Power: 3.7V
GPS Compatible: –
Altitude Hold: –
Position Hold: –
Flight Planning: –
Autonomous Flight: –
Target Tracking: –
Automatic Landing: –
Return To Home: –
Obstacle Avoidance: –
Headless Mode: –
Built-in Camera: –

Top Flight Technologies

Airborg H6 1500

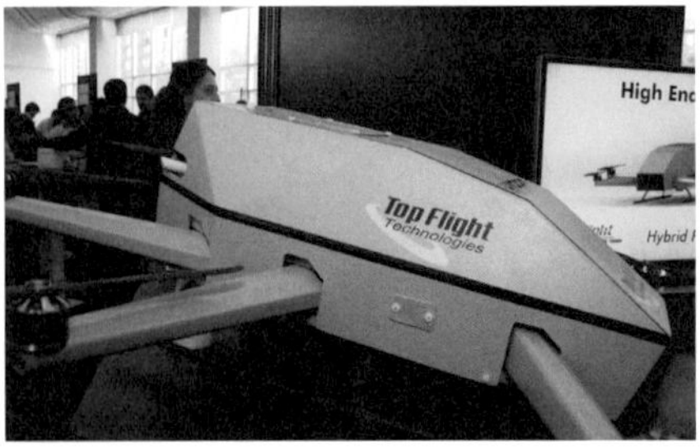

©Top Flight Technologies

Release Date: November 2014
Price: Unknown
Number of Rotors: 4
Control System: Remote Control
Applications: Reconnaissance, Surveillance
Frame Material: Aluminum
Rotor Material: Plastic
Available Colors: Grey
Max. Flight Time: 2 hours
Max. Speed: 64 km/h (40 mp/h)
Controller Frequency: 2.4 GHz
Battery Capacity: 16,000 mAh
GPS Compatible: √
Altitude Hold: √
Position Hold: √
Flight Planning: √
Autonomous Flight: √
Target Tracking: –
Automatic Landing: √
Return To Home: √
Obstacle Avoidance: –
Headless Mode: –
Built-in Camera: √

TRNDlabs

SKEYE Nano

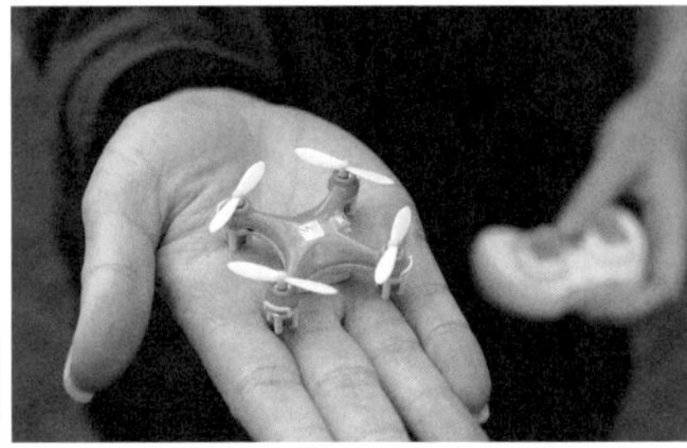

©TRNDlabs

Release Date: November 2014
Price: $40
Number of Rotors: 4
Control System: Remote Control
Applications: Recreation
Dimensions: 40 × 40 × 22 mm
Weight: 12 gr
Frame Material: Plastic
Rotor Material: Plastic
Available Colors: Orange
Max. Flight Time: 8 minutes
Max. Operating Range: 50 m
Controller Frequency: 2.4 GHz
Battery Capacity: 100 mAh
Battery Type: Lithium Polymer
Battery Power: 3.7V
GPS Compatible: –
Altitude Hold: –
Position Hold: –
Flight Planning: –
Autonomous Flight: –
Target Tracking: –
Automatic Landing: –
Return To Home: –
Obstacle Avoidance: –
Headless Mode: –
Built-in Camera: –

Thunder Tiger Robotix

Ghost+

© TTRobotix

Release Date: November 2014
Price: $1,199
Number of Rotors: 4
Control System: Remote Control
Applications: Film, Photography
Wingspan: 450 mm
Weight: 2.25 kg
Max. Payload: 810 gr
Available Colors: Black
Max. Flight Time: 25 minutes
Max. Operating Range: 1 km
Controller Frequency: 2.4 GHz
Battery Capacity: 6,000 mAh
Battery Type: Lithium Polymer 6S
Battery Power: 22.2V
GPS Compatible: √
Altitude Hold: √
Position Hold: √
Flight Planning: √
Autonomous Flight: √
Target Tracking: √
Automatic Landing: √
Return To Home: √
Obstacle Avoidance: √
Headless Mode: –
Built-in Camera: –

WLtoys

V646 mini Ufo

©WLtoys

Release Date: November 2014
Price: $18
Number of Rotors: 4
Control System: Remote Control
Applications: Recreation
Dimensions: 83 × 83 × 21 mm
Weight: 13 gr
Frame Material: Plastic
Rotor Material: Plastic
Available Colors: Black, Green
Max. Flight Time: 5 minutes
Max. Operating Range: 30 m
Controller Frequency: 2.4 GHz
Battery Capacity: 150 mAh
Battery Type: Lithium Polymer
Battery Power: 3.7V
GPS Compatible: –
Altitude Hold: –
Position Hold: –
Flight Planning: –
Autonomous Flight: –
Target Tracking: –
Automatic Landing: –
Return To Home: –
Obstacle Avoidance: –
Headless Mode: –
Built-in Camera: –

WLtoys

V676 mini Ufo

©aliexpress.com

Release Date: November 2014
Price: $17
Number of Rotors: 4
Control System: Remote Control
Applications: Recreation
Dimensions: 57 × 57 × 21 mm
Weight: 13 gr
Frame Material: Plastic
Rotor Material: Plastic
Available Colors: Red
Max. Flight Time: 5 minutes
Max. Operating Range: 30 m
Controller Frequency: 2.4 GHz
Battery Capacity: 150 mAh
Battery Type: Lithium Polymer
Battery Power: 3.7V
GPS Compatible: –
Altitude Hold: –
Position Hold: –
Flight Planning: –
Autonomous Flight: –
Target Tracking: –
Automatic Landing: –
Return To Home: –
Obstacle Avoidance: –
Headless Mode: –
Built-in Camera: –

WLtoys

V383 500 3D

©dronevisionary.com

Release Date: November 2014
Price: $320
Number of Rotors: 4
Control System: Remote Control
Applications: FPV Racing
Dimensions: 513 × 429 × 140 mm
Weight: 900 gr
Frame Material: Plastic, Carbon Fiber
Rotor Material: Plastic, Carbon Fiber
Available Colors: Red
Max. Flight Time: 8 minutes
Max. Operating Range: 500 m
Controller Frequency: 2.4 GHz
Battery Capacity: 2,200 mAh
Battery Type: Lithium Polymer
Battery Power: 14.8V
GPS Compatible: –
Altitude Hold: √
Position Hold: √
Flight Planning: –
Autonomous Flight: –
Target Tracking: –
Automatic Landing: –
Return To Home: √
Obstacle Avoidance: –
Headless Mode: –
Built-in Camera: –

Adorama

Aries Blackbird X10

Release Date: December 2014
Price: $799
Number of Rotors: 4
Control System: Remote Control, Phone/Tablet
Applications: Film, Photography
Dimensions: 584 × 584 × 140 mm
Wingspan: 450 mm
Weight: 1.4 kg
Frame Material: Plastic
Rotor Material: Carbon Fiber
Available Colors: Black
Max. Flight Time: 25 minutes
Max. Speed: 72 km/h (45 mp/h)
Max. Flying Altitude: 500 m (1,640 ft)
Max. Ascent Speed: 8 m/s
Min. Operating Temperature: 0°C
Max. Operating Temperature: 50°C
Max. Operating Range: 500 m
Controller Frequency: 2.4 GHz
Battery Capacity: 5,300 mAh
Battery Type: Lithium Polymer
Battery Power: 11V
GPS Compatible: √
Altitude Hold: √
Position Hold: √
Flight Planning: –
Autonomous Flight: –
Target Tracking: –
Automatic Landing: √
Return To Home: √
Obstacle Avoidance: –
Headless Mode: –
Built-in Camera: √ (16 megapixels)

Aeronavics

Skyjib X8

Release Date: December 2014
Price: $17.950
Number of Rotors: 8
Control System: Remote Control
Applications: Film, Photography,
Aerial Mapping, Agriculture, Surveillance
Wingspan: 1.17 m
Weight: 16 kg
Max. Payload: 5 kg
Frame Material: Carbon Fiber
Rotor Material: Carbon Fiber
Available Colors: Black
Max. Flight Time: 15 minutes
Max. Speed: 70 km/h (43,5 mp/h)
Max. Operating Range: 3 km
Controller Frequency: 2.4 GHz
Battery Capacity: 11,000 mAh
Battery Type: Lithium Polymer 6S
GPS Compatible: √
Altitude Hold: √
Position Hold: √
Flight Planning: √
Autonomous Flight: √
Target Tracking: –
Automatic Landing: √
Return To Home: √
Obstacle Avoidance: –
Headless Mode: –
Built-in Camera: –

Art-Tech

AT-100

© Art-Tech

Release Date: December 2014
Price: Unknown
Number of Rotors: 8
Control System: Flight Computer
Applications: Military Purposes, Reconnaissance, Surveillance
Wingspan: 1.2 m
Frame Material: Carbon Fiber, Plastic
Rotor Material: Carbon Fiber
Available Colors: Blue
GPS Compatible: √
Altitude Hold: √
Position Hold: √
Flight Planning: √
Autonomous Flight: √
Target Tracking: –
Automatic Landing: √
Return To Home: √
Obstacle Avoidance: –
Headless Mode: –
Built-in Camera: –

Assault

Reaper 500 3D

© Rcgroups

Release Date: December 2014
Price: $499
Number of Rotors: 4
Control System: Remote Control, Phone/Tablet
Applications: FPV Racing
Dimensions: 635 × 365 mm
Wingspan: 732 mm
Weight: 1.24 kg
Frame Material: Carbon Fiber
Rotor Material: Plastic
Motors: 500 class brushless outrunner
Available Colors: Green
Controller Frequency: 2.4 GHz
Battery Capacity: 1,550 mAh
Battery Type: Lithium Polymer 6S
Battery Power: 22.2V
Battery Weight: 250 gr
GPS Compatible: √
Altitude Hold: √
Position Hold: √
Flight Planning: √
Autonomous Flight: √
Target Tracking: –
Automatic Landing: –
Return To Home: –
Obstacle Avoidance: –
Headless Mode: –
Built-in Camera: √

Delta Drone

Delta Y

©Delta Drone

Release Date: December 2014
Price: Unknown
Number of Rotors: 1 (Fixed Wing)
Control System: Flight Computer
Applications: Aerial Mapping, Agriculture
Weight: 2 kg
Frame Material: Carbon Fiber
Available Colors: Grey
Max. Flight Time: 45 minutes
Max. Speed: 50 km/h (31 mp/h)
Max. Flying Altitude: 150 m (492 ft)
GPS Compatible: √
Altitude Hold: √
Position Hold: √
Flight Planning: √
Autonomous Flight: √
Target Tracking: √
Automatic Landing: √
Return To Home: √
Obstacle Avoidance: –
Headless Mode: –
Built-in Camera: √

Estes Rockets

Proto-X FPV

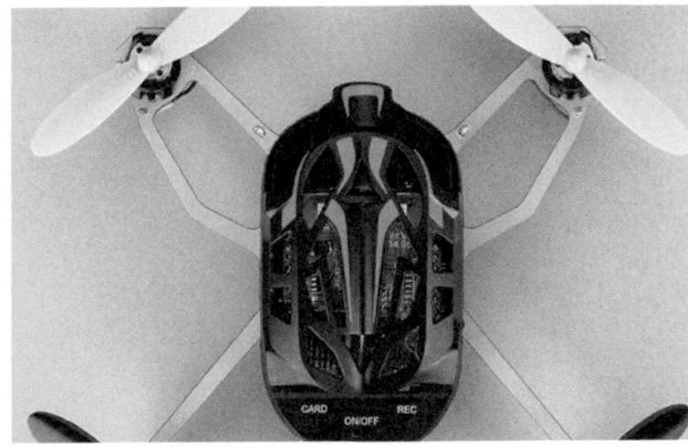

©Estes Rockets

Release Date: December 2014
Price: $55
Number of Rotors: 4
Control System: Remote Control
Applications: FPV Racing, Recreation
Wingspan: 113 mm
Weight: 58 gr
Frame Material: Plastic
Rotor Material: Plastic
Available Colors: Black
Max. Operating Range: 73 m
Controller Frequency: 2.4 GHz
Battery Capacity: 650 mAh
Battery Type: Lithium Polymer
GPS Compatible: –
Altitude Hold: √
Position Hold: √
Flight Planning: –
Autonomous Flight: –
Target Tracking: –
Automatic Landing: –
Return To Home: √
Obstacle Avoidance: –
Headless Mode: –
Built-in Camera: –

Estes Rockets

Proto-X SLT

©Estes Rockets

Release Date: December 2014
Price: $40
Number of Rotors: 4
Control System: Remote Control
Applications: Recreation
Dimensions: 45 × 45 × 23 mm
Wingspan: 50 mm
Weight: 12 gr
Frame Material: Plastic
Rotor Material: Plastic
Available Colors: Blue
Max. Flight Time: 3 minutes
Controller Frequency: 2.4 GHz
Battery Capacity: 100 mAh
Battery Type: Lithium Polymer
Battery Power: 3.7V
GPS Compatible: –
Altitude Hold: –
Position Hold: –
Flight Planning: –
Autonomous Flight: –
Target Tracking: –
Automatic Landing: –
Return To Home: –
Obstacle Avoidance: –
Headless Mode: –
Built-in Camera: –

Global Industrial and Defefense Solutions (GIDS)

Scout Mini

©armyrecognition.com

Release Date: December 2014
Price: $25,000
Number of Rotors: 1 (Fixed Wing)
Control System: Ground Control Station
Applications: Military Purposes, Inspection, Reconnaissance, Surveillance
Dimensions: 1.2 × 2 m
Wingspan: 2 m
Weight: 4.5 kg
Frame Material: Composite
Available Colors: Grey
Max. Flight Time: 60 minutes
Max. Flying Altitude: 457 m (1,500 ft)
Max. Operating Range: 10 km
Controller Frequency: 2.4 GHz
Battery Type: Lithium Polymer
GPS Compatible: √
Altitude Hold: √
Position Hold: –
Flight Planning: √
Autonomous Flight: √
Target Tracking: –
Automatic Landing: –
Return To Home: √
Obstacle Avoidance: –
Headless Mode: –
Built-in Camera: √

Immersion RC

XuGong V2

©Immersion RC

Release Date: December 2014
Price: $400
Number of Rotors: 4
Control System: Remote Control
Applications: FPV Racing
Dimensions: 240 × 300 × 140 mm
Weight: 1.5 kg
Frame Material: Carbon Fiber
Rotor Material: Plastic
Motors: 2212-920kv
Available Colors: Black
Max. Flight Time: 20 minutes
Controller Frequency: 2.4 GHz
Battery Capacity: 5,200 mAh
Battery Type: Lithium Polymer 4S
Battery Power: 11.1V
GPS Compatible: √
Altitude Hold: –
Position Hold: –
Flight Planning: √
Autonomous Flight: –
Target Tracking: –
Automatic Landing: –
Return To Home: √
Obstacle Avoidance: –
Headless Mode: –
Built-in Camera: √ (FatShark 600TVL CMOS)

JJRC

H9D

©JJRC

Release Date: December 2014
Price: $60
Number of Rotors: 4
Control System: Remote Control
Applications: Recreation
Dimensions: 315 × 315 × 95 mm
Weight: 450 gr
Frame Material: Plastic, Carbon Fiber
Rotor Material: Plastic, Carbon Fiber
Available Colors: Gold, Silver
Max. Flight Time: 8 minutes
Max. Operating Range: 150 m
Controller Frequency: 2.4 GHz
Battery Capacity: 350 mAh
Battery Type: Lithium Polymer
Battery Power: 3.7V
GPS Compatible: –
Altitude Hold: √
Position Hold: √
Flight Planning: –
Autonomous Flight: –
Target Tracking: –
Automatic Landing: –
Return To Home: √
Obstacle Avoidance: –
Headless Mode: √
Built-in Camera: √ (0.3 megapixels)

JJRC

H5C

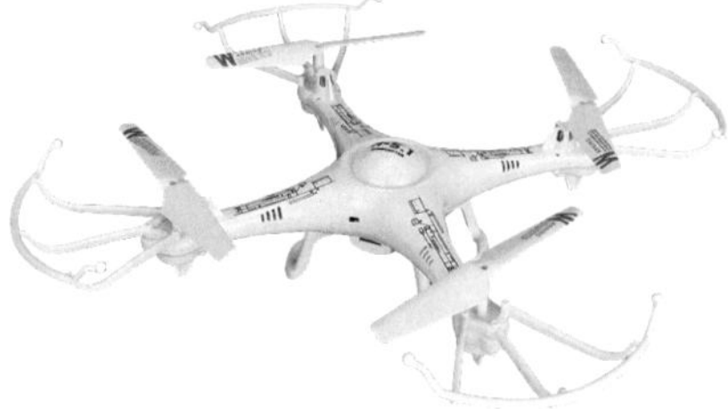
© JJRC

Release Date: December 2014
Price: $50
Number of Rotors: 4
Control System: Remote Control
Applications: Recreation
Dimensions: 310 × 310 × 80 mm
Wingspan: 438 mm
Weight: 100 gr
Frame Material: Plastic
Rotor Material: Plastic
Available Colors: White
Max. Flight Time: 8 minutes
Max. Operating Range: 100 m
Controller Frequency: 2.4 GHz
Battery Capacity: 1,100 mAh
Battery Type: Lithium Polymer
Battery Power: 3.7V
GPS Compatible: –
Altitude Hold: –
Position Hold: –
Flight Planning: –
Autonomous Flight: –
Target Tracking: –
Automatic Landing: –
Return To Home: √
Obstacle Avoidance: –
Headless Mode: √
Built-in Camera: √ (2 megapixels)

JJRC

H6C

© JJRC

Release Date: December 2014
Price: $40
Number of Rotors: 4
Control System: Remote Control
Applications: Recreation
Dimensions: 135 × 135 × 35 mm
Wingspan: 191 mm
Frame Material: Plastic
Rotor Material: Plastic
Available Colors: Black, Blue, Red
Max. Flight Time: 7 minutes
Max. Operating Range: 100 m
Controller Frequency: 2.4 GHz
Battery Capacity: 350 mAh
Battery Type: Lithium Polymer
Battery Power: 3.7V
GPS Compatible: –
Altitude Hold: –
Position Hold: –
Flight Planning: –
Autonomous Flight: –
Target Tracking: –
Automatic Landing: –
Return To Home: √
Obstacle Avoidance: –
Headless Mode: √
Built-in Camera: √ (2 megapixels)

JJRC

H7

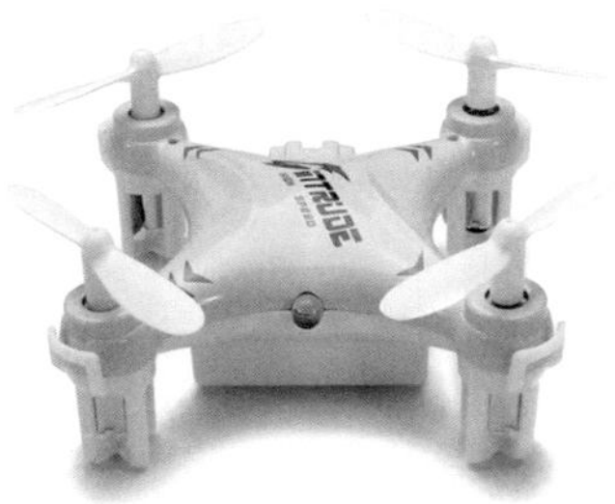

©JJRC

Release Date: December 2014
Price: $19
Number of Rotors: 4
Control System: Remote Control
Applications: Recreation
Dimensions: 45 × 45 × 12 mm
Wingspan: 64 mm
Frame Material: Plastic
Rotor Material: Plastic
Available Colors: Black, White
Max. Flight Time: 5 minutes
Max. Operating Range: 20 m
Controller Frequency: 2.4 GHz
Battery Capacity: 100 mAh
Battery Type: Lithium Polymer
Battery Power: 3.7V
GPS Compatible: –
Altitude Hold: –
Position Hold: –
Flight Planning: –
Autonomous Flight: –
Target Tracking: –
Automatic Landing: –
Return To Home: –
Obstacle Avoidance: –
Headless Mode: –
Built-in Camera: –

Meijiaxin Toys

X700C

©Meijiaxin Toys

Release Date: December 2014
Price: $80
Number of Rotors: 4
Control System: Remote Control
Applications: Recreation
Dimensions: 150 × 150 × 100 mm
Wingspan: 108 mm
Weight: 38 gr
Frame Material: Plastic
Rotor Material: Plastic
Available Colors: Orange, Red
Max. Flight Time: 6 minutes
Max. Operating Range: 100 m
Controller Frequency: 2.4 GHz
Battery Capacity: 300 mAh
Battery Type: Lithium Polymer
Battery Power: 3.7V
GPS Compatible: –
Altitude Hold: –
Position Hold: –
Flight Planning: –
Autonomous Flight: –
Target Tracking: –
Automatic Landing: –
Return To Home: –
Obstacle Avoidance: –
Headless Mode: –
Built-in Camera: √ (0.3 megapixels)

National Chung-Shan Institute of Science and Technology (NCSIST)

Cardinal II

©NCSIST

Release Date: December 2014
Price: Unknown
Number of Rotors: 1 (Fixed Wing)
Control System: Ground Control Station
Applications: Military Purposes, Inspection, Reconnaissance, Surveillance
Dimensions: 1.9 × 1.3 × 0.35 m
Wingspan: 1.3 m
Weight: 5.5 kg
Frame Material: Composite
Available Colors: Grey
Max. Flight Time: 60 minutes
Max. Speed: 55 km/h (34 mp/h)
GPS Compatible: √
Altitude Hold: √
Position Hold: –
Flight Planning: √
Autonomous Flight: √
Target Tracking: –
Automatic Landing: √
Return To Home: √
Obstacle Avoidance: –
Headless Mode: –
Built-in Camera: √

Piaggio Aerospace

P.1HH HammerHead

©Piaggio Aero

Release Date: December 2014
Price: $12 million
Number of Rotors: 2 (Fixed Wing)
Control System: Ground Control Station
Applications: Military Purposes, Inspection, Maritime Patrol, Reconnaissance, Surveillance
Dimensions: 14.4 × 15.6 × 3.98 m
Wingspan: 15.6 m
Weight: 6,146 kg
Frame Material: Composite
Motors: 2 × Pratt & Whitney Canada PT6A-66B 850 SHP
Available Colors: Grey
Max. Flight Time: 16 hours
Max. Speed: 731 km/h (454 mp/h)
Max. Flying Altitude: 13,716 m (45,000 ft)
Max. Operating Range: 8,149 km
GPS Compatible: √
Altitude Hold: √
Position Hold: –
Flight Planning: √
Autonomous Flight: √
Target Tracking: √
Automatic Landing: √
Return To Home: √
Obstacle Avoidance: √
Headless Mode: –
Built-in Camera: √

Sky-watch

Cumulus V1

© Sky-watch

Release Date: December 2014
Price: On Request
Number of Rotors: 1 (Fixed Wing)
Control System: Flight Computer, Tablet
Applications: Aerial Mapping, Agriculture, Inspection, Surveillance
Dimensions: 1.07 × 1.65 × 0.17 m
Wingspan: 1.65 m
Weight: 1.1 kg
Max. Payload: 500 gr
Frame Material: Composite
Available Colors: Grey
Max. Flight Time: 2.5 hours
Max. Speed: 108 km/h (67 mp/h)
Max. Flying Altitude: 3,500 m (11,483 ft)
Min. Operating Temperature: -20°C
Max. Operating Temperature: 45°C
Max. Operating Range: 10 km
Controller Frequency: 2.4 GHz
Battery Capacity: 8,700 mAh
Battery Type: Lithium Polymer
Battery Power: 14.8V
GPS Compatible: √
Altitude Hold: √
Position Hold: –
Flight Planning: √
Autonomous Flight: √
Target Tracking: –
Automatic Landing: √
Return To Home: √
Obstacle Avoidance: –
Headless Mode: –
Built-in Camera: √ (20.2 megapixels)

Xtim

Bionic Bird

© Bionic Bird

Release Date: December 2014
Price: $140
Number of Rotors: – (Fixed Wing)
Control System: Phone
Applications: Recreation
Dimensions: 365 × 210 × 75 mm
Wingspan: 330 mm
Weight: 910 gr
Frame Material: Plastic
Available Colors: Black
Max. Flight Time: 8 minutes
Max. Operating Range: 100 m
Controller Frequency: 2.4 GHz
Battery Capacity: 800 mAh
Battery Type: Lithium Polymer
GPS Compatible: √
Altitude Hold: √
Position Hold: √
Flight Planning: –
Autonomous Flight: –
Target Tracking: –
Automatic Landing: –
Return To Home: √
Obstacle Avoidance: –
Headless Mode: –
Built-in Camera: –

Yuneec

Typhoon Q500

© firstpersonview.co.uk

Release Date: December 2014
Price: $1,250
Number of Rotors: 4
Control System: Remote Control, Phone/Tablet
Applications: Film, Photography
Dimensions: 420 × 420 × 240 mm
Wingspan: 565 mm
Weight: 1.1 kg
Max. Payload: 600 gr
Frame Material: Aluminum
Rotor Material: Carbon Fiber
Available Colors: White
Max. Flight Time: 25 minutes
Max. Flying Altitude: 400 m (1,312 ft)
Max. Ascent Speed: 3 m/s
Max. Descent Speed: 2 m/s
Min. Operating Temperature: 0°C
Max. Operating Temperature: 50°C
Max. Operating Range: 400 m
Controller Frequency: 2.4 GHz
Battery Capacity: 5,400 mAh
Battery Type: Lithium Polymer 3S
Battery Power: 11V
Battery Weight: 40 gr
GPS Compatible: √
Altitude Hold: √
Position Hold: √
Flight Planning: √
Autonomous Flight: –
Target Tracking: √
Automatic Landing: √
Return To Home: √
Obstacle Avoidance: –
Headless Mode: –
Built-in Camera: √ (16 megapixels)

2015

Acme

Zoopa Q55 Zepto

© Acme

Release Date: January 2015
Price: $35
Number of Rotors: 4
Control System: Remote Control
Applications: Recreation
Dimensions: 50 × 50 × 25 mm
Wingspan: 55 mm
Weight: 8 gr
Frame Material: Plastic
Rotor Material: Plastic
Available Colors: Black
Max. Flight Time: 9 minutes
Max. Operating Range: 30 m
Controller Frequency: 2.4 GHz
Battery Capacity: 100 mAh
Battery Type: Lithium Polymer 1S
Battery Power: 3.7V
GPS Compatible: –
Altitude Hold: –
Position Hold: –
Flight Planning: –
Autonomous Flight: –
Target Tracking: –
Automatic Landing: –
Return To Home: –
Obstacle Avoidance: –
Headless Mode: –
Built-in Camera: √

AeroSky

C6

© thehobbyworx.com

Release Date: January 2015
Price: $599
Number of Rotors: 4
Propellers: 05H106-05
Control System: Remote Control
Applications: Film, Photography, FPV Racing
Dimensions: 600 × 600 × 350 mm
Weight: 1.15 kg
Max. Payload: 500 gr
Frame Material: Carbon Fiber
Rotor Material: Plastic
Motors: 920kv brushless
Available Colors: Black
Controller Frequency: 2.4 GHz
Battery Capacity: 2,200 mAh
Battery Type: Lithium Polymer 4S
Battery Power: 14.8V
GPS Compatible: √
Altitude Hold: √
Position Hold: √
Flight Planning: √
Autonomous Flight: √
Target Tracking: –
Automatic Landing: √
Return To Home: √
Obstacle Avoidance: –
Headless Mode: –
Built-in Camera: –

Allied Drones

EF44 Atlas

©Allied Drones

Release Date: January 2015
Price: $18,999
Number of Rotors: 4
Control System: Remote Control, Phone/Tablet
Applications: Film, Photography, Aerial Mapping, Agriculture, Surveillance
Wingspan: 1.2 m
Max. Payload: 2 kg
Frame Material: Carbon Fiber
Rotor Material: Carbon Fiber
Available Colors: Black
Max. Flight Time: 60 minutes
Controller Frequency: 2.4 GHz
GPS Compatible: √
Altitude Hold: –
Position Hold: –
Flight Planning: –
Autonomous Flight: √
Target Tracking: –
Automatic Landing: –
Return To Home: –
Obstacle Avoidance: –
Headless Mode: –
Built-in Camera: –

CATUAV

Mineos

©CATUAV

Release Date: January 2015
Price: Unknown
Number of Rotors: 1 (Fixed Wing)
Control System: Remote Control
Applications: Film, Photography, Aerial Mapping, Surveillance
Wingspan: 1.72 m
Weight: 2 kg
Max. Payload: 500 gr
Available Colors: White
Max. Flight Time: 60 minutes
Max. Speed: 110 km/h (68 mp/h)
Max. Operating Range: 15 km
GPS Compatible: √
Altitude Hold: √
Position Hold: √
Flight Planning: √
Autonomous Flight: √
Target Tracking: √
Automatic Landing: √
Return To Home: √
Obstacle Avoidance: –
Headless Mode: –
Built-in Camera: √

Flypro

PX400

© Flypro

Release Date: January 2015
Price: $2,799
Number of Rotors: 4
Control System: Remote Control, Phone/Tablet
Applications: Film, Photography, Surveillance
Dimensions: 610 × 610 × 280 mm
Wingspan: 860 mm
Weight: 2.45 kg
Max. Payload: 300 gr
Frame Material: Carbon Fiber
Rotor Material: Carbon Fiber
Available Colors: Red
Max. Flight Time: 25 minutes
Max. Speed: 72 km/h (44,7 mp/h)
Max. Flying Altitude: 1,000 m (3,280 ft)
Max. Ascent Speed: 5 m/s
Max. Descent Speed: 4 m/s
Min. Operating Temperature: -10°C
Max. Operating Temperature: 40°C
Max. Operating Range: 1 km
Controller Frequency: 2.4 GHz
Battery Capacity: 950 mAh
Battery Type: Lithium Polymer 6S
Battery Power: 22.2V
Battery Weight: 630 gr
GPS Compatible: √
Altitude Hold: √
Position Hold: √
Flight Planning: √
Autonomous Flight: √
Target Tracking: √
Automatic Landing: √
Return To Home: √
Obstacle Avoidance: √
Headless Mode: –
Built-in Camera: –

HoneyComb

AgDrone

© HoneyComb

Release Date: January 2015
Price: $15,000
Number of Rotors: 1 (Fixed Wing)
Control System: Remote Control, Tablet
Applications: Aerial Mapping, Agriculture
Wingspan: 124 mm
Frame Material: Kevlar Fiber
Available Colors: Black
Max. Flight Time: 55 minutes
Max. Speed: 82 km/h (51 mp/h)
Max. Operating Range: 4.5 km
Controller Frequency: 2.4 GHz
Battery Capacity: 8,000 mAh
Battery Type: Lithium Polymer
GPS Compatible: √
Altitude Hold: √
Position Hold: √
Flight Planning: √
Autonomous Flight: √
Target Tracking: –
Automatic Landing: –
Return To Home: √
Obstacle Avoidance: –
Headless Mode: –
Built-in Camera: √

HuaXiang

8983 Sky Walker

©HuaXiang

Release Date: January 2015
Price: $30
Number of Rotors: 4
Control System: Remote Control
Applications: Recreation
Dimensions: 146 × 146 × 35 mm
Weight: 486 gr
Frame Material: Plastic
Rotor Material: Plastic
Available Colors: Black
Max. Flight Time: 7 minutes
Max. Operating Range: 100 m
Controller Frequency: 2.4 GHz
Battery Capacity: 500 mAh
Battery Type: Lithium Polymer
Battery Power: 3.7V
GPS Compatible: –
Altitude Hold: √
Position Hold: √
Flight Planning: –
Autonomous Flight: –
Target Tracking: –
Automatic Landing: –
Return To Home: –
Obstacle Avoidance: –
Headless Mode: –
Built-in Camera: √

JJRC

JJ800

©JJRC

Release Date: January 2015
Price: $65
Number of Rotors: 4
Control System: Remote Control
Applications: Recreation
Dimensions: 135 × 135 × 52 mm
Wingspan: 191 mm
Weight: 52 gr
Frame Material: Plastic
Rotor Material: Plastic
Available Colors: Black, Silver
Max. Flight Time: 8 minutes
Max. Operating Range: 100 m
Controller Frequency: 2.4 GHz
Battery Capacity: 250 mAh
Battery Type: Lithium Polymer
Battery Power: 3.7V
GPS Compatible: –
Altitude Hold: –
Position Hold: –
Flight Planning: –
Autonomous Flight: –
Target Tracking: –
Automatic Landing: –
Return To Home: –
Obstacle Avoidance: –
Headless Mode: –
Built-in Camera: –

Revell Control

XS Nano Quad

© Revell Control

Release Date: January 2015
Price: $45
Number of Rotors: 4
Control System: Remote Control
Applications: Recreation
Dimensions: 45 × 45 × 23 mm
Weight: 12 gr
Frame Material: Plastic
Rotor Material: Plastic
Available Colors: Black, Silver
Max. Flight Time: 5 minutes
Max. Operating Range: 50 m
Controller Frequency: 2.4 GHz
Battery Capacity: 100 mAh
Battery Type: Lithium Polymer
Battery Power: 3.7V
GPS Compatible: –
Altitude Hold: –
Position Hold: –
Flight Planning: –
Autonomous Flight: –
Target Tracking: –
Automatic Landing: –
Return To Home: –
Obstacle Avoidance: –
Headless Mode: –
Built-in Camera: –

Storm Racing Drone

Storm Drone 4 mini

© SRD

Release Date: January 2015
Price: $299
Number of Rotors: 4
Control System: Remote Control
Applications: Film, Photography
Dimensions: 270 × 270 × 50 mm
Weight: 395 gr
Frame Material: Carbon Fiber
Rotor Material: Carbon Fiber
Available Colors: Black
Max. Flight Time: 19 minutes
Controller Frequency: 2.4 GHz
Battery Capacity: 2,200 mAh
Battery Type: Lithium Polymer
Battery Power: 11.1V
GPS Compatible: √
Altitude Hold: √
Position Hold: √
Flight Planning: √
Autonomous Flight: √
Target Tracking: √
Automatic Landing: √
Return To Home: √
Obstacle Avoidance: –
Headless Mode: –
Built-in Camera: –

Storm Racing Drone

Storm Drone 4 V3

©SRD

Release Date: January 2015
Price: $599
Number of Rotors: 4
Propellers: Gemfan 1038C
Control System: Remote Control
Applications: Film, Photography
Dimensions: 450 × 375 × 215 mm
Weight: 786 gr
Frame Material: Carbon Fiber
Rotor Material: Carbon Fiber
Motors: Storm M2216-800kv
Available Colors: Black
Max. Flight Time: 19 minutes
Max. Operating Range: 1.5 km
Controller Frequency: 2.4 GHz
Battery Capacity: 4,000 mAh
Battery Type: Lithium Polymer
Battery Power: 14.8V
GPS Compatible: √
Altitude Hold: √
Position Hold: √
Flight Planning: √
Autonomous Flight: √
Target Tracking: √
Automatic Landing: √
Return To Home: √
Obstacle Avoidance: –
Headless Mode: –
Built-in Camera: –

Turkish Aerospace Industries (TAI)

Anka-B

©Stéphane Pichard

Release Date: January 2015
Price: $30 million
Number of Rotors: 1 (Fixed Wing)
Control System: Ground Control Station
Applications: Military Purposes, Armed Reconnaissance, Surveillance, Target Acquisition
Dimensions: 8 × 17.3 × 3.4 m
Wingspan: 17.3 m
Weight: 1,600 kg
Frame Material: Composite
Motors: 1 × Thielert Centurion 2.0 turbocharged four-cylinder engine
Available Colors: Grey
Max. Flight Time: 26 hours
Max. Speed: 217 km/h (135 mp/h)
Max. Flying Altitude: 9,144 m (30,000 ft)
Max. Operating Range: 4,896 km
GPS Compatible: √
Altitude Hold: √
Position Hold: –
Flight Planning: √
Autonomous Flight: √
Target Tracking: √
Automatic Landing: √
Return To Home: √
Obstacle Avoidance: √
Headless Mode: –
Built-in Camera: √

UDIRC Toys

U840 mini

©UDIRC Toys

Release Date: January 2015
Price: $20
Number of Rotors: 4
Control System: Remote Control
Applications: Recreation
Dimensions: 80 × 80 × 21 mm
Wingspan: 113 mm
Weight: 13 gr
Frame Material: Plastic
Rotor Material: Plastic
Available Colors: Blue, Red, Yellow
Max. Flight Time: 5 minutes
Max. Operating Range: 30 m
Controller Frequency: 2.4 GHz
Battery Capacity: 100 mAh
Battery Type: Lithium Polymer
Battery Power: 3.7V
GPS Compatible: –
Altitude Hold: –
Position Hold: –
Flight Planning: –
Autonomous Flight: –
Target Tracking: –
Automatic Landing: –
Return To Home: –
Obstacle Avoidance: –
Headless Mode: –
Built-in Camera: –

Wltoys

V686G Explorer

©WLtoys

Release Date: January 2015
Price: $90
Number of Rotors: 4
Control System: Remote Control
Applications: Recreation
Dimensions: 320 × 320 × 93 mm
Wingspan: 453 mm
Weight: 80 gr
Frame Material: Plastic
Rotor Material: Plastic
Available Colors: White
Max. Flight Time: 10 minutes
Max. Operating Range: 150 m
Controller Frequency: 2.4 GHz
Battery Capacity: 730 mAh
Battery Type: Lithium Polymer
Battery Power: 7.3V
GPS Compatible: –
Altitude Hold: √
Position Hold: √
Flight Planning: –
Autonomous Flight: –
Target Tracking: –
Automatic Landing: –
Return To Home: √
Obstacle Avoidance: –
Headless Mode: –
Built-in Camera: √ (2 megapixels)

Acme

Zoopa Q155 Roonin

© Acme

Release Date: February 2015
Price: $60
Number of Rotors: 4
Control System: Remote Control
Applications: Recreation
Dimensions: 135 × 145 × 25 mm
Wingspan: 170 mm
Weight: 33 gr
Frame Material: Plastic
Rotor Material: Plastic
Available Colors: Black
Max. Flight Time: 9 minutes
Max. Operating Range: 60 m
Controller Frequency: 2.4 GHz
Battery Capacity: 350 mAh
Battery Type: Lithium Polymer 1S
Battery Power: 3.7V
GPS Compatible: –
Altitude Hold: –
Position Hold: –
Flight Planning: –
Autonomous Flight: –
Target Tracking: –
Automatic Landing: –
Return To Home: –
Obstacle Avoidance: –
Headless Mode: –
Built-in Camera: –

ADCOM Systems

Yabhon R

© armyrecognition.com

Release Date: February 2015
Price: Unknown
Number of Rotors: 1
Control System: Flight Computer
Applications: Military Purposes, Reconnaissance, Surveillance
Dimensions: 5 × 6.5 × 2 m
Wingspan: 6.5 m
Weight: 270 kg
Max. Payload: 280 kg
Motors: 100-hp jet turbine engine
Available Colors: Grey
Max. Flight Time: 27 hours
Max. Speed: 240 km/h (149 mp/h)
Max. Flying Altitude: 6,700 m (21,981 ft)
GPS Compatible: √
Altitude Hold: √
Position Hold: √
Flight Planning: √
Autonomous Flight: √
Target Tracking: √
Automatic Landing: √
Return To Home: √
Obstacle Avoidance: √
Headless Mode: –
Built-in Camera: √

Aerialtronics

Altura Zenith ATX8

© Acecore Technologies

Release Date: February 2015
Price: $2,000
Number of Rotors: 8
Propellers: AceCore custom
Control System: Remote Control, Tablet
Applications: Film, Photography, Agriculture, Inspection, 3D Mapping & Scanning, Surveillance
Dimensions: 600 × 600 × 570 mm
Wingspan: 600 mm
Weight: 3.75 kg
Max. Payload: 2.9 kg
Frame Material: Carbon Fiber
Rotor Material: Carbon Fiber
Available Colors: Black
Max. Flight Time: 35 minutes
Max. Speed: 72 km/h (44,7 mp/h)
Min. Operating Temperature: -10°C
Max. Operating Temperature: 50°C
Max. Operating Range: 2 km
Controller Frequency: 2.4 GHz
Battery Capacity: 20,000 mAh
Battery Type: Lithium Polymer 6S
Battery Power: 22.2V
GPS Compatible: √
Altitude Hold: √
Position Hold: √
Flight Planning: √
Autonomous Flight: √
Target Tracking: √
Automatic Landing: √
Return To Home: √
Obstacle Avoidance: –
Headless Mode: –
Built-in Camera: –

Airborne Robotics

Air 6

© Airborne Robotics

Release Date: February 2015
Price: $15,500
Number of Rotors: 6
Control System: Remote Control
Applications: Film, Photography, Aerial Mapping, Surveillance
Dimensions: 775 × 775 × 425 mm
Wingspan: 775 mm
Weight: 5 kg
Max. Payload: 1 kg
Frame Material: Aluminum
Rotor Material: Carbon Fiber
Available Colors: Black
Max. Flight Time: 18 minutes
Max. Speed: 50 km/h (31 mp/h)
Max. Operating Range: 2 km
Controller Frequency: 2.4 GHz
Battery Capacity: 7,400 mAh
Battery Type: Lithium Polymer 6S
GPS Compatible: √
Altitude Hold: √
Position Hold: √
Flight Planning: √
Autonomous Flight: √
Target Tracking: –
Automatic Landing: √
Return To Home: √
Obstacle Avoidance: –
Headless Mode: –
Built-in Camera: –

Airborne Robotics

Air 8

© Airborne Robotics

Release Date: February 2015
Price: $15,500
Number of Rotors: 8
Control System: Remote Control
Applications: Film, Photography, Aerial Mapping, Surveillance
Dimensions: 1080 × 1080 × 599 mm
Wingspan: 1080 mm
Weight: 17 kg
Max. Payload: 7 kg
Frame Material: Aluminum
Rotor Material: Carbon Fiber
Available Colors: Black
Max. Flight Time: 15 minutes
Max. Speed: 50 km/h (31 mp/h)
Max. Operating Range: 2 km
Controller Frequency: 2.4 GHz
Battery Capacity: 20,000 mAh
Battery Type: Lithium Polymer 6S
GPS Compatible: √
Altitude Hold: √
Position Hold: √
Flight Planning: √
Autonomous Flight: √
Target Tracking: –
Automatic Landing: –
Return To Home: √
Obstacle Avoidance: –
Headless Mode: –
Built-in Camera: –

Arcturus UAV

Jump 20

©Arcturus

Release Date: February 2015
Price: Unknown
Number of Rotors: 5 (Fixed Wing)
Control System: Ground Control Station
Applications: Aerial Mapping, Inspection
Dimensions: 2.9 × 5.66 m
Wingspan: 5.66 m
Weight: 79 kg
Frame Material: Composite
Motors: 190cc 4 Stroke
Available Colors: Grey
Max. Flight Time: 16 hours
Max. Speed: 133 km/h (82,6 mp/h)
Max. Flying Altitude: 4,572 m (15,000 ft)
GPS Compatible: √
Altitude Hold: √
Position Hold: √
Flight Planning: √
Autonomous Flight: √
Target Tracking: √
Automatic Landing: √
Return To Home: √
Obstacle Avoidance: –
Headless Mode: –
Built-in Camera: √

Denel Dynamics

Seeker 200

© Denel Dynamics

Release Date: February 2015
Price: Unknown
Number of Rotors: 1 (Fixed Wing)
Control System: Remote Control
Applications: Recreation
Wingspan: 6 m
Weight: 150 kg
Max. Payload: 40 kg
Frame Material: Aluminum, Carbon Fiber
Motors: Piston engine
Available Colors: Grey
Max. Flight Time: 10 hours
Max. Speed: 150 km/h (93,2 mp/h)
Max. Flying Altitude: 5,500 m (18,045 ft)
Max. Operating Range: 250 km
GPS Compatible: √
Altitude Hold: √
Position Hold: √
Flight Planning: √
Autonomous Flight: √
Target Tracking: √
Automatic Landing: √
Return To Home: √
Obstacle Avoidance: –
Headless Mode: –
Built-in Camera: √

Denel Dynamics

Seeker 400

© Denel Dynamics

Release Date: February 2015
Price: Unknown
Number of Rotors: 1 (Fixed Wing)
Control System: Flight Computer
Applications: Reconnaissance, Surveillance
Dimensions: 5.77 × 10 × 1.85 m
Wingspan: 10 m
Weight: 450 kg
Max. Payload: 100 kg
Frame Material: Aluminum, Carbon Fiber
Available Colors: Grey
Max. Flight Time: 16 hours
Max. Speed: 222 km/h (138 mp/h)
Max. Flying Altitude: 5,500 m (18,045 ft)
Max. Operating Range: 700 km
GPS Compatible: √
Altitude Hold: √
Position Hold: √
Flight Planning: √
Autonomous Flight: √
Target Tracking: √
Automatic Landing: √
Return To Home: √
Obstacle Avoidance: –
Headless Mode: –
Built-in Camera: √

Helic Max

Sky Hawkeye

Release Date: February 2015
Price: $80
Number of Rotors: 4
Control System: Remote Control, Phone/Tablet
Applications: Recreation
Dimensions: 400 × 400 × 100 mm
Frame Material: Plastic
Rotor Material: Plastic
Available Colors: White
Max. Flight Time: 5 minutes
Max. Operating Range: 50 m
Controller Frequency: 2.4 GHz
Battery Capacity: 650 mAh
Battery Type: Lithium Polymer
Battery Power: 3.7V
GPS Compatible: –
Altitude Hold: √
Position Hold: √
Flight Planning: –
Autonomous Flight: –
Target Tracking: –
Automatic Landing: –
Return To Home: –
Obstacle Avoidance: –
Headless Mode: –
Built-in Camera: √

JJRC

JJ810

© aliexpress.com

Release Date: February 2015
Price: $19
Number of Rotors: 4
Control System: Remote Control
Applications: Recreation
Dimensions: 75 × 75 × 25 mm
Wingspan: 106 mm
Frame Material: Plastic
Rotor Material: Plastic
Available Colors: Blue, Orange, White
Max. Flight Time: 5 minutes
Max. Operating Range: 30 m
Controller Frequency: 2.4 GHz
Battery Capacity: 100 mAh
Battery Type: Lithium Polymer
Battery Power: 3.7V
GPS Compatible: –
Altitude Hold: –
Position Hold: –
Flight Planning: –
Autonomous Flight: –
Target Tracking: –
Automatic Landing: –
Return To Home: –
Obstacle Avoidance: –
Headless Mode: –
Built-in Camera: –

JJRC

V686

©banggood.com

Release Date: February 2015
Price: $44
Number of Rotors: 4
Control System: Remote Control
Applications: Film, Recreation
Dimensions: 360 × 360 × 85 mm
Wingspan: 510 mm
Weight: 470 gr
Frame Material: Carbon Fiber
Rotor Material: Carbon Fiber
Available Colors: Black
Max. Flight Time: 8 minutes
Max. Flying Altitude: 150 m (492 ft)
Max. Operating Range: 150 m
Controller Frequency: 2.4 GHz
Battery Capacity: 730 mAh
Battery Type: Lithium Polymer
Battery Power: 3.7V
Battery Weight: 18 gr
GPS Compatible: –
Altitude Hold: √
Position Hold: √
Flight Planning: –
Autonomous Flight: –
Target Tracking: –
Automatic Landing: –
Return To Home: √
Obstacle Avoidance: –
Headless Mode: –
Built-in Camera: √ (2 megapixels)

Lishitoys

L6052

©aliexpress.com

Release Date: February 2015
Price: $45
Number of Rotors: 4
Control System: Remote Control
Applications: Recreation
Dimensions: 305 × 305 × 60 mm
Frame Material: Plastic
Rotor Material: Plastic
Available Colors: Blue, Red, Yellow
Max. Flight Time: 8 minutes
Max. Operating Range: 150 m
Controller Frequency: 2.4 GHz
Battery Capacity: 500 mAh
Battery Type: Lithium Polymer
Battery Power: 3.7V
GPS Compatible: –
Altitude Hold: –
Position Hold: –
Flight Planning: –
Autonomous Flight: –
Target Tracking: –
Automatic Landing: –
Return To Home: –
Obstacle Avoidance: –
Headless Mode: –
Built-in Camera: √

Nihui Model Technology

U807

Release Date: February 2015
Price: Unknown
Number of Rotors: 4
Control System: Remote Control
Applications: Recreation
Dimensions: 330 × 330 × 50 mm
Weight: 1.15 kg
Frame Material: Plastic
Rotor Material: Plastic
Available Colors: Black, White
Max. Flight Time: 12 minutes
Max. Operating Range: 150 m
Controller Frequency: 2.4 GHz
Battery Capacity: 650 mAh
Battery Type: Lithium Polymer
Battery Power: 3.7V
GPS Compatible: –
Altitude Hold: –
Position Hold: –
Flight Planning: –
Autonomous Flight: –
Target Tracking: –
Automatic Landing: –
Return To Home: √
Obstacle Avoidance: –
Headless Mode: √
Built-in Camera: –

QuadH2o

HexH2O

© quadh2o.com

Release Date: February 2015
Price: $3,795
Number of Rotors: 6
Control System: Remote Control
Applications: Film, Photography, Aerial Mapping
Dimensions: 740 × 650 × 240 mm
Weight: 1.4 kg
Max. Payload: 2 kg
Frame Material: Plastic, Carbon Fiber
Rotor Material: Carbon Fiber
Motors: 415KV
Available Colors: Yellow
Max. Flight Time: 25 minutes
Max. Speed: 56 km/h (35 mp/h)
Max. Operating Range: 1 km
Controller Frequency: 2.4 GHz
Battery Capacity: 7,000 mAh
Battery Type: Lithium Polymer 6S
Battery Power: 22.2V
Battery Weight: 900 gr
GPS Compatible: √
Altitude Hold: √
Position Hold: √
Flight Planning: √
Autonomous Flight: –
Target Tracking: √
Automatic Landing: √
Return To Home: √
Obstacle Avoidance: –
Headless Mode: –
Built-in Camera: –

Service-drone.com

G4 Eagle V2 Cargo

© service-drone.com

Release Date: February 2015
Price: $35,000
Number of Rotors: 8
Control System: Remote Control
Applications: Film, Photography, Aerial Mapping, Surveillance
Dimensions: 889 × 889 × 483 mm
Wingspan: 889 mm
Weight: 2.5 kg
Max. Payload: 2.5 kg
Frame Material: Carbon Fiber
Rotor Material: Carbon Fiber
Available Colors: Black, Yellow
Max. Flight Time: 30 minutes
Max. Speed: 59,7 km/h (37 mp/h)
Min. Operating Temperature: -10°C
Max. Operating Temperature: 40°C
Max. Operating Range: 2 km
Controller Frequency: 2.4 GHz
Battery Capacity: 2,200 mAh
Battery Type: Lithium Polymer
GPS Compatible: √
Altitude Hold: √
Position Hold: √
Flight Planning: √
Autonomous Flight: √
Target Tracking: –
Automatic Landing: –
Return To Home: √
Obstacle Avoidance: –
Headless Mode: –
Built-in Camera: –

Service-drone.com

G4 Blackbird SR

© service-drone.com

Release Date: February 2015
Price: $8,400
Number of Rotors: 4
Control System: Remote Control
Applications: Film, Photography, Aerial Mapping, Surveillance
Dimensions: 450 × 450 × 220 mm
Wingspan: 636 mm
Weight: 500 gr
Max. Payload: 1.15 kg
Frame Material: Carbon Fiber
Rotor Material: Carbon Fiber
Available Colors: Black, Yellow
Max. Flight Time: 20 minutes
Max. Speed: 49,7 km/h (30,8 mp/h)
Min. Operating Temperature: -10°C
Max. Operating Temperature: 40°C
Max. Operating Range: 1 km
Controller Frequency: 2.4 GHz
Battery Capacity: 5,200 mAh
Battery Type: Lithium Polymer
GPS Compatible: √
Altitude Hold: √
Position Hold: √
Flight Planning: √
Autonomous Flight: √
Target Tracking: –
Automatic Landing: √
Return To Home: √
Obstacle Avoidance: –
Headless Mode: –
Built-in Camera: –

Service-drone.com

G4 Skycrane

©service-drone.com

Release Date: February 2015
Price: $18,000
Number of Rotors: 8
Control System: Remote Control
Applications: Film, Photography, Aerial Mapping, Surveillance, Transport
Wingspan: 1.14 m
Weight: 3 kg
Max. Payload: 6.5 kg
Frame Material: Carbon Fiber
Rotor Material: Carbon Fiber
Available Colors: Black
Max. Flight Time: 12 minutes
Max. Speed: 50 km/h (31 mp/h)
Min. Operating Temperature: -10°C
Max. Operating Temperature: 40°C
Max. Operating Range: 2 km
Controller Frequency: 2.4 GHz
Battery Capacity: 2,200 mAh
Battery Type: Lithium Polymer
GPS Compatible: √
Altitude Hold: √
Position Hold: √
Flight Planning: √
Autonomous Flight: √
Target Tracking: –
Automatic Landing: √
Return To Home: √
Obstacle Avoidance: –
Headless Mode: –
Built-in Camera: –

Tarot RC

200 FPV Mini

©Tarot

Release Date: February 2015
Price: $175
Number of Rotors: 4
Control System: Remote Control
Applications: FPV Racing
Dimensions: 215 × 200 × 30 mm
Weight: 175 gr
Frame Material: Carbon Fiber
Rotor Material: Carbon Fiber
Motors: 1806-2280kv brushless
Available Colors: Black
Controller Frequency: 5.8 GHz
Battery Capacity: 1,300 mAh
Battery Type: Lithium Polymer
Battery Power: 11.1V
GPS Compatible: –
Altitude Hold: –
Position Hold: –
Flight Planning: –
Autonomous Flight: –
Target Tracking: –
Automatic Landing: –
Return To Home: –
Obstacle Avoidance: –
Headless Mode: –
Built-in Camera: √ (300TL)

Xiro

Xplorer

© minidronereview.com

Release Date: February 2015
Price: $400
Number of Rotors: 4
Control System: Remote Control, Phone/Tablet
Applications: Film, Photography, Aerial Mapping, Surveillance
Dimensions: 375 × 353 × 162 mm
Wingspan: 350 mm
Weight: 1.22 kg
Frame Material: Plastic
Rotor Material: Plastic
Available Colors: Black
Max. Flight Time: 25 minutes
Max. Speed: 28.8 km/h (17,8 mp/h)
Max. Flying Altitude: 120 m (394 ft)
Max. Ascent Speed: 3 m/s
Max. Descent Speed: 3 m/s
Max. Operating Range: 500 m
Controller Frequency: 2.4 GHz
Battery Capacity: 5,200 mAh
Battery Type: Lithium Polymer 3S
Battery Power: 14.8V
GPS Compatible: √
Altitude Hold: √
Position Hold: √
Flight Planning: √
Autonomous Flight: √
Target Tracking: √
Automatic Landing: √
Return To Home: √
Obstacle Avoidance: –
Headless Mode: –
Built-in Camera: –

ACSL

Mini Surveyor MS-06LA

© ACSL

Release Date: March 2015
Price: On Request
Number of Rotors: 6
Control System: Remote Control
Applications: Film, Photography, Aerial Mapping, Agriculture, Surveillance
Wingspan: 680 mm
Weight: 3 kg
Max. Payload: 6 kg
Frame Material: Plastic
Rotor Material: Plastic
Motors: Brushless 500W DC
Available Colors: Black
Max. Flight Time: 30 minutes
Max. Speed: 36 km/h (22,3 mp/h)
Max. Operating Range: 1 km
Controller Frequency: 2.4 GHz
Battery Capacity: 12,000 mAh
Battery Type: Lithium Polymer 6S
GPS Compatible: √
Altitude Hold: √
Position Hold: √
Flight Planning: √
Autonomous Flight: √
Target Tracking: √
Automatic Landing: √
Return To Home: √
Obstacle Avoidance: –
Headless Mode: –
Built-in Camera: –

Aerial Technology Int.

AgBOT

© Aerial Technology International

Release Date: March 2015
Price: $9,750
Number of Rotors: 4
Propellers: 15 inch Tri-prop
Control System: Remote Control, Phone/Tablet
Applications: Film, Photography, Aerial Mapping, Agriculture
Wingspan: 1.02 m
Weight: 4.7 kg
Frame Material: Carbon Fiber
Rotor Material: Carbon Fiber
Motors: 4012-400 kv
Available Colors: Black, Orange
Max. Flight Time: 25 minutes
Max. Speed: 61,7 km/h (38,4 mp/h)
Min. Operating Temperature: 0°C
Max. Operating Temperature: 50°C
Max. Operating Range: 26.5 km
Controller Frequency: 2.4 GHz
Battery Capacity: 6,500 mAh
Battery Type: Lithium Polymer 6S
GPS Compatible: √
Altitude Hold: √
Position Hold: √
Flight Planning: √
Autonomous Flight: √
Target Tracking: –
Automatic Landing: √
Return To Home: √
Obstacle Avoidance: –
Headless Mode: –
Built-in Camera: –

Aeromao

Aeromapper EV2

© Aeromao

Release Date: March 2015
Price: $7,990
Number of Rotors: 1 (Fixed Wing)
Control System: Remote Control, Laptop
Applications: Aerial Mapping
Wingspan: 2 m
Weight: 4.5 kg
Frame Material: Carbon Fiber, Kevlar
Available Colors: White
Max. Flight Time: 60 minutes
Max. Speed: 120 km/h (74 mp/h)
Max. Flying Altitude: 4,500 m (14,764 ft)
Controller Frequency: 2.4 GHz
Battery Type: Lithium Polymer
GPS Compatible: √
Altitude Hold: √
Position Hold: √
Flight Planning: √
Autonomous Flight: √
Target Tracking: –
Automatic Landing: –
Return To Home: √
Obstacle Avoidance: –
Headless Mode: –
Built-in Camera: √ (Sony 24 megapixels)

AeroSky

QAV 250

©xheli.com

Release Date: March 2015
Price: $365
Number of Rotors: 4
Control System: Remote Control
Applications: FPV Racing
Dimensions: 260 × 260 × 78 mm
Weight: 410 gr
Frame Material: Carbon Fiber
Rotor Material: Plastic
Motors: Outrunner brusheless
Available Colors: Black
Battery Type: Lithium Polymer
Battery Power: 11.1V
GPS Compatible: √
Altitude Hold: –
Position Hold: –
Flight Planning: –
Autonomous Flight: –
Target Tracking: –
Automatic Landing: –
Return To Home: –
Obstacle Avoidance: –
Headless Mode: –
Built-in Camera: –

AeroSky

QAV ZMR250

©xheli.com

Release Date: March 2015
Price: $435
Number of Rotors: 4
Control System: Remote Control
Applications: FPV Racing
Dimensions: 250 × 250 × 77
Weight: 390 gr
Frame Material: Carbon Fiber
Rotor Material: Plastic
Motors: Outrunner brusheless
Available Colors: Black
Battery Type: Lithium Polymer
Battery Power: 11.1V
GPS Compatible: √
Altitude Hold: –
Position Hold: –
Flight Planning: –
Autonomous Flight: –
Target Tracking: –
Automatic Landing: –
Return To Home: –
Obstacle Avoidance: –
Headless Mode: –
Built-in Camera: –

AiDrones

AiD-MC8

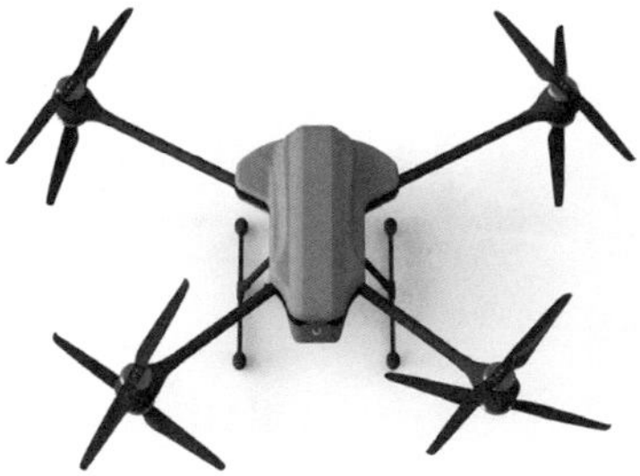
© AiDrone

Release Date: March 2015
Price: $1,800
Number of Rotors: 8
Control System: Remote Control
Applications: Film, Photography, Inspection, Reconnaissance, Surveillance
Dimensions: 0.9 × 1.1 × 0.3 m
Weight: 8 kg
Max. Payload: 5 kg
Frame Material: Carbon Fiber
Rotor Material: Carbon Fiber
Available Colors: Black, Grey, Red
Max. Flight Time: 18 minutes
Max. Speed: 40 km/h (25 mp/h)
Max. Flying Altitude: 3,000 m (9,842 ft)
Max. Ascent Speed: 4 m/s
Min. Operating Temperature: -10°C
Max. Operating Temperature: 35°C
Max. Operating Range: 3 km
Controller Frequency: 2.4 GHz
GPS Compatible: √
Altitude Hold: √
Position Hold: √
Flight Planning: √
Autonomous Flight: √
Target Tracking: √
Automatic Landing: √
Return To Home: √
Obstacle Avoidance: –
Headless Mode: –
Built-in Camera: –

Arris

X-Speed 250

© Arris Hobby

Release Date: March 2015
Price: $179
Number of Rotors: 4
Propellers: 5045 propellers
Control System: Remote Control
Applications: FPV Racing
Dimensions: 196 × 236 × 61 mm
Wingspan: 265 mm
Weight: 420 gr
Max. Payload: 159 gr
Frame Material: Carbon Fiber
Rotor Material: Plastic
Available Colors: Black
Max. Flight Time: 10 minutes
Max. Operating Range: 1 km
Controller Frequency: 5.8 GHz
Battery Capacity: 2,200 mAh
Battery Type: Lithium Polymer 3S
Battery Power: 11.1V
Battery Weight: 159 gr
GPS Compatible: √
Altitude Hold: –
Position Hold: –
Flight Planning: √
Autonomous Flight: –
Target Tracking: –
Automatic Landing: √
Return To Home: –
Obstacle Avoidance: –
Headless Mode: –
Built-in Camera: √ (700TVL)

Blade

Glimpse

©bhphotovideo.com

Release Date: March 2015
Price: $170
Number of Rotors: 4
Control System: Remote Control, Phone/Tablet
Applications: Film, Photography
Dimensions: 160 × 160 × 50 mm
Wingspan: 226 mm
Weight: 95 gr
Max. Payload: 43 gr
Frame Material: Plastic
Rotor Material: Plastic
Available Colors: White
Max. Flight Time: 8 minutes
Controller Frequency: 2.4 GHz
Battery Capacity: 500 mAh
Battery Type: Lithium Polymer 1S
Battery Power: 4V
GPS Compatible: –
Altitude Hold: –
Position Hold: –
Flight Planning: –
Autonomous Flight: –
Target Tracking: –
Automatic Landing: –
Return To Home: –
Obstacle Avoidance: –
Headless Mode: –
Built-in Camera: √

Dromida

Omnius FPV

©bhphotovideo.com

Release Date: March 2015
Price: $80
Number of Rotors: 4
Control System: Remote Control
Applications: FPV Racing
Wingspan: 238 mm
Weight: 120 gr
Frame Material: Plastic
Rotor Material: Plastic
Available Colors: Blue, Green, Red, Yellow
Max. Flight Time: 12 minutes
Max. Operating Range: 100 m
Controller Frequency: 2.4 GHz
Battery Capacity: 700 mAh
Battery Type: Lithium Polymer 1S
Battery Power: 3.7V
GPS Compatible: –
Altitude Hold: –
Position Hold: –
Flight Planning: –
Autonomous Flight: –
Target Tracking: –
Automatic Landing: –
Return To Home: –
Obstacle Avoidance: –
Headless Mode: –
Built-in Camera: √

Eachine

Q200

Release Date: March 2015
Price: $20
Number of Rotors: 4
Control System: Remote Control
Applications: FPV Racing
Dimensions: 200 mm
Weight: 40 gr
Frame Material: Carbon Fiber
Rotor Material: Plastic
Motors: Sunnysky 1306-3100kv
Available Colors: Black
Controller Frequency: 5.8 GHz
Battery Capacity: 800 mAh
Battery Type: Lithium Polymer 2S
GPS Compatible: –
Altitude Hold: –
Position Hold: –
Flight Planning: –
Autonomous Flight: –
Target Tracking: –
Automatic Landing: –
Return To Home: –
Obstacle Avoidance: –
Headless Mode: –
Built-in Camera: –

Flyingwings

Venturi Evo

© Flyingwings

Release Date: March 2015
Price: Unknown
Number of Rotors: 1 (Fixed Wing)
Control System: Remote Control
Applications: Recreation, FPV Racing
Frame Material: EPP Foam, Wood
Rotor Material: Plastic
Available Colors: White
Controller Frequency: 2.4 GHz
Battery Capacity: 3,300 mAh
Battery Type: Lithium Polymer 4S
Battery Power: 11.1V
GPS Compatible: –
Altitude Hold: –
Position Hold: –
Flight Planning: –
Autonomous Flight: –
Target Tracking: –
Automatic Landing: –
Return To Home: –
Obstacle Avoidance: –
Headless Mode: –
Built-in Camera: –

Height Tech

HT-8 C180

© Height Tech

Release Date: March 2015
Price: $36,295
Number of Rotors: 8
Control System: Remote Control
Applications: Film, Photography, Agriculture, Surveillance
Dimensions: 0.8 × 1 × 0.26 m
Weight: 2.4 kg
Max. Payload: 2.6 kg
Frame Material: Plastic, Carbon Fiber
Rotor Material: Plastic
Available Colors: Orange
Max. Flight Time: 20 minutes
Max. Speed: 65 km/h (40 mp/h)
Max. Flying Altitude: 3,900 m (12,795 ft)
Max. Ascent Speed: 8 m/s
Min. Operating Temperature: -10°C
Max. Operating Temperature: 35°C
Controller Frequency: 2.4 GHz
Battery Type: Lithium Polymer
GPS Compatible: √
Altitude Hold: √
Position Hold: √
Flight Planning: √
Autonomous Flight: √
Target Tracking: –
Automatic Landing: –
Return To Home: √
Obstacle Avoidance: –
Headless Mode: –
Built-in Camera: –

HuaXiang

8927

© HuaXiang

Release Date: March 2015
Price: $55
Number of Rotors: 4
Control System: Remote Control
Applications: Recreation
Dimensions: 376 × 376 × 53 mm
Frame Material: Plastic
Rotor Material: Plastic
Available Colors: Green, Orange, Red
Max. Flight Time: 12 minutes
Max. Operating Range: 100 m
Controller Frequency: 2.4 GHz
Battery Capacity: 600 mAh
Battery Type: Lithium Polymer
Battery Power: 3.7V
GPS Compatible: –
Altitude Hold: √
Position Hold: √
Flight Planning: –
Autonomous Flight: –
Target Tracking: –
Automatic Landing: –
Return To Home: –
Obstacle Avoidance: –
Headless Mode: –
Built-in Camera: √ (1.3 megapixels)

Izhmash Unmanned Systems

Takhion Mini-UAV

©tvzvezda.ru

Release Date: March 2015
Price: Unknown
Number of Rotors: 1 (Fixed Wing)
Control System: Ground Control Station
Applications: Military Purposes, Reconnaissance, Surveillance
Available Colors: Grey
Max. Speed: 54 km/h (33,5 mp/h)
Max. Flying Altitude: 4,000 m (13,123 ft)
Min. Operating Temperature: -30°C
Max. Operating Temperature: 40°C
Max. Operating Range: 40 km
Controller Frequency: 2.4 GHz
Battery Type: Lithium Polymer
GPS Compatible: √
Altitude Hold: √
Position Hold: –
Flight Planning: √
Autonomous Flight: √
Target Tracking: –
Automatic Landing: –
Return To Home: √
Obstacle Avoidance: –
Headless Mode: –
Built-in Camera: √

JJRC

JJ820

©JJRC

Release Date: March 2015
Price: $20
Number of Rotors: 4
Control System: Remote Control
Applications: Recreation
Dimensions: 40 × 40 mm
Wingspan: 57 mm
Frame Material: Plastic
Rotor Material: Plastic
Available Colors: Blue, Green, Orange
Max. Flight Time: 5 minutes
Max. Operating Range: 30 m
Controller Frequency: 2.4 GHz
Battery Capacity: 100 mAh
Battery Type: Lithium Polymer
Battery Power: 3.7V
GPS Compatible: –
Altitude Hold: –
Position Hold: –
Flight Planning: –
Autonomous Flight: –
Target Tracking: –
Automatic Landing: –
Return To Home: –
Obstacle Avoidance: –
Headless Mode: √
Built-in Camera: –

Lishitoys

L6039

© rcmoment.com

Release Date: March 2015
Price: $45
Number of Rotors: 4
Control System: Remote Control
Applications: Recreation
Dimensions: 298 × 298 × 65 mm
Weight: 72 gr
Frame Material: Plastic
Rotor Material: Plastic
Available Colors: White
Max. Operating Range: 100 m
Controller Frequency: 2.4 GHz
Battery Capacity: 500 mAh
Battery Type: Lithium Polymer
Battery Power: 3.7V
GPS Compatible: –
Altitude Hold: –
Position Hold: –
Flight Planning: –
Autonomous Flight: –
Target Tracking: –
Automatic Landing: –
Return To Home: –
Obstacle Avoidance: –
Headless Mode: –
Built-in Camera: √

Matternet

Matternet One

© post.ch

Release Date: March 2015
Price: $5000
Number of Rotors: 4
Control System: Remote Control
Applications: Delivery, Transport
Weight: 1 kg
Available Colors: White
Max. Flight Time: 18 minutes
Max. Speed: 65 km/h (40 mp/h)
Max. Operating Range: 15 km
Controller Frequency: 2.4 GHz
Battery Type: Lithium Polymer
GPS Compatible: √
Altitude Hold: √
Position Hold: √
Flight Planning: √
Autonomous Flight: –
Target Tracking: √
Automatic Landing: √
Return To Home: √
Obstacle Avoidance: –
Headless Mode: –
Built-in Camera: –

Revell Control

Nano Quad Cam

©Revell Control

Release Date: March 2015
Price: $65
Number of Rotors: 4
Control System: Remote Control
Applications: Recreation
Dimensions: 68 × 68 × 25 mm
Weight: 15 gr
Frame Material: Plastic
Rotor Material: Plastic
Available Colors: Black
Max. Flight Time: 5 minutes
Max. Operating Range: 30 m
Controller Frequency: 2.4 GHz
Battery Capacity: 150 mAh
Battery Type: Lithium Polymer
Battery Power: 3.7V
GPS Compatible: –
Altitude Hold: –
Position Hold: –
Flight Planning: –
Autonomous Flight: –
Target Tracking: –
Automatic Landing: –
Return To Home: –
Obstacle Avoidance: –
Headless Mode: –
Built-in Camera: √

Revell Control

Rayvore

©Revell Control

Release Date: March 2015
Price: $85
Number of Rotors: 4
Control System: Remote Control
Applications: Recreation
Dimensions: 200 × 200 × 55 mm
Weight: 105 gr
Frame Material: Plastic
Rotor Material: Plastic
Available Colors: Black, Green
Max. Flight Time: 12 minutes
Max. Operating Range: 100 m
Controller Frequency: 2.4 GHz
Battery Capacity: 700 mAh
Battery Type: Lithium Polymer
Battery Power: 3.7V
GPS Compatible: –
Altitude Hold: –
Position Hold: –
Flight Planning: –
Autonomous Flight: –
Target Tracking: –
Automatic Landing: –
Return To Home: –
Obstacle Avoidance: –
Headless Mode: –
Built-in Camera: –

Revell Control

Spot

Release Date: March 2015
Price: $70
Number of Rotors: 4
Control System: Remote Control
Applications: Recreation
Dimensions: 90 × 90 × 42 mm
Weight: 52 gr
Frame Material: Plastic
Rotor Material: Plastic
Available Colors: Black
Max. Flight Time: 7 minutes
Max. Operating Range: 50 m
Controller Frequency: 2.4 GHz
Battery Capacity: 370 mAh
Battery Type: Lithium Polymer
Battery Power: 3.7V
GPS Compatible: –
Altitude Hold: –
Position Hold: –
Flight Planning: –
Autonomous Flight: –
Target Tracking: –
Automatic Landing: –
Return To Home: –
Obstacle Avoidance: –
Headless Mode: –
Built-in Camera: √

Service-drone.com

G4 Recon One

Release Date: March 2015
Price: $10,000
Number of Rotors: 4
Control System: Remote Control
Applications: Film, Photography, Aerial Mapping, Surveillance, Transport
Wingspan: 1.14 m
Weight: 3 kg
Max. Payload: 6.5 kg
Frame Material: Carbon Fiber
Rotor Material: Carbon Fiber
Available Colors: Black, Purple, Yellow
Max. Flight Time: 12 minutes
Max. Speed: 50 km/h (31 mp/h)
Min. Operating Temperature: -10°C
Max. Operating Temperature: 40°C
Max. Operating Range: 2 km
Controller Frequency: 2.4 GHz
Battery Capacity: 2,200 mAh
Battery Type: Lithium Polymer
GPS Compatible: √
Altitude Hold: √
Position Hold: √
Flight Planning: √
Autonomous Flight: √
Target Tracking: –
Automatic Landing: –
Return To Home: √
Obstacle Avoidance: –
Headless Mode: –
Built-in Camera: –

SteadiDrone

Flare

© SteadiDrone

Release Date: March 2015
Price: $1,995
Number of Rotors: 8
Control System: Remote Control
Applications: Film, Photography
Dimensions: 440 × 645 × 180 mm
Wingspan: 677 mm
Weight: 970 gr
Max. Payload: 800 gr
Frame Material: Aluminum
Rotor Material: Carbon Fiber
Available Colors: Red
Max. Flight Time: 20 minutes
Max. Speed: 72 km/h (44,7 mp/h)
Max. Flying Altitude: 4,000 m (13,123 ft)
Max. Ascent Speed: 10 m/s
Max. Descent Speed: 7 m/s
Min. Operating Temperature: -5°C
Max. Operating Temperature: 50°C
Controller Frequency: 2.4 GHz
Battery Capacity: 5,200 mAh
Battery Type: Lithium Polymer 4S
Battery Power: 15V
Battery Weight: 520 gr
GPS Compatible: √
Altitude Hold: √
Position Hold: √
Flight Planning: √
Autonomous Flight: √
Target Tracking: –
Automatic Landing: √
Return To Home: √
Obstacle Avoidance: –
Headless Mode: –
Built-in Camera: –

Tekever

VR1 Colibri

© shephardmedia.com

Release Date: March 2015
Price: On Request
Number of Rotors: 4
Control System: Ground Control Station, Laptop
Applications: Aerial Mapping, Inspection, Surveillance
Dimensions: 500 × 500 × 200 mm
Wingspan: 2.8 kg
Frame Material: Composite
Rotor Material: Carbon Fiber
Available Colors: Black
Max. Flight Time: 30 minutes
Max. Flying Altitude: 250 m (820 ft)
Max. Operating Range: 2 km
Controller Frequency: 2.4 GHz
Battery Type: Lithium Polymer
GPS Compatible: √
Altitude Hold: √
Position Hold: √
Flight Planning: √
Autonomous Flight: √
Target Tracking: √
Automatic Landing: √
Return To Home: √
Obstacle Avoidance: –
Headless Mode: –
Built-in Camera: √

UDIRC Toys

U842-1 Lark FPV

© UDIRC Toys

Release Date: March 2015
Price: $125
Number of Rotors: 4
Control System: Remote Control
Applications: Recreation
Dimensions: 486 × 486 × 122 mm
Wingspan: 687 mm
Weight: 350 gr
Frame Material: Plastic
Rotor Material: Plastic, Carbon Fiber
Available Colors: Black, White
Max. Flight Time: 7 minutes
Max. Operating Range: 80 m
Controller Frequency: 2.4 GHz
Battery Capacity: 1,000 mAh
Battery Type: Lithium Polymer
Battery Power: 7.2V
GPS Compatible: –
Altitude Hold: √
Position Hold: √
Flight Planning: –
Autonomous Flight: –
Target Tracking: –
Automatic Landing: –
Return To Home: –
Obstacle Avoidance: –
Headless Mode: –
Built-in Camera: √

Virtualrobotix

VR Spark 350

© Daniele Avogadri

Release Date: March 2015
Price: $2,999
Number of Rotors: 4
Control System: Remote Control, Phone/Tablet
Applications: Film, Photography
Weight: 1.6 kg
Max. Payload: 500 gr
Frame Material: Plastic
Rotor Material: Plastic
Available Colors: Black
Max. Flight Time: 18 minutes
Max. Speed: 54 km/h (33,5 mp/h)
Max. Operating Range: 1.5 km
Controller Frequency: 2.4 GHz
Battery Capacity: 4,000 mAh
Battery Type: Lithium Polymer 3S
Battery Power: 14.8V
Battery Weight: 220 gr
GPS Compatible: √
Altitude Hold: √
Position Hold: √
Flight Planning: √
Autonomous Flight: –
Target Tracking: √
Automatic Landing: √
Return To Home: √
Obstacle Avoidance: –
Headless Mode: –
Built-in Camera: –

Walkera

Voyager 3

© Walkera

Release Date: March 2015
Price: $1,999
Number of Rotors: 4
Control System: Remote Control, Phone/Tablet
Applications: Film, Photography, Aerial Mapping
Dimensions: 473 × 463 × 300 mm
Wingspan: 661 mm
Weight: 3.65 kg
Max. Payload: 320 gr
Frame Material: Plastic, Carbon Fiber
Rotor Material: Carbon Fiber
Available Colors: Black, White
Max. Flight Time: 25 minutes
Max. Speed: 79 km/h (49 mp/h)
Min. Operating Temperature: -10°C
Max. Operating Temperature: 40°C
Max. Operating Range: 1 km
Controller Frequency: 2.4 GHz
Battery Capacity: 6,000 mAh
Battery Type: Lithium Polymer
Battery Power: 29.6V
GPS Compatible: √
Altitude Hold: √
Position Hold: √
Flight Planning: √
Autonomous Flight: √
Target Tracking: √
Automatic Landing: √
Return To Home: √
Obstacle Avoidance: –
Headless Mode: –
Built-in Camera: √ (4K)

XAircraft America

Xmission

© Xaircraft

Release Date: March 2015
Price: On Request
Number of Rotors: 4
Control System: Remote Control, Phone/Tablet
Applications: Inspection, Surveillance
Dimensions: 465 × 465 × 212 mm
Wingspan: 550 mm
Weight: 1.95 kg
Max. Payload: 1 kg
Frame Material: Carbon Fiber
Rotor Material: Carbon Fiber
Available Colors: Black
Max. Flight Time: 31 minutes
Max. Flight Time with Payload: 16 minutes
Max. Ascent Speed: 5 m/s
Max. Descent Speed: 4 m/s
Min. Operating Temperature: -10°C
Max. Operating Temperature: 70°C
Max. Operating Range: 5 km
Controller Frequency: 2.4 GHz
Battery Capacity: 6700 mAh
Battery Type: Lithium Polymer
Battery Power: 14.8V
Battery Weight: 500 gr
GPS Compatible: √
Altitude Hold: √
Position Hold: √
Flight Planning: √
Autonomous Flight: –
Target Tracking: √
Automatic Landing: √
Return To Home: √
Obstacle Avoidance: –
Headless Mode: –
Built-in Camera: √ (10 megapixels)

Acecore Technologies

NEO

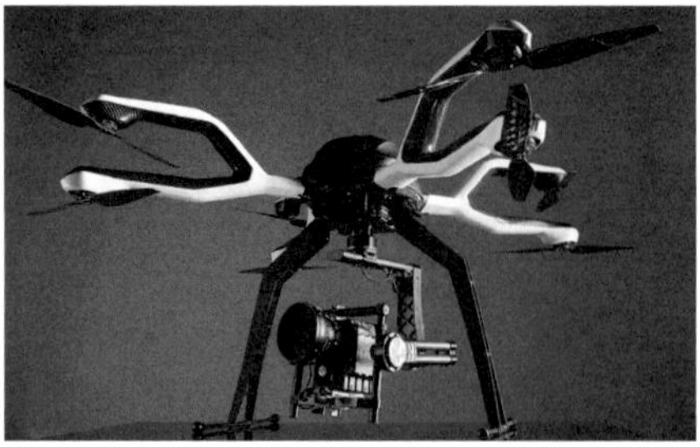

© Acecore Technologies

Release Date: April 2015
Price: $26,000
Number of Rotors: 8
Propellers: AceCore custom
Control System: Remote Control
Applications: Film, Photography, Aerial Mapping
Dimensions: 1.1 × 1.1 × 0.56 m
Weight: 6.85 kg
Max. Payload: 19 kg
Frame Material: Carbon Fiber
Rotor Material: Carbon Fiber
Available Colors: Black
Max. Flight Time: 30 minutes
Max. Speed: 72 km/h (44,7 mp/h)
Max. Flying Altitude: 2,750 m (9,022 ft)
Max. Ascent Speed: 6 m/s
Min. Operating Temperature: -15°C
Max. Operating Temperature: 60°C
Max. Operating Range: 2.75 km
Controller Frequency: 2.4 GHz
Battery Capacity: 1,500 mAh
Battery Type: Lithium Polymer
Battery Power: 7.2V
Battery Weight: 1.5 kg
GPS Compatible: √
Altitude Hold: √
Position Hold: √
Flight Planning: √
Autonomous Flight: √
Target Tracking: √
Automatic Landing: √
Return To Home: √
Obstacle Avoidance: –
Headless Mode: –
Built-in Camera: –

Acme

Zoopa Q420 Cruiser

© Acme

Release Date: April 2015
Price: $100
Number of Rotors: 4
Control System: Remote Control
Applications: Film, Photography
Dimensions: 345 × 345 × 75 mm
Wingspan: 410 mm
Weight: 100 gr
Frame Material: Plastic
Rotor Material: Plastic
Available Colors: Black
Max. Flight Time: 8 minutes
Max. Operating Range: 80 m
Controller Frequency: 2.4 GHz
Battery Capacity: 600 mAh
Battery Type: Lithium Polymer 1S
Battery Power: 3.7V
GPS Compatible: –
Altitude Hold: –
Position Hold: –
Flight Planning: –
Autonomous Flight: –
Target Tracking: –
Automatic Landing: –
Return To Home: –
Obstacle Avoidance: –
Headless Mode: –
Built-in Camera: √

AEE Technology

F100

© AEE Technologies

Release Date: April 2015
Price: $58,000
Number of Rotors: 4
Control System: Remote Control
Applications: Film, Photography, Surveillance
Dimensions: 1.55 × 1.55 × 0.24 mm
Wingspan: 2.2 m
Weight: 7.6 kg
Max. Payload: 2.5 kg
Frame Material: Carbon Fiber
Rotor Material: Carbon Fiber
Available Colors: Silver
Max. Flight Time: 60 minutes
Max. Speed: 100 km/h (62 mp/h)
Max. Flying Altitude: 1,500 m (4,921 ft)
Min. Operating Temperature: -20°C
Max. Operating Temperature: 60°C
Max. Operating Range: 10 km
Controller Frequency: 2.4 GHz
Battery Capacity: 20,000 mAh
Battery Type: Lithium Polymer
Battery Power: 22.2V
GPS Compatible: √
Altitude Hold: √
Position Hold: √
Flight Planning: √
Autonomous Flight: √
Target Tracking: –
Automatic Landing: √
Return To Home: √
Obstacle Avoidance: –
Headless Mode: –
Built-in Camera: –

AFM-Servers Ltd.

Ptero-G0

© ptero.ru

Release Date: April 2015
Price: Unknown
Number of Rotors: 1 (Fixed Wing)
Control System: Flight Computer
Applications: Military Purposes, Aerial Mapping, Reconnaissance, Surveillance
Dimensions: 2.14 × 3.13 × 0.68 m
Wingspan: 3.13 m
Weight: 15 kg
Max. Payload: 5 kg
Frame Material: Carbon Fiber, Kevlar
Motors: Saito FG-40 Fuel
Available Colors: Grey
Max. Flight Time: 8 hours
Max. Speed: 170 km/h (106 mp/h)
Max. Flying Altitude: 3,000 m (9,842 ft)
Min. Operating Temperature: -30°C
Max. Operating Temperature: 40°C
Max. Operating Range: 800 km
GPS Compatible: √
Altitude Hold: √
Position Hold: –
Flight Planning: √
Autonomous Flight: √
Target Tracking: √
Automatic Landing: –
Return To Home: √
Obstacle Avoidance: √
Headless Mode: –
Built-in Camera: –

Align

M470

©live-hobby.de

Release Date: April 2015
Price: $1,170
Number of Rotors: 4
Propellers: 7 Inch
Control System: Remote Control, Phone/Tablet
Applications: Film, Photography
Dimensions: 550 × 543 × 266 mm
Wingspan: 710 mm
Frame Material: Carbon Fiber
Rotor Material: Carbon Fiber
Motors: BL4213-370kv
Available Colors: Red
Max. Speed: 54 km/h (33,5 mp/h)
Max. Flying Altitude: 700 m (2,296 ft)
Max. Ascent Speed: 6 m/s
Min. Operating Temperature: -20°C
Max. Operating Temperature: 65°C
Controller Frequency: 2.4 GHz
Battery Capacity: 12,000 mAh
Battery Type: Lithium Polymer 6S
Battery Power: 22.2V
GPS Compatible: √
Altitude Hold: √
Position Hold: √
Flight Planning: √
Autonomous Flight: √
Target Tracking: √
Automatic Landing: √
Return To Home: √
Obstacle Avoidance: –
Headless Mode: –
Built-in Camera: –

ATyges

RPAS FV6

©ATyges

Release Date: April 2015
Price: Unknown
Number of Rotors: 6
Control System: Flight Computer
Applications: Aerial Mapping, Agriculture, Inspection, Transport
Max. Payload: 100 kg
Available Colors: Black
Max. Flight Time: 30 minutes
Battery Type: Lithium Polymer
GPS Compatible: √
Altitude Hold: √
Position Hold: √
Flight Planning: √
Autonomous Flight: √
Target Tracking: √
Automatic Landing: √
Return To Home: √
Obstacle Avoidance: √
Headless Mode: –
Built-in Camera: –

Blade

Mach25 FPV Racer

©droneshop.com

Release Date: April 2015
Price: $250
Number of Rotors: 4
Control System: Remote Control
Applications: FPV Racing
Dimensions: 210 × 210 × 80 mm
Wingspan: 250 mm
Weight: 500 gr
Frame Material: Aluminum, Carbon Fiber
Rotor Material: Plastic
Available Colors: Red
Max. Flight Time: 4 minutes
Controller Frequency: 2.4 GHz
Battery Capacity: 1,350 mAh
Battery Type: Lithium Polymer 3S
Battery Power: 11V
GPS Compatible: –
Altitude Hold: –
Position Hold: –
Flight Planning: –
Autonomous Flight: –
Target Tracking: –
Automatic Landing: –
Return To Home: –
Obstacle Avoidance: –
Headless Mode: –
Built-in Camera: √

BSK Defense

Phaethon G

©avia-pro.fr

Release Date: April 2015
Price: Unknown
Number of Rotors: 1 (Fixed Wing)
Control System: Ground Control Station
Applications: Military Purposes, Inspection, Reconnaissance, Surveillance
Dimensions: 2.2 × 3.8 × 0.8 m
Wingspan: 3.8 m
Weight: 45 kg
Frame Material: Composite
Rotor Material: Wood
Available Colors: Grey
Max. Flight Time: 180 minutes
Max. Speed: 180 km/h (112 mp/h)
Max. Flying Altitude: 4,200 m (13,780 ft)
Max. Operating Range: 100 km
GPS Compatible: √
Altitude Hold: √
Position Hold: –
Flight Planning: √
Autonomous Flight: √
Target Tracking: √
Automatic Landing: √
Return To Home: √
Obstacle Avoidance: –
Headless Mode: –
Built-in Camera: √

DJI

Phantom 3

© DJI

Release Date: April 2015
Price: $499
Number of Rotors: 4
Control System: Remote Control, Phone/Tablet
Applications: Film, Photography
Wingspan: 350 mm
Weight: 1.28 kg
Frame Material: Carbon Fiber
Rotor Material: Carbon Fiber
Available Colors: White
Max. Flight Time: 23 minutes
Max. Speed: 57 km/h (35,7 mp/h)
Max. Flying Altitude: 6,000 m (19,685 ft)
Max. Ascent Speed: 5 m/s
Max. Descent Speed: 3 m/s
Min. Operating Temperature: 0°C
Max. Operating Temperature: 40°C
Max. Operating Range: 1 km
Controller Frequency: 5.8 GHz
Battery Capacity: 4,480 mAh
Battery Type: Lithium Polymer 4S
Battery Power: 17.4V
Battery Weight: 365 gr
GPS Compatible: √
Altitude Hold: √
Position Hold: √
Flight Planning: √
Autonomous Flight: √
Target Tracking: √
Automatic Landing: √
Return To Home: √
Obstacle Avoidance: –
Headless Mode: –
Built-in Camera: √ (12.4 megapixels)

Dronetech UAV

AV-2 Pelican

© Dronetech

Release Date: April 2015
Price: Unknown
Number of Rotors: 10 (Fixed Wing)
Control System: Ground Control Station
Applications: Military Purposes, Inspection, Reconnaissance, Surveillance
Dimensions: 1.65 × 3.2 m
Wingspan: 3.2 m
Weight: 16 kg
Max. Payload: 7 kg
Frame Material: Composite
Available Colors: Grey
Max. Flight Time: 2 hours
Max. Speed: 120 km/h (74 mp/h)
Max. Flying Altitude: 3,658 m (12,000 ft)
Max. Operating Range: 100 km
GPS Compatible: √
Altitude Hold: √
Position Hold: √
Flight Planning: √
Autonomous Flight: √
Target Tracking: –
Automatic Landing: √
Return To Home: √
Obstacle Avoidance: –
Headless Mode: –
Built-in Camera: √

Huajun Toys

W609-7

©Huajun Toys

Release Date: April 2015
Price: $120
Number of Rotors: 6
Control System: Remote Control
Applications: Film, Photography
Dimensions: 630 × 690 × 210 mm
Weight: 505 gr
Frame Material: Plastic
Rotor Material: Plastic
Available Colors: Gold, Red
Max. Flight Time: 10 minutes
Max. Operating Range: 500 m
Controller Frequency: 2.4 GHz
Battery Capacity: 1,500 mAh
Battery Type: Lithium Polymer
Battery Power: 7.4V
GPS Compatible: –
Altitude Hold: √
Position Hold: √
Flight Planning: –
Autonomous Flight: –
Target Tracking: –
Automatic Landing: –
Return To Home: √
Obstacle Avoidance: –
Headless Mode: √
Built-in Camera: √ (2 megapixels)

Immersion RC

Vortex 285

©Immersion RC

Release Date: April 2015
Price: $449
Number of Rotors: 4
Control System: Remote Control
Applications: FPV Racing
Wingspan: 280 mm
Weight: 350 gr
Frame Material: Carbon Fiber
Rotor Material: Plastic
Motors: OEM 1806-2300kv
Available Colors: Black
Controller Frequency: 2.4 GHz
Battery Capacity: 1,600 mAh
Battery Type: Lithium Polymer
Battery Power: 3.7V
GPS Compatible: √
Altitude Hold: –
Position Hold: –
Flight Planning: √
Autonomous Flight: –
Target Tracking: –
Automatic Landing: –
Return To Home: √
Obstacle Avoidance: –
Headless Mode: –
Built-in Camera: √ (FatShark 600TVL CMOS)

JJRC
H12C

Release Date: April 2015
Price: $80
Number of Rotors: 4
Control System: Remote Control
Applications: Recreation
Dimensions: 310 × 310 × 85 mm
Wingspan: 438 mm
Weight: 53 gr
Frame Material: Plastic
Rotor Material: Plastic
Available Colors: Red, White
Max. Flight Time: 8 minutes
Max. Operating Range: 300 m
Controller Frequency: 2.4 GHz
Battery Capacity: 750 mAh
Battery Type: Lithium Polymer
Battery Power: 3.7V
GPS Compatible: –
Altitude Hold: –
Position Hold: –
Flight Planning: –
Autonomous Flight: –
Target Tracking: –
Automatic Landing: –
Return To Home: √
Obstacle Avoidance: –
Headless Mode: –
Built-in Camera: √ (5 megapixels)

Onyxstar
Hydra12

Release Date: April 2015
Price: $20,850
Number of Rotors: 12
Control System: Remote Control
Applications: Film, Photography, Agriculture, Inspection, Transport
Weight: 7 kg
Max. Payload: 16 kg
Frame Material: Carbon Fiber
Rotor Material: Carbon Fiber
Motors: A5050-1160 W
Available Colors: Black
Max. Flight Time: 30 minutes
GPS Compatible: √
Altitude Hold: √
Position Hold: √
Flight Planning: √
Autonomous Flight: √
Target Tracking: √
Automatic Landing: √
Return To Home: √
Obstacle Avoidance: –
Headless Mode: –
Built-in Camera: –

Perspective Robotics

Fotokite Pro

© fotokite.com

Release Date: April 2015
Price: $10,000
Number of Rotors: 4
Control System: Remote Control, Phone/Tablet
Applications: Film, Photography
Dimensions: 540 × 340 × 150 mm
Wingspan: 638 mm
Weight: 600 gr
Frame Material: Carbon Fiber
Rotor Material: Plastic
Available Colors: Blue
Max. Flight Time: 60 minutes
Max. Speed: 72 km/h (44,7 mp/h)
Max. Operating Range: 20 m
Controller Frequency: 2.4 GHz
Battery Type: Lithium Polymer 3S
Battery Power: 11V
GPS Compatible: –
Altitude Hold: √
Position Hold: √
Flight Planning: –
Autonomous Flight: –
Target Tracking: √
Automatic Landing: –
Return To Home: –
Obstacle Avoidance: –
Headless Mode: –
Built-in Camera: √

Revell Control

X-Spy 2.0

© Revell Control

Release Date: April 2015
Price: $140
Number of Rotors: 4
Control System: Remote Control
Applications: Recreation
Dimensions: 240 × 240 × 70 mm
Weight: 80 gr
Frame Material: Plastic
Rotor Material: Plastic
Available Colors: White
Max. Flight Time: 6 minutes
Max. Operating Range: 40 m
Controller Frequency: 2.4 GHz
Battery Capacity: 500 mAh
Battery Type: Lithium Polymer
Battery Power: 3.7V
GPS Compatible: –
Altitude Hold: –
Position Hold: –
Flight Planning: –
Autonomous Flight: –
Target Tracking: –
Automatic Landing: –
Return To Home: –
Obstacle Avoidance: –
Headless Mode: –
Built-in Camera: √

Revell Control

Nano Hex

Release Date: April 2015
Price: $50
Number of Rotors: 6
Control System: Remote Control
Applications: Recreation
Dimensions: 65 × 70 × 20 mm
Weight: 20 gr
Frame Material: Plastic
Rotor Material: Plastic
Available Colors: Orange, Yellow
Max. Flight Time: 7 minutes
Max. Operating Range: 30 m
Controller Frequency: 2.4 GHz
Battery Capacity: 150 mAh
Battery Type: Lithium Polymer
Battery Power: 3.7V
GPS Compatible: –
Altitude Hold: –
Position Hold: –
Flight Planning: –
Autonomous Flight: –
Target Tracking: –
Automatic Landing: –
Return To Home: –
Obstacle Avoidance: –
Headless Mode: –
Built-in Camera: –

Robodrone

Kingfisher

Release Date: April 2015
Price: $9,850
Number of Rotors: 6
Control System: Remote Control
Applications: Aerial Mapping, Agriculture, Inspection, Surveillance
Dimensions: 1.15 × 1 × 0.22 m
Weight: 12 kg
Max. Payload: 5 kg
Frame Material: Carbon Fiber
Rotor Material: Carbon Fiber
Available Colors: Grey
Max. Flight Time: 40 minutes
Max. Speed: 50 km/h (31 mp/h)
Max. Flying Altitude: 152 m (500 ft)
Max. Ascent Speed: 6 m/s
Controller Frequency: 2.4 GHz
GPS Compatible: √
Altitude Hold: √
Position Hold: √
Flight Planning: √
Autonomous Flight: √
Target Tracking: √
Automatic Landing: √
Return To Home: √
Obstacle Avoidance: –
Headless Mode: –
Built-in Camera: √

Sky Hero

Little Six

© Sky Hero

Release Date: April 2015
Price: $799
Number of Rotors: 6
Control System: Remote Control
Applications: Film, Photography, FPV Racing
Wingspan: 580 mm
Frame Material: Plastic, Carbon Fiber
Rotor Material: Plastic
Motors: Little 2806-950kv
Available Colors: White
Controller Frequency: 2.4 GHz
Battery Capacity: 3,700 mAh
Battery Type: Lithium Polymer 4S
GPS Compatible: √
Altitude Hold: √
Position Hold: √
Flight Planning: √
Autonomous Flight: –
Target Tracking: √
Automatic Landing: –
Return To Home: √
Obstacle Avoidance: –
Headless Mode: –
Built-in Camera: –

Syma Toys

X5S

© prisguide.no

Release Date: April 2015
Price: $50
Number of Rotors: 4
Control System: Remote Control
Applications: Film, Photography
Dimensions: 310 × 310 × 80 mm
Wingspan: 438 mm
Weight: 916 gr
Frame Material: Plastic
Rotor Material: Plastic
Available Colors: Black, Red, White
Max. Flight Time: 8 minutes
Max. Operating Range: 50 m
Controller Frequency: 2.4 GHz
Battery Capacity: 500 mAh
Battery Type: Lithium Polymer
Battery Power: 4V
GPS Compatible: –
Altitude Hold: –
Position Hold: –
Flight Planning: –
Autonomous Flight: –
Target Tracking: –
Automatic Landing: –
Return To Home: –
Obstacle Avoidance: –
Headless Mode: √
Built-in Camera: √ (2 megapixels)

Syma Toys

X8C Venture

©Syma Toys

Release Date: April 2015
Price: $90
Number of Rotors: 4
Control System: Remote Control
Applications: Film, Photography
Dimensions: 500 × 500 × 190 mm
Wingspan: 707 mm
Weight: 1.6 kg
Frame Material: Plastic
Rotor Material: Plastic
Available Colors: Black, Orange, White
Max. Flight Time: 7 minutes
Max. Operating Range: 50 m
Controller Frequency: 2.4 GHz
Battery Capacity: 200 mAh
Battery Type: Lithium Polymer
Battery Power: 7.4V
GPS Compatible: –
Altitude Hold: –
Position Hold: –
Flight Planning: –
Autonomous Flight: –
Target Tracking: –
Automatic Landing: –
Return To Home: –
Obstacle Avoidance: –
Headless Mode: √
Built-in Camera: √ (2 megapixels)

Syma Toys

X13 Storm

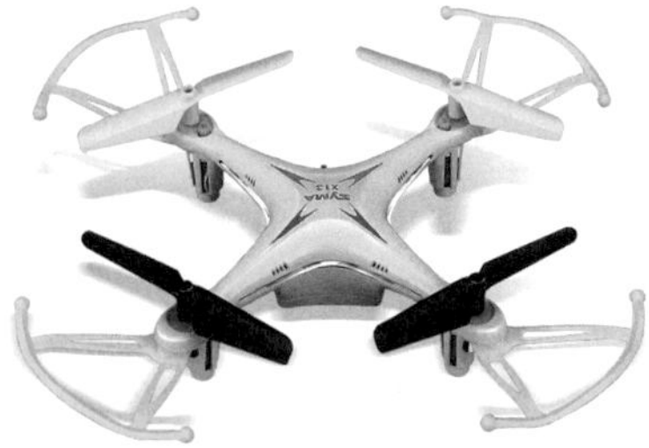

©Syma Toys

Release Date: April 2015
Price: $15
Number of Rotors: 4
Control System: Remote Control
Applications: Recreation
Dimensions: 156 × 156 × 36 mm
Wingspan: 220 mm
Weight: 445 gr
Frame Material: Plastic
Rotor Material: Plastic
Available Colors: Black, Red, White
Max. Flight Time: 8 minutes
Max. Operating Range: 50 m
Controller Frequency: 2.4 GHz
Battery Capacity: 200 mAh
Battery Type: Lithium Polymer
Battery Power: 4V
GPS Compatible: –
Altitude Hold: –
Position Hold: –
Flight Planning: –
Autonomous Flight: –
Target Tracking: –
Automatic Landing: –
Return To Home: –
Obstacle Avoidance: –
Headless Mode: –
Built-in Camera: –

UDIRC Toys

U818A-1 Discovery

Release Date: April 2015
Price: $64
Number of Rotors: 4
Control System: Remote Control
Applications: Recreation
Dimensions: 330 × 340 × 60 mm
Wingspan: 475 mm
Weight: 346 gr
Frame Material: Plastic
Rotor Material: Plastic, Carbon Fiber
Available Colors: Black, Blue
Max. Flight Time: 8 minutes
Max. Operating Range: 100 m
Controller Frequency: 2.4 GHz
Battery Capacity: 350 mAh
Battery Type: Lithium Polymer
Battery Power: 7V
GPS Compatible: –
Altitude Hold: √
Position Hold: √
Flight Planning: –
Autonomous Flight: –
Target Tracking: –
Automatic Landing: –
Return To Home: –
Obstacle Avoidance: –
Headless Mode: –
Built-in Camera: √

UDIRC Toys

U27 Free Loop

Release Date: April 2015
Price: $36
Number of Rotors: 4
Control System: Remote Control
Applications: Recreation
Dimensions: 139 × 139 × 35.5 mm
Wingspan: 195 mm
Weight: 54 gr
Frame Material: Plastic
Rotor Material: Plastic, Carbon Fiber
Available Colors: Black
Max. Flight Time: 7 minutes
Max. Operating Range: 80 m
Controller Frequency: 2.4 GHz
Battery Capacity: 450 mAh
Battery Type: Lithium Polymer
Battery Power: 3.7V
GPS Compatible: –
Altitude Hold: √
Position Hold: √
Flight Planning: –
Autonomous Flight: –
Target Tracking: –
Automatic Landing: –
Return To Home: –
Obstacle Avoidance: –
Headless Mode: –
Built-in Camera: –

Wingsland Technology

Scarlet minivet

Release Date: April 2015
Price: $449
Number of Rotors: 4
Control System: Remote Control, Phone/Tablet
Applications: Film, Photography
Dimensions: 303 × 303 × 176 mm
Wingspan: 390 mm
Weight: 1.09 kg
Frame Material: Plastic
Rotor Material: Plastic
Available Colors: Orange
Max. Flight Time: 25 minutes
Max. Speed: 64,8 km/h (40 mp/h)
Max. Ascent Speed: 3 m/s
Max. Descent Speed: 2 m/s
Max. Operating Range: 1.5 km
Controller Frequency: 2.4 GHz
Battery Capacity: 5,200 mAh
Battery Type: Lithium Polymer 3S
Battery Power: 14.7V
GPS Compatible: √
Altitude Hold: √
Position Hold: √
Flight Planning: √
Autonomous Flight: –
Target Tracking: –
Automatic Landing: √
Return To Home: √
Obstacle Avoidance: –
Headless Mode: –
Built-in Camera: √ (4 megapixels)

3D Robotics

Solo

Release Date: May 2015
Price: $800
Number of Rotors: 4
Propellers: PP11A
Control System: Remote Control, Phone/Tablet
Applications: Film, Photography
Wingspan: 460 mm
Weight: 1.5 kg
Max. Payload: 420 gr
Frame Material: Plastic
Rotor Material: Glass-reinforced Nylon
Motors: 880 k
Available Colors: Black
Max. Flight Time: 25 minutes
Max. Flight Time with Payload: 20 minutes
Max. Speed: 89 km/h (55 mp/h)
Max. Flying Altitude: 122 m (400 ft)
Max. Ascent Speed: 10 m/s
Min. Operating Temperature: 0°C
Max. Operating Temperature: 45°C
Max. Operating Range: 800 m
Controller Frequency: 2.4 GHz
Battery Capacity: 5,200 mAh
Battery Type: Lithium Polymer
Battery Power: 15V
Battery Weight: 500 gr
GPS Compatible: √
Altitude Hold: √
Position Hold: √
Flight Planning: √
Autonomous Flight: √
Target Tracking: √
Automatic Landing: √
Return To Home: √
Obstacle Avoidance: –
Headless Mode: –
Built-in Camera: –

AeroSky

QAV 280

Release Date: May 2015
Price: $445
Number of Rotors: 4
Control System: Remote Control
Applications: FPV Racing
Dimensions: 280 × 280 × 92 mm
Weight: 450 gr
Frame Material: Carbon Fiber
Rotor Material: Carbon Fiber
Motors: AeroSky MC2206-1900kv
Available Colors: Black
Battery Type: Lithium Polymer
Battery Power: 11.1V
GPS Compatible: √
Altitude Hold: –
Position Hold: –
Flight Planning: –
Autonomous Flight: –
Target Tracking: –
Automatic Landing: –
Return To Home: –
Obstacle Avoidance: –
Headless Mode: –
Built-in Camera: –

Blade

Chroma

© bhphotovideo.com

Release Date: May 2015
Price: $349
Number of Rotors: 4
Control System: Remote Control
Applications: Film, Photography
Wingspan: 400 mm
Weight: 1.3 kg
Frame Material: Plastic
Rotor Material: Plastic
Motors: 780KV Brushless Outrunner
Available Colors: White
Max. Flight Time: 30 minutes
Max. Speed: 65 km/h (40 mp/h)
Max. Operating Range: 400 m
Controller Frequency: 2.4 GHz
Battery Capacity: 5,400 mAh
Battery Type: Lithium Polymer 3S
Battery Power: 11V
GPS Compatible: √
Altitude Hold: –
Position Hold: √
Flight Planning: √
Autonomous Flight: –
Target Tracking: √
Automatic Landing: √
Return To Home: √
Obstacle Avoidance: –
Headless Mode: –
Built-in Camera: –

Bormatec

Ninox

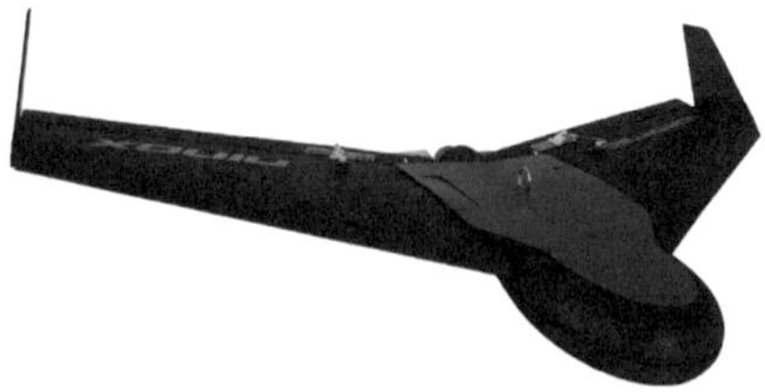

© Bormatec

Release Date: May 2015
Price: $165
Number of Rotors: 1 (Fixed Wing)
Propellers: PP11A
Control System: Remote Control, Phone/Tablet
Applications: Aerial Mapping
Dimensions: 0.6 × 1 m
Wingspan: 1 m
Weight: 700 gr
Max. Payload: 200 gr
Frame Material: EPP, Coroplast
Available Colors: Black
Max. Flight Time: 30 minutes
Controller Frequency: 2.4 GHz
Battery Capacity: 2,200 mAh
Battery Type: Lithium Polymer 3S
GPS Compatible: √
Altitude Hold: √
Position Hold: √
Flight Planning: √
Autonomous Flight: √
Target Tracking: √
Automatic Landing: √
Return To Home: √
Obstacle Avoidance: –
Headless Mode: –
Built-in Camera: √

Cheerson

CX-22

© Alibaba.com

Release Date: May 2015
Price: $650
Number of Rotors: 4
Control System: Remote Control, Phone/Tablet
Applications: Surveillance
Dimensions: 360 × 360 × 200 mm
Wingspan: 590 mm
Weight: 700 gr
Max. Payload: 500 gr
Frame Material: Plastic
Rotor Material: Plastic
Available Colors: White
Max. Flight Time: 25 minutes
Max. Flying Altitude: 800 m (2,624 ft)
Max. Operating Range: 1 km
Controller Frequency: 5.8 GHz
Battery Capacity: 5,400 mAh
Battery Type: Lithium Polymer 3S
Battery Power: 11.1V
GPS Compatible: √
Altitude Hold: √
Position Hold: √
Flight Planning: –
Autonomous Flight: –
Target Tracking: √
Automatic Landing: √
Return To Home: √
Obstacle Avoidance: –
Headless Mode: √
Built-in Camera: √

Cheerson

CX-31

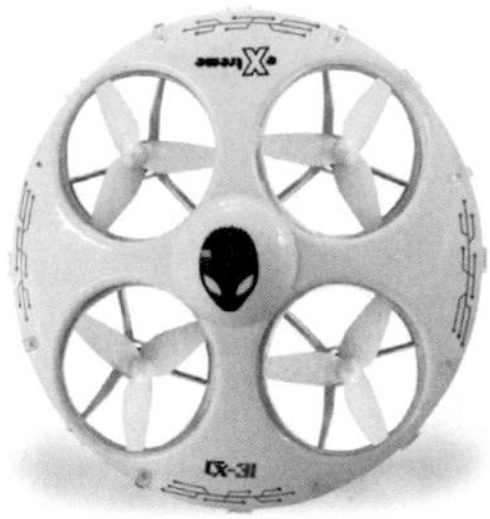

Release Date: May 2015
Price: $30
Number of Rotors: 4
Control System: Remote Control
Applications: Recreation
Dimensions: 120 × 120 × 50 mm
Wingspan: 120 mm
Weight: 52 gr
Frame Material: Plastic
Rotor Material: Plastic
Available Colors: Silver, White
Max. Flight Time: 6 minutes
Max. Operating Range: 30 m
Controller Frequency: 2.4 GHz
Battery Capacity: 500 mAh
Battery Type: Lithium Polymer
Battery Power: 3.7V
GPS Compatible: –
Altitude Hold: √
Position Hold: √
Flight Planning: –
Autonomous Flight: –
Target Tracking: –
Automatic Landing: –
Return To Home: –
Obstacle Avoidance: –
Headless Mode: –
Built-in Camera: –

DaMing

Spy 007

© aliexpress.com

Release Date: May 2015
Price: $36
Number of Rotors: 4
Control System: Remote Control, Phone/Tablet
Applications: Film, Photography, Recreation
Dimensions: 205 × 185 × 70 mm
Weight: 750 gr
Frame Material: Plastic
Rotor Material: Plastic
Available Colors: Black, White
Max. Flight Time: 8 minutes
Max. Operating Range: 150 m
Controller Frequency: 2.4 GHz
Battery Capacity: 400 mAh
Battery Type: Lithium Polymer
Battery Power: 7.4V
GPS Compatible: –
Altitude Hold: √
Position Hold: √
Flight Planning: –
Autonomous Flight: –
Target Tracking: –
Automatic Landing: –
Return To Home: √
Obstacle Avoidance: –
Headless Mode: –
Built-in Camera: √ (2 megapixels)

Delair-Tech

DT26X

© Delair-Tech

Release Date: May 2015
Price: $12,000
Number of Rotors: 1 (Fixed Wing)
Control System: Flight Computer
Applications: Reconnaissance, Surveillance
Dimensions: 1.6 m
Wingspan: 3.3 m
Weight: 15 kg
Max. Payload: 4 kg
Frame Material: Carbon Fiber
Available Colors: Black
Max. Flight Time: 2.5 hours
Max. Speed: 50 km/h (31 mp/h)
Max. Flying Altitude: 3,000 m (9,842 ft)
Min. Operating Temperature: -20°C
Max. Operating Temperature: 50°C
Max. Operating Range: 150 km
Controller Frequency: 2.4 GHz
GPS Compatible: √
Altitude Hold: √
Position Hold: √
Flight Planning: √
Autonomous Flight: √
Target Tracking: √
Automatic Landing: √
Return To Home: √
Obstacle Avoidance: –
Headless Mode: –
Built-in Camera: √

DYS

BG-250

© banggood.com

Release Date: May 2015
Price: $190
Number of Rotors: 4
Control System: Remote Control
Applications: FPV Racing
Dimensions: 225 × 204 × 65 mm
Wingspan: 250 mm
Weight: 290 gr
Frame Material: Carbon Fiber
Rotor Material: Carbon Fiber
Motors: BE1806-2300kv
Available Colors: Black
Controller Frequency: 5.8 GHz
Battery Capacity: 2,800 mAh
Battery Type: Lithium Polymer 3S
Battery Power: 11.1V
GPS Compatible: √
Altitude Hold: –
Position Hold: –
Flight Planning: –
Autonomous Flight: –
Target Tracking: –
Automatic Landing: –
Return To Home: –
Obstacle Avoidance: –
Headless Mode: –
Built-in Camera: –

Eachine

H8 Mini

Release Date: May 2015
Price: $15
Number of Rotors: 4
Control System: Remote Control
Applications: Recreation
Dimensions: 135 × 135 × 28 mm
Wingspan: 155 mm
Weight: 20 gr
Frame Material: Plastic
Rotor Material: Plastic
Available Colors: Black, White
Max. Flight Time: 7 minutes
Max. Operating Range: 30 m
Controller Frequency: 2.4 GHz
Battery Capacity: 150 mAh
Battery Type: Lithium Polymer 1S
Battery Power: 3.7V
GPS Compatible: –
Altitude Hold: –
Position Hold: –
Flight Planning: –
Autonomous Flight: –
Target Tracking: –
Automatic Landing: –
Return To Home: √
Obstacle Avoidance: –
Headless Mode: –
Built-in Camera: –

Fineco

FX-4V

Release Date: May 2015
Price: $20
Number of Rotors: 4
Control System: Remote Control
Applications: Recreation
Dimensions: 125 × 125 × 33 mm
Weight: 12 gr
Frame Material: Plastic
Rotor Material: Plastic
Available Colors: Red, White
Max. Flight Time: 8 minutes
Max. Operating Range: 80 m
Controller Frequency: 2.4 GHz
Battery Capacity: 300 mAh
Battery Type: Lithium Polymer
Battery Power: 3.7V
GPS Compatible: –
Altitude Hold: –
Position Hold: –
Flight Planning: –
Autonomous Flight: –
Target Tracking: –
Automatic Landing: –
Return To Home: √
Obstacle Avoidance: –
Headless Mode: –
Built-in Camera: √

Hobby Lord

Alpha Y3

© rcuniverse.com

Release Date: May 2015
Price: $399
Number of Rotors: 6
Control System: Remote Control
Applications: Film, Photography
Wingspan: 750 mm
Weight: 3 kg
Max. Payload: 2.5 kg
Frame Material: Carbon Fiber
Rotor Material: Carbon Fiber
Motors: ST4708 brushless
Available Colors: White, Yellow
Max. Flight Time: 25 minutes
Controller Frequency: 2.4 GHz
Battery Capacity: 15,000 mAh
Battery Type: Lithium Polymer 6S
Battery Power: 22.2V
GPS Compatible: √
Altitude Hold: √
Position Hold: √
Flight Planning: √
Autonomous Flight: √
Target Tracking: –
Automatic Landing: √
Return To Home: √
Obstacle Avoidance: –
Headless Mode: –
Built-in Camera: –

HSE

AG-V8A Plus

© HSE

Release Date: May 2015
Price: On Request
Number of Rotors: 8
Control System: Remote Control
Applications: Agriculture
Wingspan: 1.63 kg
Weight: 48 kg
Frame Material: Carbon Fiber
Available Colors: Orange
Max. Flight Time: 35 minutes
Max. Speed: 35,4 km/h (22 mp/h)
Controller Frequency: 2.4 GHz
Battery Type: Lithium Polymer
GPS Compatible: √
Altitude Hold: √
Position Hold: √
Flight Planning: √
Autonomous Flight: √
Target Tracking: –
Automatic Landing: √
Return To Home: √
Obstacle Avoidance: –
Headless Mode: –
Built-in Camera: –

Huajun Toys

W609-8 Pathfinder

©Huajun Toys

Release Date: May 2015
Price: $100
Number of Rotors: 6
Control System: Remote Control
Applications: Film, Photography
Dimensions: 540 × 470 × 210 mm
Weight: 515 gr
Frame Material: Aluminum
Rotor Material: Plastic
Available Colors: Blue, Red
Max. Flight Time: 10 minutes
Max. Speed: 15 km/h (9 mp/h)
Max. Operating Range: 500 m
Controller Frequency: 2.4 GHz
Battery Capacity: 1,500 mAh
Battery Type: Lithium Polymer
Battery Power: 7.4V
GPS Compatible: –
Altitude Hold: √
Position Hold: √
Flight Planning: –
Autonomous Flight: –
Target Tracking: –
Automatic Landing: –
Return To Home: √
Obstacle Avoidance: –
Headless Mode: √
Built-in Camera: √ (2 megapixels)

Huajun Toys

W609-9

©Huajun Toys

Release Date: May 2015
Price: $65
Number of Rotors: 6
Control System: Remote Control
Applications: Recreation
Dimensions: 320 × 280 × 115 mm
Weight: 147 gr
Frame Material: Plastic
Rotor Material: Plastic
Available Colors: Blue, Gold
Max. Flight Time: 7 minutes
Max. Operating Range: 100 m
Controller Frequency: 2.4 GHz
Battery Capacity: 650 mAh
Battery Type: Lithium Polymer
Battery Power: 3.7V
GPS Compatible: –
Altitude Hold: √
Position Hold: √
Flight Planning: –
Autonomous Flight: –
Target Tracking: –
Automatic Landing: –
Return To Home: –
Obstacle Avoidance: –
Headless Mode: –
Built-in Camera: √ (0.3 megapixels)

Huajun Toys

W609-10

© Huajun Toys

Release Date: May 2015
Price: $65
Number of Rotors: 6
Control System: Remote Control
Applications: Recreation
Dimensions: 320 × 280 × 115 mm
Weight: 147 gr
Frame Material: Plastic
Rotor Material: Plastic
Available Colors: Blue, Red
Max. Flight Time: 10 minutes
Max. Operating Range: 100 m
Controller Frequency: 2.4 GHz
Battery Capacity: 650 mAh
Battery Type: Lithium Polymer
Battery Power: 3.7V
GPS Compatible: –
Altitude Hold: √
Position Hold: √
Flight Planning: –
Autonomous Flight: –
Target Tracking: –
Automatic Landing: –
Return To Home: √
Obstacle Avoidance: –
Headless Mode: √
Built-in Camera: √ (2 megapixels)

HuaXiang

8923 Smart

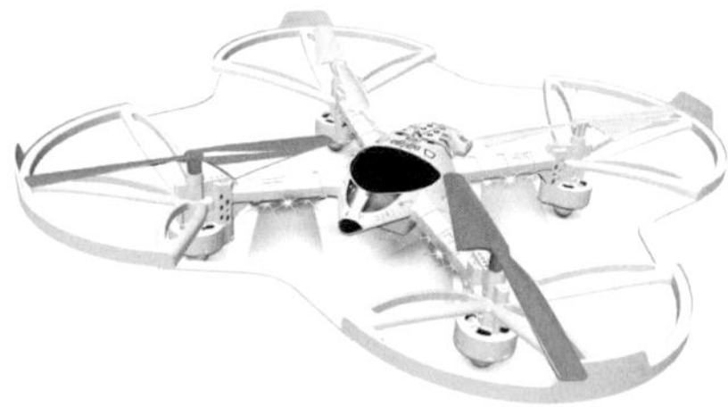

© HuaXiang

Release Date: May 2015
Price: $65
Number of Rotors: 4
Control System: Remote Control
Applications: Recreation
Dimensions: 305 × 305 × 80 mm
Frame Material: Plastic
Rotor Material: Plastic
Available Colors: White
Max. Flight Time: 8 minutes
Max. Operating Range: 200 m
Controller Frequency: 2.4 GHz
Battery Capacity: 600 mAh
Battery Type: Lithium Polymer
Battery Power: 3.7V
GPS Compatible: –
Altitude Hold: √
Position Hold: √
Flight Planning: –
Autonomous Flight: –
Target Tracking: –
Automatic Landing: –
Return To Home: –
Obstacle Avoidance: –
Headless Mode: –
Built-in Camera: √

JJRC

H8D

© JJRC

Release Date: May 2015
Price: $115
Number of Rotors: 4
Control System: Remote Control
Applications: Recreation
Dimensions: 330 × 330 × 110 mm
Wingspan: 467 mm
Weight: 1.2 kg
Max. Payload: 34 gr
Frame Material: Plastic
Rotor Material: Plastic
Available Colors: Red
Max. Flight Time: 8 minutes
Max. Operating Range: 300 m
Controller Frequency: 2.4 GHz
Battery Capacity: 500 mAh
Battery Type: Lithium Polymer
Battery Power: 7.4V
GPS Compatible: –
Altitude Hold: √
Position Hold: √
Flight Planning: –
Autonomous Flight: –
Target Tracking: –
Automatic Landing: –
Return To Home: √
Obstacle Avoidance: –
Headless Mode: √
Built-in Camera: √ (2 megapixels)

JJRC

JJ1000

© aliexpress.com

Release Date: May 2015
Price: $15
Number of Rotors: 4
Control System: Remote Control
Applications: Recreation
Dimensions: 80 × 80 × 30 mm
Wingspan: 113 mm
Frame Material: Plastic
Rotor Material: Plastic
Available Colors: White
Max. Flight Time: 8 minutes
Max. Operating Range: 80 m
Controller Frequency: 2.4 GHz
Battery Capacity: 250 mAh
Battery Type: Lithium Polymer
Battery Power: 7.4V
GPS Compatible: –
Altitude Hold: √
Position Hold: √
Flight Planning: –
Autonomous Flight: –
Target Tracking: –
Automatic Landing: –
Return To Home: –
Obstacle Avoidance: –
Headless Mode: √
Built-in Camera: √ (2 megapixels)

Kaiser Baas

Alpha

©techdaily.com

Release Date: May 2015
Price: $149
Number of Rotors: 4
Control System: Remote Control
Applications: Film, Photography
Dimensions: 310 × 310 mm
Wingspan: 330 mm
Weight: 109 gr
Frame Material: Plastic
Rotor Material: Plastic
Available Colors: Black
Max. Flight Time: 7 minutes
Max. Operating Range: 60 m
Controller Frequency: 2.4 GHz
Battery Capacity: 550 mAh
Battery Type: Lithium Polymer
GPS Compatible: –
Altitude Hold: √
Position Hold: √
Flight Planning: –
Autonomous Flight: –
Target Tracking: –
Automatic Landing: –
Return To Home: –
Obstacle Avoidance: –
Headless Mode: –
Built-in Camera: √ (2 megapixels)

Meijiaxin Toys

X400 V2

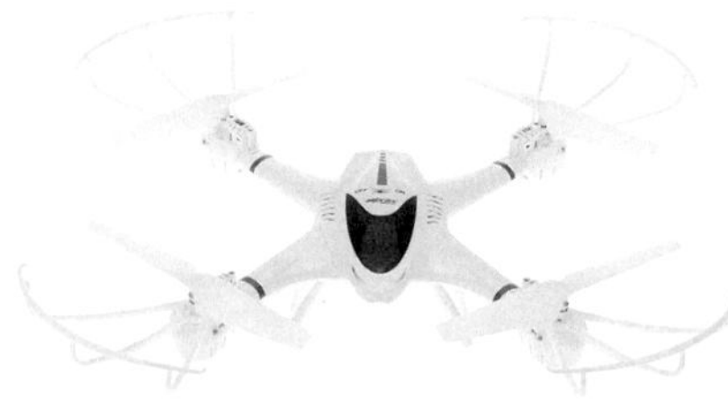

©aliexpress.com

Release Date: May 2015
Price: $45
Number of Rotors: 4
Control System: Remote Control
Applications: Recreation
Dimensions: 320 × 310 × 70 mm
Wingspan: 445 mm
Weight: 112 gr
Frame Material: Plastic
Rotor Material: Plastic
Available Colors: Red, White
Max. Flight Time: 12 minutes
Max. Operating Range: 100 m
Controller Frequency: 2.4 GHz
Battery Capacity: 750 mAh
Battery Type: Lithium Polymer
Battery Power: 3.7V
GPS Compatible: –
Altitude Hold: –
Position Hold: –
Flight Planning: –
Autonomous Flight: –
Target Tracking: –
Automatic Landing: –
Return To Home: –
Obstacle Avoidance: –
Headless Mode: √
Built-in Camera: –

RC Logger

RC Eye One NovaX 350

Release Date: May 2015
Price: $570
Number of Rotors: 4
Control System: Remote Control, Phone/Tablet
Applications: Film, Photography
Wingspan: 350 mm
Weight: 808 gr
Max. Payload: 350 gr
Frame Material: Carbon Fiber
Rotor Material: Plastic
Available Colors: Black
Max. Flight Time: 20 minutes
Max. Operating Range: 1 km
Controller Frequency: 2.4 GHz
Battery Capacity: 5,200 mAh
Battery Type: Lithium Polymer 3S
Battery Power: 11.1V
GPS Compatible: √
Altitude Hold: √
Position Hold: √
Flight Planning: √
Autonomous Flight: √
Target Tracking: √
Automatic Landing: –
Return To Home: √
Obstacle Avoidance: –
Headless Mode: –
Built-in Camera: –

Syma Toys

X5SW

© Syma Toys

Release Date: May 2015
Price: $75
Number of Rotors: 4
Control System: Remote Control, Phone/Tablet
Applications: Film, Photography
Dimensions: 315 × 315 × 105 mm
Wingspan: 445 mm
Weight: 967 gr
Frame Material: Plastic
Rotor Material: Plastic
Available Colors: Black, White
Max. Flight Time: 6 minutes
Max. Operating Range: 50 m
Controller Frequency: 2.4 GHz
Battery Capacity: 500 mAh
Battery Type: Lithium Polymer
Battery Power: 4V
GPS Compatible: –
Altitude Hold: –
Position Hold: –
Flight Planning: –
Autonomous Flight: –
Target Tracking: –
Automatic Landing: –
Return To Home: –
Obstacle Avoidance: –
Headless Mode: √
Built-in Camera: √ (0.3 megapixels)

UDIRC Toys

U839 Nano

©UDIRC Toys

Release Date: May 2015
Price: $90
Number of Rotors: 4
Control System: Remote Control
Applications: Recreation
Dimensions: 89 × 89 × 29 mm
Wingspan: 125 mm
Weight: 20 gr
Frame Material: Plastic
Rotor Material: Plastic
Available Colors: Green
Max. Flight Time: 6 minutes
Max. Operating Range: 100 m
Controller Frequency: 2.4 GHz
Battery Capacity: 150 mAh
Battery Type: Lithium Polymer
Battery Power: 3.7V
GPS Compatible: –
Altitude Hold: √
Position Hold: √
Flight Planning: –
Autonomous Flight: –
Target Tracking: –
Automatic Landing: –
Return To Home: –
Obstacle Avoidance: –
Headless Mode: –
Built-in Camera: –

Walkera

QR X350 Premium

©Walkera

Release Date: May 2015
Price: $1,499
Number of Rotors: 4
Control System: Remote Control, Phone/Tablet
Applications: Film, Photography, FPV Racing
Dimensions: 303 × 303 × 176 mm
Wingspan: 428 mm
Weight: 1.65 kg
Max. Payload: 200 gr
Frame Material: Plastic, Carbon Fiber
Rotor Material: Plastic
Available Colors: White
Max. Flight Time: 25 minutes
Max. Speed: 72 km/h (44,7 mp/h)
Max. Operating Range: 1.5 km
Controller Frequency: 2.4 GHz
Battery Capacity: 3,000 mAh
Battery Type: Lithium Polymer 8S
Battery Power: 29.6V
GPS Compatible: √
Altitude Hold: √
Position Hold: √
Flight Planning: √
Autonomous Flight: –
Target Tracking: –
Automatic Landing: √
Return To Home: √
Obstacle Avoidance: –
Headless Mode: –
Built-in Camera: –

WLtoys

V606 Explorers

Release Date: May 2015
Price: $145
Number of Rotors: 4
Control System: Remote Control
Applications: Recreation
Dimensions: 51 × 51 × 10 mm
Weight: 100 gr
Frame Material: Plastic, Foam
Rotor Material: Plastic
Available Colors: Black
Max. Flight Time: 10 minutes
Max. Operating Range: 150 m
Controller Frequency: 2.4 GHz
Battery Capacity: 730 mAh
Battery Type: Lithium Polymer
Battery Power: 3.7V
GPS Compatible: –
Altitude Hold: √
Position Hold: √
Flight Planning: –
Autonomous Flight: –
Target Tracking: –
Automatic Landing: –
Return To Home: √
Obstacle Avoidance: –
Headless Mode: –
Built-in Camera: –

WLtoys

Q202 Aircraft Carrier

Release Date: May 2015
Price: $60
Number of Rotors: 4
Control System: Remote Control
Applications: Recreation
Dimensions: 640 × 640 × 140 mm
Weight: 24 gr
Frame Material: Plastic, Foam
Rotor Material: Plastic
Available Colors: Black
Max. Flight Time: 8 minutes
Max. Operating Range: 150 m
Controller Frequency: 2.4 GHz
Battery Capacity: 850 mAh
Battery Type: Lithium Polymer
Battery Power: 3.7V
GPS Compatible: –
Altitude Hold: √
Position Hold: √
Flight Planning: –
Autonomous Flight: –
Target Tracking: –
Automatic Landing: –
Return To Home: √
Obstacle Avoidance: –
Headless Mode: –
Built-in Camera: –

XK Innovations
Alien X250

Release Date: May 2015
Price: $44
Number of Rotors: 4
Control System: Remote Control
Applications: Recreation
Dimensions: 194 × 194 × 61 mm
Weight: 100 gr
Frame Material: Plastic
Rotor Material: Plastic
Available Colors: Black
Max. Flight Time: 12 minutes
Max. Operating Range: 300 m
Controller Frequency: 2.4 GHz
Battery Capacity: 800 mAh
Battery Type: Lithium Polymer
Battery Power: 3.7V
GPS Compatible: –
Altitude Hold: √
Position Hold: √
Flight Planning: –
Autonomous Flight: –
Target Tracking: –
Automatic Landing: –
Return To Home: √
Obstacle Avoidance: –
Headless Mode: –
Built-in Camera: –

XK Innovations
Air Dancer X350

Release Date: May 2015
Price: $250
Number of Rotors: 4
Control System: Remote Control
Applications: Recreation
Dimensions: 305 × 305 × 210 mm
Wingspan: 376 mm
Weight: 770 gr
Frame Material: Plastic
Rotor Material: Plastic
Available Colors: White
Max. Flight Time: 18 minutes
Max. Operating Range: 1 km
Controller Frequency: 2.4 GHz
Battery Capacity: 2,200 mAh
Battery Type: Lithium Polymer
Battery Power: 11.1V
GPS Compatible: –
Altitude Hold: √
Position Hold: √
Flight Planning: –
Autonomous Flight: –
Target Tracking: –
Automatic Landing: √
Return To Home: √
Obstacle Avoidance: –
Headless Mode: –
Built-in Camera: –

XK Innovations

Detect X380

©aliexpress.com

Release Date: May 2015
Price: $200
Number of Rotors: 4
Control System: Remote Control
Applications: Film, Photography
Dimensions: 305 × 305 × 210 mm
Wingspan: 376 mm
Weight: 950 gr
Frame Material: Plastic, Carbon Fiber
Rotor Material: Plastic, Carbon Fiber
Available Colors: Black, White
Max. Flight Time: 30 minutes
Max. Operating Range: 1 km
Controller Frequency: 2.4 GHz
Battery Capacity: 5,400 mAh
Battery Type: Lithium Polymer
Battery Power: 11.1V
GPS Compatible: √
Altitude Hold: √
Position Hold: √
Flight Planning: –
Autonomous Flight: √
Target Tracking: –
Automatic Landing: √
Return To Home: √
Obstacle Avoidance: –
Headless Mode: –
Built-in Camera: –

Acme

Zoopa Q Crosswave 290

©Acme

Release Date: June 2015
Price: $550
Number of Rotors: 4
Propellers: Typ 6030, 4 × 6
Control System: Remote Control
Applications: FPV Racing
Dimensions: 235 × 235 × 68 mm
Wingspan: 278 mm
Weight: 570 gr
Frame Material: Carbon Fiber
Rotor Material: Plastic
Motors: 22Zero4 brushless
Available Colors: White
Max. Flight Time: 12 minutes
Max. Speed: 54 km/h (33,5 mp/h)
Max. Operating Range: 700 m
Controller Frequency: 2.4 GHz
Battery Capacity: 2,200 mAh
Battery Type: Lithium Polymer 3S
Battery Power: 11.1V
GPS Compatible: –
Altitude Hold: –
Position Hold: –
Flight Planning: –
Autonomous Flight: –
Target Tracking: –
Automatic Landing: –
Return To Home: –
Obstacle Avoidance: –
Headless Mode: –
Built-in Camera: √ (FlyCamOne Eco HD)

Acme

Zoopa Q165 Riot

© Acme

Release Date: June 2015
Price: $69
Number of Rotors: 4
Control System: Remote Control
Applications: Recreation
Dimensions: 160 × 160 × 40 mm
Wingspan: 210 mm
Weight: 44 gr
Frame Material: Plastic
Rotor Material: Plastic
Available Colors: Black
Max. Flight Time: 8 minutes
Max. Operating Range: 60 m
Controller Frequency: 2.4 GHz
Battery Capacity: 350 mAh
Battery Type: Lithium Polymer 1S
Battery Power: 3.7V
GPS Compatible: –
Altitude Hold: –
Position Hold: –
Flight Planning: –
Autonomous Flight: –
Target Tracking: –
Automatic Landing: –
Return To Home: –
Obstacle Avoidance: –
Headless Mode: –
Built-in Camera: –

Aerix Drones

Nano Drone

© Aerix Drones

Release Date: June 2015
Price: $30
Number of Rotors: 4
Control System: Remote Control, Phone
Applications: Recreation
Dimensions: 88 × 88 × 25 mm
Weight: 14 gr
Frame Material: Plastic
Rotor Material: Plastic
Motors: 3mm brushless
Available Colors: Black, Blue, Green, Orange, White
Max. Flight Time: 7 minutes
Max. Operating Range: 30 m
Controller Frequency: 2.4 GHz
Battery Capacity: 100 mAh
Battery Type: Lithium Polymer
Battery Power: 3.7V
GPS Compatible: –
Altitude Hold: √
Position Hold: –
Flight Planning: –
Autonomous Flight: –
Target Tracking: –
Automatic Landing: –
Return To Home: –
Obstacle Avoidance: –
Headless Mode: –
Built-in Camera: –

Airbus

Talarion

© air-cosmos.com

Release Date: June 2015
Price: $120 million
Number of Rotors: – (Fixed Wing)
Control System: Ground Control Station
Applications: Military Purposes, Intelligence, Reconnaissance, Surveillance
Dimensions: 12 × 28 × 4.5 m
Wingspan: 28 m
Weight: 7,000 kg
Max. Payload: 800 gr
Frame Material: Composite
Motors: 2 × turbojet engine
Available Colors: Grey
Max. Flight Time: 20 hours
Max. Speed: 630 km/h (391 mp/h)
Max. Flying Altitude: 15,000 m (49,213 ft)
Max. Operating Range: 16,000 km
GPS Compatible: √
Altitude Hold: √
Position Hold: –
Flight Planning: √
Autonomous Flight: √
Target Tracking: √
Automatic Landing: √
Return To Home: √
Obstacle Avoidance: √
Headless Mode: –
Built-in Camera: √

DJI

Matrice 100

© DJI

Release Date: June 2015
Price: $3,299
Number of Rotors: 4
Propellers: DJI 1345s
Control System: Remote Control, Phone/Tablet
Applications: Film, Photography
Wingspan: 650 mm
Weight: 2.43 kg
Max. Payload: 1 kg
Frame Material: Carbon Fiber
Rotor Material: Plastic
Motors: DJI 3510
Available Colors: Black
Max. Flight Time: 33 minutes
Max. Flight Time with Payload: 13 minutes
Max. Speed: 79 km/h (49 mp/h)
Max. Ascent Speed: 5 m/s
Max. Descent Speed: 4 m/s
Min. Operating Temperature: -10°C
Max. Operating Temperature: 40°C
Max. Operating Range: 5 km
Controller Frequency: 2.4 GHz
Battery Capacity: 4,500 mAh
Battery Type: Lithium Polymer 6S
Battery Power: 22.2V
Battery Weight: 595 gr
GPS Compatible: √
Altitude Hold: √
Position Hold: √
Flight Planning: √
Autonomous Flight: –
Target Tracking: √
Automatic Landing: –
Return To Home: √
Obstacle Avoidance: –
Headless Mode: –
Built-in Camera: –

Eachine

3D X4

©helicomicro.com

Release Date: June 2015
Price: $50
Number of Rotors: 4
Control System: Remote Control
Applications: Recreation
Dimensions: 173 × 173 × 36 mm
Frame Material: Plastic
Rotor Material: Plastic
Available Colors: Black, Red, White, Yellow
Max. Flight Time: 7 minutes
Max. Flying Altitude: 60 m (197 ft)
Max. Operating Range: 60 m
Controller Frequency: 2.4 GHz
Battery Capacity: 350 mAh
Battery Type: Lithium Polymer 1S
Battery Power: 4V
GPS Compatible: –
Altitude Hold: –
Position Hold: –
Flight Planning: –
Autonomous Flight: –
Target Tracking: –
Automatic Landing: –
Return To Home: –
Obstacle Avoidance: –
Headless Mode: –
Built-in Camera: –

Elbit Systems

Skylark I-LEX

©Israel Defense Forces/Zev Marmorstein

Release Date: June 2015
Price: Unknown
Number of Rotors: 1 (Fixed Wing)
Control System: Flight Computer
Applications: Military Purposes, Reconnaissance, Surveillance
Wingspan: 3 m
Weight: 7.5 kg
Max. Payload: 1.2 kg
Available Colors: Grey
Max. Flight Time: 3 hours
Max. Flying Altitude: 5,000 m (16,404 ft)
Max. Operating Range: 40 km
GPS Compatible: √
Altitude Hold: √
Position Hold: –
Flight Planning: √
Autonomous Flight: √
Target Tracking: –
Automatic Landing: √
Return To Home: √
Obstacle Avoidance: √
Headless Mode: –
Built-in Camera: √

Fineco
FX-3

© Fineco

Release Date: June 2015
Price: $18
Number of Rotors: 4
Control System: Remote Control
Applications: Recreation
Dimensions: 160 × 160 × 80 mm
Weight: 30 gr
Frame Material: Plastic
Rotor Material: Plastic
Available Colors: Green
Max. Flight Time: 12 minutes
Max. Operating Range: 30 m
Controller Frequency: 2.4 GHz
Battery Capacity: 380 mAh
Battery Type: Lithium Polymer
Battery Power: 3.7V
GPS Compatible: –
Altitude Hold: –
Position Hold: –
Flight Planning: –
Autonomous Flight: –
Target Tracking: –
Automatic Landing: –
Return To Home: √
Obstacle Avoidance: –
Headless Mode: –
Built-in Camera: –

Flytrex
Sky

© Flytrex

Release Date: June 2015
Price: $649
Number of Rotors: 4
Control System: Phone/Tablet
Applications: Film, Photography, Delivery
Wingspan: 470 mm
Weight: 1.25 kg
Max. Payload: 2 kg
Frame Material: Carbon Fiber
Rotor Material: Carbon Fiber
Available Colors: Black
Max. Flight Time: 30 minutes
Max. Speed: 72 km/h (44,7 mp/h)
Max. Ascent Speed: 5 m/s
Max. Descent Speed: 3 m/s
Min. Operating Temperature: 0°C
Max. Operating Temperature: 45°C
Max. Operating Range: 11 km
Controller Frequency: 2.4 GHz
Battery Capacity: 7,000 mAh
Battery Type: Lithium Polymer 3S
GPS Compatible: √
Altitude Hold: √
Position Hold: √
Flight Planning: √
Autonomous Flight: √
Target Tracking: –
Automatic Landing: √
Return To Home: √
Obstacle Avoidance: –
Headless Mode: –
Built-in Camera: –

FreeFly Systems

Alta 6

© coptercraft.com

Release Date: June 2015
Price: $11,995
Number of Rotors: 6
Control System: Remote Control
Applications: Film, Photography
Wingspan: 1.13 m
Weight: 4.5 kg
Max. Payload: 6.8 kg
Frame Material: Carbon Fiber
Rotor Material: Carbon Fiber, Balsa Core
Motors: Direct Drive 3-Phase PMAC Outrunner
Available Colors: Black
Controller Frequency: 2.4 GHz
Battery Capacity: 20,000 mAh
Battery Type: Lithium Polymer 6S
Battery Power: 22.2V
GPS Compatible: √
Altitude Hold: √
Position Hold: √
Flight Planning: √
Autonomous Flight: √
Target Tracking: √
Automatic Landing: √
Return To Home: √
Obstacle Avoidance: √
Headless Mode: –
Built-in Camera: –

Fuqi Toys

FQ777-124 Pocket Drone

© Fuqi Toys

Release Date: June 2015
Price: $14
Number of Rotors: 4
Control System: Remote Control
Applications: Recreation
Dimensions: 70 × 70 × 30 mm
Wingspan: 42 mm
Weight: 13 gr
Frame Material: Plastic
Rotor Material: Plastic
Available Colors: Black, Blue, Red, White
Max. Flight Time: 6 minutes
Max. Operating Range: 50 m
Controller Frequency: 2.4 GHz
Battery Capacity: 100 mAh
Battery Type: Lithium Polymer
Battery Power: 3.7V
GPS Compatible: –
Altitude Hold: –
Position Hold: –
Flight Planning: –
Autonomous Flight: –
Target Tracking: –
Automatic Landing: –
Return To Home: –
Obstacle Avoidance: –
Headless Mode: –
Built-in Camera: –

Helic Max

Sky Cruiser

© Helic Max

Release Date: June 2015
Price: $45
Number of Rotors: 4
Control System: Remote Control, Phone/Tablet
Applications: Recreation
Dimensions: 620 × 440 × 110 mm
Frame Material: Plastic
Rotor Material: Plastic
Available Colors: Black, White
Max. Flight Time: 5 minutes
Max. Operating Range: 50 m
Controller Frequency: 2.4 GHz
Battery Capacity: 700 mAh
Battery Type: Lithium Polymer
Battery Power: 7.4V
GPS Compatible: –
Altitude Hold: √
Position Hold: √
Flight Planning: –
Autonomous Flight: –
Target Tracking: –
Automatic Landing: –
Return To Home: –
Obstacle Avoidance: –
Headless Mode: –
Built-in Camera: √

Hesel EU

Splash Drone

© Hesel EU

Release Date: June 2015
Price: $999
Number of Rotors: 4
Control System: Remote Control
Applications: Film, Photography
Dimensions: 500 × 500 mm
Frame Material: Plastic
Rotor Material: Carbon Fiber
Motors: 650kv brushless
Available Colors: Red, White
Controller Frequency: 5.8 GHz
Battery Type: Lithium Polymer
GPS Compatible: √
Altitude Hold: √
Position Hold: √
Flight Planning: √
Autonomous Flight: √
Target Tracking: √
Automatic Landing: √
Return To Home: √
Obstacle Avoidance: –
Headless Mode: –
Built-in Camera: √

HiSky Hobby

HMX280

© helicomicro.com

Release Date: June 2015
Price: $170
Number of Rotors: 4
Control System: Remote Control
Applications: Recreation, FPV Racing
Wingspan: 280 mm
Weight: 420 gr
Frame Material: Plastic
Rotor Material: Plastic
Motors: 1818-1300kv
Available Colors: Black
Max. Flight Time: 10 minutes
Max. Speed: 100 km/h (62 mp/h)
Max. Operating Range: 400 m
Controller Frequency: 2.4 GHz
Battery Capacity: 1,350 mAh
Battery Type: Lithium Polymer 4S
Battery Power: 11.1V
GPS Compatible: –
Altitude Hold: √
Position Hold: √
Flight Planning: –
Autonomous Flight: –
Target Tracking: –
Automatic Landing: –
Return To Home: –
Obstacle Avoidance: –
Headless Mode: –
Built-in Camera: –

HuaXiang

8933V Exceed

© HuaXiang

Release Date: June 2015
Price: $45
Number of Rotors: 4
Control System: Remote Control
Applications: Recreation
Dimensions: 305 × 305 × 80 mm
Frame Material: Plastic
Rotor Material: Plastic
Available Colors: White
Max. Flight Time: 8 minutes
Max. Operating Range: 200 m
Controller Frequency: 2.4 GHz
Battery Capacity: 600 mAh
Battery Type: Lithium Polymer
Battery Power: 3.7V
GPS Compatible: –
Altitude Hold: √
Position Hold: √
Flight Planning: –
Autonomous Flight: –
Target Tracking: –
Automatic Landing: –
Return To Home: –
Obstacle Avoidance: –
Headless Mode: –
Built-in Camera: √ (0.3 megapixels)

Jin Xing Da

398

© Jin Xing Da

Release Date: June 2015
Price: $30
Number of Rotors: 4
Control System: Remote Control
Applications: Recreation
Dimensions: 145 × 145 × 25 mm
Frame Material: Plastic
Rotor Material: Plastic
Available Colors: Blue, Red
Max. Flight Time: 6 minutes
Max. Operating Range: 100 m
Controller Frequency: 2.4 GHz
Battery Capacity: 500 mAh
Battery Type: Lithium Polymer
Battery Power: 3.7V
GPS Compatible: –
Altitude Hold: –
Position Hold: –
Flight Planning: –
Autonomous Flight: –
Target Tracking: –
Automatic Landing: –
Return To Home: –
Obstacle Avoidance: –
Headless Mode: –
Built-in Camera: –

JJRC

H15

© JJRC

Release Date: June 2015
Price: $155
Number of Rotors: 4
Control System: Remote Control
Applications: Recreation
Dimensions: 520 × 520 × 160 mm
Wingspan: 734 mm
Frame Material: Plastic
Rotor Material: Plastic
Available Colors: Orange, White
Max. Flight Time: 8 minutes
Max. Operating Range: 150 m
Controller Frequency: 2.4 GHz
Battery Capacity: 850 mAh
Battery Type: Lithium Polymer
Battery Power: 7.4V
GPS Compatible: –
Altitude Hold: –
Position Hold: –
Flight Planning: –
Autonomous Flight: –
Target Tracking: –
Automatic Landing: –
Return To Home: –
Obstacle Avoidance: –
Headless Mode: –
Built-in Camera: √ (2 megapixels)

JJRC

H13C

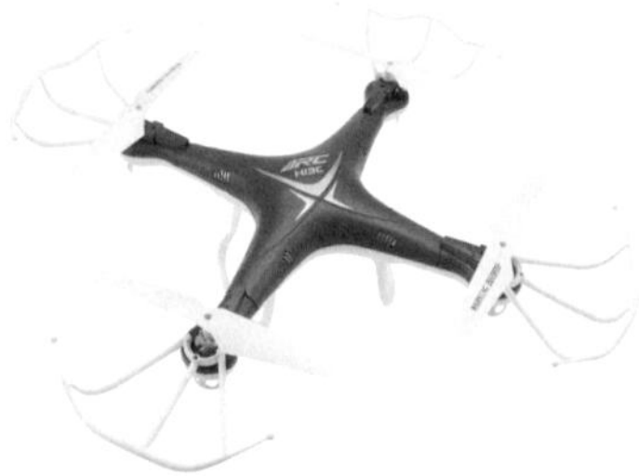

© JJRC

Release Date: June 2015
Price: $37
Number of Rotors: 4
Control System: Remote Control
Applications: Recreation
Dimensions: 315 × 315 × 80 mm
Wingspan: 445 mm
Frame Material: Plastic
Rotor Material: Plastic
Available Colors: Blue, Green
Max. Flight Time: 8 minutes
Max. Operating Range: 150 m
Controller Frequency: 2.4 GHz
Battery Capacity: 650 mAh
Battery Type: Lithium Polymer
Battery Power: 3.7V
GPS Compatible: –
Altitude Hold: –
Position Hold: –
Flight Planning: –
Autonomous Flight: –
Target Tracking: –
Automatic Landing: –
Return To Home: –
Obstacle Avoidance: –
Headless Mode: √
Built-in Camera: √ (2 megapixels)

JJRC

JJ669

© aliexpress.com

Release Date: June 2015
Price: $119
Number of Rotors: 4
Control System: Remote Control
Applications: Recreation
Dimensions: 540 × 540 × 165 mm
Wingspan: 467 mm
Frame Material: Plastic
Rotor Material: Plastic
Available Colors: Red
Max. Flight Time: 8 minutes
Max. Operating Range: 150 m
Controller Frequency: 2.4 GHz
Battery Capacity: 650 mAh
Battery Type: Lithium Polymer
Battery Power: 7V
GPS Compatible: –
Altitude Hold: –
Position Hold: –
Flight Planning: –
Autonomous Flight: –
Target Tracking: –
Automatic Landing: –
Return To Home: √
Obstacle Avoidance: –
Headless Mode: –
Built-in Camera: √ (2 megapixels)

JJRC

JJ830

© JJRC

Release Date: June 2015
Price: $35
Number of Rotors: 4
Control System: Remote Control
Applications: Recreation
Dimensions: 130 × 130 × 35 mm
Wingspan: 189 mm
Frame Material: Plastic
Rotor Material: Plastic
Available Colors: Black, Blue, Red
Max. Flight Time: 5 minutes
Max. Operating Range: 150 m
Controller Frequency: 2.4 GHz
Battery Capacity: 250 mAh
Battery Type: Lithium Polymer
Battery Power: 3.7V
GPS Compatible: –
Altitude Hold: –
Position Hold: –
Flight Planning: –
Autonomous Flight: –
Target Tracking: –
Automatic Landing: –
Return To Home: –
Obstacle Avoidance: –
Headless Mode: √
Built-in Camera: –

Meijiaxin Toys

X300C

© aliexpress.com

Release Date: June 2015
Price: $55
Number of Rotors: 4
Control System: Remote Control, Phone/Tablet
Applications: Recreation
Dimensions: 240 × 240 × 50 mm
Wingspan: 339 mm
Weight: 102 gr
Frame Material: Plastic
Rotor Material: Plastic
Available Colors: Silver, White
Max. Flight Time: 9 minutes
Max. Operating Range: 100 m
Controller Frequency: 2.4 GHz
Battery Capacity: 750 mAh
Battery Type: Lithium Polymer
Battery Power: 3.7V
GPS Compatible: –
Altitude Hold: –
Position Hold: –
Flight Planning: –
Autonomous Flight: –
Target Tracking: –
Automatic Landing: –
Return To Home: –
Obstacle Avoidance: –
Headless Mode: –
Built-in Camera: √ (0.3 megapixels)

Meijiaxin Toys

X500

© aliexpress.com

Release Date: June 2015
Price: $15
Number of Rotors: 4
Control System: Remote Control
Applications: Recreation
Dimensions: 315 × 285 × 85 mm
Wingspan: 425 mm
Frame Material: Plastic
Rotor Material: Plastic
Available Colors: Black, Gold
Max. Flight Time: 10 minutes
Max. Operating Range: 100 m
Controller Frequency: 2.4 GHz
Battery Capacity: 750 mAh
Battery Type: Lithium Polymer
Battery Power: 3.7V
GPS Compatible: –
Altitude Hold: –
Position Hold: –
Flight Planning: –
Autonomous Flight: –
Target Tracking: –
Automatic Landing: –
Return To Home: √
Obstacle Avoidance: –
Headless Mode: –
Built-in Camera: √ (0.3 megapixels)

Meijiaxin Toys

X600

© aliexpress.com

Release Date: June 2015
Price: $60
Number of Rotors: 6
Control System: Remote Control, Phone/Tablet
Applications: Recreation
Dimensions: 385 × 425 × 60 mm
Wingspan: 573 mm
Weight: 193 gr
Frame Material: Plastic
Rotor Material: Plastic
Available Colors: Black, White
Max. Flight Time: 12 minutes
Max. Operating Range: 100 m
Controller Frequency: 2.4 GHz
Battery Capacity: 750 mAh
Battery Type: Lithium Polymer
Battery Power: 7.4V
GPS Compatible: –
Altitude Hold: –
Position Hold: –
Flight Planning: –
Autonomous Flight: –
Target Tracking: –
Automatic Landing: –
Return To Home: √
Obstacle Avoidance: –
Headless Mode: √
Built-in Camera: –

Meijiaxin Toys

X800

©Meijiaxin Toys

Release Date: June 2015
Price: $30
Number of Rotors: 6
Control System: Remote Control
Applications: Recreation
Dimensions: 210 × 200 × 25 mm
Wingspan: 290 mm
Weight: 71 gr
Frame Material: Plastic
Rotor Material: Plastic
Available Colors: Black, White
Max. Flight Time: 10 minutes
Max. Operating Range: 100 m
Controller Frequency: 2.4 GHz
Battery Capacity: 750 mAh
Battery Type: Lithium Polymer
Battery Power: 3.7V
GPS Compatible: –
Altitude Hold: √
Position Hold: √
Flight Planning: –
Autonomous Flight: –
Target Tracking: –
Automatic Landing: –
Return To Home: √
Obstacle Avoidance: –
Headless Mode: –
Built-in Camera: –

Monstertronic

Invader V2

©Monstertronic

Release Date: June 2015
Price: $339
Number of Rotors: 4
Control System: Remote Control
Applications: Recreation, FPV Racing
Dimensions: 200 × 195 × 150 mm
Wingspan: 270 mm
Weight: 320 gr
Frame Material: Plastic
Rotor Material: Plastic
Available Colors: White
Controller Frequency: 2.4 GHz
Battery Capacity: 850 mAh
Battery Type: Lithium Polymer 2S
Battery Power: 14.8V
GPS Compatible: √
Altitude Hold: √
Position Hold: √
Flight Planning: –
Autonomous Flight: –
Target Tracking: –
Automatic Landing: –
Return To Home: √
Obstacle Avoidance: –
Headless Mode: –
Built-in Camera: –

Parrot

Airborne Cargo

©Parrot

Release Date: June 2015
Price: $110
Number of Rotors: 4
Control System: Phone/Tablet
Applications: Film, Photography, Recreation
Dimensions: 230 × 210 × 48 mm
Wingspan: 258 mm
Weight: 54 gr
Frame Material: Plastic
Rotor Material: Plastic
Available Colors: White, Yellow
Max. Flight Time: 9 minutes
Max. Speed: 32 km/h (20 mp/h)
Max. Ascent Speed: 5 m/s
Max. Descent Speed: 6 m/s
Max. Operating Range: 20 m
Controller Frequency: 2.4 GHz
Battery Capacity: 550 mAh
Battery Type: Lithium Polymer 3S
GPS Compatible: –
Altitude Hold: √
Position Hold: √
Flight Planning: –
Autonomous Flight: –
Target Tracking: –
Automatic Landing: √
Return To Home: –
Obstacle Avoidance: –
Headless Mode: –
Built-in Camera: √

Parrot

Airborne Night

©Parrot

Release Date: June 2015
Price: $145
Number of Rotors: 4
Control System: Phone/Tablet
Applications: Film, Photography, Recreation
Dimensions: 150 × 150 × 38 mm
Wingspan: 258 mm
Weight: 54 gr
Frame Material: Plastic
Rotor Material: Plastic
Available Colors: Black, Blue, Red
Max. Flight Time: 9 minutes
Max. Speed: 18 km/h (11,2 mp/h)
Max. Flying Altitude: 4 m (13 ft)
Max. Operating Range: 20 m
Controller Frequency: 2.4 GHz
Battery Capacity: 550 mAh
Battery Type: Lithium Polymer 3S
GPS Compatible: –
Altitude Hold: –
Position Hold: –
Flight Planning: –
Autonomous Flight: –
Target Tracking: –
Automatic Landing: –
Return To Home: –
Obstacle Avoidance: –
Headless Mode: –
Built-in Camera: √ (0.3 megapixels)

RoboFlight

RF-70

©diydrones.com

Release Date: June 2015
Price: $7,495
Number of Rotors: 1 (Fixed Wing)
Control System: Remote Control
Applications: Agriculture, Aerial Mapping
Wingspan: 1.8 m
Weight: 3 kg
Max. Payload: 2 kg
Frame Material: Foam
Available Colors: White
Max. Flight Time: 60 minutes
Controller Frequency: 2.4 GHz
Battery Capacity: 21,000 mAh
Battery Type: Lithium Polymer 3S
Battery Power: 12.6V
GPS Compatible: √
Altitude Hold: √
Position Hold: √
Flight Planning: √
Autonomous Flight: √
Target Tracking: √
Automatic Landing: √
Return To Home: √
Obstacle Avoidance: –
Headless Mode: –
Built-in Camera: √

Storm Racing Drone

SRD 260 Pro V2

©SRD

Release Date: June 2015
Price: $499
Number of Rotors: 6
Control System: Remote Control
Applications: FPV Racing
Dimensions: 240 × 218 × 60 mm
Weight: 546 gr
Frame Material: Carbon Fiber
Rotor Material: Carbon Fiber
Available Colors: Black, Red, White
Max. Operating Range: 1.5 km
Controller Frequency: 5.8 GHz
Battery Capacity: 1,800 mAh
Battery Type: Lithium Polymer
Battery Power: 14.8V
GPS Compatible: –
Altitude Hold: –
Position Hold: –
Flight Planning: –
Autonomous Flight: –
Target Tracking: –
Automatic Landing: –
Return To Home: –
Obstacle Avoidance: –
Headless Mode: –
Built-in Camera: √ (700TVL)

Storm Racing Drone

Storm Drone 8

©SRD

Release Date: June 2015
Price: $669
Number of Rotors: 8
Propellers: HQ 6035
Control System: Remote Control
Applications: Film, Photography
Dimensions: 410 × 410 × 210 mm
Weight: 940 gr
Frame Material: Carbon Fiber
Rotor Material: Carbon Fiber
Motors: Storm M2204-2300kv
Available Colors: Black
Max. Flight Time: 15 minutes
Max. Operating Range: 1.5 km
Controller Frequency: 2.4 GHz
Battery Capacity: 5,500 mAh
Battery Type: Lithium Polymer
Battery Power: 11.1V
GPS Compatible: √
Altitude Hold: √
Position Hold: √
Flight Planning: √
Autonomous Flight: √
Target Tracking: √
Automatic Landing: √
Return To Home: √
Obstacle Avoidance: –
Headless Mode: –
Built-in Camera: –

SteadiDrone

Vader

©SteadiDrone

Release Date: June 2015
Price: $19,995
Number of Rotors: 8
Control System: Remote Control
Applications: Film, Photography, Agriculture, Aerial Mapping, Inspection, Surveillance
Dimensions: 970 × 962 × 565 mm
Wingspan: 1320 mm
Weight: 3.8 kg
Max. Payload: 4.3 kg
Frame Material: Carbon Fiber
Rotor Material: Carbon Fiber
Available Colors: Red
Max. Flight Time: 50 minutes
Max. Speed: 72 km/h (44,7 mp/h)
Max. Ascent Speed: 9 m/s
Max. Descent Speed: 6 m/s
Min. Operating Temperature: -5°C
Max. Operating Temperature: 50°C
Max. Operating Range: 1.5 km
Controller Frequency: 2.4 GHz
Battery Capacity: 10,000 mAh
Battery Type: Lithium Polymer 5S
Battery Power: 22V
GPS Compatible: √
Altitude Hold: √
Position Hold: √
Flight Planning: √
Autonomous Flight: √
Target Tracking: –
Automatic Landing: √
Return To Home: √
Obstacle Avoidance: –
Headless Mode: –
Built-in Camera: –

Uvision

Hero-30

Release Date: June 2015
Price: Unknown
Number of Rotors: 1 (Fixed Wing)
Control System: Flight Computer
Applications: Loitering munition
Dimensions: 800 × 350 × 350 mm
Wingspan: 350 mm
Weight: 3 kg
Max. Payload: 0.5 kg
Available Colors: White
Max. Flight Time: 30 minutes
Max. Speed: 185 km/h (115 mp/h)
Max. Flying Altitude: 3,100 m (10,170 ft)
Max. Operating Range: 40 km
GPS Compatible: √
Altitude Hold: √
Position Hold: –
Flight Planning: √
Autonomous Flight: √
Target Tracking: √
Automatic Landing: –
Return To Home: –
Obstacle Avoidance: √
Headless Mode: –
Built-in Camera: √

Uvision

Hero-70

Release Date: June 2015
Price: Unknown
Number of Rotors: 1 (Fixed Wing)
Control System: Flight Computer
Applications: Loitering munition
Dimensions: 1 × 0.4 × 0.4 m
Wingspan: 0.4 m
Weight: 7 kg
Max. Payload: 1.2 kg
Available Colors: White
Max. Flight Time: 45 minutes
Max. Speed: 150 km/h (93,2 mp/h)
Max. Flying Altitude: 3,100 m (10,170 ft)
Max. Operating Range: 40 km
GPS Compatible: √
Altitude Hold: √
Position Hold: –
Flight Planning: √
Autonomous Flight: √
Target Tracking: √
Automatic Landing: –
Return To Home: –
Obstacle Avoidance: √
Headless Mode: –
Built-in Camera: √

Uvision

Hero-120

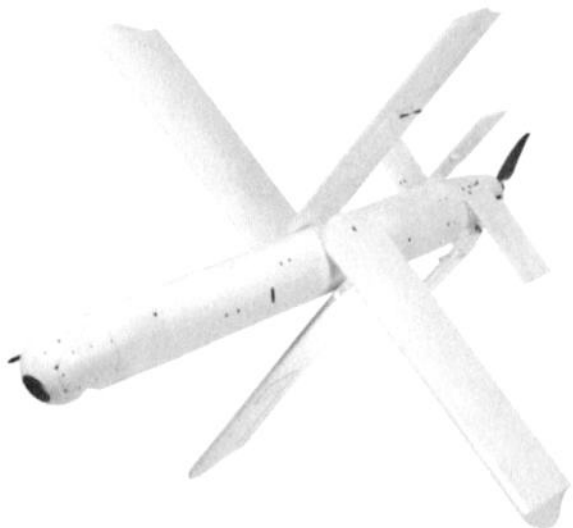

© Uvision

Release Date: June 2015
Price: Unknown
Number of Rotors: 1 (Fixed Wing)
Control System: Flight Computer
Applications: Loitering munition
Dimensions: 1.5 × 0.6 × 0.6 m
Wingspan: 0.6 m
Weight: 12.5 kg
Max. Payload: 3.5 kg
Available Colors: White
Max. Flight Time: 60 minutes
Max. Speed: 120 km/h (74 mp/h)
Max. Flying Altitude: 3,100 m (10,170 ft)
Max. Operating Range: 40 km
GPS Compatible: √
Altitude Hold: √
Position Hold: –
Flight Planning: √
Autonomous Flight: √
Target Tracking: √
Automatic Landing: –
Return To Home: –
Obstacle Avoidance: √
Headless Mode: –
Built-in Camera: √

Uvision

Hero-250

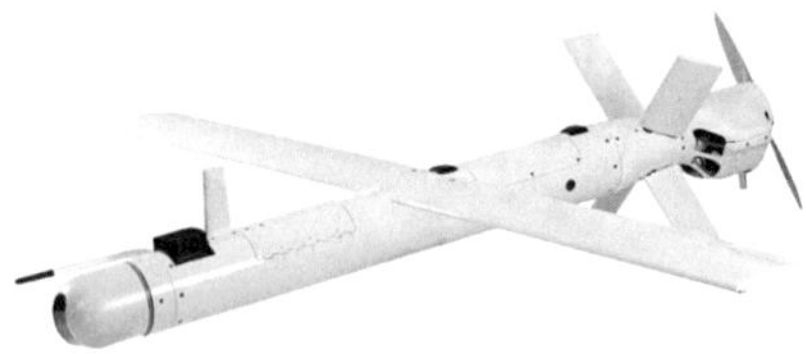

© Uvision

Release Date: June 2015
Price: Unknown
Number of Rotors: 1 (Fixed Wing)
Control System: Flight Computer
Applications: Loitering munition
Dimensions: 1.8 × 1.2 × 0.35 m
Wingspan: 1.2 m
Weight: 25 kg
Max. Payload: 5 kg
Available Colors: White
Max. Flight Time: 3 hours
Max. Speed: 150 km/h (93,2 mp/h)
Max. Flying Altitude: 14,000 m (45,931 ft)
Max. Operating Range: 150 km
GPS Compatible: √
Altitude Hold: √
Position Hold: –
Flight Planning: √
Autonomous Flight: √
Target Tracking: √
Automatic Landing: –
Return To Home: –
Obstacle Avoidance: √
Headless Mode: –
Built-in Camera: √

Uvision

Hero-400

Release Date: June 2015
Price: Unknown
Number of Rotors: 1 (Fixed Wing)
Control System: Flight Computer
Applications: Loitering munition
Dimensions: 2.2 × 1.5 × 0.4 m
Wingspan: 1.5 m
Weight: 40 kg
Max. Payload: 8 kg
Available Colors: White
Max. Flight Time: 4 hours
Max. Speed: 120 km/h (74 mp/h)
Max. Flying Altitude: 14,000 m (45,931 ft)
Max. Operating Range: 150 km
GPS Compatible: √
Altitude Hold: √
Position Hold: –
Flight Planning: √
Autonomous Flight: √
Target Tracking: √
Automatic Landing: –
Return To Home: –
Obstacle Avoidance: √
Headless Mode: –
Built-in Camera: √

Uvision

Hero-900

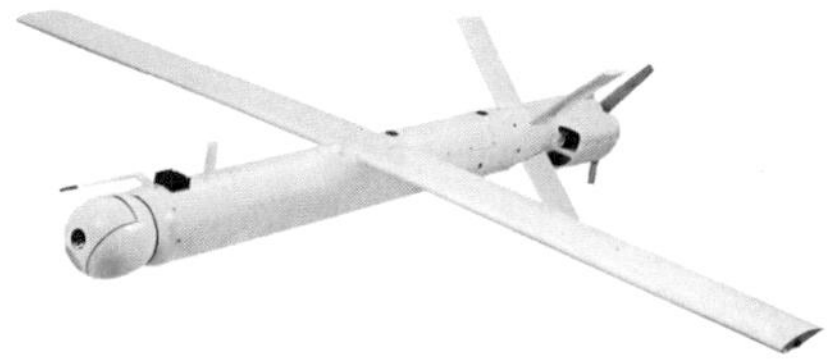

Release Date: June 2015
Price: Unknown
Number of Rotors: 1 (Fixed Wing)
Control System: Flight Computer
Applications: Loitering munition
Dimensions: 2.5 × 1.8 × 0.5 m
Wingspan: 1.8 m
Weight: 97 kg
Max. Payload: 20 kg
Available Colors: White
Max. Flight Time: 7 hours
Max. Speed: 120 km/h (74 mp/h)
Max. Flying Altitude: 14,000 m (45,931 ft)
Max. Operating Range: 250 km
GPS Compatible: √
Altitude Hold: √
Position Hold: –
Flight Planning: √
Autonomous Flight: √
Target Tracking: √
Automatic Landing: –
Return To Home: –
Obstacle Avoidance: √
Headless Mode: –
Built-in Camera: √

Volt Robotics

Venture

© Volt Robotics

Release Date: June 2015
Price: Unknown
Number of Rotors: 4
Control System: Remote Control
Applications: Aerial Mapping, Agriculture, Inspection, Surveillance
Dimensions: 838 × 838 × 247 mm
Weight: 2 kg
Max. Payload: 1 kg
Frame Material: Carbon Fiber
Available Colors: Black
Max. Flight Time: 30 minutes
Max. Speed: 30 km/h (18,6 mp/h)
Max. Operating Range: 10 km
Controller Frequency: 2.4 GHz
Battery Type: Lithium Polymer
GPS Compatible: √
Altitude Hold: √
Position Hold: √
Flight Planning: √
Autonomous Flight: √
Target Tracking: –
Automatic Landing: √
Return To Home: √
Obstacle Avoidance: √
Headless Mode: –
Built-in Camera: √ (20 megapixels)

Yi Zhan Toys

Tarantula X6

© koopjedrone.nl

Release Date: June 2015
Price: $55
Number of Rotors: 4
Control System: Remote Control
Applications: Film, Photography, Recreation
Dimensions: 480 × 480 × 160 mm
Wingspan: 679 mm
Weight: 310 gr
Frame Material: Plastic
Rotor Material: Plastic
Available Colors: Blue, White
Max. Flight Time: 7 minutes
Max. Operating Range: 300 m
Controller Frequency: 2.4 GHz
Battery Capacity: 1,200 mAh
Battery Type: Lithium Polymer
Battery Power: 7.4V
GPS Compatible: √
Altitude Hold: –
Position Hold: –
Flight Planning: –
Autonomous Flight: –
Target Tracking: –
Automatic Landing: –
Return To Home: √
Obstacle Avoidance: –
Headless Mode: –
Built-in Camera: √ (5 megapixels)

Acme

Zoopa Q550 Evolution

Release Date: July 2015
Price: $425
Number of Rotors: 4
Control System: Remote Control
Applications: Film, Photography
Dimensions: 245 × 245 × 210 mm
Wingspan: 350 mm
Weight: 900 gr
Frame Material: Plastic
Rotor Material: Plastic
Available Colors: Black
Max. Flight Time: 15 minutes
Max. Speed: 54 km/h (33,5 mp/h)
Max. Ascent Speed: 6 m/s
Max. Operating Range: 1 km
Controller Frequency: 2.4 GHz
Battery Capacity: 2,200 mAh
Battery Type: Lithium Polymer 3S
Battery Power: 11.1V
GPS Compatible: √
Altitude Hold: √
Position Hold: √
Flight Planning: √
Autonomous Flight: √
Target Tracking: –
Automatic Landing: √
Return To Home: –
Obstacle Avoidance: –
Headless Mode: –
Built-in Camera: –

AEE Technology

Toruk AP9

Release Date: July 2015
Price: $349
Number of Rotors: 4
Control System: Remote Control, Phone/Tablet
Applications: Film, Photography
Wingspan: 450 mm
Weight: 920 gr
Frame Material: Plastic
Rotor Material: Plastic
Motors:
Available Colors: White
Max. Flight Time: 25 minutes
Max. Speed: 72 km/h (44,7 mp/h)
Max. Ascent Speed: 6 m/s
Max. Descent Speed: 4 m/s
Max. Operating Range: 500 m
Controller Frequency: 2.4 GHz
Battery Capacity: 5,300 mAh
Battery Type: Lithium Polymer
Battery Power: 11.1V
GPS Compatible: √
Altitude Hold: √
Position Hold: √
Flight Planning: –
Autonomous Flight: –
Target Tracking: –
Automatic Landing: √
Return To Home: √
Obstacle Avoidance: –
Headless Mode: –
Built-in Camera: √ (16 megapixels)

AeroSky

X350

© raidentech.com

Release Date: July 2015
Price: $569
Number of Rotors: 4
Propellers: 8045
Control System: Remote Control
Applications: Film, Photography
Wingspan: 350 mm
Weight: 900 gr
Max. Payload: 500 gr
Frame Material: Plastic
Rotor Material: Plastic
Motors: 2312-1000kv
Available Colors: White
Max. Flight Time: 10 minutes
Max. Speed: 36 km/h (22,3 mp/h)
Max. Ascent Speed: 3 m/s
Max. Descent Speed: 3 m/s
Min. Operating Temperature: -10°C
Max. Operating Temperature: 50°C
Controller Frequency: 2.4 GHz
Battery Capacity: 2,800 mAh
Battery Type: Lithium Polymer 3S
Battery Power: 11.1V
GPS Compatible: √
Altitude Hold: √
Position Hold: √
Flight Planning: –
Autonomous Flight: √
Target Tracking: –
Automatic Landing: √
Return To Home: √
Obstacle Avoidance: –
Headless Mode: –
Built-in Camera: –

Autel Robotics

X-Star

© Autel Robotics

Release Date: July 2015
Price: $699
Number of Rotors: 4
Control System: Remote Control, Phone/Tablet
Applications: Film, Photography
Wingspan: 352 mm
Weight: 1.4 kg
Max. Payload: 180 gr
Frame Material: Plastic
Rotor Material: Plastic
Available Colors: Orange, White
Max. Flight Time: 25 minutes
Max. Speed: 57 km/h (35,7 mp/h)
Max. Ascent Speed: 6 m/s
Min. Operating Temperature: 0°C
Max. Operating Temperature: 50°C
Max. Operating Range: 800 m
Controller Frequency: 2.4 GHz
Battery Capacity: 6,400 mAh
Battery Type: Lithium Polymer
Battery Power: 14.8V
GPS Compatible: –
Altitude Hold: √
Position Hold: √
Flight Planning: –
Autonomous Flight: –
Target Tracking: √
Automatic Landing: √
Return To Home: √
Obstacle Avoidance: –
Headless Mode: –
Built-in Camera: –

Blade

Inductrix

© bhphotovideo.com

Release Date: July 2015
Price: $50
Number of Rotors: 4
Control System: Remote Control
Applications: Recreation
Dimensions: 83 × 83 × 28 mm
Wingspan: 117 mm
Weight: 19 gr
Frame Material: Plastic
Rotor Material: Plastic
Available Colors: Blue, Red
Max. Flight Time: 8 minutes
Controller Frequency: 2.4 GHz
Battery Capacity: 150 mAh
Battery Type: Lithium Polymer 1S
Battery Power: 3.7V
GPS Compatible: –
Altitude Hold: –
Position Hold: –
Flight Planning: –
Autonomous Flight: –
Target Tracking: –
Automatic Landing: –
Return To Home: –
Obstacle Avoidance: –
Headless Mode: –
Built-in Camera: –

EMAX

Nighthawk 280 Pro

© rctech.de

Release Date: July 2015
Price: $258
Number of Rotors: 4
Propellers: 6045
Control System: Remote Control
Applications: FPV Racing
Dimensions: 280 mm
Weight: 440 gr
Frame Material: Carbon Fiber
Rotor Material: Plastic
Motors: EMAX 2204-2300kv
Available Colors: Black
Controller Frequency: 5.8 GHz
Battery Capacity: 2,200 mAh
Battery Type: Lithium Polymer 3S
GPS Compatible: √
Altitude Hold: –
Position Hold: –
Flight Planning: –
Autonomous Flight: –
Target Tracking: –
Automatic Landing: –
Return To Home: –
Obstacle Avoidance: –
Headless Mode: –
Built-in Camera: √ (700TVL)

Fineco

FX-7Ci

©costway.com

Release Date: July 2015
Price: $60
Number of Rotors: 4
Control System: Remote Control
Applications: Recreation
Dimensions: 450 × 450 × 105 mm
Weight: 450 gr
Frame Material: Plastic
Rotor Material: Plastic
Available Colors: White
Max. Flight Time: 12 minutes
Max. Operating Range: 100 m
Controller Frequency: 2.4 GHz
Battery Capacity: 1,000 mAh
Battery Type: Lithium Polymer
Battery Power: 3.7V
GPS Compatible: –
Altitude Hold: √
Position Hold: √
Flight Planning: –
Autonomous Flight: –
Target Tracking: –
Automatic Landing: –
Return To Home: √
Obstacle Avoidance: –
Headless Mode: –
Built-in Camera: √ (2 megapixels)

Fineco

FX-2

©gearbest.com

Release Date: July 2015
Price: $25
Number of Rotors: 4
Control System: Remote Control
Applications: Recreation
Dimensions: 40 × 40 × 24 mm
Frame Material: Plastic
Rotor Material: Plastic
Available Colors: Orange
Max. Flight Time: 8 minutes
Max. Operating Range: 80 m
Controller Frequency: 2.4 GHz
Battery Capacity: 100 mAh
Battery Type: Lithium Polymer
Battery Power: 3.7V
GPS Compatible: –
Altitude Hold: –
Position Hold: –
Flight Planning: –
Autonomous Flight: –
Target Tracking: –
Automatic Landing: –
Return To Home: –
Obstacle Avoidance: –
Headless Mode: –
Built-in Camera: –

Foxtech

Devourer130 V2

Release Date: July 2015
Price: $6,220
Number of Rotors: 8
Propellers: Supreme 2880 Pro
Control System: Remote Control
Applications: Film, Photography
Dimensions: 1.2 × 1.2 × 0.73 m
Wingspan: 1.3 m
Weight: 7.2 kg
Max. Payload: 16 kg
Frame Material: Aluminum, Carbon Fiber
Rotor Material: Plastic
Motors: T-Motor U8 kV170
Available Colors: Black
Max. Flight Time: 48 minutes
Controller Frequency: 2.4 GHz
Battery Capacity: 22,000 mAh
Battery Type: Lithium Polymer 6S
GPS Compatible: √
Altitude Hold: √
Position Hold: √
Flight Planning: √
Autonomous Flight: √
Target Tracking: –
Automatic Landing: √
Return To Home: √
Obstacle Avoidance: –
Headless Mode: –
Built-in Camera: –

FUAV

Seraphi Pro

Release Date: July 2015
Price: $419
Number of Rotors: 4
Control System: Remote Control, Phone/Tablet
Applications: Film, Photography
Weight: 1.35 kg
Frame Material: Carbon Fiber
Rotor Material: Carbon Fiber
Available Colors: Black, White
Max. Flight Time: 15 minutes
Max. Speed: 57,6 km/h (35,7 mp/h)
Min. Operating Temperature: -10°C
Max. Operating Temperature: 40°C
Controller Frequency: 2.4 GHz
Battery Capacity: 5,200 mAh
Battery Type: Lithium Polymer 6S
Battery Power: 11.1V
Battery Weight: 472 gr
GPS Compatible: √
Altitude Hold: √
Position Hold: √
Flight Planning: –
Autonomous Flight: –
Target Tracking: –
Automatic Landing: √
Return To Home: √
Obstacle Avoidance: –
Headless Mode: –
Built-in Camera: √ (Sony EXMOR 16.7 megapixel)

Helic Max

Sky Roller

© Helic Max

Release Date: July 2015
Price: $38
Number of Rotors: 4
Control System: Remote Control
Applications: Recreation
Weight: 22 gr
Frame Material: Plastic
Rotor Material: Plastic
Available Colors: Green
Max. Flight Time: 5 minutes
Max. Operating Range: 30 m
Controller Frequency: 2.4 GHz
Battery Capacity: 100 mAh
Battery Type: Lithium Polymer
Battery Power: 3.7V
GPS Compatible: –
Altitude Hold: √
Position Hold: √
Flight Planning: –
Autonomous Flight: –
Target Tracking: –
Automatic Landing: –
Return To Home: –
Obstacle Avoidance: –
Headless Mode: –
Built-in Camera: –

Hobbyzone

Zugo

© Horizon Hobby

Release Date: July 2015
Price: $60
Number of Rotors: 4
Control System: Remote Control
Applications: Recreation
Dimensions: 130 × 130 × 38 mm
Weight: 52 gr
Frame Material: Plastic
Rotor Material: Plastic
Available Colors: Black
Max. Flight Time: 7 minutes
Controller Frequency: 2.4 GHz
Battery Capacity: 380 mAh
Battery Type: Lithium Polymer 1S
Battery Power: 3.7V
GPS Compatible: –
Altitude Hold: √
Position Hold: √
Flight Planning: √
Autonomous Flight: –
Target Tracking: –
Automatic Landing: –
Return To Home: –
Obstacle Avoidance: –
Headless Mode: –
Built-in Camera: √ (2 megapixels)

HuaXiang

8963 Camera

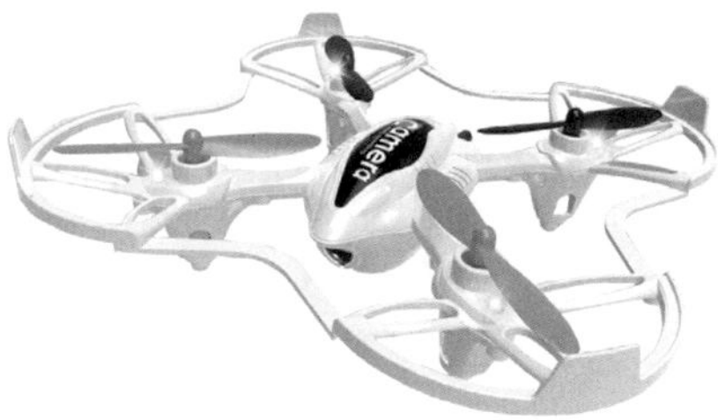

© HuaXiang

Release Date: July 2015
Price: $45
Number of Rotors: 4
Control System: Remote Control
Applications: Recreation
Dimensions: 130 × 130 × 40 mm
Weight: 542 gr
Frame Material: Plastic
Rotor Material: Plastic
Available Colors: White
Max. Flight Time: 7 minutes
Max. Operating Range: 100 m
Controller Frequency: 2.4 GHz
Battery Capacity: 500 mAh
Battery Type: Lithium Polymer
Battery Power: 3.7V
GPS Compatible: –
Altitude Hold: √
Position Hold: √
Flight Planning: –
Autonomous Flight: –
Target Tracking: –
Automatic Landing: –
Return To Home: –
Obstacle Avoidance: –
Headless Mode: –
Built-in Camera: √

Jin Xing Da

502

© aliexpress.com

Release Date: July 2015
Price: $20
Number of Rotors: 4
Control System: Remote Control
Applications: Recreation
Dimensions: 30 × 30 × 20 mm
Frame Material: Plastic
Rotor Material: Plastic
Available Colors: Blue
Max. Flight Time: 5 minutes
Max. Operating Range: 15 m
Controller Frequency: 2.4 GHz
Battery Capacity: 75 mAh
Battery Type: Lithium Polymer
Battery Power: 3.7V
GPS Compatible: –
Altitude Hold: –
Position Hold: –
Flight Planning: –
Autonomous Flight: –
Target Tracking: –
Automatic Landing: –
Return To Home: √
Obstacle Avoidance: –
Headless Mode: √
Built-in Camera: –

JJRC

H8 Mini

© dhgate.com

Release Date: July 2015
Price: $15
Number of Rotors: 4
Control System: Remote Control
Applications: Recreation
Dimensions: 110 × 110 × 30 mm
Wingspan: 156 mm
Frame Material: Plastic
Rotor Material: Plastic
Available Colors: Black, White
Max. Flight Time: 5 minutes
Max. Operating Range: 150 m
Controller Frequency: 2.4 GHz
Battery Type: Lithium Polymer
Battery Power: 3.7V
GPS Compatible: –
Altitude Hold: –
Position Hold: –
Flight Planning: –
Autonomous Flight: –
Target Tracking: –
Automatic Landing: –
Return To Home: √
Obstacle Avoidance: –
Headless Mode: –
Built-in Camera: –

JJRC

JJ850

© specsout.com

Release Date: July 2015
Price: $18
Number of Rotors: 4
Control System: Remote Control
Applications: Recreation
Dimensions: 165 × 165 × 36 mm
Wingspan: 165 mm
Weight: 39 gr
Frame Material: Plastic
Rotor Material: Plastic
Available Colors: Red
Max. Flight Time: 7 minutes
Max. Operating Range: 100 m
Controller Frequency: 2.4 GHz
Battery Capacity: 300 mAh
Battery Type: Lithium Polymer
Battery Power: 3.7V
GPS Compatible: –
Altitude Hold: √
Position Hold: √
Flight Planning: –
Autonomous Flight: –
Target Tracking: –
Automatic Landing: –
Return To Home: √
Obstacle Avoidance: –
Headless Mode: √
Built-in Camera: –

Meijiaxin Toys

X705C

© Meijiaxin Toys

Release Date: July 2015
Price: $70
Number of Rotors: 4
Control System: Remote Control, Phone/Tablet
Applications: Recreation
Dimensions: 315 × 315 × 120 mm
Wingspan: 445 mm
Weight: 750 gr
Frame Material: Plastic
Rotor Material: Plastic
Available Colors: White
Max. Flight Time: 9 minutes
Max. Operating Range: 100 m
Controller Frequency: 2.4 GHz
Battery Capacity: 550 mAh
Battery Type: Lithium Polymer
Battery Power: 3.7V
GPS Compatible: –
Altitude Hold: √
Position Hold: √
Flight Planning: –
Autonomous Flight: –
Target Tracking: –
Automatic Landing: –
Return To Home: √
Obstacle Avoidance: –
Headless Mode: –
Built-in Camera: –

Meijiaxin Toys

X900

© gearbest.com

Release Date: July 2015
Price: $19
Number of Rotors: 4
Control System: Remote Control
Applications: Recreation
Dimensions: 86 × 75 × 20 mm
Wingspan: 114 mm
Weight: 22 gr
Frame Material: Plastic
Rotor Material: Plastic
Motors: Coreless Motor
Available Colors: Black, White
Max. Flight Time: 5 minutes
Max. Operating Range: 100 m
Controller Frequency: 2.4 GHz
Battery Capacity: 180 mAh
Battery Type: Lithium Polymer
Battery Power: 3.7V
GPS Compatible: –
Altitude Hold: –
Position Hold: –
Flight Planning: –
Autonomous Flight: –
Target Tracking: –
Automatic Landing: –
Return To Home: –
Obstacle Avoidance: –
Headless Mode: –
Built-in Camera: –

Parrot

Hydrofoil

© Parrot

Release Date: July 2015
Price: $160
Number of Rotors: 4
Control System: Phone/Tablet
Applications: Film, Photography, Recreation
Dimensions: 340 × 320 × 140 mm
Wingspan: 180 mm
Weight: 245 gr
Frame Material: Plastic
Rotor Material: Plastic
Available Colors: Black
Max. Flight Time: 9 minutes
Max. Speed: 18 km/h (11,2 mp/h)
Max. Operating Range: 20 m
Controller Frequency: 2.4 GHz
Battery Capacity: 550 mAh
Battery Type: Lithium Polymer 3S
GPS Compatible: –
Altitude Hold: –
Position Hold: –
Flight Planning: –
Autonomous Flight: –
Target Tracking: –
Automatic Landing: –
Return To Home: –
Obstacle Avoidance: –
Headless Mode: –
Built-in Camera: √ (3 megapixels)

Robodrone

Sparrow

© Robodrone

Release Date: July 2015
Price: $55,700
Number of Rotors: 4
Control System: Remote Control
Applications: Reconnaissance, Surveillance
Dimensions: 1100 × 1100 × 220 mm
Wingspan: 813 mm
Weight: 12 kg
Max. Payload: 5 kg
Frame Material: Carbon Fiber
Rotor Material: Carbon Fiber
Available Colors: Black, Green
Max. Flight Time: 60 minutes
Max. Speed: 40 km/h (25 mp/h)
Max. Flying Altitude: 152 m (500 ft)
Max. Ascent Speed: 6 m/s
Controller Frequency: 2.4 GHz
Battery Capacity: 36,000 mAh
Battery Type: Lithium Polymer 8S
Battery Power: 29V
GPS Compatible: √
Altitude Hold: √
Position Hold: √
Flight Planning: √
Autonomous Flight: √
Target Tracking: √
Automatic Landing: √
Return To Home: √
Obstacle Avoidance: –
Headless Mode: –
Built-in Camera: –

Sci.Aero

CyberQuad Maxi

©dronelife.com

Release Date: July 2015
Price: $37,000
Number of Rotors: 4
Control System: Remote Control, Phone/Tablet
Applications: Film, Photography, Agriculture, Inspection, Surveillance
Dimensions: 560 × 640 × 250 mm
Wingspan: 850 mm
Weight: 1.7 kg
Max. Payload: 800 gr
Frame Material: Plastic
Rotor Material: Plastic
Available Colors: Orange
Max. Flight Time: 25 minutes
Max. Speed: 60 km/h (37 mp/h)
Max. Ascent Speed: 7 m/s
Max. Operating Range: 500 m
Controller Frequency: 2.4 GHz
GPS Compatible: √
Altitude Hold: √
Position Hold: √
Flight Planning: √
Autonomous Flight: √
Target Tracking: –
Automatic Landing: –
Return To Home: √
Obstacle Avoidance: –
Headless Mode: –
Built-in Camera: √

Sky Hero

Anakin

© Sky Hero

Release Date: July 2015
Price: $499
Number of Rotors: 4
Control System: Remote Control
Applications: FPV Racing
Wingspan: 200 mm
Weight: 116 gr
Frame Material: Plastic, Carbon Fiber
Rotor Material: Plastic
Available Colors: Black, White
Max. Flight Time: 10 minutes
Controller Frequency: 2.4 GHz
Battery Capacity: 2,200 mAh
Battery Type: Lithium Polymer 3S
GPS Compatible: –
Altitude Hold: –
Position Hold: –
Flight Planning: –
Autonomous Flight: –
Target Tracking: –
Automatic Landing: –
Return To Home: –
Obstacle Avoidance: –
Headless Mode: –
Built-in Camera: √

Tekever

AR3 Net Ray

© air-cosmos.com

Release Date: July 2015
Price: On Request
Number of Rotors: 1 (Fixed Wing)
Control System: Ground Control Station
Applications: Military Purposes, Maritime Patrol, Reconnaissance, Surveillance
Dimensions: 1.4 × 3.2 m
Wingspan: 3.2 m
Weight: 14 kg
Max. Payload: 8 kg
Frame Material: Composite
Available Colors: Black
Max. Flight Time: 10 hours
Max. Speed: 140 km/h (87 mp/h)
Max. Operating Range: 80 km
GPS Compatible: √
Altitude Hold: √
Position Hold: √
Flight Planning: √
Autonomous Flight: √
Target Tracking: –
Automatic Landing: –
Return To Home: √
Obstacle Avoidance: √
Headless Mode: –
Built-in Camera: √

TRNDlabs

SKEYE Mini

© TRNDlabs

Release Date: July 2015
Price: $100
Number of Rotors: 4
Control System: Remote Control
Applications: Recreation
Dimensions: 88 × 88 × 23 mm
Weight: 37 gr
Frame Material: Plastic
Rotor Material: Plastic
Available Colors: Orange
Max. Flight Time: 8 minutes
Max. Operating Range: 50 m
Controller Frequency: 2.4 GHz
Battery Capacity: 250 mAh
Battery Type: Lithium Polymer
Battery Power: 3.7V
GPS Compatible: –
Altitude Hold: –
Position Hold: –
Flight Planning: –
Autonomous Flight: –
Target Tracking: –
Automatic Landing: –
Return To Home: –
Obstacle Avoidance: –
Headless Mode: –
Built-in Camera: √ (2 megapixels)

UDIRC Toys
U842 Falcon

Release Date: July 2015
Price: $64
Number of Rotors: 4
Control System: Remote Control
Applications: Recreation
Dimensions: 486 × 486 × 122 mm
Wingspan: 685 mm
Weight: 346 gr
Frame Material: Plastic
Rotor Material: Plastic, Carbon Fiber
Available Colors: Black, White
Max. Flight Time: 8 minutes
Max. Operating Range: 80 m
Controller Frequency: 2.4 GHz
Battery Capacity: 1,000 mAh
Battery Type: Lithium Polymer
Battery Power: 7.2V
GPS Compatible: –
Altitude Hold: √
Position Hold: √
Flight Planning: –
Autonomous Flight: –
Target Tracking: –
Automatic Landing: –
Return To Home: –
Obstacle Avoidance: –
Headless Mode: –
Built-in Camera: √

UDIRC Toys
U843 multi-skywalker

Release Date: July 2015
Price: $55
Number of Rotors: 4
Control System: Remote Control
Applications: Recreation
Dimensions: 110 × 110 × 43 mm
Wingspan: 177 mm
Weight: 52 gr
Frame Material: Plastic
Rotor Material: Plastic
Available Colors: Green, Orange
Max. Flight Time: 7 minutes
Max. Operating Range: 30 m
Controller Frequency: 2.4 GHz
Battery Capacity: 450 mAh
Battery Type: Lithium Polymer
Battery Power: 3.7V
GPS Compatible: –
Altitude Hold: –
Position Hold: –
Flight Planning: –
Autonomous Flight: –
Target Tracking: –
Automatic Landing: –
Return To Home: –
Obstacle Avoidance: –
Headless Mode: –
Built-in Camera: –

Walkera

QR X900

© Walkera

Release Date: July 2015
Price: $4,399
Number of Rotors: 6
Control System: Remote Control
Applications: Film, Photography, Aerial Mapping, Surveillance
Dimensions: 983 × 853 × 687 mm
Wingspan: 1300 mm
Weight: 7 kg
Max. Payload: 3 kg
Frame Material: Carbon Fiber
Rotor Material: Carbon Fiber
Available Colors: Black
Max. Flight Time: 25 minutes
Max. Speed: 79 km/h (49 mp/h)
Min. Operating Temperature: -10°C
Max. Operating Temperature: 40°C
Max. Operating Range: 1 km
Controller Frequency: 2.4 GHz
Battery Capacity: 16,000 mAh
Battery Type: Lithium Polymer 6S
Battery Power: 22.2V
GPS Compatible: √
Altitude Hold: √
Position Hold: √
Flight Planning: √
Autonomous Flight: –
Target Tracking: –
Automatic Landing: –
Return To Home: √
Obstacle Avoidance: –
Headless Mode: –
Built-in Camera: –

Xamen Technologies

LE 4-8 X Dual

© Xamen Technologies

Release Date: July 2015
Price: On Request
Number of Rotors: 4
Control System: Ground Control Stration, Remote Control
Applications: Film, Photography, Aerial Mapping
Dimensions: 310 × 245 × 300 mm
Weight: 3 kg
Max. Payload: 3 kg
Frame Material: Carbon Fiber
Rotor Material: Carbon Fiber
Available Colors: White
Max. Flight Time: 20 minutes
Max. Speed: 43 km/h (26,8 mp/h)
Max. Flying Altitude: 1,500 m (4,921 ft)
Min. Operating Temperature: -15°C
Max. Operating Temperature: 55°C
Max. Operating Range: 4 km
Controller Frequency: 2.4 GHz
Battery Type: Lithium Polymer
GPS Compatible: √
Altitude Hold: √
Position Hold: √
Flight Planning: √
Autonomous Flight: √
Target Tracking: –
Automatic Landing: √
Return To Home: √
Obstacle Avoidance: –
Headless Mode: –
Built-in Camera: –

Yuneec

Typhoon Q500 4K

© Yuneec

Release Date: July 2015
Price: $1,000
Number of Rotors: 4
Control System: Remote Control, Phone/Tablet
Applications: Film, Photography
Dimensions: 500 × 420 × 240 mm
Wingspan: 565 mm
Weight: 1.1 kg
Max. Payload: 600 gr
Frame Material: Carbon Fiber
Rotor Material: Carbon Fiber
Available Colors: Black
Max. Flight Time: 25 minutes
Max. Speed: 29 km/h (18 mp/h)
Max. Flying Altitude: 122 m (400 ft)
Max. Ascent Speed: 3 m/s
Max. Descent Speed: 2 m/s
Min. Operating Temperature: -5°C
Max. Operating Temperature: 80°C
Max. Operating Range: 800 m
Controller Frequency: 2.4 GHz
Battery Capacity: 5,400 mAh
Battery Type: Lithium Polymer 3S
Battery Power: 11V
GPS Compatible: √
Altitude Hold: √
Position Hold: √
Flight Planning: –
Autonomous Flight: –
Target Tracking: √
Automatic Landing: √
Return To Home: √
Obstacle Avoidance: –
Headless Mode: –
Built-in Camera: √ (12 megapixels)

Ares

Spidex

© Ares-RC

Release Date: August 2015
Price: $35
Number of Rotors: 4
Control System: Remote Control, Phone/Tablet
Applications: Recreation
Dimensions: 162 × 162 × 33 mm
Wingspan: 102 mm
Weight: 20 gr
Frame Material: Plastic
Rotor Material: Plastic
Available Colors: Green, Orange
Controller Frequency: 2.4 GHz
Battery Capacity: 180 mAh
Battery Type: Lithium Polymer 1S
Battery Power: 3.7V
GPS Compatible: –
Altitude Hold: –
Position Hold: √
Flight Planning: –
Autonomous Flight: –
Target Tracking: –
Automatic Landing: –
Return To Home: –
Obstacle Avoidance: –
Headless Mode: –
Built-in Camera: –

China Aerospace Science and Technology Corporation (CASC)

CH-3 Rainbow

© weibo.com/dskaogu

Release Date: August 2015
Price: Unknown
Number of Rotors: 1 (Fixed Wing)
Control System: Ground Control Station
Applications: Military Purposes, Armed Reconnaissance, Surveillance
Wingspan: 8 m
Max. Payload: 80 kg
Frame Material: Composite
Available Colors: Green
Max. Flight Time: 12 hours
Max. Flying Altitude: 4,000 m (13,123 ft)
Max. Operating Range: 960 km
GPS Compatible: √
Altitude Hold: √
Position Hold: –
Flight Planning: √
Autonomous Flight: √
Target Tracking: √
Automatic Landing: √
Return To Home: √
Obstacle Avoidance: √
Headless Mode: –
Built-in Camera: √

Dromida

Vista UAV

© Dromida

Release Date: August 2015
Price: $80
Number of Rotors: 4
Control System: Remote Control
Applications: Recreation
Wingspan: 251 mm
Weight: 122 gr
Frame Material: Plastic
Rotor Material: Plastic
Available Colors: Blue, Green, Red, White
Max. Flight Time: 15 minutes
Controller Frequency: 2.4 GHz
Battery Capacity: 850 mAh
Battery Type: Lithium Polymer 1S
Battery Power: 3.7V
GPS Compatible: –
Altitude Hold: –
Position Hold: –
Flight Planning: √
Autonomous Flight: –
Target Tracking: –
Automatic Landing: –
Return To Home: –
Obstacle Avoidance: –
Headless Mode: –
Built-in Camera: –

Eachine

H8 3D Mini

© banggood.com

Release Date: August 2015
Price: $40
Number of Rotors: 4
Control System: Remote Control
Applications: Recreation
Dimensions: 124 × 124 × 81 mm
Frame Material: Plastic
Rotor Material: Plastic
Available Colors: Blue, Green, Orange
Max. Flight Time: 8 minutes
Max. Operating Range: 30 m
Controller Frequency: 2.4 GHz
Battery Capacity: 220 mAh
Battery Type: Lithium Polymer 1S
Battery Power: 3V
GPS Compatible: –
Altitude Hold: –
Position Hold: –
Flight Planning: –
Autonomous Flight: –
Target Tracking: –
Automatic Landing: –
Return To Home: √
Obstacle Avoidance: –
Headless Mode: –
Built-in Camera: –

Eachine

H7

© banggood.com

Release Date: August 2015
Price: $35
Number of Rotors: 4
Control System: Remote Control
Applications: Recreation
Dimensions: 95 × 95 × 27 mm
Weight: 14 gr
Frame Material: Plastic
Rotor Material: Plastic
Available Colors: Green, Silver
Max. Flight Time: 6 minutes
Max. Operating Range: 20 m
Controller Frequency: 2.4 GHz
Battery Capacity: 150 mAh
Battery Type: Lithium Polymer 1S
Battery Power: 3V
GPS Compatible: –
Altitude Hold: √
Position Hold: √
Flight Planning: –
Autonomous Flight: –
Target Tracking: –
Automatic Landing: –
Return To Home: –
Obstacle Avoidance: –
Headless Mode: –
Built-in Camera: –

EscaDrone

Squad3

© EscaDrone

Release Date: August 2015
Price: $6,480
Number of Rotors: 4
Control System: Remote Control
Applications: Aerial Mapping, Agriculture, Inspection, Surveillance
Dimensions: 660 × 660 × 480 mm
Weight: 1.8 kg
Max. Payload: 950 gr
Frame Material: Aluminum, Carbon Fiber
Rotor Material: Carbon Fiber
Available Colors: Black
Max. Flight Time: 30 minutes
Max. Speed: 29 km/h (18 mp/h)
Controller Frequency: 2.4 GHz
Battery Capacity: 8,000 mAh
Battery Type: Lithium Polymer 6S
GPS Compatible: √
Altitude Hold: √
Position Hold: √
Flight Planning: √
Autonomous Flight: √
Target Tracking: –
Automatic Landing: √
Return To Home: √
Obstacle Avoidance: –
Headless Mode: –
Built-in Camera: –

eTurbine

TB 250

© rc-quadcopter.de

Release Date: August 2015
Price: $180
Number of Rotors: 4
Control System: Remote Control
Applications: FPV Racing
Dimensions: 230 × 210 × 55 mm
Wingspan: 250 mm
Weight: 391 gr
Frame Material: Aluminum
Rotor Material: Plastic
Available Colors: Black
Max. Flight Time: 10 minutes
Max. Operating Range: 1 km
Controller Frequency: 5.8 GHz
Battery Capacity: 1,500 mAh
Battery Type: Lithium Polymer 3S
Battery Power: 11.1V
GPS Compatible: –
Altitude Hold: √
Position Hold: √
Flight Planning: –
Autonomous Flight: –
Target Tracking: –
Automatic Landing: –
Return To Home: –
Obstacle Avoidance: –
Headless Mode: –
Built-in Camera: √

Extreme Fliers

Microdrone 3.0

© Extreme Fliers

Release Date: August 2015
Price: $180
Number of Rotors: 4
Control System: Remote Control, Phone/Tablet
Applications: FPV Racing
Dimensions: 145 × 145 × 50 mm
Wingspan: 185 mm
Weight: 56 gr
Max. Payload: 35 gr
Frame Material: Plastic
Rotor Material: Plastic
Available Colors: Black
Max. Flight Time: 8 minutes
Max. Speed: 72 km/h (44,7 mp/h)
Max. Operating Range: 90 m
Controller Frequency: 2.4 GHz
Battery Capacity: 450 mAh
Battery Type: Lithium Polymer
GPS Compatible: √
Altitude Hold: –
Position Hold: –
Flight Planning: √
Autonomous Flight: –
Target Tracking: –
Automatic Landing: –
Return To Home: √
Obstacle Avoidance: –
Headless Mode: –
Built-in Camera: √

Fineco

FX-21

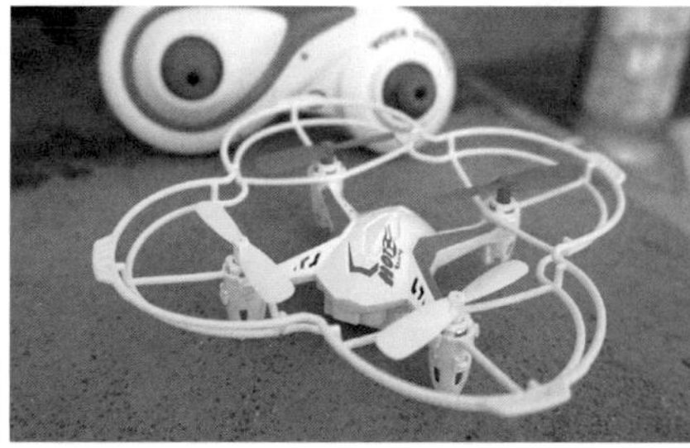

© Fineco

Release Date: August 2015
Price: $30
Number of Rotors: 4
Control System: Remote Control
Applications: Recreation
Dimensions: 125 × 125 × 33 mm
Weight: 15 gr
Frame Material: Plastic
Rotor Material: Plastic
Available Colors: Red, White
Max. Flight Time: 10 minutes
Max. Operating Range: 80 m
Controller Frequency: 2.4 GHz
Battery Capacity: 150 mAh
Battery Type: Lithium Polymer
Battery Power: 3.7V
GPS Compatible: –
Altitude Hold: –
Position Hold: –
Flight Planning: –
Autonomous Flight: –
Target Tracking: –
Automatic Landing: –
Return To Home: √
Obstacle Avoidance: –
Headless Mode: –
Built-in Camera: √

Fuqi Toys

FQ777-956C Sirius

© Fuqi Toys

Release Date: August 2015
Price: $54
Number of Rotors: 6
Control System: Remote Control
Applications: Recreation
Dimensions: 210 × 210 × 57 mm
Wingspan: 297 mm
Weight: 66 gr
Frame Material: Plastic
Rotor Material: Plastic
Available Colors: Black, White
Max. Flight Time: 7 minutes
Max. Operating Range: 80 m
Controller Frequency: 2.4 GHz
Battery Capacity: 550 mAh
Battery Type: Lithium Polymer
Battery Power: 3.7V
GPS Compatible: –
Altitude Hold: √
Position Hold: –
Flight Planning: –
Autonomous Flight: –
Target Tracking: –
Automatic Landing: –
Return To Home: –
Obstacle Avoidance: –
Headless Mode: –
Built-in Camera: √ (2 megapixels)

HeliMax

Voltage 500 3D

© rotordronemag.com

Release Date: August 2015
Price: $430
Number of Rotors: 4
Control System: Remote Control
Applications: Recreation, FPV Racing
Wingspan: 500 mm
Weight: 900 gr
Frame Material: Carbon Fiber
Rotor Material: Carbon Fiber
Available Colors: Red
Max. Flight Time: 15 minutes
Max. Operating Range: 1 km
Controller Frequency: 2.4 GHz
Battery Capacity: 4,000 mAh
Battery Type: Lithium Polymer 3S
Battery Power: 11.1V
GPS Compatible: –
Altitude Hold: –
Position Hold: –
Flight Planning: –
Autonomous Flight: –
Target Tracking: –
Automatic Landing: –
Return To Home: √
Obstacle Avoidance: –
Headless Mode: –
Built-in Camera: –

Hobby King

Black Widow 260 FPV

©Hobby King

Release Date: August 2015
Price: $170
Number of Rotors: 4
Control System: Remote Control
Applications: FPV Racing
Dimensions: 260 mm
Frame Material: Carbon Fiber
Rotor Material: Plastic
Motors: 2204-2300kv brushless
Available Colors: Black
Controller Frequency: 2.4 GHz
Battery Capacity: 1,300 mAh
Battery Type: Lithium Polymer 3S
Battery Power: 11.1V
GPS Compatible: √
Altitude Hold: √
Position Hold: √
Flight Planning: –
Autonomous Flight: –
Target Tracking: –
Automatic Landing: –
Return To Home: √
Obstacle Avoidance: –
Headless Mode: –
Built-in Camera: √

JJRC

H20

©aliexpress.com

Release Date: August 2015
Price: $19
Number of Rotors: 6
Control System: Remote Control
Applications: Recreation
Dimensions: 110 × 100 × 25 mm
Wingspan: 149 mm
Weight: 110 gr
Frame Material: Plastic
Rotor Material: Plastic
Available Colors: Gold, Red
Max. Flight Time: 7 minutes
Max. Operating Range: 30 m
Controller Frequency: 2.4 GHz
Battery Capacity: 150 mAh
Battery Type: Lithium Polymer
Battery Power: 3.7V
GPS Compatible: –
Altitude Hold: √
Position Hold: √
Flight Planning: –
Autonomous Flight: –
Target Tracking: –
Automatic Landing: –
Return To Home: √
Obstacle Avoidance: –
Headless Mode: –
Built-in Camera: –

JJRC

H6D

© helicomicro.com

Release Date: August 2015
Price: $90
Number of Rotors: 4
Control System: Remote Control
Applications: Recreation
Dimensions: 110 × 105 × 35 mm
Wingspan: 152 mm
Frame Material: Plastic
Rotor Material: Plastic
Available Colors: Black
Max. Flight Time: 7 minutes
Max. Operating Range: 300 m
Controller Frequency: 2.4 GHz
Battery Capacity: 350 mAh
Battery Type: Lithium Polymer
Battery Power: 3.7V
GPS Compatible: –
Altitude Hold: √
Position Hold: √
Flight Planning: –
Autonomous Flight: –
Target Tracking: –
Automatic Landing: –
Return To Home: √
Obstacle Avoidance: –
Headless Mode: –
Built-in Camera: √ (2 megapixels)

JJRC

H10

© JJRC

Release Date: August 2015
Price: $25
Number of Rotors: 4
Control System: Remote Control
Applications: Recreation
Dimensions: 300 × 300 × 100 mm
Wingspan: 424 mm
Frame Material: Plastic
Rotor Material: Plastic
Available Colors: Blue, Orange
Max. Flight Time: 8 minutes
Max. Operating Range: 100 m
Controller Frequency: 2.4 GHz
Battery Capacity: 150 mAh
Battery Type: Lithium Polymer
Battery Power: 3.7V
GPS Compatible: –
Altitude Hold: –
Position Hold: –
Flight Planning: –
Autonomous Flight: –
Target Tracking: –
Automatic Landing: –
Return To Home: √
Obstacle Avoidance: –
Headless Mode: √
Built-in Camera: √ (2 megapixels)

JJRC

JJ200

© aliexpress.com

Release Date: August 2015
Price: $69
Number of Rotors: 4
Control System: Remote Control
Applications: Recreation
Dimensions: 260 × 228 × 228 mm
Weight: 91 gr
Frame Material: Plastic
Rotor Material: Plastic
Available Colors: Orange
Max. Flight Time: 6 minutes
Max. Operating Range: 100 m
Controller Frequency: 2.4 GHz
Battery Capacity: 380 mAh
Battery Type: Lithium Polymer
Battery Power: 3.7V
GPS Compatible: –
Altitude Hold: –
Position Hold: –
Flight Planning: –
Autonomous Flight: –
Target Tracking: –
Automatic Landing: –
Return To Home: –
Obstacle Avoidance: –
Headless Mode: –
Built-in Camera: √ (2 megapixels)

JJRC

H18

© aliexpress.com

Release Date: August 2015
Price: $19
Number of Rotors: 4
Control System: Remote Control
Applications: Recreation
Dimensions: 85 × 85 × 25 mm
Wingspan: 85 mm
Weight: 24 gr
Frame Material: Plastic
Rotor Material: Plastic
Available Colors: Black, Orange
Max. Flight Time: 6 minutes
Max. Operating Range: 20 m
Controller Frequency: 2.4 GHz
Battery Capacity: 180 mAh
Battery Type: Lithium Polymer
Battery Power: 3.7V
GPS Compatible: –
Altitude Hold: √
Position Hold: √
Flight Planning: –
Autonomous Flight: –
Target Tracking: –
Automatic Landing: –
Return To Home: –
Obstacle Avoidance: –
Headless Mode: –
Built-in Camera: √ (2 megapixels)

Keyshare US
Glint

© firstquadcopter.com

Release Date: August 2015
Price: Unknown
Number of Rotors: 4
Control System: Remote Control
Applications: Film, Photography
Wingspan: 450 mm
Frame Material: Plastic
Rotor Material: Plastic
Available Colors: Black, Blue, Silver, White
Max. Flight Time: 25 minutes
Max. Speed: 36 km/h (22,3 mp/h)
Max. Flying Altitude: 5,000 m (16,404 ft)
Min. Operating Temperature: -10°C
Max. Operating Temperature: 40°C
Max. Operating Range: 1.2 km
Controller Frequency: 5.8 GHz
Battery Capacity: 5,300 mAh
Battery Type: Lithium Polymer
Battery Power: 14.8V
GPS Compatible: √
Altitude Hold: √
Position Hold: √
Flight Planning: √
Autonomous Flight: √
Target Tracking: –
Automatic Landing: √
Return To Home: √
Obstacle Avoidance: –
Headless Mode: –
Built-in Camera: –

Lishitoys
L6053

© aliexpress.com

Release Date: August 2015
Price: $30
Number of Rotors: 4
Control System: Remote Control
Applications: Recreation
Dimensions: 215 × 215 × 50 mm
Weight: 65 gr
Frame Material: Plastic
Rotor Material: Plastic
Available Colors: Red
Max. Flight Time: 5 minutes
Max. Operating Range: 50 m
Controller Frequency: 2.4 GHz
Battery Capacity: 380 mAh
Battery Type: Lithium Polymer
Battery Power: 3.7V
GPS Compatible: –
Altitude Hold: –
Position Hold: –
Flight Planning: –
Autonomous Flight: –
Target Tracking: –
Automatic Landing: –
Return To Home: √
Obstacle Avoidance: –
Headless Mode: √
Built-in Camera: –

MOLA

Tourist 1

©drohnenstore24.de

Release Date: August 2015
Price: $599
Number of Rotors: 4
Control System: Remote Control, Phone/Tablet
Applications: Film, Photography
Weight: 700 gr
Frame Material: Carbon Fiber, Plastic
Rotor Material: Carbon Fiber
Available Colors: Black, White
Max. Flight Time: 20 minutes
Controller Frequency: 2.4 GHz
GPS Compatible: √
Altitude Hold: –
Position Hold: –
Flight Planning: √
Autonomous Flight: √
Target Tracking: √
Automatic Landing: √
Return To Home: √
Obstacle Avoidance: √
Headless Mode: –
Built-in Camera: √ (12 megapixels)

Monstertronic

Sirius Quattro

©Monstertronic

Release Date: August 2015
Price: $699
Number of Rotors: 4
Control System: Remote Control
Applications: Film, Photography, Recreation
Dimensions: 700 × 420 × 320 mm
Wingspan: 600 mm
Weight: 1.5 kg
Frame Material: Plastic
Rotor Material: Plastic
Available Colors: Black
Controller Frequency: 2.4 GHz
Battery Capacity: 850 mAh
Battery Type: Lithium Polymer
GPS Compatible: √
Altitude Hold: √
Position Hold: √
Flight Planning: √
Autonomous Flight: √
Target Tracking: –
Automatic Landing: √
Return To Home: √
Obstacle Avoidance: –
Headless Mode: –
Built-in Camera: –

Muvi

X drone

©Muvi

Release Date: August 2015
Price: $1,000
Number of Rotors: 4
Control System: Remote Control, Phone/Tablet
Applications: Film, Photography, Aerial Mapping
Dimensions: 432 × 432 × 241 mm
Wingspan: 450 mm
Weight: 1.4 kg
Frame Material: Plastic
Rotor Material: Plastic
Available Colors: Black
Max. Flight Time: 20 minutes
Max. Speed: 72 km/h (44,7 mp/h)
Max. Flying Altitude: 100 m (328 ft)
Max. Ascent Speed: 8 m/s
Max. Descent Speed: 5 m/s
Min. Operating Temperature: 0°C
Max. Operating Temperature: 50°C
Max. Operating Range: 1 km
Controller Frequency: 2.4 GHz
Battery Capacity: 5,300 mAh
Battery Type: Lithium Polymer 4S
GPS Compatible: √
Altitude Hold: √
Position Hold: √
Flight Planning: √
Autonomous Flight: –
Target Tracking: √
Automatic Landing: √
Return To Home: √
Obstacle Avoidance: –
Headless Mode: –
Built-in Camera: √ (16 megapixels)

Precision Drone

Scout

©Precision Drone

Release Date: August 2015
Price: Unknown
Number of Rotors: 6
Control System: Remote Control
Applications: Film, Photography, Aerial Mapping, Agriculture
Frame Material: Carbon Fiber
Rotor Material: Carbon Fiber
Available Colors: Black
Controller Frequency: 2.4 GHz
GPS Compatible: √
Altitude Hold: √
Position Hold: √
Flight Planning: √
Autonomous Flight: √
Target Tracking: –
Automatic Landing: –
Return To Home: √
Obstacle Avoidance: –
Headless Mode: –
Built-in Camera: –

Revell Control

Proto Quad

©Revell Control

Release Date: August 2015
Price: $45
Number of Rotors: 4
Control System: Remote Control
Applications: Recreation
Dimensions: 28 × 28 × 19 mm
Weight: 7 gr
Frame Material: Plastic
Rotor Material: Plastic
Available Colors: Black, Green, Red, White
Max. Flight Time: 5 minutes
Max. Operating Range: 20 m
Controller Frequency: 2.4 GHz
Battery Capacity: 80 mAh
Battery Type: Lithium Polymer
Battery Power: 3.7V
GPS Compatible: –
Altitude Hold: –
Position Hold: –
Flight Planning: –
Autonomous Flight: –
Target Tracking: –
Automatic Landing: –
Return To Home: –
Obstacle Avoidance: –
Headless Mode: –
Built-in Camera: –

Syma Toys

X8G

©hitoys.org

Release Date: August 2015
Price: $125
Number of Rotors: 4
Control System: Remote Control
Applications: Film, Photography, Recreation
Dimensions: 500 × 500 × 190 mm
Wingspan: 707 mm
Weight: 1.6 kg
Frame Material: Plastic
Rotor Material: Plastic
Available Colors: Silver
Max. Flight Time: 12 minutes
Max. Operating Range: 100 m
Controller Frequency: 2.4 GHz
Battery Capacity: 2,000 mAh
Battery Type: Lithium Polymer
Battery Power: 7.4V
GPS Compatible: –
Altitude Hold: –
Position Hold: –
Flight Planning: –
Autonomous Flight: –
Target Tracking: –
Automatic Landing: –
Return To Home: –
Obstacle Avoidance: –
Headless Mode: –
Built-in Camera: √ (5 megapixels)

Trimble

ZX5 multirotor

© Trimble

Release Date: August 2015
Price: On Request
Number of Rotors: 6
Control System: Remote Control
Applications: Film, Photography, Surveillance
Dimensions: 850 × 490 mm
Wingspan: 981 mm
Weight: 2.7 kg
Max. Payload: 2.3 kg
Frame Material: Carbon Fiber
Rotor Material: Carbon Fiber
Available Colors: Yellow
Max. Flight Time: 20 minutes
Max. Speed: 36 km/h (22,3 mp/h)
Max. Flying Altitude: 3,000 m (9,842 ft)
Max. Operating Range: 750 m
Controller Frequency: 2.4 GHz
Battery Capacity: 6,600 mAh
Battery Type: Lithium Polymer
Battery Power: 14.8V
GPS Compatible: √
Altitude Hold: √
Position Hold: √
Flight Planning: √
Autonomous Flight: –
Target Tracking: √
Automatic Landing: √
Return To Home: √
Obstacle Avoidance: –
Headless Mode: –
Built-in Camera: √ (16 megapixels)

UAV Solutions

Phoenix 30

© UAV Solutions

Release Date: August 2015
Price: On Request
Number of Rotors: 4
Control System: Ground Control Station, Laptop, Tablet
Applications: Reconnaissance, Surveillance
Dimensions: 508 × 508 mm
Weight: 4.5 kg
Max. Payload: 900 gr
Frame Material: Carbon Fiber, Kevlar
Available Colors: White
Max. Flight Time: 30 minutes
Max. Flying Altitude: 3,048 m (10,000 ft)
Controller Frequency: 2.4 GHz
Battery Capacity: 16,000 mAh
Battery Type: Lithium Polymer
GPS Compatible: √
Altitude Hold: √
Position Hold: √
Flight Planning: √
Autonomous Flight: √
Target Tracking: √
Automatic Landing: √
Return To Home: √
Obstacle Avoidance: √
Headless Mode: –
Built-in Camera: √

Walkera

Runner 250R

© Walkera

Release Date: August 2015
Price: $260
Number of Rotors: 4
Control System: Remote Control
Applications: FPV Racing
Dimensions: 236 × 205 × 102 mm
Wingspan: 304 mm
Weight: 446 gr
Max. Payload: 320 gr
Frame Material: Carbon Fiber
Rotor Material: Plastic
Available Colors: Black
Max. Flight Time: 12 minutes
Max. Speed: 65 km/h (40 mp/h)
Max. Operating Range: 1 km
Controller Frequency: 2.4 GHz
Battery Capacity: 2,200 mAh
Battery Type: Lithium Polymer 3S
Battery Power: 11.1V
GPS Compatible: –
Altitude Hold: –
Position Hold: –
Flight Planning: –
Autonomous Flight: –
Target Tracking: –
Automatic Landing: –
Return To Home: –
Obstacle Avoidance: –
Headless Mode: –
Built-in Camera: √ (800TVL)

WLtoys

Q212 Spaceship

© Wltoys

Release Date: August 2015
Price: $100
Number of Rotors: 4
Control System: Remote Control
Applications: Recreation
Dimensions: 320 × 320 × 100 mm
Wingspan: 453 mm
Weight: 90 gr
Frame Material: Plastic
Rotor Material: Plastic
Available Colors: Red
Max. Flight Time: 12 minutes
Max. Operating Range: 150 m
Controller Frequency: 2.4 GHz
Battery Capacity: 1,200 mAh
Battery Type: Lithium Polymer
Battery Power: 7.4V
GPS Compatible: –
Altitude Hold: √
Position Hold: √
Flight Planning: –
Autonomous Flight: –
Target Tracking: –
Automatic Landing: –
Return To Home: √
Obstacle Avoidance: –
Headless Mode: –
Built-in Camera: √ (2 megapixels)

Acme

Zoopa Q650 Razor

© Acme

Release Date: September 2015
Price: $100
Number of Rotors: 4
Control System: Remote Control
Applications: Film, Photography
Dimensions: 520 × 520 × 100 mm
Wingspan: 630 mm
Weight: 335 gr
Frame Material: Carbon Fiber, Foam
Rotor Material: Plastic
Available Colors: Black
Max. Flight Time: 10 minutes
Max. Operating Range: 150 m
Controller Frequency: 2.4 GHz
Battery Capacity: 850 mAh
Battery Type: Lithium Polymer 2S
Battery Power: 7.4V
GPS Compatible: –
Altitude Hold: –
Position Hold: –
Flight Planning: –
Autonomous Flight: –
Target Tracking: –
Automatic Landing: –
Return To Home: –
Obstacle Avoidance: –
Headless Mode: –
Built-in Camera: –

Airhogs

Helix Video Drone

© Airhogs

Release Date: September 2015
Price: $110
Number of Rotors: 4
Control System: Remote Control
Applications: Recreation
Dimensions: 305 × 305 × 135 mm
Weight: 250 gr
Frame Material: Foam
Rotor Material: Plastic
Available Colors: Black, Red
Max. Flight Time: 15 minutes
Max. Operating Range: 76 m
Controller Frequency: 2.4 GHz
Battery Capacity: 450 mAh
Battery Type: Lithium Polymer
Battery Power: 3.7V
GPS Compatible: –
Altitude Hold: √
Position Hold: √
Flight Planning: –
Autonomous Flight: –
Target Tracking: –
Automatic Landing: –
Return To Home: –
Obstacle Avoidance: –
Headless Mode: –
Built-in Camera: √

Airhogs

Millenium Falcon Quad

Release Date: September 2015
Price: $70
Number of Rotors: 4
Control System: Remote Control
Applications: Recreation
Dimensions: 240 × 178 × 38 mm
Weight: 66 gr
Frame Material: Foam
Rotor Material: Plastic
Available Colors: White
Max. Flight Time: 5 minutes
Controller Frequency: 2.4 GHz
Battery Capacity: 140 mAh
Battery Type: Lithium Ion
Battery Power: 3.7V
GPS Compatible: –
Altitude Hold: –
Position Hold: –
Flight Planning: –
Autonomous Flight: –
Target Tracking: –
Automatic Landing: –
Return To Home: –
Obstacle Avoidance: –
Headless Mode: –
Built-in Camera: –

C-Astral

Atlas UAS

©C-Astral

Release Date: September 2015
Price: Unknown
Number of Rotors: 1 (Fixed Wing)
Control System: Flight Computer
Applications: Aerial Mapping, Agriculture, Surveillance
Wingspan: 1.55 m
Weight: 1.55 kg
Max. Payload: 300 gr
Frame Material: Carbon Fiber, Kevlar
Available Colors: White
Max. Flight Time: 60 minutes
Max. Speed: 60 km/h (37 mp/h)
Min. Operating Temperature: -25°C
Max. Operating Temperature: 45°C
Max. Operating Range: 40 km
Battery Type: Lithium Polymer
Battery Weight: 450 gr
GPS Compatible: √
Altitude Hold: √
Position Hold: √
Flight Planning: √
Autonomous Flight: √
Target Tracking: √
Automatic Landing: √
Return To Home: √
Obstacle Avoidance: –
Headless Mode: –
Built-in Camera: √ (24.3 megapixels)

Carrera RC

Quadrocopter Video Next

© bol.com

Release Date: September 2015
Price: $120
Number of Rotors: 4
Control System: Remote Control
Applications: Recreation
Dimensions: 350 × 350 mm
Frame Material: Plastic
Rotor Material: Plastic
Available Colors: Red
Max. Flight Time: 7 minutes
Controller Frequency: 2.4 GHz
Battery Capacity: 1,000 mAh
Battery Type: Lithium Polymer
Battery Power: 7V
GPS Compatible: –
Altitude Hold: √
Position Hold: √
Flight Planning: –
Autonomous Flight: –
Target Tracking: –
Automatic Landing: √
Return To Home: √
Obstacle Avoidance: –
Headless Mode: √
Built-in Camera: √ (2 megapixels)

Cheerson

CX-10C

© hobygaga.com

Release Date: September 2015
Price: $15
Number of Rotors: 4
Control System: Remote Control
Applications: Recreation
Dimensions: 62 × 62 × 20 mm
Wingspan: 65 mm
Weight: 15 gr
Frame Material: Plastic
Rotor Material: Plastic
Available Colors: Black, Orange
Max. Flight Time: 5 minutes
Max. Operating Range: 30 m
Controller Frequency: 2.4 GHz
Battery Capacity: 120 mAh
Battery Type: Lithium Polymer
Battery Power: 3.7V
GPS Compatible: –
Altitude Hold: √
Position Hold: √
Flight Planning: –
Autonomous Flight: –
Target Tracking: –
Automatic Landing: –
Return To Home: –
Obstacle Avoidance: –
Headless Mode: –
Built-in Camera: √ (0.3 megapixels)

Chengdu Aircraft Industry Group (CAIG)

Wing Loong II

Release Date: September 2015
Price: Unknown
Number of Rotors: 1 (Fixed Wing)
Control System: Ground Control Station
Applications: Military Purposes, Armed Reconnaissance, Surveillance
Dimensions: 11 × 20.5 × 4.1 m
Wingspan: 20.5 m
Weight: 4,200 kg
Max. Payload: 480 kg
Frame Material: Composite
Available Colors: Grey
Max. Flight Time: 20 hours
Max. Speed: 370 km/h (230 mp/h)
Max. Flying Altitude: 9,144 m (30,000 ft)
Max. Operating Range: 1,500 km
GPS Compatible: √
Altitude Hold: √
Position Hold: –
Flight Planning: √
Autonomous Flight: √
Target Tracking: √
Automatic Landing: √
Return To Home: √
Obstacle Avoidance: √
Headless Mode: –
Built-in Camera: √

China Aerospace Science and Technology Corporation (CASC)

CH-91

© Chinese Military Aviation

Release Date: September 2015
Price: Unknown
Number of Rotors: 1 (Fixed Wing)
Control System: Ground Control Station
Applications: Military Purposes, Reconnaissance, Surveillance
Dimensions: 3.5 × 4.4 × 1.4 m
Wingspan: 4.4 m
Weight: 110 kg
Frame Material: Composite
Available Colors: White
Max. Flight Time: 5 hours
Max. Speed: 180 km/h (112 mp/h)
Max. Flying Altitude: 4,500 m (14,764 ft)
Max. Operating Range: 150 km
GPS Compatible: √
Altitude Hold: √
Position Hold: –
Flight Planning: √
Autonomous Flight: √
Target Tracking: √
Automatic Landing: –
Return To Home: √
Obstacle Avoidance: √
Headless Mode: –
Built-in Camera: √

Dromida

Vista FPV

© Dromida

Release Date: September 2015
Price: $160
Number of Rotors: 4
Control System: Remote Control
Applications: FPV Racing
Wingspan: 251 mm
Weight: 136 gr
Frame Material: Plastic
Rotor Material: Plastic
Available Colors: Blue, Green, Orange, Yellow
Max. Flight Time: 15 minutes
Controller Frequency: 2.4 GHz
Battery Capacity: 850 mAh
Battery Type: Lithium Polymer 1S
Battery Power: 3.7V
GPS Compatible: –
Altitude Hold: –
Position Hold: –
Flight Planning: √
Autonomous Flight: –
Target Tracking: –
Automatic Landing: –
Return To Home: –
Obstacle Avoidance: –
Headless Mode: –
Built-in Camera: √

Flytronic

Manta VTOL

© miltechmag.com

Release Date: September 2015
Price: Unknown
Number of Rotors: 2 (Fixed Wing)
Control System: Ground Control Station
Applications: Military Purposes, Maritime Patrol, Reconnaissance, Surveillance
Dimensions: 3.3 × 6.3 m
Wingspan: 6.3 m
Weight: 75 kg
Max. Payload: 40 kg
Frame Material: Composite
Available Colors: Grey
Max. Speed: 150 km/h (93,2 mp/h)
Max. Operating Range: 180 km
GPS Compatible: √
Altitude Hold: √
Position Hold: √
Flight Planning: √
Autonomous Flight: √
Target Tracking: –
Automatic Landing: √
Return To Home: √
Obstacle Avoidance: –
Headless Mode: –
Built-in Camera: √

Fuqi Toys

FQ777-AF957 The Gazer

Release Date: September 2015
Price: $87
Number of Rotors: 4
Control System: Remote Control
Applications: Recreation, FPV Racing
Dimensions: 340 × 340 × 80 mm
Weight: 129 gr
Frame Material: Plastic
Rotor Material: Plastic
Available Colors: Black, Green
Max. Flight Time: 7 minutes
Max. Operating Range: 150 m
Controller Frequency: 2.4 GHz
Battery Capacity: 650 mAh
Battery Type: Lithium Polymer
Battery Power: 3.7V
GPS Compatible: –
Altitude Hold: –
Position Hold: –
Flight Planning: –
Autonomous Flight: –
Target Tracking: –
Automatic Landing: –
Return To Home: –
Obstacle Avoidance: –
Headless Mode: –
Built-in Camera: √ (2 megapixels)

Jin Xing Da

509G

Release Date: September 2015
Price: $80
Number of Rotors: 4
Control System: Remote Control
Applications: Film, Photography
Dimensions: 316 × 316 × 106 mm
Frame Material: Plastic
Rotor Material: Plastic
Available Colors: Black
Max. Flight Time: 10 minutes
Max. Operating Range: 150 m
Controller Frequency: 2.4 GHz
Battery Capacity: 600 mAh
Battery Type: Lithium Polymer
Battery Power: 3.7V
GPS Compatible: –
Altitude Hold: √
Position Hold: √
Flight Planning: –
Autonomous Flight: –
Target Tracking: –
Automatic Landing: –
Return To Home: √
Obstacle Avoidance: –
Headless Mode: –
Built-in Camera: √ (2 megapixels)

JJRC

H5P

© aliexpress.com

Release Date: September 2015
Price: $50
Number of Rotors: 4
Control System: Remote Control
Applications: Recreation
Dimensions: 310 × 310 × 85 mm
Wingspan: 438 mm
Frame Material: Plastic
Rotor Material: Plastic
Available Colors: Black, Green, White
Max. Flight Time: 14 minutes
Max. Operating Range: 80 m
Controller Frequency: 2.4 GHz
Battery Capacity: 1,100 mAh
Battery Type: Lithium Polymer
Battery Power: 3.7V
GPS Compatible: –
Altitude Hold: –
Position Hold: –
Flight Planning: –
Autonomous Flight: –
Target Tracking: –
Automatic Landing: –
Return To Home: √
Obstacle Avoidance: –
Headless Mode: –
Built-in Camera: √ (2 megapixels)

Martin UAV

Super Bat DA-50

© Martin UAV

Release Date: September 2015
Price: $150,000
Number of Rotors: 1 (Fixed Wing)
Control System: Flight Computer
Applications: Aerial Mapping, Border Patrol, Inspection, Surveillance
Dimensions: 1.6 × 2.6 m
Wingspan: 2.6 m
Weight: 21.3 kg
Max. Payload: 2 kg
Frame Material: Aluminum, Carbon Fiber, Kevlar
Motors: 50cc two-stroke engine
Available Colors: Grey
Max. Flight Time: 10 hours
Max. Speed: 120 km/h (74 mp/h)
Max. Flying Altitude: 3,048 m (10,000 ft)
Max. Operating Range: 640 km
Controller Frequency: 2.4 GHz
GPS Compatible: √
Altitude Hold: √
Position Hold: –
Flight Planning: √
Autonomous Flight: √
Target Tracking: √
Automatic Landing: –
Return To Home: √
Obstacle Avoidance: √
Headless Mode: –
Built-in Camera: √

Meijiaxin Toys

X101

Release Date: September 2015
Price: $80
Number of Rotors: 4
Control System: Remote Control
Applications: Recreation
Dimensions: 505 × 500 × 180 mm
Wingspan: 711 mm
Weight: 420 gr
Frame Material: Plastic
Rotor Material: Plastic
Available Colors: White
Max. Flight Time: 10 minutes
Max. Operating Range: 150 m
Controller Frequency: 2.4 GHz
Battery Capacity: 1,200 mAh
Battery Type: Lithium Polymer
Battery Power: 3.7V
GPS Compatible: –
Altitude Hold: √
Position Hold: √
Flight Planning: –
Autonomous Flight: –
Target Tracking: –
Automatic Landing: –
Return To Home: √
Obstacle Avoidance: –
Headless Mode: –
Built-in Camera: –

MOLA

Fanatic 1

Release Date: September 2015
Price: Unknown
Number of Rotors: 4
Control System: Remote Control, Phone/Tablet
Applications: Film, Photography
Weight: 900 gr
Frame Material: Plastic
Rotor Material: Plastic
Available Colors: Black, Desert, Forest, Navy, Snow, White
Controller Frequency: 2.4 GHz
GPS Compatible: √
Altitude Hold: √
Position Hold: √
Flight Planning: √
Autonomous Flight: √
Target Tracking: √
Automatic Landing: √
Return To Home: √
Obstacle Avoidance: √
Headless Mode: √
Built-in Camera: √

Silent Falcon UAS

Silent Falcon

© Silent Falcon

Release Date: September 2015
Price: $250,000
Number of Rotors: 1 (Fixed Wing)
Control System: Remote Control
Applications: Military Purposes, Aerial Mapping, Reconnaissance, Surveillance
Dimensions: 1.9 × 4.4 m
Wingspan: 4.4 m
Weight: 14.5 kg
Frame Material: Carbon Fiber
Rotor Material: Carbon Fiber
Available Colors: Grey
Max. Flight Time: 5 hours
Max. Speed: 90 km/h (56 mp/h)
Max. Flying Altitude: 6,000 m (19,685 ft)
Max. Operating Range: 15 km
Controller Frequency: 2.4 GHz
Battery Type: Lithium Polymer, Solar Energy
GPS Compatible: √
Altitude Hold: √
Position Hold: √
Flight Planning: √
Autonomous Flight: √
Target Tracking: √
Automatic Landing: –
Return To Home: √
Obstacle Avoidance: –
Headless Mode: –
Built-in Camera: √ (38 megapixels)

Tarot RC

250 FPV

© Tarot

Release Date: September 2015
Price: $180
Number of Rotors: 4
Control System: Remote Control
Applications: FPV Racing
Dimensions: 230 × 210 × 40 mm
Weight: 190 gr
Frame Material: Carbon Fiber
Rotor Material: Carbon Fiber
Motors: 1806-2280kv brushless
Available Colors: Black
Controller Frequency: 5.8 GHz
Battery Capacity: 1,300 mAh
Battery Type: Lithium Polymer
Battery Power: 11.1V
GPS Compatible: –
Altitude Hold: –
Position Hold: –
Flight Planning: –
Autonomous Flight: –
Target Tracking: –
Automatic Landing: –
Return To Home: –
Obstacle Avoidance: –
Headless Mode: –
Built-in Camera: √ (300TL)

Tekever

AR5 Life Ray Evolution

© aeroexpo.online

Release Date: September 2015
Price: On Request
Number of Rotors: 2 (Fixed Wing)
Control System: Ground Control Station
Applications: Military Purposes, Maritime Patrol, Reconnaissance, Search & Rescue, Surveillance
Dimensions: 4 × 7.3 m
Wingspan: 7.3 m
Weight: 180 kg
Max. Payload: 50 kg
Frame Material: Composite
Rotor Material: Wood
Available Colors: Grey
Max. Flight Time: 20 hours
Max. Speed: 100 km/h (62 mp/h)
Max. Operating Range: 80 km
GPS Compatible: √
Altitude Hold: √
Position Hold: √
Flight Planning: √
Autonomous Flight: √
Target Tracking: –
Automatic Landing: √
Return To Home: √
Obstacle Avoidance: √
Headless Mode: –
Built-in Camera: √

Tobyrich

X-spy

© hitoys.org

Release Date: September 2015
Price: $100
Number of Rotors: 4
Control System: Remote Control
Applications: Recreation
Dimensions: 42 × 42 × 20 mm
Weight: 15 gr
Frame Material: Plastic
Rotor Material: Plastic
Available Colors: Black
Max. Flight Time: 7 minutes
Max. Operating Range: 30 m
Controller Frequency: 2.4 GHz
Battery Capacity: 100 mAh
Battery Type: Lithium Polymer
Battery Power: 3.7V
GPS Compatible: –
Altitude Hold: –
Position Hold: –
Flight Planning: –
Autonomous Flight: –
Target Tracking: –
Automatic Landing: –
Return To Home: –
Obstacle Avoidance: –
Headless Mode: –
Built-in Camera: √ (0.3 megapixels)

TRNDlabs

SKEYE Pico

©TRNDlabs

Release Date: September 2015
Price: $55
Number of Rotors: 4
Control System: Remote Control
Applications: Recreation
Dimensions: 22 × 22 × 19 mm
Weight: 7 gr
Frame Material: Plastic
Rotor Material: Plastic
Available Colors: Orange
Max. Flight Time: 8 minutes
Max. Operating Range: 50 m
Controller Frequency: 2.4 GHz
Battery Capacity: 80 mAh
Battery Type: Lithium Polymer
Battery Power: 3.7V
GPS Compatible: –
Altitude Hold: –
Position Hold: –
Flight Planning: –
Autonomous Flight: –
Target Tracking: –
Automatic Landing: –
Return To Home: –
Obstacle Avoidance: –
Headless Mode: –
Built-in Camera: –

Ucon System

Remo M-002

©Ucon System

Release Date: September 2015
Price: Unknown
Number of Rotors: 1 (Fixed Wing)
Control System: Ground Control Station
Applications: Aerial Mapping, Agriculture, Inspection, Surveillance, Target Acquisition
Dimensions: 1.44 × 1.8 m
Wingspan: 1.8 m
Weight: 3.5 kg
Frame Material: Composite
Available Colors: White
Max. Flight Time: 90 minutes
Max. Speed: 80 km/h (49,7 mp/h)
Max. Operating Range: 10 km
GPS Compatible: √
Altitude Hold: √
Position Hold: –
Flight Planning: √
Autonomous Flight: √
Target Tracking: –
Automatic Landing: –
Return To Home: √
Obstacle Avoidance: –
Headless Mode: –
Built-in Camera: √

WLtoys

Q282J Spaceship

© WLtoys

Release Date: September 2015
Price: $38
Number of Rotors: 6
Control System: Remote Control
Applications: Recreation
Dimensions: 137 × 130 × 50 mm
Weight: 50 gr
Frame Material: Plastic
Rotor Material: Plastic
Available Colors: White
Max. Flight Time: 6 minutes
Max. Operating Range: 80 m
Controller Frequency: 2.4 GHz
Battery Capacity: 500 mAh
Battery Type: Lithium Polymer
Battery Power: 3.7V
GPS Compatible: –
Altitude Hold: –
Position Hold: –
Flight Planning: –
Autonomous Flight: –
Target Tracking: –
Automatic Landing: –
Return To Home: –
Obstacle Avoidance: –
Headless Mode: –
Built-in Camera: √

Xtreem

Mini Stealth Drone

© Swann

Release Date: September 2015
Price: $50
Number of Rotors: 4
Control System: Remote Control
Applications: Recreation
Dimensions: 152 × 152 × 37 mm
Wingspan: 215 mm
Weight: 37 gr
Frame Material: Plastic
Rotor Material: Plastic
Available Colors: Black
Max. Flight Time: 7 minutes
Max. Operating Range: 30 m
Controller Frequency: 2.4 GHz
Battery Capacity: 150 mAh
Battery Type: Lithium Polymer
Battery Power: 3.7V
GPS Compatible: –
Altitude Hold: –
Position Hold: –
Flight Planning: –
Autonomous Flight: –
Target Tracking: –
Automatic Landing: –
Return To Home: –
Obstacle Avoidance: –
Headless Mode: –
Built-in Camera: –

Yuneec

Tornado H920

© rcatmosphere.com

Release Date: September 2015
Price: $4,999
Number of Rotors: 6
Control System: Remote Control, Phone/Tablet
Applications: Film, Photography
Dimensions: 797 × 920 × 461 mm
Wingspan: 920 mm
Weight: 5 kg
Max. Payload: 1.6 kg
Frame Material: Carbon Fiber
Rotor Material: Carbon Fiber
Available Colors: Black
Max. Flight Time: 24 minutes
Max. Speed: 75,6 km/h (47 mp/h)
Max. Flying Altitude: 4,000 m (13,123 ft)
Max. Ascent Speed: 5 m/s
Max. Descent Speed: 3 m/s
Min. Operating Temperature: -5°C
Max. Operating Temperature: 80°C
Max. Operating Range: 2 km
Controller Frequency: 2.4 GHz
Battery Capacity: 8,800 mAh
Battery Type: Lithium Polymer 6S
Battery Power: 22.2V
GPS Compatible: √
Altitude Hold: √
Position Hold: √
Flight Planning: √
Autonomous Flight: √
Target Tracking: √
Automatic Landing: –
Return To Home: √
Obstacle Avoidance: –
Headless Mode: –
Built-in Camera: –

Acme

Zoopa Q600 Mantis

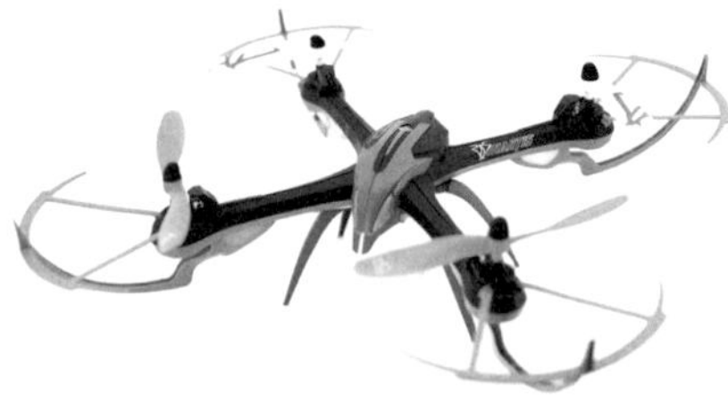

© Acme

Release Date: October 2015
Price: $110
Number of Rotors: 4
Control System: Remote Control
Applications: Film, Photography
Dimensions: 485 × 485 × 150 mm
Wingspan: 585 mm
Weight: 290 gr
Frame Material: Plastic
Rotor Material: Plastic
Available Colors: Black
Max. Flight Time: 7 minutes
Max. Operating Range: 300 m
Controller Frequency: 2.4 GHz
Battery Capacity: 1,200 mAh
Battery Type: Lithium Polymer 2S
Battery Power: 7.4V
GPS Compatible: –
Altitude Hold: –
Position Hold: –
Flight Planning: –
Autonomous Flight: –
Target Tracking: –
Automatic Landing: –
Return To Home: –
Obstacle Avoidance: –
Headless Mode: –
Built-in Camera: –

AEE Technology

Toruk AP11

©AEE Technologies

Release Date: October 2015
Price: $699
Number of Rotors: 4
Control System: Remote Control, Phone/Tablet
Applications: Film, Photography
Dimensions: 370 × 370 × 140 mm
Wingspan: 450 mm
Weight: 1.55 kg
Max. Payload: 500 gr
Frame Material: Plastic
Rotor Material: Plastic
Available Colors: White
Max. Flight Time: 25 minutes
Max. Speed: 70 km/h (44 mp/h)
Max. Flying Altitude: 300 m (984 ft)
Max. Ascent Speed: 6 m/s
Min. Operating Temperature: 0°C
Max. Operating Temperature: 50°C
Max. Operating Range: 500 m
Controller Frequency: 2.4 GHz
Battery Capacity: 6,800 mAh
Battery Type: Lithium Polymer
Battery Power: 11V
Battery Weight: 300 gr
GPS Compatible: √
Altitude Hold: √
Position Hold: √
Flight Planning: –
Autonomous Flight: –
Target Tracking: –
Automatic Landing: √
Return To Home: √
Obstacle Avoidance: –
Headless Mode: –
Built-in Camera: –

Aerix Drones

Turbo-X

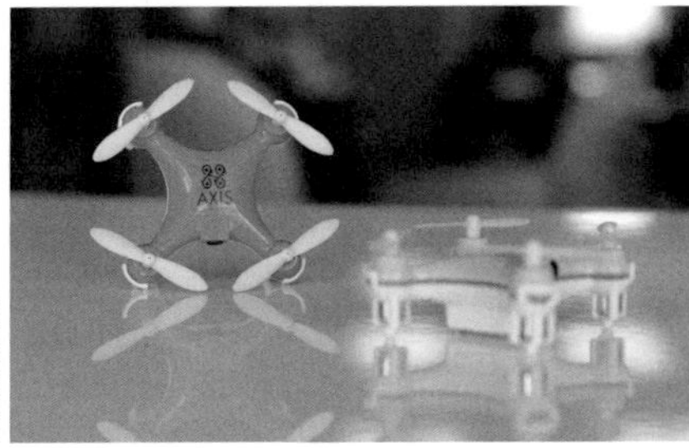

©Aerix Drones

Release Date: October 2015
Price: $30
Number of Rotors: 4
Control System: Remote Control, Phone/Tablet
Applications: Recreation
Dimensions: 42 × 42 × 25 mm
Weight: 17 gr
Frame Material: Plastic
Rotor Material: Plastic
Available Colors: Blue, Green, Orange, Red
Max. Flight Time: 7 minutes
Max. Operating Range: 30 m
Controller Frequency: 2.4 GHz
Battery Capacity: 100 mAh
Battery Type: Lithium Ion
Battery Power: 3.7V
GPS Compatible: –
Altitude Hold: √
Position Hold: –
Flight Planning: –
Autonomous Flight: –
Target Tracking: –
Automatic Landing: –
Return To Home: –
Obstacle Avoidance: –
Headless Mode: –
Built-in Camera: √

Airnamics

R5

© Airnamics

Release Date: October 2015
Price: Unknown
Number of Rotors: 4
Control System: Flight Computer
Applications: Film, Photography, Aerial Mapping, Inspection, Surveillance
Dimensions: 3.7 × 2.7 × 0.7 m
Weight: 60 kg
Frame Material: Composite
Available Colors: White
Max. Flight Time: 30 minutes
Max. Speed: 130 km/h (80 mp/h)
GPS Compatible: √
Altitude Hold: √
Position Hold: √
Flight Planning: √
Autonomous Flight: √
Target Tracking: √
Automatic Landing: √
Return To Home: √
Obstacle Avoidance: √
Headless Mode: –
Built-in Camera: √

Ares

Spidex 3D

© Ares-RC

Release Date: October 2015
Price: $40
Number of Rotors: 4
Control System: Remote Control, Phone/Tablet
Applications: Recreation
Dimensions: 170 × 170 × 45 mm
Wingspan: 99 mm
Weight: 32 gr
Frame Material: Plastic
Rotor Material: Plastic
Available Colors: Blue, Red
Controller Frequency: 2.4 GHz
Battery Capacity: 30 mAh
Battery Type: Lithium Polymer 1S
Battery Power: 3.7V
GPS Compatible: –
Altitude Hold: –
Position Hold: √
Flight Planning: –
Autonomous Flight: –
Target Tracking: –
Automatic Landing: –
Return To Home: –
Obstacle Avoidance: –
Headless Mode: –
Built-in Camera: –

Cheerson

CX-33S

© Cheerson Hobby

Release Date: October 2015
Price: $90
Number of Rotors: 6
Control System: Remote Control
Applications: Film, Photography, Recreation
Dimensions: 310 × 345 × 85 mm
Wingspan: 360 mm
Weight: 164 gr
Frame Material: Plastic
Rotor Material: Plastic
Available Colors: White
Max. Flight Time: 7 minutes
Max. Operating Range: 80 m
Controller Frequency: 2.4 GHz
Battery Capacity: 450 mAh
Battery Type: Lithium Polymer
Battery Power: 7.4V
GPS Compatible: –
Altitude Hold: √
Position Hold: √
Flight Planning: –
Autonomous Flight: –
Target Tracking: –
Automatic Landing: √
Return To Home: –
Obstacle Avoidance: –
Headless Mode: –
Built-in Camera: √ (1 megapixels)

Drone America

Savant

© Drone America

Release Date: October 2015
Price: On Request
Number of Rotors: 1 (Fixed Wing)
Control System: Laptop/Tablet
Applications: Aerial Mapping, Agriculture, Search & Rescue, Surveillance
Dimensions: 1.7 × 3.6 m
Wingspan: 3.6 m
Weight: 4.5 kg
Max. Payload: 3.7 kg
Frame Material: Carbon Fiber
Rotor Material: Carbon Fiber
Available Colors: White
Max. Flight Time: 2 hours
Max. Speed: 136 km/h (85 mp/h)
Max. Flying Altitude: 3,650 m (11,975 ft)
Controller Frequency: 2.4 GHz
GPS Compatible: –
Altitude Hold: –
Position Hold: –
Flight Planning: √
Autonomous Flight: –
Target Tracking: –
Automatic Landing: –
Return To Home: –
Obstacle Avoidance: –
Headless Mode: –
Built-in Camera: √

Eachine

Racer 250

Release Date: October 2015
Price: $400
Number of Rotors: 4
Control System: Remote Control
Applications: FPV Racing
Dimensions: 220 × 233 × 50 mm
Wingspan: 280 mm
Weight: 400 gr
Frame Material: Carbon Fiber
Rotor Material: Plastic
Available Colors: Black
Max. Flight Time: 14 minutes
Min. Operating Temperature: -10°C
Max. Operating Temperature: 40°C
Max. Operating Range: 30 m
Controller Frequency: 2.4 GHz
Battery Capacity: 1,300 mAh
Battery Type: Lithium Polymer 3S
Battery Power: 11.1V
GPS Compatible: √
Altitude Hold: –
Position Hold: –
Flight Planning: –
Autonomous Flight: –
Target Tracking: –
Automatic Landing: –
Return To Home: √
Obstacle Avoidance: –
Headless Mode: –
Built-in Camera: √

Estes Rockets

Proto-N

Release Date: October 2015
Price: $35
Number of Rotors: 4
Control System: Remote Control
Applications: Recreation
Dimensions: 22 × 22 × 19 mm
Wingspan: 31 mm
Weight: 8 gr
Frame Material: Plastic
Rotor Material: Plastic
Available Colors: Black, Green, Red, White
Max. Flight Time: 5 minutes
Controller Frequency: 2.4 GHz
Battery Capacity: 80 mAh
Battery Type: Lithium Polymer
Battery Power: 3.7V
GPS Compatible: –
Altitude Hold: √
Position Hold: √
Flight Planning: –
Autonomous Flight: –
Target Tracking: –
Automatic Landing: –
Return To Home: √
Obstacle Avoidance: –
Headless Mode: –
Built-in Camera: –

Fuqi Toys

FQ777-955C Scorpius

©Fuqi Toys

Release Date: October 2015
Price: $44
Number of Rotors: 4
Control System: Remote Control
Applications: Recreation, FPV Racing
Dimensions: 300 × 300 × 90 mm
Weight: 118 gr
Frame Material: Plastic
Rotor Material: Plastic
Available Colors: Black, White
Max. Flight Time: 7 minutes
Max. Operating Range: 100 m
Controller Frequency: 2.4 GHz
Battery Capacity: 600 mAh
Battery Type: Lithium Polymer
Battery Power: 3.7V
GPS Compatible: –
Altitude Hold: –
Position Hold: –
Flight Planning: –
Autonomous Flight: –
Target Tracking: –
Automatic Landing: –
Return To Home: –
Obstacle Avoidance: –
Headless Mode: –
Built-in Camera: √ (2 megapixels)

Gear2Play

Nano Smart Drone

©droneshop.nl

Release Date: October 2015
Price: $65
Number of Rotors: 4
Control System: Remote Control
Applications: Recreation
Dimensions: 70 × 70 mm
Frame Material: Plastic
Rotor Material: Plastic
Available Colors: Black, Blue, White
Max. Flight Time: 8 minutes
Max. Operating Range: 30 m
Controller Frequency: 2.4 GHz
GPS Compatible: –
Altitude Hold: –
Position Hold: –
Flight Planning: –
Autonomous Flight: –
Target Tracking: –
Automatic Landing: –
Return To Home: √
Obstacle Avoidance: –
Headless Mode: √
Built-in Camera: √ (0.3 megapixels)

Huajun Toys

W609-5

© Huajun Toys

Release Date: October 2015
Price: $24
Number of Rotors: 6
Control System: Remote Control
Applications: Recreation
Dimensions: 80 × 80 × 26 mm
Weight: 22 gr
Frame Material: Plastic
Rotor Material: Plastic
Available Colors: Black, Blue
Max. Flight Time: 5 minutes
Max. Operating Range: 50 m
Controller Frequency: 2.4 GHz
Battery Capacity: 150 mAh
Battery Type: Lithium Polymer
Battery Power: 3.7V
GPS Compatible: –
Altitude Hold: √
Position Hold: √
Flight Planning: –
Autonomous Flight: –
Target Tracking: –
Automatic Landing: –
Return To Home: –
Obstacle Avoidance: –
Headless Mode: –
Built-in Camera: –

Idea-Fly

Mars-350

© Idea-Fly

Release Date: October 2015
Price: $600
Number of Rotors: 4
Control System: Remote Control
Applications: Film, Photography
Wingspan: 350 mm
Weight: 1.1 kg
Max. Payload: 440 gr
Frame Material: Plastic
Rotor Material: Plastic
Available Colors: White
Max. Flight Time: 25 minutes
Max. Speed: 29 km/h (18 mp/h)
Max. Ascent Speed: 10 m/s
Max. Descent Speed: 5 m/s
Max. Operating Range: 1 km
Controller Frequency: 2.4 GHz
Battery Capacity: 5,400 mAh
Battery Type: Lithium Polymer 3S
Battery Power: 11.1V
Battery Weight: 314 gr
GPS Compatible: √
Altitude Hold: √
Position Hold: √
Flight Planning: √
Autonomous Flight: √
Target Tracking: √
Automatic Landing: √
Return To Home: √
Obstacle Avoidance: –
Headless Mode: –
Built-in Camera: –

Izhmash Unmanned Systems

Granat-1

© i-korotchenko.livejournal.com

Release Date: October 2015
Price: Unknown
Number of Rotors: 1 (Fixed Wing)
Control System: Ground Control Station
Applications: Military Purposes, Reconnaissance, Surveillance
Wingspan: 820 mm
Weight: 2.4 kg
Frame Material: Composite
Available Colors: Grey
Max. Flight Time: 75 minutes
Max. Flying Altitude: 3,500 m (11,483 ft)
Min. Operating Temperature: -30°C
Max. Operating Temperature: 40°C
Max. Operating Range: 10 km
Controller Frequency: 2.4 GHz
Battery Type: Lithium Polymer
GPS Compatible: √
Altitude Hold: √
Position Hold: –
Flight Planning: √
Autonomous Flight: √
Target Tracking: –
Automatic Landing: –
Return To Home: √
Obstacle Avoidance: –
Headless Mode: –
Built-in Camera: √

JJRC

H11D

© JJRC

Release Date: October 2015
Price: $80
Number of Rotors: 4
Control System: Remote Control
Applications: Recreation
Dimensions: 330 × 330 × 85 mm
Wingspan: 467 mm
Frame Material: Plastic
Rotor Material: Plastic
Available Colors: Orange
Max. Flight Time: 12 minutes
Max. Operating Range: 120 m
Controller Frequency: 2.4 GHz
Battery Capacity: 1,100 mAh
Battery Type: Lithium Polymer
Battery Power: 3.7V
GPS Compatible: –
Altitude Hold: –
Position Hold: –
Flight Planning: –
Autonomous Flight: –
Target Tracking: –
Automatic Landing: –
Return To Home: √
Obstacle Avoidance: –
Headless Mode: –
Built-in Camera: √ (2 megapixels)

JJRC

H21

© gearbest.com

Release Date: October 2015
Price: $28
Number of Rotors: 6
Control System: Remote Control
Applications: Recreation
Dimensions: 255 × 180 × 35 mm
Wingspan: 312 mm
Frame Material: Plastic
Rotor Material: Plastic
Available Colors: Black, White
Max. Flight Time: 6 minutes
Max. Operating Range: 100 m
Controller Frequency: 2.4 GHz
Battery Capacity: 500 mAh
Battery Type: Lithium Polymer
Battery Power: 3.7V
GPS Compatible: –
Altitude Hold: √
Position Hold: √
Flight Planning: –
Autonomous Flight: –
Target Tracking: –
Automatic Landing: –
Return To Home: √
Obstacle Avoidance: –
Headless Mode: –
Built-in Camera: √ (2 megapixels)

JJRC

H12W

© droneflyersguide.com

Release Date: October 2015
Price: $70
Number of Rotors: 4
Control System: Remote Control, Phone/Tablet
Applications: Recreation
Dimensions: 310 × 310 × 90 mm
Frame Material: Plastic
Rotor Material: Plastic
Available Colors: Orange
Max. Flight Time: 8 minutes
Max. Operating Range: 60 m
Controller Frequency: 2.4 GHz
Battery Capacity: 750 mAh
Battery Type: Lithium Polymer
Battery Power: 3.7V
GPS Compatible: –
Altitude Hold: –
Position Hold: –
Flight Planning: –
Autonomous Flight: –
Target Tracking: –
Automatic Landing: –
Return To Home: –
Obstacle Avoidance: –
Headless Mode: –
Built-in Camera: √ (2 megapixels)

NASA

Prandtl-D No.3

© Nasa

Release Date: October 2015
Price: Unknown
Number of Rotors: 1 (Fixed Wing)
Control System: Flight Computer
Applications: Reconnaissance, Surveillance
Wingspan: 7.62 m
Weight: 14 kg
Max. Payload: 200 gr
Frame Material: Carbon Fiber
Available Colors: White
Max. Speed: 33.7 km/h (21 mp/h)
Controller Frequency: 2.4 GHz
GPS Compatible: √
Altitude Hold: √
Position Hold: √
Flight Planning: √
Autonomous Flight: –
Target Tracking: √
Automatic Landing: √
Return To Home: √
Obstacle Avoidance: –
Headless Mode: –
Built-in Camera: √ (10 megapixels)

Rapid Composites

Bullray Heavy Aero

© unmannedsystemstechnology.com

Release Date: October 2015
Price: Unknown
Number of Rotors: 4
Control System: Ground Control Station, Remote Control
Applications: Military Purposes, Armed Reconnaissance, Surveillance
Dimensions: 1.8 × 1.8 × 0.6 m
Weight: 16 kg
Max. Payload: 9 kg
Available Colors: Black
Max. Flight Time: 20 minutes
Max. Speed: 112 km/h (70 mp/h)
Controller Frequency: 2.4 GHz
Battery Capacity: 44,000 mAh
Battery Type: Lithium Polymer
Battery Weight: 5.4 kg
GPS Compatible: √
Altitude Hold: √
Position Hold: √
Flight Planning: √
Autonomous Flight: √
Target Tracking: √
Automatic Landing: √
Return To Home: √
Obstacle Avoidance: –
Headless Mode: –
Built-in Camera: √

Revell Control

Nano Pocket

© Revell Control

Release Date: October 2015
Price: $32
Number of Rotors: 4
Control System: Remote Control
Applications: Recreation
Dimensions: 40 × 40 × 20 mm
Weight: 13 gr
Frame Material: Plastic
Rotor Material: Plastic
Available Colors: Blue, Orange
Max. Flight Time: 6 minutes
Max. Operating Range: 25 m
Controller Frequency: 2.4 GHz
Battery Capacity: 100 mAh
Battery Type: Lithium Polymer
Battery Power: 3.7V
GPS Compatible: –
Altitude Hold: –
Position Hold: –
Flight Planning: –
Autonomous Flight: –
Target Tracking: –
Automatic Landing: –
Return To Home: –
Obstacle Avoidance: –
Headless Mode: –
Built-in Camera: –

Walkera

Runner 250 Pro

© Walkera

Release Date: October 2015
Price: $359
Number of Rotors: 4
Control System: Remote Control
Applications: FPV Racing
Dimensions: 221 × 205 × 117 mm
Wingspan: 312 mm
Weight: 464 gr
Max. Payload: 320 gr
Frame Material: Carbon Fiber
Rotor Material: Plastic
Motors: WK-WS-28-014A brushless
Available Colors: Black
Max. Flight Time: 12 minutes
Max. Speed: 39,6 km/h (24,6 mp/h)
Min. Operating Temperature: -10°C
Max. Operating Temperature: 40°C
Max. Operating Range: 1 km
Controller Frequency: 5.8 GHz
Battery Capacity: 2,200 mAh
Battery Type: Lithium Polymer 3S
Battery Power: 11.1V
GPS Compatible: √
Altitude Hold: √
Position Hold: √
Flight Planning: √
Autonomous Flight: –
Target Tracking: –
Automatic Landing: –
Return To Home: √
Obstacle Avoidance: –
Headless Mode: –
Built-in Camera: √ (800TVL)

WLtoys

Q222 Spaceship

© maisondudrone.com

Release Date: October 2015
Price: $85
Number of Rotors: 4
Control System: Remote Control
Applications: Recreation
Dimensions: 170 × 170 × 97 mm
Weight: 100 gr
Frame Material: Plastic
Rotor Material: Plastic
Available Colors: Black, Red
Max. Flight Time: 7 minutes
Max. Operating Range: 150 m
Controller Frequency: 2.4 GHz
Battery Capacity: 730 mAh
Battery Type: Lithium Polymer
Battery Power: 3.7V
GPS Compatible: –
Altitude Hold: √
Position Hold: √
Flight Planning: –
Autonomous Flight: –
Target Tracking: –
Automatic Landing: –
Return To Home: √
Obstacle Avoidance: –
Headless Mode: √
Built-in Camera: √ (2 megapixels)

Xinlin Shiye

X118 FPV

© aliexpress.com

Release Date: October 2015
Price: $135
Number of Rotors: 4
Control System: Remote Control
Applications: Recreation, FPV Racing
Dimensions: 300 × 300 × 90 mm
Weight: 138 gr
Frame Material: Plastic
Rotor Material: Plastic
Available Colors: White
Max. Flight Time: 6 minutes
Max. Operating Range: 100 m
Controller Frequency: 2.4 GHz
Battery Capacity: 850 mAh
Battery Type: Lithium Polymer
Battery Power: 3.7V
GPS Compatible: –
Altitude Hold: –
Position Hold: –
Flight Planning: –
Autonomous Flight: –
Target Tracking: –
Automatic Landing: –
Return To Home: √
Obstacle Avoidance: –
Headless Mode: √
Built-in Camera: √

XK Innovations

X260

© thedronefiles.net

Release Date: October 2015
Price: $75
Number of Rotors: 4
Control System: Remote Control, Phone/Tablet
Applications: Film, Photography, Recreation
Dimensions: 356 × 350 × 86 mm
Wingspan: 245 mm
Weight: 120 gr
Frame Material: Plastic
Rotor Material: Plastic
Available Colors: Black
Max. Flight Time: 12 minutes
Max. Operating Range: 300 m
Controller Frequency: 2.4 GHz
Battery Capacity: 730 mAh
Battery Type: Lithium Polymer
Battery Power: 3.7V
GPS Compatible: –
Altitude Hold: √
Position Hold: √
Flight Planning: –
Autonomous Flight: –
Target Tracking: –
Automatic Landing: √
Return To Home: √
Obstacle Avoidance: –
Headless Mode: √
Built-in Camera: √ (2 megapixels)

Aerix Drones

Aerius

© newatlas.com/Stanley Goodner

Release Date: November 2015
Price: $35
Number of Rotors: 4
Control System: Remote Control, Phone
Applications: Recreation
Dimensions: 30 × 30 × 20 mm
Weight: 7 gr
Frame Material: Plastic
Rotor Material: Plastic
Motors: 3 mm brushless
Available Colors: Black, Blue, Green, Orange
Max. Flight Time: 5 minutes
Max. Operating Range: 20 m
Controller Frequency: 2.4 GHz
Battery Capacity: 85 mAh
Battery Type: Lithium Ion
GPS Compatible: –
Altitude Hold: √
Position Hold: –
Flight Planning: –
Autonomous Flight: –
Target Tracking: –
Automatic Landing: –
Return To Home: –
Obstacle Avoidance: –
Headless Mode: –
Built-in Camera: –

Airhogs

Helix Ion

©Airhogs

Release Date: November 2015
Price: $50
Number of Rotors: 4
Control System: Remote Control
Applications: Recreation
Dimensions: 89 × 89 × 30,5 mm
Frame Material: Foam
Rotor Material: Plastic
Available Colors: Black
Max. Flight Time: 6 minutes
Controller Frequency: 2.4 GHz
Battery Capacity: 140 mAh
Battery Type: Lithium Ion
Battery Power: 3.7V
GPS Compatible: –
Altitude Hold: –
Position Hold: –
Flight Planning: –
Autonomous Flight: –
Target Tracking: –
Automatic Landing: –
Return To Home: –
Obstacle Avoidance: –
Headless Mode: –
Built-in Camera: –

Align

MR25

©Align

Release Date: November 2015
Price: $399
Number of Rotors: 4
Propellers: 5045/6040 inch propeller
Control System: Remote Control, Phone/Tablet
Applications: FPV Racing
Dimensions: 199 × 196 × 73 mm
Wingspan: 250 mm
Weight: 410 gr
Frame Material: Carbon Fiber
Rotor Material: Carbon Fiber
Motors: 2300kv
Available Colors: Green, Red, Yellow, White
Max. Flight Time: 9 minutes
Min. Operating Temperature: -5°C
Max. Operating Temperature: 45°C
Controller Frequency: 2.4 GHz
Battery Capacity: 1,800 mAh
Battery Type: Lithium Polymer 3S
Battery Power: 11.1V
GPS Compatible: √
Altitude Hold: √
Position Hold: √
Flight Planning: –
Autonomous Flight: –
Target Tracking: –
Automatic Landing: √
Return To Home: –
Obstacle Avoidance: –
Headless Mode: –
Built-in Camera: √

DaMing

Spy 003

© aliexpress.com

Release Date: November 2015
Price: $20
Number of Rotors: 4
Control System: Remote Control, Phone/Tablet
Applications: Recreation
Dimensions: 140 × 140 × 40 mm
Wingspan: 198 mm
Weight: 39 gr
Frame Material: Plastic
Rotor Material: Plastic
Available Colors: Black
Max. Flight Time: 8 minutes
Max. Operating Range: 100 m
Controller Frequency: 2.4 GHz
Battery Capacity: 300 mAh
Battery Type: Lithium Polymer
Battery Power: 3.7V
GPS Compatible: –
Altitude Hold: √
Position Hold: √
Flight Planning: –
Autonomous Flight: –
Target Tracking: –
Automatic Landing: –
Return To Home: √
Obstacle Avoidance: –
Headless Mode: –
Built-in Camera: √ (2 megapixels)

Drone America

Ariel

© RC Groups

Release Date: November 2015
Price: On Request
Number of Rotors: 2 (Fixed Wing)
Control System: Ground Control Station, Laptop,
Applications: Aerial Mapping, Inspection, Maritime Patrol, Maritime Research, Search & Rescue
Dimensions: 4.8 × 6.8 m
Wingspan: 6.8 m
Weight: 294 kg
Max. Payload: 158.7 kg
Frame Material: Composite
Available Colors: White
Max. Flight Time: 10 hours
Max. Speed: 161 km/h (100 mp/h)
Max. Flying Altitude: 5,486 m (18,000 ft)
Max. Operating Range: 1,050 km
GPS Compatible: √
Altitude Hold: √
Position Hold: √
Flight Planning: √
Autonomous Flight: √
Target Tracking: √
Automatic Landing: √
Return To Home: √
Obstacle Avoidance: √
Headless Mode: –
Built-in Camera: –

DYS

X-Ray V2

© DYS

Release Date: November 2015
Price: $115
Number of Rotors: 4
Control System: Remote Control
Applications: FPV Racing
Dimensions: 131 × 131 × 38 mm
Wingspan: 160 mm
Weight: 110 gr
Frame Material: Carbon Fiber
Rotor Material: Carbon Fiber
Motors: 1104-4000kv
Available Colors: Black
Controller Frequency: 5.8 GHz
Battery Capacity: 1,000 mAh
Battery Type: Lithium Polymer 3S
GPS Compatible: –
Altitude Hold: –
Position Hold: –
Flight Planning: –
Autonomous Flight: –
Target Tracking: –
Automatic Landing: –
Return To Home: –
Obstacle Avoidance: –
Headless Mode: –
Built-in Camera: √

DYS

X230 FPV

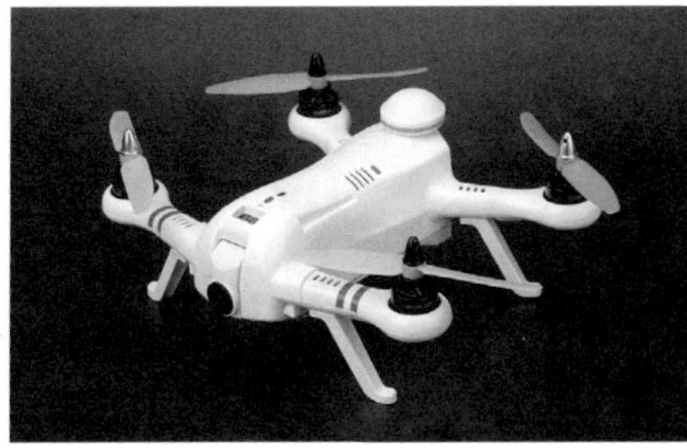
© RC Groups

Release Date: November 2015
Price: $330
Number of Rotors: 4
Control System: Remote Control
Applications: FPV Racing
Dimensions: 225 × 204 × 65 mm
Wingspan: 250 mm
Weight: 500 gr
Frame Material: Carbon Fiber
Rotor Material: Carbon Fiber
Available Colors: Black
Max. Flight Time: 10 minutes
Max. Speed: 108 km/h (67 mp/h)
Max. Flying Altitude: 500 m (1,640 ft)
Max. Ascent Speed: 5 m/s
Max. Descent Speed: 4 m/s
Min. Operating Temperature: -10°C
Max. Operating Temperature: 40°C
Max. Operating Range: 1 km
Controller Frequency: 5.8 GHz
Battery Capacity: 2,200 mAh
Battery Type: Lithium Polymer 3S
Battery Power: 11.1V
GPS Compatible: √
Altitude Hold: –
Position Hold: –
Flight Planning: –
Autonomous Flight: –
Target Tracking: –
Automatic Landing: –
Return To Home: –
Obstacle Avoidance: –
Headless Mode: –
Built-in Camera: √ (3.4 megapixels)

Eachine

H8C Mini

© banggood.com

Release Date: November 2015
Price: $60
Number of Rotors: 4
Control System: Remote Control
Applications: Recreation
Dimensions: 180 × 180 × 55 mm
Wingspan: 255 mm
Frame Material: Plastic
Rotor Material: Plastic
Available Colors: White
Max. Flight Time: 8 minutes
Max. Operating Range: 100 m
Controller Frequency: 2.4 GHz
Battery Capacity: 250 mAh
Battery Type: Lithium Polymer 1S
Battery Power: 3.7V
GPS Compatible: –
Altitude Hold: –
Position Hold: –
Flight Planning: –
Autonomous Flight: –
Target Tracking: –
Automatic Landing: –
Return To Home: √
Obstacle Avoidance: –
Headless Mode: √
Built-in Camera: √ (2 megapixels)

EHANG

Ghost 2.0

© dhgate.com

Release Date: November 2015
Price: $599
Number of Rotors: 4
Control System: Remote Control, Phone/Tablet
Applications: Film, Photography, FPV Racing
Wingspan: 350 mm
Weight: 1.15 kg
Frame Material: Plastic
Rotor Material: Plastic
Available Colors: Black
Max. Flight Time: 25 minutes
Max. Speed: 60 km/h (37 mp/h)
Max. Ascent Speed: 2,5 m/s
Max. Descent Speed: 1,5 m/s
Min. Operating Temperature: -10°C
Max. Operating Temperature: 40°C
Max. Operating Range: 1 km
Controller Frequency: 2.4 GHz
Battery Capacity: 4,500 mAh
Battery Type: Lithium Polymer 4S
Battery Power: 15V
Battery Weight: 400 gr
GPS Compatible: √
Altitude Hold: √
Position Hold: √
Flight Planning: √
Autonomous Flight: √
Target Tracking: √
Automatic Landing: √
Return To Home: √
Obstacle Avoidance: √
Headless Mode: –
Built-in Camera: –

Event38

E386

© Event38

Release Date: November 2015
Price: $6,899
Number of Rotors: 1 (Fixed Wing)
Control System: Flight Computer, Laptop
Applications: Aerial Mapping, Agriculture, Surveillance
Dimensions: 1.3 m
Wingspan: 1.9 m
Weight: 2.75 kg
Max. Payload: 500 gr
Frame Material: Foam, Carbon Fiber
Rotor Material: Plastic
Available Colors: White
Max. Flight Time: 75 minutes
Max. Speed: 52 km/h (29 mp/h)
Max. Flying Altitude: 3,960 m (12,992 ft)
Min. Operating Temperature: -20°C
Max. Operating Temperature: 50°C
Max. Operating Range: 10 km
Controller Frequency: 2.4 GHz
GPS Compatible: √
Altitude Hold: √
Position Hold: √
Flight Planning: √
Autonomous Flight: √
Target Tracking: √
Automatic Landing: √
Return To Home: √
Obstacle Avoidance: √
Headless Mode: –
Built-in Camera: √

Flairics

Inspector S

© Flairics

Release Date: November 2015
Price: On Request
Number of Rotors: 8
Control System: Remote Control
Applications: Film, Photography, Aerial Mapping, Inspection, Surveillance
Frame Material: Plastic, Carbon Fiber
Rotor Material: Carbon Fiber
Available Colors: Orange
Controller Frequency: 2.4 GHz
GPS Compatible: √
Altitude Hold: √
Position Hold: √
Flight Planning: √
Autonomous Flight: √
Target Tracking: –
Automatic Landing: √
Return To Home: √
Obstacle Avoidance: –
Headless Mode: –
Built-in Camera: –

Gear2Play

Sky

Release Date: November 2015
Price: $55
Number of Rotors: 4
Control System: Remote Control
Applications: Recreation
Dimensions: 300 × 300 mm
Weight: 70 gr
Frame Material: Plastic
Rotor Material: Plastic
Available Colors: Grey
Max. Flight Time: 8 minutes
Max. Operating Range: 30 m
Controller Frequency: 2.4 GHz
Battery Capacity: 250 mAh
Battery Type: Lithium Polymer
Battery Power: 3.7V
GPS Compatible: –
Altitude Hold: –
Position Hold: –
Flight Planning: –
Autonomous Flight: –
Target Tracking: –
Automatic Landing: –
Return To Home: √
Obstacle Avoidance: –
Headless Mode: √
Built-in Camera: √ (0.3 megapixels)

Hubsan

X4 Pro FPV H109S

Release Date: November 2015
Price: $499
Number of Rotors: 4
Control System: Remote Control
Applications: Film, Photography
Dimensions: 300 × 300 × 205 mm
Wingspan: 424 mm
Weight: 1.4 kg
Max. Payload: 650 gr
Frame Material: Plastic
Rotor Material: Plastic
Available Colors: Black
Max. Flight Time: 23 minutes
Max. Speed: 21,6 km/h (13,4 mp/h)
Max. Flying Altitude: 250 m (820 ft)
Max. Ascent Speed: 4 m/s
Max. Descent Speed: 4 m/s
Min. Operating Temperature: -10°C
Max. Operating Temperature: 40°C
Max. Operating Range: 1 km
Controller Frequency: 2.4 GHz
Battery Capacity: 7,000 mAh
Battery Type: Lithium Polymer
Battery Power: 11.1V
Battery Weight: 384 gr
GPS Compatible: –
Altitude Hold: √
Position Hold: √
Flight Planning: √
Autonomous Flight: √
Target Tracking: –
Automatic Landing: √
Return To Home: √
Obstacle Avoidance: –
Headless Mode: √
Built-in Camera: √ (3 megapixels)

Hubsan
X4 Plus H107D+

© Hubsan

Release Date: November 2015
Price: $175
Number of Rotors: 4
Control System: Remote Control
Applications: Recreation, FPV Racing
Wingspan: 140 mm
Frame Material: Plastic
Rotor Material: Plastic
Available Colors: White
Max. Flight Time: 7 minutes
Max. Flying Altitude: 100 m (328 ft)
Max. Operating Range: 100 m
Controller Frequency: 2.4 GHz
Battery Capacity: 380 mAh
Battery Type: Lithium Polymer
Battery Power: 3.7V
GPS Compatible: –
Altitude Hold: –
Position Hold: –
Flight Planning: –
Autonomous Flight: –
Target Tracking: –
Automatic Landing: –
Return To Home: –
Obstacle Avoidance: –
Headless Mode: –
Built-in Camera: √ (0.3 megapixels)

Immersion RC
Vortex 250 Pro

©Immersion RC

Release Date: November 2015
Price: $499
Number of Rotors: 4
Propellers: OEM Gemfan 5045 GF Blunt-Nose
Control System: Remote Control
Applications: FPV Racing
Wingspan: 250 mm
Weight: 415 gr
Frame Material: Carbon Fiber
Rotor Material: Plastic
Motors: 2204-2300kv
Available Colors: Black
Controller Frequency: 2.4 GHz
Battery Capacity: 5,400 mAh
Battery Type: Lithium Polymer 3S
Battery Power: 11.1V
GPS Compatible: √
Altitude Hold: –
Position Hold: –
Flight Planning: √
Autonomous Flight: –
Target Tracking: –
Automatic Landing: –
Return To Home: √
Obstacle Avoidance: –
Headless Mode: –
Built-in Camera: √ (FatShark 700TVL V2 CMOS)

Jin Xing Da

509V

© aliexpress.com

Release Date: November 2015
Price: $65
Number of Rotors: 4
Control System: Remote Control, Phone/Tablet
Applications: Film, Photography
Dimensions: 316 × 316 × 106 mm
Frame Material: Plastic
Rotor Material: Plastic
Available Colors: White
Max. Flight Time: 6 minutes
Max. Operating Range: 80 m
Controller Frequency: 2.4 GHz
Battery Capacity: 600 mAh
Battery Type: Lithium Polymer
Battery Power: 3.7V
GPS Compatible: –
Altitude Hold: √
Position Hold: √
Flight Planning: –
Autonomous Flight: –
Target Tracking: –
Automatic Landing: –
Return To Home: √
Obstacle Avoidance: –
Headless Mode: –
Built-in Camera: √ (2 megapixels)

JJRC

H22

© banggood.com

Release Date: November 2015
Price: $20
Number of Rotors: 4
Control System: Remote Control
Applications: Recreation
Dimensions: 100 × 100 × 20 mm
Wingspan: 168 mm
Weight: 270 gr
Frame Material: Plastic
Rotor Material: Plastic
Available Colors: Blue, Red
Max. Flight Time: 7 minutes
Max. Operating Range: 30 m
Controller Frequency: 2.4 GHz
Battery Capacity: 370 mAh
Battery Type: Lithium Polymer
Battery Power: 3.7V
GPS Compatible: –
Altitude Hold: –
Position Hold: –
Flight Planning: –
Autonomous Flight: –
Target Tracking: –
Automatic Landing: –
Return To Home: √
Obstacle Avoidance: –
Headless Mode: √
Built-in Camera: –

JTT UAV

T50

Release Date: November 2015
Price: $5,400
Number of Rotors: 4
Control System: Remote Control
Applications: Film, Photography, Recreation
Dimensions: 400 × 300 × 200 mm
Weight: 2.7 kg
Max. Payload: 650 gr
Frame Material: Plastic
Rotor Material: Plastic
Available Colors: Black, White
Max. Flight Time: 30 minutes
Max. Speed: 54 km/h (33,5 mp/h)
Max. Flying Altitude: 1,500 m (4,921 ft)
Max. Ascent Speed: 5 m/s
Max. Descent Speed: 4 m/s
Min. Operating Temperature: -10°C
Max. Operating Temperature: 40°C
Max. Operating Range: 1.5 km
Controller Frequency: 2.4 GHz
Battery Capacity: 10,000 mAh
Battery Type: Lithium Polymer 4S
GPS Compatible: √
Altitude Hold: √
Position Hold: √
Flight Planning: √
Autonomous Flight: √
Target Tracking: √
Automatic Landing: √
Return To Home: √
Obstacle Avoidance: –
Headless Mode: √
Built-in Camera: √ (1080P HD)

Monstertronic

Sky Eagle

© Monstertronic

Release Date: November 2015
Price: $70
Number of Rotors: 4
Control System: Remote Control
Applications: Recreation
Dimensions: 320 × 320 × 95 mm
Weight: 93 gr
Frame Material: Plastic
Rotor Material: Plastic
Available Colors: White
Controller Frequency: 2.4 GHz
Battery Type: Lithium Polymer 1S
GPS Compatible: –
Altitude Hold: √
Position Hold: √
Flight Planning: –
Autonomous Flight: –
Target Tracking: –
Automatic Landing: –
Return To Home: √
Obstacle Avoidance: –
Headless Mode: –
Built-in Camera: √

Monstertronic

Sky Bot

© Monstertronic

Release Date: November 2015
Price: $100
Number of Rotors: 4
Control System: Remote Control
Applications: Recreation
Dimensions: 335 × 335 × 75 mm
Weight: 132 gr
Frame Material: Plastic
Rotor Material: Plastic
Available Colors: Black
Max. Flight Time: 8 minutes
Controller Frequency: 2.4 GHz
Battery Capacity: 650 mAh
Battery Type: Lithium Polymer 1S
Battery Power: 3.7V
GPS Compatible: –
Altitude Hold: √
Position Hold: √
Flight Planning: –
Autonomous Flight: –
Target Tracking: –
Automatic Landing: –
Return To Home: √
Obstacle Avoidance: –
Headless Mode: –
Built-in Camera: √

Monstertronic

Sky Hawk

© Monstertronic

Release Date: November 2015
Price: $190
Number of Rotors: 4
Control System: Remote Control
Applications: Recreation
Dimensions: 310 × 310 × 90 mm
Weight: 135 gr
Frame Material: Plastic
Rotor Material: Plastic
Available Colors: Green, Red
Max. Flight Time: 6 minutes
Controller Frequency: 2.4 GHz
Battery Capacity: 650 mAh
Battery Type: Lithium Polymer 1S
Battery Power: 3.7V
GPS Compatible: –
Altitude Hold: √
Position Hold: √
Flight Planning: –
Autonomous Flight: –
Target Tracking: –
Automatic Landing: –
Return To Home: √
Obstacle Avoidance: –
Headless Mode: –
Built-in Camera: √

Novadem

NX110m

Release Date: November 2015
Price: $50,000
Number of Rotors: 4
Control System: Remote Control
Applications: Inspection, Military Purposes, Reconnaissance, Surveillance
Dimensions: 1.1 × 1.1 × 0.14 m
Weight: 1.7 kg
Max. Payload: 300 gr
Frame Material: Plastic
Rotor Material: Plastic
Available Colors: White
Max. Flight Time: 20 minutes
Max. Speed: 43 km/h (26,8 mp/h)
Max. Flying Altitude: 2,200 m (7218 ft)
Min. Operating Temperature: -10°C
Max. Operating Temperature: 35°C
Max. Operating Range: 1 km
GPS Compatible: √
Altitude Hold: √
Position Hold: √
Flight Planning: √
Autonomous Flight: √
Target Tracking: √
Automatic Landing: √
Return To Home: √
Obstacle Avoidance: –
Headless Mode: –
Built-in Camera: √

Onyxstar

Fox-C8 XT

Release Date: November 2015
Price: $17,500
Number of Rotors: 8
Control System: Remote Control
Applications: Film, Photography, Agriculture, Aerial Mapping, Inspection, Surveillance
Weight: 10 kg
Max. Payload: 5 kg
Frame Material: Carbon Fiber
Rotor Material: Carbon Fiber
Available Colors: Black
Max. Flight Time: 35 minutes
GPS Compatible: √
Altitude Hold: √
Position Hold: √
Flight Planning: √
Autonomous Flight: √
Target Tracking: √
Automatic Landing: √
Return To Home: √
Obstacle Avoidance: –
Headless Mode: –
Built-in Camera: –

Prodrone

Byrd

© Prodrone

Release Date: November 2015
Price: $950
Number of Rotors: 4
Control System: Remote Control
Applications: Film, Photography
Dimensions: 273 × 223 × 107 mm
Wingspan: 622 mm
Weight: 1.8 kg
Max. Payload: 2 kg
Frame Material: Plastic, Carbon Fiber
Rotor Material: Carbon Fiber
Available Colors: Black, White
Max. Flight Time: 29 minutes
Max. Speed: 54 km/h (33,5 mp/h)
Max. Flying Altitude: 152 m (500 ft)
Min. Operating Temperature: -10°C
Max. Operating Temperature: 50°C
Max. Operating Range: 2 km
Controller Frequency: 2.4 GHz
Battery Capacity: 6,700 mAh
Battery Type: Lithium Polymer
Battery Power: 14.8V
GPS Compatible: √
Altitude Hold: √
Position Hold: √
Flight Planning: √
Autonomous Flight: –
Target Tracking: √
Automatic Landing: √
Return To Home: √
Obstacle Avoidance: –
Headless Mode: –
Built-in Camera: –

Revell Control

Backflip 3D

© Revell Control

Release Date: November 2015
Price: $79
Number of Rotors: 4
Control System: Remote Control
Applications: Recreation
Dimensions: 140 × 140 × 35 mm
Weight: 54 gr
Frame Material: Plastic
Rotor Material: Plastic
Available Colors: Black
Max. Flight Time: 7 minutes
Max. Operating Range: 40 m
Controller Frequency: 2.4 GHz
Battery Capacity: 450 mAh
Battery Type: Lithium Polymer
Battery Power: 3.7V
GPS Compatible: –
Altitude Hold: –
Position Hold: –
Flight Planning: –
Autonomous Flight: –
Target Tracking: –
Automatic Landing: –
Return To Home: –
Obstacle Avoidance: –
Headless Mode: –
Built-in Camera: –

Revell Control

Formula Q

©Revell Control

Release Date: November 2015
Price: $120
Number of Rotors: 4
Control System: Remote Control
Applications: Recreation
Dimensions: 480 × 480 × 100 mm
Weight: 220 gr
Frame Material: Plastic
Rotor Material: Plastic
Available Colors: White
Max. Flight Time: 10 minutes
Max. Operating Range: 50 m
Controller Frequency: 2.4 GHz
Battery Capacity: 850 mAh
Battery Type: Lithium Polymer
Battery Power: 7.4V
GPS Compatible: –
Altitude Hold: –
Position Hold: –
Flight Planning: –
Autonomous Flight: –
Target Tracking: –
Automatic Landing: –
Return To Home: –
Obstacle Avoidance: –
Headless Mode: √
Built-in Camera: √

Revell Control

Funtic

©Revell Control

Release Date: November 2015
Price: $50
Number of Rotors: 4
Control System: Remote Control
Applications: Recreation
Dimensions: 290 × 290 × 70 mm
Weight: 100 gr
Frame Material: Plastic
Rotor Material: Plastic
Available Colors: White
Max. Flight Time: 7 minutes
Max. Operating Range: 50 m
Controller Frequency: 2.4 GHz
Battery Capacity: 500 mAh
Battery Type: Lithium Polymer
Battery Power: 3.7V
GPS Compatible: –
Altitude Hold: –
Position Hold: –
Flight Planning: –
Autonomous Flight: –
Target Tracking: –
Automatic Landing: –
Return To Home: –
Obstacle Avoidance: –
Headless Mode: –
Built-in Camera: √

Storm Racing Drone

SRD Lumenier QAV210

© SRD

Release Date: November 2015
Price: $589
Number of Rotors: 4
Control System: Remote Control
Applications: FPV Racing
Dimensions: 162 × 205 × 55 mm
Weight: 336 gr
Frame Material: Carbon Fiber
Rotor Material: Carbon Fiber
Motors: Storm M2204-UC 2300kv
Available Colors: Black
Max. Operating Range: 1.5 km
Controller Frequency: 5.8 GHz
Battery Capacity: 1,300 mAh
Battery Type: Lithium Polymer
Battery Power: 14.8V
GPS Compatible: –
Altitude Hold: –
Position Hold: –
Flight Planning: –
Autonomous Flight: –
Target Tracking: –
Automatic Landing: –
Return To Home: –
Obstacle Avoidance: –
Headless Mode: –
Built-in Camera: √ (600TVL)

Syma Toys

X18

© supplymicro.com

Release Date: November 2015
Price: $20
Number of Rotors: 4
Control System: Remote Control
Applications: Recreation
Dimensions: 100 × 100 × 95 mm
Frame Material: Plastic
Rotor Material: Plastic
Available Colors: Orange, Red
Max. Flight Time: 5 minutes
Max. Operating Range: 30 m
Controller Frequency: 2.4 GHz
Battery Capacity: 200 mAh
Battery Type: Lithium Polymer
Battery Power: 3.7V
GPS Compatible: –
Altitude Hold: –
Position Hold: –
Flight Planning: –
Autonomous Flight: –
Target Tracking: –
Automatic Landing: –
Return To Home: –
Obstacle Avoidance: –
Headless Mode: –
Built-in Camera: –

Thales

Spy'Ranger

© usinenouvelle.com/Gilles Cohen

Release Date: November 2015
Price: Unknown
Number of Rotors: 1 (Fixed Wing)
Control System: Ground Control Station
Applications: Military Purposes, Reconnaissance, Surveillance
Wingspan: 3.8 m
Weight: 14 kg
Max. Payload: 1.2 kg
Frame Material: Composite
Available Colors: Green
Max. Flight Time: 180 minutes
Max. Speed: 90 km/h (56 mp/h)
Max. Flying Altitude: 4,500 m (14,764 ft)
Max. Operating Range: 30 km
GPS Compatible: √
Altitude Hold: √
Position Hold: –
Flight Planning: √
Autonomous Flight: √
Target Tracking: –
Automatic Landing: –
Return To Home: √
Obstacle Avoidance: –
Headless Mode: –
Built-in Camera: √

Traxxas

Aton

© Traxxas

Release Date: November 2015
Price: $319
Number of Rotors: 4
Control System: Remote Control, Phone/Tablet
Applications: Film, Photography
Dimensions: 262 × 262 × 220 mm
Wingspan: 370 mm
Frame Material: Composite
Rotor Material: Plastic
Available Colors: Red
Max. Flight Time: 20 minutes
Controller Frequency: 2.4 GHz
Battery Capacity: 3,000 mAh
Battery Type: Lithium Polymer 3S
Battery Power: 11V
GPS Compatible: √
Altitude Hold: √
Position Hold: √
Flight Planning: √
Autonomous Flight: –
Target Tracking: √
Automatic Landing: √
Return To Home: √
Obstacle Avoidance: –
Headless Mode: –
Built-in Camera: –

Traxxas

Aton Plus

Release Date: November 2015
Price: $399
Number of Rotors: 4
Control System: Remote Control, Phone/Tablet
Applications: Film, Photography
Dimensions: 262 × 262 × 220 mm
Wingspan: 370 mm
Frame Material: Composite
Rotor Material: Plastic
Available Colors: Red
Max. Flight Time: 28 minutes
Controller Frequency: 2.4 GHz
Battery Capacity: 5,000 mAh
Battery Type: Lithium Polymer 3S
Battery Power: 11V
GPS Compatible: √
Altitude Hold: √
Position Hold: √
Flight Planning: √
Autonomous Flight: –
Target Tracking: √
Automatic Landing: √
Return To Home: √
Obstacle Avoidance: –
Headless Mode: –
Built-in Camera: –

WLtoys

DQ242G

Release Date: November 2015
Price: $85
Number of Rotors: 4
Control System: Remote Control
Applications: Recreation
Dimensions: 120 × 120 × 42 mm
Weight: 54 gr
Frame Material: Plastic
Rotor Material: Plastic
Available Colors: Black
Max. Flight Time: 6 minutes
Max. Operating Range: 50 m
Controller Frequency: 2.4 GHz
Battery Capacity: 380 mAh
Battery Type: Lithium Polymer
Battery Power: 3.7V
GPS Compatible: –
Altitude Hold: √
Position Hold: √
Flight Planning: –
Autonomous Flight: –
Target Tracking: –
Automatic Landing: –
Return To Home: √
Obstacle Avoidance: –
Headless Mode: √
Built-in Camera: √ (0.3 megapixels)

Xtreem

Gravity Pursuit

©Swann

Release Date: November 2015
Price: $299
Number of Rotors: 4
Control System: Remote Control
Applications: Recreation
Dimensions: 500 × 500 × 190 mm
Wingspan: 707 mm
Weight: 610 gr
Frame Material: Plastic
Rotor Material: Plastic
Available Colors: Blue
Max. Flight Time: 12 minutes
Max. Operating Range: 30 m
Controller Frequency: 2.4 GHz
Battery Capacity: 2,000 mAh
Battery Type: Lithium Polymer
Battery Power: 7.4V
GPS Compatible: –
Altitude Hold: –
Position Hold: –
Flight Planning: –
Autonomous Flight: –
Target Tracking: –
Automatic Landing: –
Return To Home: –
Obstacle Avoidance: –
Headless Mode: –
Built-in Camera: –

Xtreem

Electro-Max Eye

©Swann

Release Date: November 2015
Price: $69
Number of Rotors: 4
Control System: Remote Control
Applications: Recreation
Dimensions: 152 × 152 × 37 mm
Wingspan: 215 mm
Weight: 37 gr
Frame Material: Plastic
Rotor Material: Plastic
Available Colors: Black
Max. Flight Time: 7 minutes
Max. Operating Range: 30 m
Controller Frequency: 2.4 GHz
Battery Capacity: 250 mAh
Battery Type: Lithium Polymer
Battery Power: 3.7V
GPS Compatible: –
Altitude Hold: –
Position Hold: –
Flight Planning: –
Autonomous Flight: –
Target Tracking: –
Automatic Landing: –
Return To Home: –
Obstacle Avoidance: –
Headless Mode: –
Built-in Camera: √

Xtreem

Atom II Lightning Fast Mini

© Swann

Release Date: November 2015
Price: $40
Number of Rotors: 4
Control System: Remote Control
Applications: Recreation
Dimensions: 77 × 77 × 27 mm
Weight: 15 gr
Frame Material: Plastic
Rotor Material: Plastic
Available Colors: Blue
Max. Flight Time: 4 minutes
Max. Operating Range: 30 m
Controller Frequency: 2.4 GHz
Battery Capacity: 150 mAh
Battery Type: Lithium Polymer
Battery Power: 3V
GPS Compatible: –
Altitude Hold: –
Position Hold: –
Flight Planning: –
Autonomous Flight: –
Target Tracking: –
Automatic Landing: –
Return To Home: –
Obstacle Avoidance: –
Headless Mode: –
Built-in Camera: –

Xtreem

Quadforce

© bhphotovideo.com

Release Date: November 2015
Price: $90
Number of Rotors: 4
Control System: Remote Control
Applications: Recreation
Dimensions: 315 × 315 × 76 mm
Wingspan: 445 mm
Weight: 93 gr
Frame Material: Plastic
Rotor Material: Plastic
Available Colors: Black, Orange
Max. Flight Time: 8 minutes
Max. Operating Range: 100 m
Controller Frequency: 2.4 GHz
Battery Capacity: 500 mAh
Battery Type: Lithium Polymer
Battery Power: 4V
GPS Compatible: –
Altitude Hold: –
Position Hold: –
Flight Planning: –
Autonomous Flight: –
Target Tracking: –
Automatic Landing: –
Return To Home: –
Obstacle Avoidance: –
Headless Mode: –
Built-in Camera: √

Arris

X-Speed 280

©Arris Hobby

Release Date: December 2015
Price: $285
Number of Rotors: 4
Propellers: 5045 propellers
Control System: Remote Control
Applications: FPV Racing
Dimensions: 245 × 213 × 112 mm
Wingspan: 274 mm
Weight: 420 gr
Max. Payload: 159 gr
Frame Material: Carbon Fiber
Rotor Material: Plastic
Motors: Arris 2205-2300kv brushless
Available Colors: Black, Green, Orange, Pink, White, Yellow
Controller Frequency: 2.4 GHz
Battery Capacity: 2,200 mAh
Battery Type: Lithium Polymer 4S
Battery Power: 11.1V
Battery Weight: 159 gr
GPS Compatible: √
Altitude Hold: –
Position Hold: –
Flight Planning: √
Autonomous Flight: –
Target Tracking: –
Automatic Landing: √
Return To Home: –
Obstacle Avoidance: –
Headless Mode: –
Built-in Camera: √ (Sony 700TVL)

Cheerson

CX-32S Falcon

© banggood.com

Release Date: December 2015
Price: $90
Number of Rotors: 4
Control System: Remote Control
Applications: Film, Photography, Recreation
Dimensions: 340 × 340 × 80 mm
Wingspan: 380 mm
Weight: 130 gr
Max. Payload: 44 gr
Frame Material: Plastic
Rotor Material: Plastic
Available Colors: Black, White
Max. Flight Time: 7 minutes
Max. Operating Range: 80 m
Controller Frequency: 2.4 GHz
Battery Capacity: 360 mAh
Battery Type: Lithium Polymer
Battery Power: 7.4V
GPS Compatible: –
Altitude Hold: √
Position Hold: √
Flight Planning: –
Autonomous Flight: –
Target Tracking: –
Automatic Landing: √
Return To Home: –
Obstacle Avoidance: –
Headless Mode: –
Built-in Camera: √ (2 megapixels)

Cheerson

CX-Stars

© Ingenious-ingenuity.com

Release Date: December 2015
Price: $15
Number of Rotors: 4
Control System: Remote Control
Applications: Recreation
Dimensions: 22 × 22 × 20 mm
Wingspan: 30 mm
Weight: 7.5 gr
Frame Material: Plastic
Rotor Material: Plastic
Available Colors: Black, Blue, Green, Orange
Max. Flight Time: 5 minutes
Controller Frequency: 2.4 GHz
Battery Capacity: 80 mAh
Battery Type: Lithium Polymer
Battery Power: 3.7V
GPS Compatible: –
Altitude Hold: –
Position Hold: –
Flight Planning: –
Autonomous Flight: –
Target Tracking: –
Automatic Landing: –
Return To Home: –
Obstacle Avoidance: –
Headless Mode: –
Built-in Camera: –

HeliMax

Form 500

© Helicomicro

Release Date: December 2015
Price: $300
Number of Rotors: 4
Control System: Remote Control
Applications: Film, Photography
Dimensions: 345 × 345 × 235 mm
Wingspan: 500 mm
Weight: 1.75 kg
Max. Payload: 1 kg
Frame Material: Plastic
Rotor Material: Plastic
Available Colors: Yellow
Max. Flight Time: 15 minutes
Controller Frequency: 2.4 GHz
Battery Capacity: 5,000 mAh
Battery Type: Lithium Polymer 3S
Battery Power: 11.1V
GPS Compatible: √
Altitude Hold: √
Position Hold: –
Flight Planning: –
Autonomous Flight: –
Target Tracking: –
Automatic Landing: √
Return To Home: √
Obstacle Avoidance: –
Headless Mode: √
Built-in Camera: –

HiSky Hobby

Talon 280

Release Date: December 2015
Price: $260
Number of Rotors: 4
Control System: Remote Control
Applications: FPV Racing
Dimensions: 350 × 120 mm
Wingspan: 280 mm
Weight: 450 gr
Frame Material: Plastic
Rotor Material: Plastic
Motors: 1818-1300kv
Available Colors: Green
Max. Flight Time: 10 minutes
Max. Speed: 100 km/h (62 mp/h)
Max. Operating Range: 400 m
Controller Frequency: 5.8 GHz
Battery Capacity: 1,350 mAh
Battery Type: Lithium Polymer 4S
Battery Power: 11.1V
GPS Compatible: √
Altitude Hold: √
Position Hold: √
Flight Planning: –
Autonomous Flight: –
Target Tracking: –
Automatic Landing: –
Return To Home: √
Obstacle Avoidance: –
Headless Mode: –
Built-in Camera: √ (13 megapixels)

Jin Xing Da

509W

Release Date: December 2015
Price: $55
Number of Rotors: 4
Control System: Remote Control, Phone/Tablet
Applications: Film, Photography
Dimensions: 299 × 295 × 102 mm
Wingspan: 227 mm
Weight: 138 gr
Frame Material: Plastic
Rotor Material: Plastic
Available Colors: Black
Max. Flight Time: 6 minutes
Max. Operating Range: 50 m
Controller Frequency: 2.4 GHz
Battery Capacity: 600 mAh
Battery Type: Lithium Polymer
Battery Power: 3.7V
GPS Compatible: –
Altitude Hold: √
Position Hold: √
Flight Planning: –
Autonomous Flight: –
Target Tracking: –
Automatic Landing: –
Return To Home: √
Obstacle Avoidance: –
Headless Mode: –
Built-in Camera: √ (0.3 megapixels)

JJRC

H26D

© JJRC

Release Date: December 2015
Price: $80
Number of Rotors: 4
Control System: Phone/Tablet
Applications: Film, Photography, Recreation
Dimensions: 455 × 455 × 180 mm
Wingspan: 636 mm
Frame Material: Plastic, Carbon Fiber
Rotor Material: Plastic
Available Colors: Green, White
Max. Flight Time: 7 minutes
Max. Operating Range: 300 m
Controller Frequency: 2.4 GHz
Battery Capacity: 1,200 mAh
Battery Type: Lithium Polymer
Battery Power: 7.4V
GPS Compatible: –
Altitude Hold: –
Position Hold: –
Flight Planning: –
Autonomous Flight: –
Target Tracking: –
Automatic Landing: –
Return To Home: √
Obstacle Avoidance: –
Headless Mode: √
Built-in Camera: √ (3 megapixels)

JJRC

H11C

© banggood.com

Release Date: December 2015
Price: $45
Number of Rotors: 4
Control System: Remote Control
Applications: Recreation
Dimensions: 330 × 330 × 85 mm
Wingspan: 467 mm
Frame Material: Plastic
Rotor Material: Plastic
Available Colors: Black, Blue, White
Max. Flight Time: 10 minutes
Max. Operating Range: 120 m
Controller Frequency: 2.4 GHz
Battery Capacity: 1,100 mAh
Battery Type: Lithium Polymer
Battery Power: 3.7V
GPS Compatible: –
Altitude Hold: –
Position Hold: –
Flight Planning: –
Autonomous Flight: –
Target Tracking: –
Automatic Landing: –
Return To Home: √
Obstacle Avoidance: –
Headless Mode: –
Built-in Camera: √ (2 megapixels)

JJRC

H5M

© JJRC

Release Date: December 2015
Price: $35
Number of Rotors: 4
Control System: Remote Control
Applications: Recreation
Dimensions: 290 × 290 × 54 mm
Weight: 108 gr
Frame Material: Plastic
Rotor Material: Plastic
Available Colors: Green, Red
Max. Flight Time: 8 minutes
Max. Operating Range: 80 m
Controller Frequency: 2.4 GHz
Battery Capacity: 600 mAh
Battery Type: Lithium Polymer
Battery Power: 3.7V
GPS Compatible: –
Altitude Hold: √
Position Hold: √
Flight Planning: –
Autonomous Flight: –
Target Tracking: –
Automatic Landing: –
Return To Home: √
Obstacle Avoidance: –
Headless Mode: –
Built-in Camera: √ (2 megapixels)

JJRC

X1

© aliexpress.com

Release Date: December 2015
Price: $100
Number of Rotors: 4
Control System: Remote Control
Applications: Recreation
Dimensions: 328 × 325 × 53 mm
Wingspan: 192 mm
Frame Material: Plastic
Rotor Material: Plastic
Available Colors: Black, Blue, Green
Max. Flight Time: 15 minutes
Max. Operating Range: 400 m
Controller Frequency: 2.4 GHz
Battery Capacity: 1,300 mAh
Battery Type: Lithium Polymer
Battery Power: 7.4V
GPS Compatible: –
Altitude Hold: √
Position Hold: √
Flight Planning: –
Autonomous Flight: –
Target Tracking: –
Automatic Landing: –
Return To Home: √
Obstacle Avoidance: –
Headless Mode: –
Built-in Camera: √ (2 megapixels)

JJRC

H26

© banggood.com

Release Date: December 2015
Price: $45
Number of Rotors: 4
Control System: Remote Control
Applications: Recreation
Dimensions: 450 × 450 × 180 mm
Wingspan: 467 mm
Frame Material: Plastic
Rotor Material: Plastic
Available Colors: Green, White
Max. Flight Time: 7 minutes
Max. Operating Range: 300 m
Controller Frequency: 2.4 GHz
Battery Capacity: 1,200 mAh
Battery Type: Lithium Polymer
Battery Power: 7.4V
GPS Compatible: –
Altitude Hold: √
Position Hold: √
Flight Planning: –
Autonomous Flight: –
Target Tracking: –
Automatic Landing: –
Return To Home: √
Obstacle Avoidance: –
Headless Mode: √
Built-in Camera: –

JJRC

H26W

© banggood.com

Release Date: December 2015
Price: $80
Number of Rotors: 4
Control System: Remote Control, Phone/Tablet
Applications: Film, Photography, Recreation
Dimensions: 455 × 455 × 180 mm
Weight: 1 kg
Frame Material: Plastic
Rotor Material: Plastic
Available Colors: Green, White
Max. Flight Time: 7 minutes
Max. Operating Range: 300 m
Controller Frequency: 2.4 GHz
Battery Capacity: 1,200 mAh
Battery Type: Lithium Polymer
Battery Power: 7.4V
GPS Compatible: –
Altitude Hold: √
Position Hold: √
Flight Planning: –
Autonomous Flight: –
Target Tracking: –
Automatic Landing: –
Return To Home: √
Obstacle Avoidance: –
Headless Mode: –
Built-in Camera: √ (2 megapixels)

Nine Eagles

Galaxy Visitor 7

©Nine Eagles

Release Date: December 2015
Price: $269
Number of Rotors: 4
Control System: Remote Control, Phone/Tablet
Applications: Agriculture, Inspection, Surveillance
Dimensions: 199 × 199 × 66 mm
Wingspan: 241 mm
Weight: 130 gr
Frame Material: Plastic
Rotor Material: Plastic
Available Colors: White
Max. Flight Time: 10 minutes
Max. Operating Range: 100 m
Controller Frequency: 2.4 GHz
Battery Capacity: 700 mAh
Battery Type: Lithium Polymer
Battery Power: 3.7V
GPS Compatible: –
Altitude Hold: √
Position Hold: √
Flight Planning: –
Autonomous Flight: –
Target Tracking: –
Automatic Landing: –
Return To Home: √
Obstacle Avoidance: –
Headless Mode: –
Built-in Camera: √

Parrot

Bebop 2

©Parrot

Release Date: December 2015
Price: $550
Number of Rotors: 4
Control System: Remote Control, Phone/Tablet
Applications: Film, Photography
Dimensions: 328 × 382 × 89 mm
Weight: 500 gr
Frame Material: Plastic, Carbon Fiber
Rotor Material: Plastic
Available Colors: Red, White
Max. Flight Time: 25 minutes
Max. Speed: 64 km/h (40 mp/h)
Max. Ascent Speed: 6 m/s
Max. Operating Range: 300 m
Controller Frequency: 2.4 GHz
Battery Capacity: 2,700 mAh
Battery Type: Lithium Polymer 3S
GPS Compatible: √
Altitude Hold: √
Position Hold: √
Flight Planning: √
Autonomous Flight: √
Target Tracking: √
Automatic Landing: √
Return To Home: √
Obstacle Avoidance: –
Headless Mode: –
Built-in Camera: √ (14 megapixels)

Squadrone Systems

Hexo Plus

© Hexo Plus

Release Date: December 2015
Price: $1,099
Number of Rotors: 6
Control System: Remote Control, Phone/Tablet
Applications: Film, Photography
Dimensions: 435 × 410 × 244 mm
Weight: 1.7 kg
Frame Material: Plastic, Carbon Fiber
Rotor Material: Carbon Fiber
Available Colors: Blue
Max. Flight Time: 15 minutes
Max. Speed: 70 km/h (44 mp/h)
Max. Flying Altitude: 4,800 m (15,748 ft)
Min. Operating Temperature: -10°C
Max. Operating Temperature: 50°C
Controller Frequency: 2.4 GHz
Battery Capacity: 5,000 mAh
Battery Type: Lithium Polymer 5S
Battery Power: 11.1V
GPS Compatible: √
Altitude Hold: √
Position Hold: √
Flight Planning: √
Autonomous Flight: √
Target Tracking: √
Automatic Landing: √
Return To Home: √
Obstacle Avoidance: √
Headless Mode: √
Built-in Camera: –

Storm Racing Drone

SRD 370 Naza V2

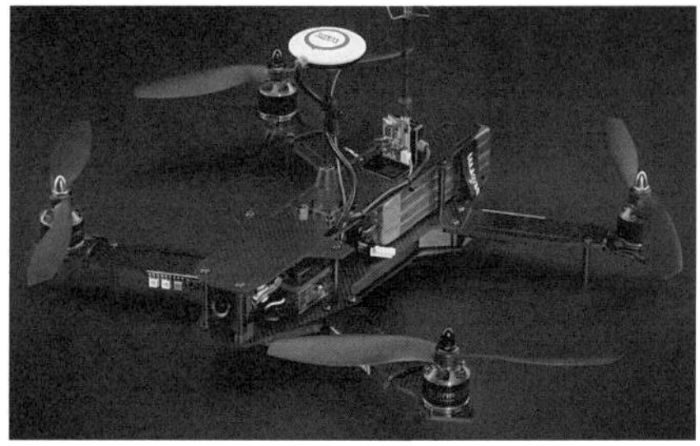

© SRD

Release Date: December 2015
Price: $699
Number of Rotors: 4
Control System: Remote Control
Applications: FPV Racing
Dimensions: 325 × 283 × 72 mm
Weight: 742 gr
Frame Material: Carbon Fiber
Rotor Material: Carbon Fiber
Available Colors: Black
Max. Flight Time: 19 minutes
Max. Operating Range: 1.5 km
Controller Frequency: 5.8 GHz
Battery Capacity: 4,000 mAh
Battery Type: Lithium Polymer
Battery Power: 14.8V
GPS Compatible: –
Altitude Hold: –
Position Hold: –
Flight Planning: –
Autonomous Flight: –
Target Tracking: –
Automatic Landing: –
Return To Home: –
Obstacle Avoidance: –
Headless Mode: –
Built-in Camera: √ (700TVL)

Trident LLC

Carbon Flyer

Release Date: December 2015
Price: $149
Number of Rotors: 2 (Fixed Wing)
Control System: Remote Control
Applications: Recreation
Wingspan: 370 mm
Frame Material: Carbon Fiber
Rotor Material: Carbon Fiber
Available Colors: Black
Max. Flight Time: 7 minutes
Max. Operating Range: 70 m
Controller Frequency: 2.4 GHz
Battery Capacity: 150 mAh
Battery Type: Lithium Polymer
Battery Power: 3.7V
GPS Compatible: –
Altitude Hold: –
Position Hold: –
Flight Planning: –
Autonomous Flight: –
Target Tracking: –
Automatic Landing: –
Return To Home: –
Obstacle Avoidance: –
Headless Mode: –
Built-in Camera: √

WLtoys

Flying 3D X6 Plus

Release Date: December 2015
Price: $300
Number of Rotors: 4
Control System: Remote Control
Applications: Film, Photography, FPV Racing
Dimensions: 195 × 195 × 150 mm
Weight: 331 gr
Frame Material: Plastic
Rotor Material: Plastic
Available Colors: White
Max. Flight Time: 14 minutes
Max. Operating Range: 1 km
Controller Frequency: 5.8 GHz
Battery Capacity: 1,000 mAh
Battery Type: Lithium Polymer
Battery Power: 7.4V
GPS Compatible: √
Altitude Hold: √
Position Hold: √
Flight Planning: √
Autonomous Flight: √
Target Tracking: –
Automatic Landing: √
Return To Home: √
Obstacle Avoidance: –
Headless Mode: –
Built-in Camera: √ (720P HD)

xFold

Dragon X12

© bhphotovideo.com

Release Date: December 2015
Price: $31,599
Number of Rotors: 12
Control System: Remote Control
Applications: Film, Photography, Inspection, Search & Rescue
Dimensions: 1.36 × 1.56 × 0.81 m
Weight: 19.4 kg
Frame Material: Carbon Fiber
Motors: xFold custom 120kv
Available Colors: Black
Max. Flight Time: 25 minutes
Controller Frequency: 2.4 GHz
Battery Capacity: 22,000 mAh
Battery Type: Lithium Polymer 6S
GPS Compatible: √
Altitude Hold: √
Position Hold: √
Flight Planning: √
Autonomous Flight: √
Target Tracking: –
Automatic Landing: √
Return To Home: √
Obstacle Avoidance: –
Headless Mode: –
Built-in Camera: –

XK Innovations

Aircam X500

© gearbest.com

Release Date: December 2015
Price: $350
Number of Rotors: 4
Control System: Remote Control
Applications: Film, Photography, Recreation
Dimensions: 430 × 400 × 208 mm
Wingspan: 500 mm
Weight: 1.3 kg
Max. Payload: 500 gr
Frame Material: Carbon Fiber
Rotor Material: Carbon Fiber
Available Colors: Red
Max. Flight Time: 30 minutes
Max. Operating Range: 1 km
Controller Frequency: 2.4 GHz
Battery Capacity: 5,400 mAh
Battery Type: Lithium Polymer 4S
Battery Power: 14.8V
GPS Compatible: √
Altitude Hold: √
Position Hold: √
Flight Planning: –
Autonomous Flight: √
Target Tracking: –
Automatic Landing: √
Return To Home: √
Obstacle Avoidance: –
Headless Mode: √
Built-in Camera: √

Yi Zhan Toys

i6S

©rcdronenews.com

Release Date: December 2015
Price: $40
Number of Rotors: 6
Control System: Remote Control
Applications: Recreation
Dimensions: 95 × 95 × 80 mm
Weight: 39 gr
Frame Material: Plastic
Rotor Material: Plastic
Available Colors: Black, White
Max. Flight Time: 6 minutes
Max. Operating Range: 50 m
Controller Frequency: 2.4 GHz
Battery Capacity: 450 mAh
Battery Type: Lithium Polymer
Battery Power: 3.7V
GPS Compatible: –
Altitude Hold: √
Position Hold: √
Flight Planning: –
Autonomous Flight: –
Target Tracking: –
Automatic Landing: –
Return To Home: √
Obstacle Avoidance: –
Headless Mode: –
Built-in Camera: √ (2 megapixels)

Yuneec

Typhoon G

©bhphotovideo.com

Release Date: December 2015
Price: $499
Number of Rotors: 4
Control System: Remote Control, Phone/Tablet
Applications: Film, Photography
Dimensions: 500 × 420 × 240 mm
Wingspan: 565 mm
Weight: 1.1 kg
Max. Payload: 600 gr
Frame Material: Carbon Fiber
Rotor Material: Carbon Fiber
Available Colors: Black
Max. Flight Time: 25 minutes
Max. Speed: 29 km/h (18 mp/h)
Max. Flying Altitude: 122 m (400 ft)
Max. Ascent Speed: 3 m/s
Max. Descent Speed: 2 m/s
Min. Operating Temperature: 0°C
Max. Operating Temperature: 40°C
Max. Operating Range: 800 m
Controller Frequency: 2.4 GHz
Battery Capacity: 5,400 mAh
Battery Type: Lithium Polymer 3S
Battery Power: 11V
GPS Compatible: √
Altitude Hold: √
Position Hold: √
Flight Planning: –
Autonomous Flight: –
Target Tracking: √
Automatic Landing: √
Return To Home: √
Obstacle Avoidance: –
Headless Mode: –
Built-in Camera: –

Zerotech

Guardian Z10

Release Date: December 2015
Price: On Request
Number of Rotors: 4
Control System: Flight Computer, Laptop
Applications: Aerial Mapping, Agriculture
Dimensions: 1.3 × 1.3 × 0.45 m
Wingspan: 1.2 m
Weight: 14 kg
Max. Payload: 10 kg
Frame Material: Carbon Fiber
Available Colors: Black
Max. Flight Time: 15 minutes
Controller Frequency: 2.4 GHz
Battery Capacity: 16,000 mAh
Battery Type: Lithium Polymer 12S
GPS Compatible: √
Altitude Hold: √
Position Hold: √
Flight Planning: √
Autonomous Flight: √
Target Tracking: –
Automatic Landing: √
Return To Home: √
Obstacle Avoidance: –
Headless Mode: –
Built-in Camera: –

2016

Action Drone

AD2

Release Date: January 2016
Price: $6,727
Number of Rotors: 4
Propellers: 18 inch carbon fiber hollow core
Control System: Remote Control, Tablet
Applications: Aerial Mapping, Agriculture, Surveillance
Wingspan: 950 mm
Max. Payload: 5.5 kg
Frame Material: Carbon Fiber
Rotor Material: Carbon Fiber
Available Colors: Red
Max. Flight Time: 45 minutes
Max. Speed: 50 km/h (31 mp/h)
Max. Flying Altitude: 600 m (1,968 ft)
Max. Operating Range: 1.6 km
Controller Frequency: 2.4 GHz
Battery Capacity: 17,000 mAh
Battery Type: Lithium Polymer 6S
GPS Compatible: √
Altitude Hold: √
Position Hold: √
Flight Planning: √
Autonomous Flight: √
Target Tracking: –
Automatic Landing: √
Return To Home: –
Obstacle Avoidance: –
Headless Mode: –
Built-in Camera: –

Aerial Technology Int.

Thor X4

Release Date: January 2016
Price: $17,899
Number of Rotors: 4
Propellers: 18 inch
Control System: Remote Control, Phone/Tablet
Applications: Film, Photography
Wingspan: 1.48 m
Weight: 15.5 kg
Max. Payload: 4.5 kg
Frame Material: Carbon Fiber, Aluminum
Rotor Material: Carbon Fiber
Motors: 5215-435kv
Available Colors: Black
Max. Flight Time: 15 minutes
Max. Speed: 107 km/h (67 mp/h)
Max. Operating Range: 26.7 km
Controller Frequency: 2.4 GHz
Battery Capacity: 22,000 mAh
Battery Type: Lithium Polymer 6S
Battery Power: 12V
GPS Compatible: √
Altitude Hold: –
Position Hold: –
Flight Planning: –
Autonomous Flight: –
Target Tracking: –
Automatic Landing: –
Return To Home: –
Obstacle Avoidance: –
Headless Mode: –
Built-in Camera: –

Aerix Drones

Vidius

©Aerix Drones

Release Date: January 2016
Price: $95
Number of Rotors: 4
Control System: Remote Control, Phone/Tablet
Applications: Film, Photography
Dimensions: 43 × 43 × 25 mm
Weight: 17 gr
Frame Material: Plastic
Rotor Material: Plastic
Available Colors: Black
Max. Flight Time: 7 minutes
Max. Operating Range: 30 m
Controller Frequency: 2.4 GHz
Battery Capacity: 150 mAh
Battery Type: Lithium Polymer
Battery Power: 3.7V
GPS Compatible: –
Altitude Hold: √
Position Hold: √
Flight Planning: –
Autonomous Flight: –
Target Tracking: –
Automatic Landing: –
Return To Home: –
Obstacle Avoidance: –
Headless Mode: –
Built-in Camera: √

Ares

Shadow 240

©Ares-RC

Release Date: January 2016
Price: $100
Number of Rotors: 4
Control System: Remote Control, Phone/Tablet
Applications: Recreation
Wingspan: 140 mm
Weight: 110 gr
Frame Material: Plastic
Rotor Material: Plastic
Motors: 8520 coreless
Available Colors: Black
Battery Capacity: 780 mAh
Battery Type: Lithium Polymer 1S
Battery Power: 3.7V
GPS Compatible: –
Altitude Hold: –
Position Hold: √
Flight Planning: –
Autonomous Flight: –
Target Tracking: –
Automatic Landing: –
Return To Home: –
Obstacle Avoidance: –
Headless Mode: –
Built-in Camera: –

Arris

M700

© Droneboomer

Release Date: January 2016
Price: $349
Number of Rotors: 6
Control System: Remote Control
Applications: Film, Photography, Surveillance
Wingspan: 710 mm
Weight: 1.6 kg
Frame Material: Carbon Fiber
Rotor Material: Plastic
Available Colors: Black
Controller Frequency: 2.4 GHz
Battery Type: Lithium Polymer 6S
GPS Compatible: √
Altitude Hold: √
Position Hold: √
Flight Planning: √
Autonomous Flight: √
Target Tracking: √
Automatic Landing: √
Return To Home: √
Obstacle Avoidance: –
Headless Mode: –
Built-in Camera: –

Autel Robotics

X-Star Premium

© Autel Robotics

Release Date: January 2016
Price: $899
Number of Rotors: 4
Control System: Remote Control, Phone/Tablet
Applications: Film, Photography
Wingspan: 352 mm
Weight: 1.4 kg
Max. Payload: 180 gr
Frame Material: Plastic
Rotor Material: Plastic
Available Colors: Orange, White
Max. Flight Time: 25 minutes
Max. Speed: 57 km/h (35,7 mp/h)
Max. Ascent Speed: 6 m/s
Min. Operating Temperature: 0°C
Max. Operating Temperature: 50°C
Max. Operating Range: 2 km
Controller Frequency: 5.8 GHz
Battery Capacity: 4,900 mAh
Battery Type: Lithium Polymer
Battery Power: 14.8V
GPS Compatible: √
Altitude Hold: √
Position Hold: √
Flight Planning: √
Autonomous Flight: √
Target Tracking: √
Automatic Landing: √
Return To Home: √
Obstacle Avoidance: –
Headless Mode: –
Built-in Camera: –

Blade

Zeyrok

© Horizon Hobby

Release Date: January 2016
Price: $160
Number of Rotors: 4
Control System: Remote Control, Phone/Tablet
Applications: Film, Photography
Dimensions: 220 × 200 × 45 mm
Weight: 107 gr
Frame Material: Plastic, Carbon Fiber
Rotor Material: Plastic
Available Colors: Green, Yellow
Max. Flight Time: 10 minutes
Controller Frequency: 2.4 GHz
Battery Capacity: 750 mAh
Battery Type: Lithium Polymer 1S
GPS Compatible: –
Altitude Hold: –
Position Hold: –
Flight Planning: –
Autonomous Flight: –
Target Tracking: –
Automatic Landing: –
Return To Home: –
Obstacle Avoidance: –
Headless Mode: –
Built-in Camera: √ (1.3 megapixels)

Blade

Inductrix 200

© Horizon Hobby

Release Date: January 2016
Price: $250
Number of Rotors: 4
Control System: Remote Control
Applications: FPV Racing
Dimensions: 155 × 155 × 45 mm
Weight: 185 gr
Frame Material: Plastic
Rotor Material: Plastic
Motors: Brushless
Available Colors: Black
Max. Flight Time: 8 minutes
Controller Frequency: 2.4 GHz
Battery Capacity: 800 mAh
Battery Type: Lithium Polymer 3S
Battery Power: 11.1V
GPS Compatible: –
Altitude Hold: –
Position Hold: –
Flight Planning: –
Autonomous Flight: –
Target Tracking: –
Automatic Landing: –
Return To Home: –
Obstacle Avoidance: –
Headless Mode: –
Built-in Camera: √

Cheerson

CX-10W

© banggood.com

Release Date: January 2016
Price: $25
Number of Rotors: 4
Control System: Phone/Tablet
Applications: Recreation
Dimensions: 42 × 42 × 25 mm
Wingspan: 45 mm
Weight: 17 gr
Frame Material: Plastic
Rotor Material: Plastic
Available Colors: Gold, Pink, Silver
Max. Flight Time: 4 minutes
Max. Operating Range: 30 m
Controller Frequency: 2.4 GHz
Battery Capacity: 150 mAh
Battery Type: Lithium Polymer
Battery Power: 3.7V
GPS Compatible: –
Altitude Hold: √
Position Hold: √
Flight Planning: –
Autonomous Flight: –
Target Tracking: –
Automatic Landing: –
Return To Home: –
Obstacle Avoidance: –
Headless Mode: –
Built-in Camera: √ (0.3 megapixels)

Cheerson

CX-35 Leopard

© geekbuying.com

Release Date: January 2016
Price: $110
Number of Rotors: 4
Control System: Remote Control, Phone/Tablet
Applications: Film, Photography
Dimensions: 450 × 450 × 170 mm
Wingspan: 636 mm
Weight: 480 gr
Max. Payload: 50 gr
Frame Material: Plastic
Rotor Material: Plastic
Available Colors: Black
Max. Flight Time: 7 minutes
Max. Operating Range: 100 m
Controller Frequency: 2.4 GHz
Battery Capacity: 1,300 mAh
Battery Type: Lithium Polymer
Battery Power: 7.4V
GPS Compatible: –
Altitude Hold: √
Position Hold: √
Flight Planning: –
Autonomous Flight: –
Target Tracking: –
Automatic Landing: √
Return To Home: √
Obstacle Avoidance: –
Headless Mode: –
Built-in Camera: –

DJI

Agras MG-1

© DJI

Release Date: January 2016
Price: ± $15,000
Number of Rotors: 8
Control System: Remote Control
Applications: Agriculture
Dimensions: 1.47 × 1.47 × 0.48 m
Wingspan: 1.52 m
Weight: 8.8 kg
Max. Payload: 10 kg
Frame Material: Carbon Fiber
Rotor Material: Plastic
Motors: 130 rpm/V kv
Available Colors: Black
Max. Flight Time: 24 minutes
Max. Flight Time with Payload: 10 minutes
Max. Speed: 79 km/h (49 mp/h)
Min. Operating Temperature: 0°C
Max. Operating Temperature: 40°C
Max. Operating Range: 1 km
Controller Frequency: 2.4 GHz
GPS Compatible: √
Altitude Hold: √
Position Hold: √
Flight Planning: √
Autonomous Flight: √
Target Tracking: √
Automatic Landing: √
Return To Home: √
Obstacle Avoidance: –
Headless Mode: –
Built-in Camera: –

Dromida

Verso

© hobbyohio.com

Release Date: January 2016
Price: $40
Number of Rotors: 4
Control System: Remote Control
Applications: Recreation
Dimensions: 200 × 200 × 111 mm
Wingspan: 110 mm
Weight: 34 gr
Frame Material: Plastic
Rotor Material: Plastic
Available Colors: Blue, Green, Red, Yellow
Max. Flight Time: 5 minutes
Controller Frequency: 2.4 GHz
Battery Capacity: 350 mAh
Battery Type: Lithium Polymer 1S
Battery Power: 3.7V
GPS Compatible: –
Altitude Hold: –
Position Hold: –
Flight Planning: √
Autonomous Flight: –
Target Tracking: –
Automatic Landing: –
Return To Home: –
Obstacle Avoidance: –
Headless Mode: –
Built-in Camera: –

Eachine

Assassin 180 FPV

© Eachine

Release Date: January 2016
Price: $750
Number of Rotors: 4
Control System: Remote Control
Applications: FPV Racing
Dimensions: 128 × 180 × 74 mm
Wingspan: 180 mm
Weight: 288 gr
Frame Material: Carbon Fiber
Rotor Material: Plastic
Motors: B1806-2300kv
Available Colors: Black
Max. Flight Time: 9 minutes
Controller Frequency: 5.8 GHz
Battery Capacity: 1,000 mAh
Battery Type: Lithium Polymer 3S
Battery Power: 11.1V
Battery Weight: 24 gr
GPS Compatible: √
Altitude Hold: √
Position Hold: √
Flight Planning: –
Autonomous Flight: –
Target Tracking: –
Automatic Landing: –
Return To Home: –
Obstacle Avoidance: –
Headless Mode: –
Built-in Camera: √

Eachine

Falcon 250

© Eachine

Release Date: January 2016
Price: $250
Number of Rotors: 4
Control System: Remote Control
Applications: FPV Racing
Weight: 400 gr
Frame Material: Carbon Fiber
Rotor Material: Plastic
Motors: B1806-2300kv
Available Colors: Black
Max. Speed: 150 km/h (93 mp/h)
Controller Frequency: 5.8 GHz
Battery Capacity: 1,500 mAh
Battery Type: Lithium Polymer 3S
Battery Power: 11.1V
GPS Compatible: √
Altitude Hold: –
Position Hold: –
Flight Planning: –
Autonomous Flight: –
Target Tracking: –
Automatic Landing: –
Return To Home: √
Obstacle Avoidance: –
Headless Mode: –
Built-in Camera: √ (700TVL)

Eachine

H99D

© banggood.com

Release Date: January 2016
Price: $90
Number of Rotors: 4
Control System: Remote Control
Applications: Recreation
Dimensions: 230 × 205 × 75 mm
Frame Material: Plastic
Rotor Material: Plastic
Available Colors: Green
Max. Flight Time: 6 minutes
Max. Operating Range: 60 m
Controller Frequency: 2.4 GHz
Battery Capacity: 380 mAh
Battery Type: Lithium Polymer 1S
Battery Power: 3.7V
GPS Compatible: –
Altitude Hold: –
Position Hold: –
Flight Planning: –
Autonomous Flight: –
Target Tracking: –
Automatic Landing: –
Return To Home: √
Obstacle Avoidance: –
Headless Mode: –
Built-in Camera: √ (3 megapixels)

Easy Aerial

Easy Drone XL

© dronesworld.net

Release Date: January 2016
Price: $1,995
Number of Rotors: 4
Control System: Remote Control, Flight Computer
Applications: Film, Photography, Agriculture, Surveillance
Wingspan: 700 mm
Max. Payload: 1.4 kg
Frame Material: Carbon Fiber
Rotor Material: Plastic
Available Colors: Black
Max. Flight Time: 45 minutes
Controller Frequency: 2.4 GHz
GPS Compatible: √
Altitude Hold: √
Position Hold: √
Flight Planning: √
Autonomous Flight: √
Target Tracking: √
Automatic Landing: √
Return To Home: √
Obstacle Avoidance: –
Headless Mode: –
Built-in Camera: –

Fineco

FX-6Ci

© Fineco

Release Date: January 2016
Price: $45
Number of Rotors: 4
Control System: Remote Control
Applications: Recreation
Dimensions: 305 × 305 × 75 mm
Weight: 90 gr
Frame Material: Plastic
Rotor Material: Plastic
Available Colors: White
Max. Flight Time: 8 minutes
Max. Operating Range: 100 m
Controller Frequency: 2.4 GHz
Battery Capacity: 500 mAh
Battery Type: Lithium Polymer
Battery Power: 3.7V
GPS Compatible: –
Altitude Hold: √
Position Hold: √
Flight Planning: –
Autonomous Flight: –
Target Tracking: –
Automatic Landing: –
Return To Home: √
Obstacle Avoidance: –
Headless Mode: –
Built-in Camera: √ (2 megapixels)

Fuqi Toys

FQ777-126 Spider

© Fuqi Toys

Release Date: January 2016
Price: $37
Number of Rotors: 6
Control System: Remote Control
Applications: Recreation
Dimensions: 108 × 98 × 25 mm
Frame Material: Plastic
Rotor Material: Plastic
Available Colors: Black, Red, White
Max. Flight Time: 8 minutes
Max. Operating Range: 50 m
Controller Frequency: 2.4 GHz
Battery Capacity: 220 mAh
Battery Type: Lithium Polymer
Battery Power: 3.7V
GPS Compatible: –
Altitude Hold: –
Position Hold: –
Flight Planning: –
Autonomous Flight: –
Target Tracking: –
Automatic Landing: –
Return To Home: –
Obstacle Avoidance: –
Headless Mode: –
Built-in Camera: √ (2 megapixels)

Fuqi Toys

FQ777-954W FPV

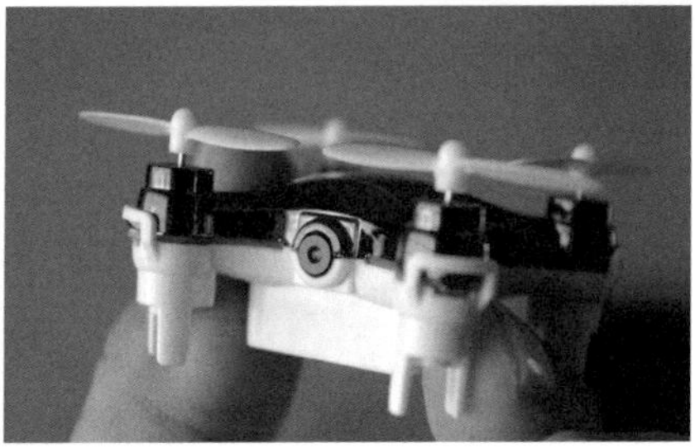

© Fuqi Toys

Release Date: January 2016
Price: $36
Number of Rotors: 4
Control System: Remote Control
Applications: Recreation, FPV Racing
Dimensions: 90 × 40 × 30 mm
Wingspan: 70 mm
Weight: 21 gr
Frame Material: Plastic
Rotor Material: Plastic
Available Colors: Black, Red
Max. Flight Time: 7 minutes
Max. Operating Range: 50 m
Controller Frequency: 2.4 GHz
Battery Capacity: 150 mAh
Battery Type: Lithium Polymer
Battery Power: 3.7V
GPS Compatible: –
Altitude Hold: –
Position Hold: –
Flight Planning: –
Autonomous Flight: –
Target Tracking: –
Automatic Landing: –
Return To Home: –
Obstacle Avoidance: –
Headless Mode: –
Built-in Camera: √ (0.3 megapixels)

HuaXiang

8925 Explorer

© HuaXiang

Release Date: January 2016
Price: $30
Number of Rotors: 4
Control System: Remote Control
Applications: Recreation
Dimensions: 194 × 215 × 57 mm
Weight: 22 gr
Frame Material: Plastic
Rotor Material: Plastic
Available Colors: White
Max. Flight Time: 6 minutes
Max. Operating Range: 80 m
Controller Frequency: 2.4 GHz
Battery Capacity: 550 mAh
Battery Type: Lithium Polymer
Battery Power: 3.7V
GPS Compatible: –
Altitude Hold: √
Position Hold: √
Flight Planning: –
Autonomous Flight: –
Target Tracking: –
Automatic Landing: –
Return To Home: –
Obstacle Avoidance: –
Headless Mode: –
Built-in Camera: √ (0.3 megapixels)

Hubsan

X4 Cam H501C

© Hubsan

Release Date: January 2016
Price: $169
Number of Rotors: 4
Control System: Remote Control
Applications: Film, Photography
Dimensions: 220 × 220 × 70 mm
Wingspan: 311 mm
Weight: 435 gr
Frame Material: Plastic
Rotor Material: Plastic
Available Colors: Black
Max. Flight Time: 20 minutes
Max. Operating Range: 200 m
Controller Frequency: 2.4 GHz
Battery Capacity: 2,700 mAh
Battery Type: Lithium Polymer
Battery Power: 7.4V
Battery Weight: 384 gr
GPS Compatible: –
Altitude Hold: √
Position Hold: √
Flight Planning: √
Autonomous Flight: √
Target Tracking: –
Automatic Landing: √
Return To Home: √
Obstacle Avoidance: –
Headless Mode: –
Built-in Camera: √ (3 megapixels)

Hubsan

X4 Cam Plus H107C+

© Hubsan

Release Date: January 2016
Price: $80
Number of Rotors: 4
Control System: Remote Control
Applications: Recreation, FPV Racing
Dimensions: 135 × 135 × 38 mm
Wingspan: 146 mm
Frame Material: Plastic
Rotor Material: Plastic
Available Colors: Black, Blue, Red, White
Max. Flight Time: 7 minutes
Max. Operating Range: 100 m
Controller Frequency: 2.4 GHz
Battery Capacity: 380 mAh
Battery Type: Lithium Polymer
Battery Power: 3.7V
GPS Compatible: –
Altitude Hold: –
Position Hold: –
Flight Planning: –
Autonomous Flight: –
Target Tracking: –
Automatic Landing: –
Return To Home: –
Obstacle Avoidance: –
Headless Mode: –
Built-in Camera: √ (2 megapixels)

JJRC

H6W

Release Date: January 2016
Price: $40
Number of Rotors: 4
Control System: Remote Control, Phone/Tablet
Applications: Recreation
Dimensions: 138 × 134 × 29 mm
Weight: 53 gr
Frame Material: Plastic
Rotor Material: Plastic
Available Colors: Red
Max. Flight Time: 7 minutes
Max. Operating Range: 300 m
Controller Frequency: 2.4 GHz
Battery Capacity: 350 mAh
Battery Type: Lithium Polymer
Battery Power: 3.7V
GPS Compatible: –
Altitude Hold: √
Position Hold: √
Flight Planning: –
Autonomous Flight: –
Target Tracking: –
Automatic Landing: –
Return To Home: √
Obstacle Avoidance: –
Headless Mode: –
Built-in Camera: √ (2 megapixels)

JJRC

H25G

Release Date: January 2016
Price: $125
Number of Rotors: 4
Control System: Remote Control
Applications: Recreation
Dimensions: 500 × 500 × 165 mm
Wingspan: 707 mm
Frame Material: Plastic
Rotor Material: Plastic
Available Colors: Red, Silver
Max. Flight Time: 10 minutes
Max. Operating Range: 200 m
Controller Frequency: 2.4 GHz
Battery Capacity: 1,100 mAh
Battery Type: Lithium Polymer
Battery Power: 3.7V
GPS Compatible: –
Altitude Hold: √
Position Hold: √
Flight Planning: –
Autonomous Flight: –
Target Tracking: –
Automatic Landing: –
Return To Home: √
Obstacle Avoidance: –
Headless Mode: –
Built-in Camera: √ (2 megapixels)

Lishitoys
L6055

Release Date: January 2016
Price: $75
Number of Rotors: 4
Control System: Remote Control
Applications: Recreation, FPV Racing
Dimensions: 300 × 300 × 80 mm
Frame Material: Plastic
Rotor Material: Plastic
Available Colors: Black
Max. Flight Time: 10 minutes
Max. Operating Range: 100 m
Controller Frequency: 2.4 GHz
Battery Capacity: 900 mAh
Battery Type: Lithium Polymer
GPS Compatible: –
Altitude Hold: –
Position Hold: –
Flight Planning: –
Autonomous Flight: –
Target Tracking: –
Automatic Landing: –
Return To Home: –
Obstacle Avoidance: –
Headless Mode: –
Built-in Camera: √ (2 megapixels)

Meijiaxin Toys
X902

Release Date: January 2016
Price: $18
Number of Rotors: 4
Control System: Remote Control
Applications: Recreation
Dimensions: 90 × 90 × 30 mm
Wingspan: 127 mm
Weight: 20 gr
Frame Material: Plastic
Rotor Material: Plastic
Available Colors: Black
Max. Flight Time: 6 minutes
Max. Operating Range: 100 m
Controller Frequency: 2.4 GHz
Battery Capacity: 180 mAh
Battery Type: Lithium Polymer
Battery Power: 3.7V
GPS Compatible: –
Altitude Hold: –
Position Hold: –
Flight Planning: –
Autonomous Flight: –
Target Tracking: –
Automatic Landing: –
Return To Home: –
Obstacle Avoidance: –
Headless Mode: –
Built-in Camera: –

MMC

F4 Precision Agriculture Drone

©MMC

Release Date: January 2016
Price: $1,450
Number of Rotors: 4
Control System: Remote Control
Applications: Agriculture
Dimensions: 700 × 540 × 600 mm
Weight: 9 kg
Max. Payload: 10 kg
Frame Material: Carbon Fiber
Rotor Material: Carbon Fiber
Motors: Q9XL KV120
Available Colors: Black
Max. Flight Time: 20 minutes
Max. Speed: 43 km/h (26,8 mp/h)
Min. Operating Temperature: -10°C
Max. Operating Temperature: 50°C
Max. Operating Range: 1.5 km
Controller Frequency: 2.4 GHz
Battery Capacity: 16,000 mAh
Battery Type: Lithium Polymer 6S
Battery Power: 22.2V
GPS Compatible: √
Altitude Hold: √
Position Hold: √
Flight Planning: √
Autonomous Flight: √
Target Tracking: √
Automatic Landing: √
Return To Home: √
Obstacle Avoidance: –
Headless Mode: –
Built-in Camera: –

MMC

F6 Precision Agriculture Drone

©MMC

Release Date: January 2016
Price: $1,500
Number of Rotors: 6
Control System: Remote Control
Applications: Agriculture
Dimensions: 700 × 540 × 600 mm
Weight: 9 kg
Max. Payload: 10 kg
Frame Material: Carbon Fiber
Rotor Material: Carbon Fiber
Motors: Q9XL KV120
Available Colors: Black
Max. Flight Time: 22 minutes
Max. Speed: 43 km/h (26,8 mp/h)
Min. Operating Temperature: -10°C
Max. Operating Temperature: 50°C
Max. Operating Range: 1.5 km
Controller Frequency: 2.4 GHz
Battery Capacity: 16,000 mAh
Battery Type: Lithium Polymer 6S
Battery Power: 22.2V
GPS Compatible: √
Altitude Hold: √
Position Hold: √
Flight Planning: √
Autonomous Flight: √
Target Tracking: √
Automatic Landing: √
Return To Home: √
Obstacle Avoidance: –
Headless Mode: –
Built-in Camera: –

Mota

Giga 6000

© bhphotovideo.com

Release Date: January 2016
Price: $975
Number of Rotors: 4
Control System: Remote Control
Applications: Film, Photography
Dimensions: 317 × 317 × 229 mm
Weight: 680 gr
Frame Material: Plastic
Rotor Material: Plastic
Available Colors: Black
Max. Flight Time: 15 minutes
Max. Operating Range: 300 m
Controller Frequency: 2.4 GHz
Battery Capacity: 2,700 mAh
Battery Type: Lithium Polymer
Battery Power: 11.1V
GPS Compatible: √
Altitude Hold: √
Position Hold: √
Flight Planning: √
Autonomous Flight: –
Target Tracking: –
Automatic Landing: –
Return To Home: √
Obstacle Avoidance: –
Headless Mode: –
Built-in Camera: –

Mota

Giga 8000

© bhphotovideo.com

Release Date: January 2016
Price: $1,350
Number of Rotors: 4
Control System: Remote Control
Applications: Film, Photography
Dimensions: 398 × 398 × 215 mm
Frame Material: Plastic
Rotor Material: Plastic
Available Colors: Black
Max. Flight Time: 25 minutes
Max. Flying Altitude: 1,000 m (3,280 ft)
Max. Operating Range: 300 m
Controller Frequency: 2.4 GHz
Battery Capacity: 5,400 mAh
Battery Type: Lithium Polymer
Battery Power: 11.1V
GPS Compatible: √
Altitude Hold: √
Position Hold: √
Flight Planning: √
Autonomous Flight: –
Target Tracking: √
Automatic Landing: –
Return To Home: √
Obstacle Avoidance: –
Headless Mode: –
Built-in Camera: √ (1080P HD)

Mota

Jetjat Live-W

©amazon.com

Release Date: January 2016
Price: $69
Number of Rotors: 4
Control System: Remote Control
Applications: Recreation
Dimensions: 150 × 150 × 45 mm
Frame Material: Plastic
Rotor Material: Plastic
Available Colors: Black
Max. Flight Time: 6 minutes
Max. Operating Range: 30 m
Controller Frequency: 2.4 GHz
Battery Capacity: 540 mAh
Battery Type: Lithium Polymer
GPS Compatible: –
Altitude Hold: √
Position Hold: –
Flight Planning: –
Autonomous Flight: –
Target Tracking: –
Automatic Landing: –
Return To Home: √
Obstacle Avoidance: –
Headless Mode: –
Built-in Camera: √ (720P HD)

Mota

Jetjat Nano-C

©bhphotovideo.com

Release Date: January 2016
Price: $50
Number of Rotors: 4
Control System: Remote Control
Applications: Recreation
Dimensions: 60 × 60 × 20 mm
Frame Material: Plastic
Rotor Material: Plastic
Available Colors: Black
Max. Flight Time: 4 minutes
Max. Operating Range: 30 m
Controller Frequency: 2.4 GHz
Battery Type: Lithium Polymer
GPS Compatible: –
Altitude Hold: –
Position Hold: –
Flight Planning: –
Autonomous Flight: –
Target Tracking: –
Automatic Landing: –
Return To Home: –
Obstacle Avoidance: –
Headless Mode: –
Built-in Camera: √

Mota

Pro Live 4000

©amazon.com

Release Date: January 2016
Price: $299
Number of Rotors: 4
Control System: Remote Control
Applications: Recreation
Dimensions: 305 × 345 × 83 mm
Weight: 816 gr
Frame Material: Plastic
Rotor Material: Plastic
Available Colors: Black
Max. Flight Time: 7 minutes
Max. Operating Range: 80 m
Controller Frequency: 2.4 GHz
Battery Capacity: 360 mAh
Battery Type: Lithium Polymer
GPS Compatible: –
Altitude Hold: √
Position Hold: –
Flight Planning: –
Autonomous Flight: –
Target Tracking: –
Automatic Landing: –
Return To Home: √
Obstacle Avoidance: –
Headless Mode: –
Built-in Camera: √ (720P HD)

Mota

Pro Live 5000

©Mota

Release Date: January 2016
Price: $399
Number of Rotors: 6
Control System: Remote Control
Applications: Recreation
Dimensions: 340 × 340 × 58 mm
Weight: 170 gr
Frame Material: Plastic
Rotor Material: Plastic
Available Colors: Black
Max. Flight Time: 7 minutes
Max. Operating Range: 80 m
Controller Frequency: 2.4 GHz
Battery Capacity: 360 mAh
Battery Type: Lithium Polymer
GPS Compatible: –
Altitude Hold: √
Position Hold: –
Flight Planning: –
Autonomous Flight: –
Target Tracking: –
Automatic Landing: –
Return To Home: √
Obstacle Avoidance: –
Headless Mode: –
Built-in Camera: √ (720P HD)

Revell Control

Orbix

© crowdtox.com

Release Date: January 2016
Price: $39
Number of Rotors: 4
Control System: Remote Control
Applications: Recreation
Dimensions: 80 × 80 × 25 mm
Weight: 18 gr
Frame Material: Plastic
Rotor Material: Plastic
Available Colors: Red
Max. Flight Time: 6 minutes
Max. Operating Range: 25 m
Controller Frequency: 2.4 GHz
Battery Capacity: 110 mAh
Battery Type: Lithium Polymer
Battery Power: 3.7V
GPS Compatible: –
Altitude Hold: –
Position Hold: –
Flight Planning: –
Autonomous Flight: –
Target Tracking: –
Automatic Landing: –
Return To Home: –
Obstacle Avoidance: –
Headless Mode: –
Built-in Camera: –

Skyrc

Sokar FPV

© Skyrc

Release Date: January 2016
Price: $399
Number of Rotors: 4
Control System: Remote Control
Applications: FPV Racing
Dimensions: 220 × 173 × 74 mm
Frame Material: Plastic
Rotor Material: Plastic
Available Colors: Red
Max. Flight Time: 12 minutes
Max. Operating Range: 500 m
Controller Frequency: 5.8 GHz
Battery Capacity: 2,000 mAh
Battery Type: Lithium Polymer 3S
GPS Compatible: –
Altitude Hold: –
Position Hold: –
Flight Planning: –
Autonomous Flight: –
Target Tracking: –
Automatic Landing: –
Return To Home: –
Obstacle Avoidance: –
Headless Mode: –
Built-in Camera: √ (0.3 megapixels)

TRNDlabs

SKEYE Hexa

© TRNDlabs

Release Date: January 2016
Price: $45
Number of Rotors: 6
Control System: Remote Control
Applications: Recreation
Dimensions: 130 × 130 × 33 mm
Weight: 30 gr
Frame Material: Plastic
Rotor Material: Plastic
Available Colors: Orange
Max. Flight Time: 8 minutes
Max. Operating Range: 50 m
Controller Frequency: 2.4 GHz
Battery Capacity: 250 mAh
Battery Type: Lithium Polymer
Battery Power: 3.7V
GPS Compatible: –
Altitude Hold: –
Position Hold: –
Flight Planning: –
Autonomous Flight: –
Target Tracking: –
Automatic Landing: –
Return To Home: –
Obstacle Avoidance: –
Headless Mode: –
Built-in Camera: –

UDIRC Toys

U818A Discovery WiFi

© bhphotovideo.com

Release Date: January 2016
Price: $75
Number of Rotors: 4
Control System: Remote Control
Applications: Recreation
Dimensions: 330 × 340 × 60 mm
Wingspan: 475 mm
Weight: 346 gr
Frame Material: Plastic
Rotor Material: Plastic, Carbon Fiber
Available Colors: Black, Blue
Max. Flight Time: 8 minutes
Max. Operating Range: 100 m
Controller Frequency: 2.4 GHz
Battery Capacity: 500 mAh
Battery Type: Lithium Polymer
Battery Power: 4V
GPS Compatible: –
Altitude Hold: √
Position Hold: √
Flight Planning: –
Autonomous Flight: –
Target Tracking: –
Automatic Landing: –
Return To Home: √
Obstacle Avoidance: –
Headless Mode: –
Built-in Camera: √ (2 megapixels)

Walkera

F210

© dronewish.com

Release Date: January 2016
Price: $369
Number of Rotors: 4
Control System: Remote Control
Applications: FPV Racing
Dimensions: 182 × 182 × 103 mm
Wingspan: 312 mm
Weight: 370 gr
Frame Material: Carbon Fiber
Rotor Material: Plastic
Available Colors: Black
Max. Flight Time: 9 minutes
Min. Operating Temperature: -10°C
Max. Operating Temperature: 40°C
Max. Operating Range: 800 m
Controller Frequency: 2.4 GHz
Battery Capacity: 1,300 mAh
Battery Type: Lithium Polymer 4S
Battery Power: 14V
GPS Compatible: –
Altitude Hold: –
Position Hold: –
Flight Planning: –
Autonomous Flight: –
Target Tracking: –
Automatic Landing: –
Return To Home: –
Obstacle Avoidance: –
Headless Mode: –
Built-in Camera: √

Walkera

Furious 320

© banggood.com

Release Date: January 2016
Price: $489
Number of Rotors: 4
Control System: Remote Control, Phone/Tablet
Applications: FPV Racing
Dimensions: 328 × 254 × 118 mm
Weight: 810 gr
Frame Material: Carbon Fiber
Rotor Material: Plastic, Carbon Fiber
Available Colors: White
Max. Flight Time: 10 minutes
Max. Speed: 120 km/h (74,5 mp/h)
Min. Operating Temperature: -10°C
Max. Operating Temperature: 40°C
Max. Operating Range: 1 km
Controller Frequency: 2.4 GHz
Battery Capacity: 2,600 mAh
Battery Type: Lithium Polymer 4S
Battery Power: 14V
GPS Compatible: √
Altitude Hold: –
Position Hold: –
Flight Planning: –
Autonomous Flight: –
Target Tracking: –
Automatic Landing: –
Return To Home: √
Obstacle Avoidance: –
Headless Mode: –
Built-in Camera: √ (12 megapixels)

Wingsland Technology

K3

©gearbest.com

Release Date: January 2016
Price: $1,469
Number of Rotors: 4
Control System: Remote Control, Phone/Tablet
Applications: Film, Photography
Wingspan: 430 mm
Weight: 4 kg
Frame Material: Plastic
Rotor Material: Plastic
Available Colors: Orange
Max. Flight Time: 20 minutes
Max. Speed: 43 km/h (27 mp/h)
Max. Ascent Speed: 2.5 m/s
Max. Descent Speed: 2.5 m/s
Min. Operating Temperature: -10°C
Max. Operating Temperature: 40°C
Max. Operating Range: 2 km
Controller Frequency: 2.4 GHz
Battery Capacity: 6,200 mAh
Battery Type: Lithium Polymer 4S
Battery Power: 16.8V
Battery Weight: 520 gr
GPS Compatible: √
Altitude Hold: √
Position Hold: √
Flight Planning: √
Autonomous Flight: –
Target Tracking: –
Automatic Landing: √
Return To Home: √
Obstacle Avoidance: –
Headless Mode: –
Built-in Camera: √ (16 megapixels)

Wingsland Technology

M5

©rcmoment.com

Release Date: January 2016
Price: $599
Number of Rotors: 4
Control System: Remote Control, Phone/Tablet
Applications: Film, Photography
Wingspan: 275 mm
Weight: 4 kg
Frame Material: Plastic
Rotor Material: Plastic
Available Colors: Orange
Max. Flight Time: 12 minutes
Max. Speed: 43 km/h (27 mp/h)
Max. Ascent Speed: 6 m/s
Max. Descent Speed: 3 m/s
Min. Operating Temperature: -10°C
Max. Operating Temperature: 40°C
Max. Operating Range: 2 km
Controller Frequency: 2.4 GHz
Battery Capacity: 1,850 mAh
Battery Type: Lithium Polymer 4S
Battery Power: 16.8V
GPS Compatible: √
Altitude Hold: √
Position Hold: √
Flight Planning: √
Autonomous Flight: –
Target Tracking: –
Automatic Landing: √
Return To Home: √
Obstacle Avoidance: –
Headless Mode: –
Built-in Camera: √ (13 megapixels)

Xiro

Xplorer 2

©Xiro

Release Date: January 2016
Price: $1,499
Number of Rotors: 4
Control System: Remote Control, Phone/Tablet
Applications: Film, Photography
Dimensions: 500 × 500 × 190 mm
Wingspan: 350 mm
Weight: 1.4 kg
Frame Material: Plastic, Carbon Fiber
Rotor Material: Plastic
Available Colors: Black
Max. Flight Time: 25 minutes
Max. Speed: 29 km/h (18 mp/h)
Max. Flying Altitude: 122 m (400 ft)
Max. Ascent Speed: 3 m/s
Max. Descent Speed: 3 m/s
Max. Operating Range: 1 km
Controller Frequency: 2.4 GHz
Battery Capacity: 5,300 mAh
Battery Type: Lithium Polymer 4S
Battery Power: 14.8V
GPS Compatible: √
Altitude Hold: √
Position Hold: √
Flight Planning: √
Autonomous Flight: √
Target Tracking: √
Automatic Landing: √
Return To Home: √
Obstacle Avoidance: –
Headless Mode: –
Built-in Camera: √ (12 megapixels)

Xiro

Xplorer Mini

©Xiro

Release Date: January 2016
Price: $400
Number of Rotors: 4
Control System: Remote Control, Phone/Tablet
Applications: Film, Photography
Dimensions: 320 × 320 × 55 mm
Wingspan: 225 mm
Weight: 410 gr
Frame Material: Plastic
Rotor Material: Plastic
Available Colors: Black, White
Max. Flight Time: 15 minutes
Max. Speed: 29 km/h (18 mp/h)
Max. Flying Altitude: 50 m (164 ft)
Max. Ascent Speed: 3 m/s
Max. Descent Speed: 2 m/s
Min. Operating Temperature: -10°C
Max. Operating Temperature: 50°C
Max. Operating Range: 100 m
Controller Frequency: 2.4 GHz
Battery Capacity: 1,400 mAh
Battery Type: Lithium Polymer 3S
GPS Compatible: √
Altitude Hold: √
Position Hold: √
Flight Planning: –
Autonomous Flight: √
Target Tracking: √
Automatic Landing: √
Return To Home: √
Obstacle Avoidance: √
Headless Mode: –
Built-in Camera: √ (13 megapixels)

Yi Zhan Toys
i7H

© rcdronenews.com

Release Date: January 2016
Price: $80
Number of Rotors: 4
Control System: Remote Control, Phone/Tablet
Applications: Recreation
Dimensions: 335 × 335 × 145 mm
Weight: 290 gr
Frame Material: Plastic
Rotor Material: Plastic
Available Colors: Black, Gold
Max. Flight Time: 7 minutes
Max. Operating Range: 300 m
Controller Frequency: 2.4 GHz
Battery Capacity: 850 mAh
Battery Type: Lithium Polymer
Battery Power: 7.4V
GPS Compatible: –
Altitude Hold: √
Position Hold: √
Flight Planning: –
Autonomous Flight: –
Target Tracking: –
Automatic Landing: –
Return To Home: √
Obstacle Avoidance: –
Headless Mode: –
Built-in Camera: √ (5 megapixels)

Yi Zhan Toys
i8H

© Yi Zhan Toys

Release Date: January 2016
Price: $80
Number of Rotors: 4
Control System: Remote Control, Phone/Tablet
Applications: Recreation
Dimensions: 480 × 480 × 160 mm
Weight: 290 gr
Frame Material: Plastic
Rotor Material: Plastic
Available Colors: Black
Max. Flight Time: 8 minutes
Max. Operating Range: 300 m
Controller Frequency: 2.4 GHz
Battery Capacity: 1,200 mAh
Battery Type: Lithium Polymer
Battery Power: 7.4V
GPS Compatible: –
Altitude Hold: √
Position Hold: √
Flight Planning: –
Autonomous Flight: –
Target Tracking: –
Automatic Landing: –
Return To Home: √
Obstacle Avoidance: –
Headless Mode: √
Built-in Camera: √ (2 megapixels)

QuestUAV

Q-100 Datahawk

© QuestUAV

Release Date: January 2016
Price: On Request
Number of Rotors: 1 (Fixed Wing)
Control System: Flight Computer
Applications: Aerial Mapping, Inspection
Dimensions: 0.75 × 1.25 × 0.13 m
Wingspan: 1.25 m
Weight: 1.95 kg
Frame Material: EPP Foam
Rotor Material: Carbon Fiber
Available Colors: Grey
Max. Flight Time: 60 minutes
Max. Speed: 64 km/h (40 mp/h)
Max. Flying Altitude: 2,500 m (8,200 ft)
Max. Operating Range: 30 km
Controller Frequency: 2.4 GHz
Battery Capacity: 8,000 mAh
Battery Type: Lithium Polymer
Battery Power: 11.1V
GPS Compatible: √
Altitude Hold: √
Position Hold: √
Flight Planning: √
Autonomous Flight: –
Target Tracking: √
Automatic Landing: √
Return To Home: √
Obstacle Avoidance: –
Headless Mode: –
Built-in Camera: √ (20 megapixels)

Airhogs

Helix Race

© Airhogs

Release Date: February 2016
Price: $60
Number of Rotors: 4
Control System: Remote Control
Applications: Recreation
Dimensions: 355 × 305 × 76 mm
Weight: 440 gr
Frame Material: Foam
Rotor Material: Plastic
Available Colors: Blue, Red
Controller Frequency: 2.4 GHz
Battery Type: Lithium Polymer
GPS Compatible: –
Altitude Hold: –
Position Hold: –
Flight Planning: –
Autonomous Flight: –
Target Tracking: –
Automatic Landing: –
Return To Home: –
Obstacle Avoidance: –
Headless Mode: –
Built-in Camera: –

Ares

Crossfire Racer

© Ares-RC

Release Date: February 2016
Price: $350
Number of Rotors: 4
Control System: Remote Control
Applications: FPV Racing
Dimensions: 280 mm
Weight: 560 gr
Frame Material: Carbon Fiber
Rotor Material: Plastic
Motors: 2204-2300kv brushless
Available Colors: Blue, Orange, Red
Max. Flight Time: 7 minutes
Controller Frequency: 2.4 GHz
Battery Capacity: 1,350 mAh
Battery Type: Lithium Polymer 3S
GPS Compatible: √
Altitude Hold: –
Position Hold: –
Flight Planning: –
Autonomous Flight: –
Target Tracking: –
Automatic Landing: –
Return To Home: –
Obstacle Avoidance: –
Headless Mode: –
Built-in Camera: √

ATyges

RPAS FV1

© ATyges

Release Date: February 2016
Price: Unknown
Number of Rotors: 1 (Fixed Wing)
Control System: Flight Computer
Applications: Aerial Mapping, Inspection
Weight: 3 kg
Max. Payload: 2 kg
Frame Material: EPO Foam
Available Colors: Black
Max. Flight Time: 180 minutes
Battery Type: Lithium Polymer
GPS Compatible: √
Altitude Hold: √
Position Hold: √
Flight Planning: √
Autonomous Flight: √
Target Tracking: –
Automatic Landing: –
Return To Home: √
Obstacle Avoidance: –
Headless Mode: –
Built-in Camera: √

Bee Rotor

Victory 230

© BeeRotor

Release Date: February 2016
Price: $240
Number of Rotors: 4
Control System: Remote Control, Phone/Tablet
Applications: FPV Racing
Wingspan: 230 mm
Weight: 368 gr
Frame Material: Carbon Fiber
Rotor Material: Plastic
Motors: FR 2306-2100kv
Available Colors: Black
Controller Frequency: 2.4 GHz
Battery Type: Lithium Polymer
GPS Compatible: √
Altitude Hold: –
Position Hold: –
Flight Planning: –
Autonomous Flight: –
Target Tracking: –
Automatic Landing: –
Return To Home: –
Obstacle Avoidance: –
Headless Mode: –
Built-in Camera: √ (BeeRotor 700TVL)

Cheerson

CX-91 Jumper

© Cheerson Hobby

Release Date: February 2016
Price: $260
Number of Rotors: 4
Control System: Remote Control
Applications: FPV Racing
Dimensions: 240 × 250 × 80 mm
Wingspan: 240 mm
Weight: 420 gr
Frame Material: Carbon Fiber
Rotor Material: Carbon Fiber
Motors: Brushless 1806
Available Colors: Black
Max. Flight Time: 12 minutes
Min. Operating Temperature: -10°C
Max. Operating Temperature: 40°C
Max. Operating Range: 300 m
Controller Frequency: 5.8 GHz
Battery Capacity: 16,000 mAh
Battery Type: Lithium Ion 3S
Battery Power: 11.1V
GPS Compatible: –
Altitude Hold: –
Position Hold: –
Flight Planning: –
Autonomous Flight: –
Target Tracking: –
Automatic Landing: –
Return To Home: –
Obstacle Avoidance: –
Headless Mode: –
Built-in Camera: √

Eachine

H99W

© banggood.com

Release Date: February 2016
Price: $79
Number of Rotors: 4
Control System: Remote Control, Phone/Tablet
Applications: Recreation
Dimensions: 230 × 205 × 75 mm
Frame Material: Plastic
Rotor Material: Plastic
Available Colors: Blue
Max. Flight Time: 6 minutes
Max. Operating Range: 60 m
Controller Frequency: 2.4 GHz
Battery Capacity: 380 mAh
Battery Type: Lithium Polymer 1S
Battery Power: 3.7V
GPS Compatible: –
Altitude Hold: –
Position Hold: –
Flight Planning: –
Autonomous Flight: –
Target Tracking: –
Automatic Landing: –
Return To Home: √
Obstacle Avoidance: –
Headless Mode: –
Built-in Camera: √ (2 megapixels)

Elbit Systems

Skylark 3

© Elbit Systems

Release Date: February 2016
Price: $1.5 million
Number of Rotors: 1 (Fixed Wing)
Control System: Flight Computer
Applications: Military Purposes, Border Patrol, Reconnaissance, Surveillance
Wingspan: 4.7 m
Weight: 40 kg
Max. Payload: 10 kg
Available Colors: Grey
Max. Flight Time: 6 hours
Max. Flying Altitude: 5,000 m (16,404 ft)
Max. Operating Range: 100 km
GPS Compatible: √
Altitude Hold: √
Position Hold: –
Flight Planning: √
Autonomous Flight: √
Target Tracking: –
Automatic Landing: √
Return To Home: √
Obstacle Avoidance: √
Headless Mode: –
Built-in Camera: √

Fly Sight

F250

© Fly Sight

Release Date: February 2016
Price: $789
Number of Rotors: 4
Control System: Remote Control
Applications: FPV Racing
Dimensions: 250 mm
Weight: 1.26 kg
Frame Material: Carbon Fiber
Rotor Material: Carbon Fiber
Available Colors: Black
Max. Flight Time: 14 minutes
Max. Speed: 40 km/h (25 mp/h)
Max. Descent Speed: 5 m/s
Max. Operating Range: 1 km
Controller Frequency: 5.8 GHz
Battery Capacity: 2,200 mAh
Battery Type: Lithium Polymer 3S
Battery Power: 11V
Battery Weight: 187 gr
GPS Compatible: –
Altitude Hold: –
Position Hold: –
Flight Planning: –
Autonomous Flight: –
Target Tracking: –
Automatic Landing: –
Return To Home: –
Obstacle Avoidance: –
Headless Mode: –
Built-in Camera: √ (700TVL)

Helic Max

Sky Phantom

© Helic Max

Release Date: February 2016
Price: $25
Number of Rotors: 4
Control System: Remote Control, Phone/Tablet
Applications: Recreation
Dimensions: 280 × 280 × 80 mm
Frame Material: Plastic
Rotor Material: Plastic
Available Colors: Red, White
Max. Flight Time: 5 minutes
Max. Operating Range: 50 m
Controller Frequency: 2.4 GHz
Battery Capacity: 500 mAh
Battery Type: Lithium Polymer
Battery Power: 3.7V
GPS Compatible: –
Altitude Hold: √
Position Hold: √
Flight Planning: –
Autonomous Flight: –
Target Tracking: –
Automatic Landing: –
Return To Home: –
Obstacle Avoidance: –
Headless Mode: –
Built-in Camera: √

Hobby King

KingKong 260 FPV

Release Date: February 2016
Price: $120
Number of Rotors: 4
Control System: Remote Control
Applications: FPV Racing
Dimensions: 260 mm
Weight: 138 gr
Frame Material: Carbon Fiber
Rotor Material: Plastic
Motors: 2300kv brushless
Available Colors: Green, Red
Controller Frequency: 5.8 GHz
Battery Capacity: 2,200 mAh
Battery Type: Lithium Polymer 3S
Battery Power: 11.1V
GPS Compatible: √
Altitude Hold: √
Position Hold: √
Flight Planning: –
Autonomous Flight: –
Target Tracking: –
Automatic Landing: –
Return To Home: √
Obstacle Avoidance: –
Headless Mode: –
Built-in Camera: √ (1000TVL)

Hubsan

X4 Air H501S

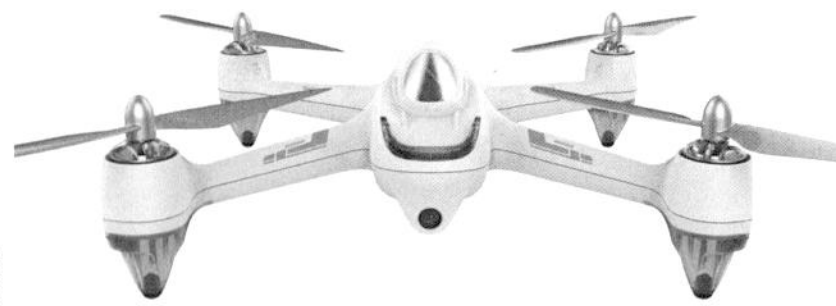

Release Date: February 2016
Price: $250
Number of Rotors: 4
Control System: Remote Control
Applications: Film, Photography
Dimensions: 220 × 220 × 70 mm
Wingspan: 311 mm
Weight: 410 gr
Frame Material: Plastic
Rotor Material: Plastic
Available Colors: Black, White
Max. Flight Time: 20 minutes
Max. Operating Range: 300 m
Controller Frequency: 2.4 GHz
Battery Capacity: 2,300 mAh
Battery Type: Lithium Polymer
Battery Power: 7.4V
Battery Weight: 105 gr
GPS Compatible: √
Altitude Hold: √
Position Hold: √
Flight Planning: –
Autonomous Flight: –
Target Tracking: √
Automatic Landing: √
Return To Home: √
Obstacle Avoidance: –
Headless Mode: –
Built-in Camera: √

Hubsan

X4 Desire H502S

© Hubsan

Release Date: February 2016
Price: $299
Number of Rotors: 4
Control System: Remote Control
Applications: Film, Photography
Dimensions: 165 × 165 × 60 mm
Wingspan: 233mm
Weight: 140 gr
Frame Material: Plastic
Rotor Material: Plastic
Available Colors: White
Max. Flight Time: 12 minutes
Max. Operating Range: 200 m
Controller Frequency: 2.4 GHz
Battery Capacity: 610 mAh
Battery Type: Lithium Polymer 2S
Battery Power: 7.4V
Battery Weight: 35 gr
GPS Compatible: √
Altitude Hold: √
Position Hold: √
Flight Planning: –
Autonomous Flight: –
Target Tracking: √
Automatic Landing: √
Return To Home: √
Obstacle Avoidance: –
Headless Mode: –
Built-in Camera: √

Intuitive Aerial

Aerigon MK II

© Intuitive Aerial

Release Date: February 2016
Price: On Request
Number of Rotors: 12
Control System: Remote Control
Applications: Broadcasting, Film, Photography
Wingspan: 1.51 m
Weight: 8.1 kg
Max. Payload: 16 kg
Frame Material: Carbon Fiber
Rotor Material: Carbon Fiber
Available Colors: Black, White
Max. Flight Time: 15 minutes
Max. Flying Altitude: 5,500 m (18,045 ft)
Max. Operating Range: 1 km
Controller Frequency: 2.4 GHz
Battery Capacity: 10,000 mAh
Battery Type: Lithium Polymer
Battery Power: 22.2V
GPS Compatible: √
Altitude Hold: √
Position Hold: √
Flight Planning: √
Autonomous Flight: √
Target Tracking: –
Automatic Landing: √
Return To Home: √
Obstacle Avoidance: √
Headless Mode: –
Built-in Camera: –

Israel Aerospace Industries (IAI)

FE-Panther

© IAI

Release Date: February 2016
Price: $4 million
Number of Rotors: 4 (Fixed Wing)
Control System: Flight Computer
Applications: Maritime Patrol, Reconnaissance, Surveillance
Weight: 67 kg
Max. Payload: 6 kg
Available Colors: Grey
Controller Frequency: 2.4 GHz
GPS Compatible: √
Altitude Hold: √
Position Hold: √
Flight Planning: √
Autonomous Flight: √
Target Tracking: –
Automatic Landing: √
Return To Home: √
Obstacle Avoidance: –
Headless Mode: –
Built-in Camera: √

Israel Aerospace Industries (IAI)

Harpy NG

© Wikipedia / Jastrow

Release Date: February 2016
Price: $4.5 million
Number of Rotors: 1 (Fixed Wing)
Control System: Ground Control Station
Applications: Loitering Munition
Dimensions: 2.7 × 2.1 m
Wingspan: 2.1 m
Weight: 135 kg
Max. Payload: 32 kg
Frame Material: Composite
Motors: 1 × UEL AR731 Wankel rotary engine
Available Colors: Grey
Max. Speed: 185 km/h (115 mp/h)
Max. Operating Range: 500 km
GPS Compatible: √
Altitude Hold: √
Position Hold: –
Flight Planning: √
Autonomous Flight: √
Target Tracking: √
Automatic Landing: –
Return To Home: –
Obstacle Avoidance: √
Headless Mode: –
Built-in Camera: √

Israel Aerospace Industries (IAI)

Green Dragon

© IAI

Release Date: February 2016
Price: $2.5 million
Number of Rotors: 1 (Fixed Wing)
Control System: Tablet
Applications: Loitering Munition
Dimensions: 1.6 × 1.7 m
Wingspan: 1.7 m
Weight: 16 kg
Max. Payload: 3 kg
Frame Material: Composite
Available Colors: Grey
Max. Flight Time: 75 minutes
Max. Speed: 204 km/h (126,5 mp/h)
Max. Flying Altitude: 1,524 m (5,000 ft)
Max. Operating Range: 40 km
GPS Compatible: √
Altitude Hold: √
Position Hold: –
Flight Planning: √
Autonomous Flight: √
Target Tracking: √
Automatic Landing: –
Return To Home: √
Obstacle Avoidance: √
Headless Mode: –
Built-in Camera: √

Israel Aerospace Industries (IAI)

Rotem-L

© reddit.com

Release Date: February 2016
Price: Unknown
Number of Rotors: 4
Control System: Tablet
Applications: Loitering Munition, Target Acquisition
Weight: 5.8 kg
Max. Payload: 1.2 kg
Frame Material: Composite
Available Colors: Grey
Max. Flight Time: 45 minutes
Max. Speed: 92 km/h (57 mp/h)
Max. Flying Altitude: 1,524 m (5,000 ft)
Max. Operating Range: 10 km
GPS Compatible: √
Altitude Hold: √
Position Hold: √
Flight Planning: √
Autonomous Flight: √
Target Tracking: √
Automatic Landing: √
Return To Home: √
Obstacle Avoidance: √
Headless Mode: –
Built-in Camera: √

JJRC

H29

©tmart.com

Release Date: February 2016
Price: $35
Number of Rotors: 4
Control System: Remote Control
Applications: Recreation
Dimensions: 285 × 280 × 95 mm
Weight: 100 gr
Frame Material: Plastic
Rotor Material: Plastic
Available Colors: Blue, Green, Red
Max. Flight Time: 8 minutes
Max. Operating Range: 150 m
Controller Frequency: 2.4 GHz
Battery Capacity: 450 mAh
Battery Type: Lithium Polymer
Battery Power: 3.7V
GPS Compatible: –
Altitude Hold: √
Position Hold: √
Flight Planning: –
Autonomous Flight: –
Target Tracking: –
Automatic Landing: –
Return To Home: √
Obstacle Avoidance: –
Headless Mode: √
Built-in Camera: –

JJRC

H20W

©aliexpress.com

Release Date: February 2016
Price: $40
Number of Rotors: 6
Control System: Remote Control, Phone/Tablet
Applications: Recreation
Dimensions: 115 × 110 × 30 mm
Weight: 35 gr
Frame Material: Plastic
Rotor Material: Plastic
Available Colors: Blue, Green
Max. Flight Time: 6 minutes
Max. Operating Range: 20 m
Controller Frequency: 2.4 GHz
Battery Capacity: 2,800 mAh
Battery Type: Lithium Polymer
Battery Power: 3.7V
GPS Compatible: –
Altitude Hold: √
Position Hold: √
Flight Planning: –
Autonomous Flight: –
Target Tracking: –
Automatic Landing: –
Return To Home: √
Obstacle Avoidance: –
Headless Mode: –
Built-in Camera: √ (2 megapixels)

JJRC

H30C

© aliexpress.com

Release Date: February 2016
Price: $30
Number of Rotors: 4
Control System: Remote Control
Applications: Recreation
Dimensions: 85 × 85 × 33 mm
Frame Material: Plastic
Rotor Material: Plastic
Available Colors: Black, Red
Max. Flight Time: 7 minutes
Max. Operating Range: 30 m
Controller Frequency: 2.4 GHz
Battery Capacity: 150 mAh
Battery Type: Lithium Polymer
Battery Power: 3.7V
GPS Compatible: –
Altitude Hold: –
Position Hold: –
Flight Planning: –
Autonomous Flight: –
Target Tracking: –
Automatic Landing: –
Return To Home: √
Obstacle Avoidance: –
Headless Mode: √
Built-in Camera: √ (2 megapixels)

JJRC

H29G

© JJRC

Release Date: February 2016
Price: $70
Number of Rotors: 4
Control System: Remote Control
Applications: Film, Photography, Recreation
Dimensions: 340 × 280 × 100 mm
Weight: 250 gr
Frame Material: Plastic
Rotor Material: Plastic
Available Colors: Black, Blue, Green, Red
Max. Flight Time: 8 minutes
Max. Operating Range: 150 m
Controller Frequency: 2.4 GHz
Battery Capacity: 450 mAh
Battery Type: Lithium Polymer
Battery Power: 3.7V
GPS Compatible: –
Altitude Hold: √
Position Hold: √
Flight Planning: –
Autonomous Flight: –
Target Tracking: –
Automatic Landing: –
Return To Home: √
Obstacle Avoidance: –
Headless Mode: –
Built-in Camera: √ (2 megapixels)

Kaiser Baas

Delta

© techdaily.com

Release Date: February 2016
Price: $499
Number of Rotors: 4
Control System: Remote Control
Applications: Film, Photography
Dimensions: 290 × 290 × 185 mm
Weight: 810 gr
Frame Material: Plastic
Rotor Material: Plastic
Available Colors: Black
Max. Flight Time: 17 minutes
Max. Speed: 54 km/h (33,5 mp/h)
Max. Ascent Speed: 6 m/s
Max. Operating Range: 400 m
Controller Frequency: 2.4 GHz
Battery Capacity: 2,200 mAh
Battery Type: Lithium Polymer 3S
Battery Power: 11.1V
GPS Compatible: √
Altitude Hold: √
Position Hold: √
Flight Planning: –
Autonomous Flight: –
Target Tracking: –
Automatic Landing: –
Return To Home: √
Obstacle Avoidance: –
Headless Mode: –
Built-in Camera: –

Monstertronic

Sky Bolt

© Monstertronic

Release Date: February 2016
Price: $250
Number of Rotors: 4
Control System: Remote Control
Applications: Recreation
Dimensions: 335 × 335 × 75 mm
Frame Material: Plastic
Rotor Material: Plastic
Available Colors: Black, White
Max. Flight Time: 8 minutes
Controller Frequency: 2.4 GHz
Battery Type: Lithium Polymer 1S
Battery Power: 3.7V
GPS Compatible: –
Altitude Hold: √
Position Hold: √
Flight Planning: –
Autonomous Flight: –
Target Tracking: –
Automatic Landing: –
Return To Home: √
Obstacle Avoidance: –
Headless Mode: –
Built-in Camera: √

Monstertronic

Sky Runner

©Monstertronic

Release Date: February 2016
Price: $70
Number of Rotors: 4
Control System: Remote Control
Applications: Recreation
Dimensions: 350 × 350 × 100 mm
Frame Material: Plastic
Rotor Material: Plastic
Available Colors: Black
Controller Frequency: 2.4 GHz
Battery Capacity: 550 mAh
Battery Type: Lithium Polymer 1S
Battery Power: 3.7V
GPS Compatible: –
Altitude Hold: √
Position Hold: √
Flight Planning: –
Autonomous Flight: –
Target Tracking: –
Automatic Landing: –
Return To Home: √
Obstacle Avoidance: –
Headless Mode: –
Built-in Camera: √

Robot Aviation

Skyrobot FX450

©Robot Aviation

Release Date: February 2016
Price: On Request
Number of Rotors: 1 (Fixed Wing)
Control System: Ground Control Station
Applications: Aerial Mapping, Inspection, Surveillance, Transport
Dimensions: 4 × 7.2 m
Wingspan: 7.2 m
Weight: 80 kg
Max. Payload: 100 kg
Frame Material: Composite
Available Colors: White
Max. Flight Time: 20 hours
Max. Speed: 216 km/h (134 mp/h)
Max. Flying Altitude: 4,000 m (13,123 ft)
Max. Operating Range: 2,000 km
GPS Compatible: √
Altitude Hold: √
Position Hold: –
Flight Planning: √
Autonomous Flight: √
Target Tracking: –
Automatic Landing: √
Return To Home: √
Obstacle Avoidance: –
Headless Mode: –
Built-in Camera: √

Sky Hero

Anakin Club Racer

© Sky Hero

Release Date: February 2016
Price: $299
Number of Rotors: 4
Control System: Remote Control
Applications: FPV Racing
Dimensions: 211 × 200 × 60 mm
Wingspan: 215 mm
Weight: 410 gr
Frame Material: Plastic, Carbon Fiber
Rotor Material: Plastic
Available Colors: Blue, Green, Red, White
Max. Flight Time: 12 minutes
Controller Frequency: 2.4 GHz
Battery Type: Lithium Polymer
GPS Compatible: –
Altitude Hold: –
Position Hold: –
Flight Planning: –
Autonomous Flight: –
Target Tracking: –
Automatic Landing: –
Return To Home: –
Obstacle Avoidance: –
Headless Mode: –
Built-in Camera: √

Storm Racing Drone

SRD 210

© SRD

Release Date: February 2016
Price: $499
Number of Rotors: 4
Control System: Remote Control
Applications: FPV Racing
Dimensions: 172 × 210 × 60 mm
Weight: 368 gr
Frame Material: Carbon Fiber
Rotor Material: Carbon Fiber
Motors: Storm M1306-3100kv
Available Colors: Black
Max. Operating Range: 1.5 km
Controller Frequency: 5.8 GHz
Battery Capacity: 1,300 mAh
Battery Type: Lithium Polymer
Battery Power: 14.8V
GPS Compatible: –
Altitude Hold: –
Position Hold: –
Flight Planning: –
Autonomous Flight: –
Target Tracking: –
Automatic Landing: –
Return To Home: –
Obstacle Avoidance: –
Headless Mode: –
Built-in Camera: √ (700TVL)

Storm Racing Drone

SRD 250 V3

©SRD

Release Date: February 2016
Price: $499
Number of Rotors: 4
Control System: Remote Control
Applications: FPV Racing
Dimensions: 220 × 200 × 98 mm
Weight: 414 gr
Frame Material: Carbon Fiber
Rotor Material: Carbon Fiber
Available Colors: Black, Red, White
Max. Operating Range: 1.5 km
Controller Frequency: 5.8 GHz
Battery Capacity: 1,300 mAh
Battery Type: Lithium Polymer
Battery Power: 14.8V
GPS Compatible: –
Altitude Hold: –
Position Hold: –
Flight Planning: –
Autonomous Flight: –
Target Tracking: –
Automatic Landing: –
Return To Home: –
Obstacle Avoidance: –
Headless Mode: –
Built-in Camera: √ (700TVL)

Storm Racing Drone

SRD Ninja 250

©SRD

Release Date: February 2016
Price: $499
Number of Rotors: 4
Control System: Remote Control
Applications: FPV Racing
Dimensions: 220 × 200 × 98 mm
Weight: 308 gr
Frame Material: Carbon Fiber
Rotor Material: Carbon Fiber
Available Colors: Black, Red, White
Max. Operating Range: 1.5 km
Controller Frequency: 5.8 GHz
Battery Capacity: 1,000 mAh
Battery Type: Lithium Polymer
Battery Power: 14.8V
GPS Compatible: –
Altitude Hold: –
Position Hold: –
Flight Planning: –
Autonomous Flight: –
Target Tracking: –
Automatic Landing: –
Return To Home: –
Obstacle Avoidance: –
Headless Mode: –
Built-in Camera: √ (700TVL)

Syma Toys

X54HC

© thedronefiles.org

Release Date: February 2016
Price: $65
Number of Rotors: 4
Control System: Remote Control
Applications: Film, Photography, Recreation
Dimensions: 370 × 370 × 100 mm
Weight: 900 gr
Frame Material: Plastic
Rotor Material: Plastic
Available Colors: Black, White
Max. Flight Time: 7 minutes
Max. Operating Range: 30 m
Controller Frequency: 2.4 GHz
Battery Capacity: 2,000 mAh
Battery Type: Lithium Polymer
Battery Power: 7.4V
GPS Compatible: –
Altitude Hold: –
Position Hold: –
Flight Planning: –
Autonomous Flight: –
Target Tracking: –
Automatic Landing: –
Return To Home: –
Obstacle Avoidance: –
Headless Mode: –
Built-in Camera: √ (2 megapixels)

Team Blacksheep

Vendetta

© Team BlackSheep

Release Date: February 2016
Price: $499
Number of Rotors: 4
Control System: Remote Control
Applications: FPV Racing
Dimensions: 240 mm
Weight: 410 gr
Frame Material: Carbon Fiber
Rotor Material: Carbon Fiber
Available Colors: Black
Max. Flight Time: 10 minutes
Controller Frequency: 5.8 GHz
Battery Capacity: 1,700 mAh
Battery Type: Lithium Polymer 3S
GPS Compatible: –
Altitude Hold: –
Position Hold: –
Flight Planning: –
Autonomous Flight: –
Target Tracking: –
Automatic Landing: –
Return To Home: –
Obstacle Avoidance: –
Headless Mode: –
Built-in Camera: √

Virtualrobotix

VR Mapper 300

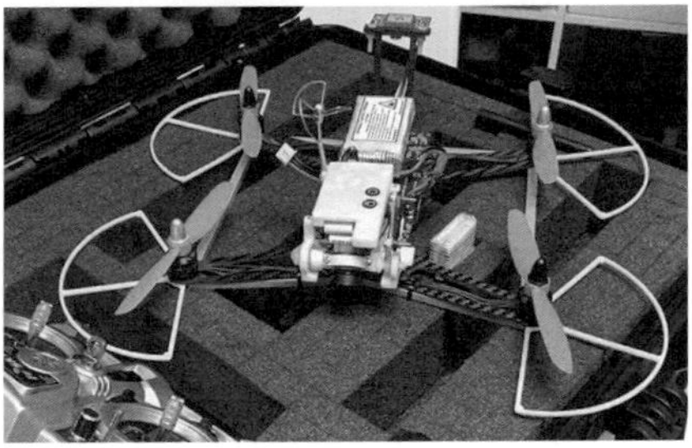

Release Date: February 2016
Price: $1,695
Number of Rotors: 4
Control System: Remote Control
Applications: Aerial Mapping, Inspection, Surveillance
Dimensions: 215 × 220 × 85 mm
Weight: 300 gr
Frame Material: Carbon Fiber
Rotor Material: Carbon Fiber
Motors: Class 1306
Available Colors: Grey
Max. Flight Time: 10 minutes
Controller Frequency: 2.4 GHz
Battery Capacity: 850 mAh
Battery Type: Lithium Polymer 3S
Battery Power: 11.1V
GPS Compatible: √
Altitude Hold: √
Position Hold: √
Flight Planning: √
Autonomous Flight: √
Target Tracking: –
Automatic Landing: √
Return To Home: √
Obstacle Avoidance: –
Headless Mode: –
Built-in Camera: √

Aeronavics

Navi

Release Date: March 2016
Price: On Request
Number of Rotors: 4
Control System: Remote Control
Applications: Aerial Mapping, Agriculture, Surveillance
Weight: 5 kg
Max. Payload: 1.2 kg
Frame Material: Carbon Fiber
Rotor Material: Carbon Fiber
Available Colors: Yellow, White
Max. Flight Time: 35 minutes
Max. Speed: 90 km/h (55 mp/h)
Controller Frequency: 2.4 GHz
GPS Compatible: √
Altitude Hold: √
Position Hold: √
Flight Planning: √
Autonomous Flight: √
Target Tracking: –
Automatic Landing: √
Return To Home: √
Obstacle Avoidance: –
Headless Mode: –
Built-in Camera: –

AgEagle

RX60

©AgEagle

Release Date: March 2016
Price: $17,500
Number of Rotors: 1 (Fixed Wing)
Control System: Remote Control, Phone/Tablet
Applications: Aerial Mapping, Agriculture
Wingspan: 1.37 m
Weight: 3.2 kg
Frame Material: Carbon Fiber
Rotor Material: Carbon Fiber
Available Colors: White
Max. Flight Time: 60 minutes
Max. Speed: 79 km/h (49 mp/h)
Max. Ascent Speed: 5 m/s
Max. Descent Speed: 2 m/s
Min. Operating Temperature: 0°C
Max. Operating Temperature: 75°C
Controller Frequency: 2.4 GHz
Battery Capacity: 5,500 mAh
Battery Type: Lithium Polymer
Battery Power: 14.8V
GPS Compatible: √
Altitude Hold: √
Position Hold: √
Flight Planning: √
Autonomous Flight: √
Target Tracking: –
Automatic Landing: –
Return To Home: √
Obstacle Avoidance: –
Headless Mode: –
Built-in Camera: √

All Tech

KWT-X6L

©made-in-china.com

Release Date: March 2016
Price: $19,999
Number of Rotors: 6
Control System: Ground Control Station
Applications: Military Purposes, Aerial Mapping, Inspection, Reconnaissance, Surveillance
Wingspan: 1.6 m
Weight: 11.5 kg
Max. Payload: 5 kg
Frame Material: Carbon Fiber
Rotor Material: Carbon Fiber
Available Colors: Black, White
Max. Flight Time: 65 minutes
Max. Speed: 54 km/h (33,5 mp/h)
Max. Flying Altitude: 5,000 m (16,404 ft)
Max. Operating Range: 10 km
Controller Frequency: 2.4 GHz
Battery Type: Lithium Polymer
GPS Compatible: √
Altitude Hold: √
Position Hold: √
Flight Planning: √
Autonomous Flight: √
Target Tracking: –
Automatic Landing: √
Return To Home: √
Obstacle Avoidance: –
Headless Mode: –
Built-in Camera: √

Ares
X:Bolt 250

© Ares-RC

Release Date: March 2016
Price: $500
Number of Rotors: 4
Control System: Remote Control
Applications: FPV Racing
Dimensions: 203 × 228 × 53 mm
Wingspan: 260 mm
Frame Material: Carbon Fiber
Rotor Material: Plastic
Motors: 2206-2250kv brushless
Available Colors: Black
Controller Frequency: 2.4 GHz
Battery Capacity: 2,200 mAh
Battery Type: Lithium Polymer 4S
GPS Compatible: –
Altitude Hold: –
Position Hold: –
Flight Planning: –
Autonomous Flight: –
Target Tracking: –
Automatic Landing: –
Return To Home: –
Obstacle Avoidance: –
Headless Mode: –
Built-in Camera: √ (600TVL)

Blade
Nano QX2 FPV

© Helicomicro

Release Date: March 2016
Price: $160
Number of Rotors: 4
Control System: Remote Control
Applications: FPV Racing
Dimensions: 130 × 130 × 50 mm
Weight: 52 gr
Frame Material: Plastic
Rotor Material: Plastic
Available Colors: Black
Max. Flight Time: 8 minutes
Controller Frequency: 2.4 GHz
Battery Capacity: 750 mAh
Battery Type: Lithium Polymer 3S
Battery Power: 3.7V
GPS Compatible: –
Altitude Hold: –
Position Hold: –
Flight Planning: –
Autonomous Flight: –
Target Tracking: –
Automatic Landing: –
Return To Home: –
Obstacle Avoidance: –
Headless Mode: √
Built-in Camera: √

DJI

Phantom 4

©bhphotovideo.com

Release Date: March 2016
Price: $1,399
Number of Rotors: 4
Control System: Remote Control, Phone/Tablet
Applications: Film, Photography
Wingspan: 350 mm
Weight: 1.38 kg
Max. Payload: 462 gr
Frame Material: Plastic, Magnesium
Rotor Material: Carbon Fiber
Available Colors: Black, White
Max. Flight Time: 28 minutes
Max. Speed: 72 km/h (44,7 mp/h)
Max. Flying Altitude: 6,000 m (19,685 ft)
Max. Ascent Speed: 6 m/s
Max. Descent Speed: 4 m/s
Min. Operating Temperature: 0°C
Max. Operating Temperature: 40°C
Max. Operating Range: 5 km
Controller Frequency: 2.4 GHz
Battery Capacity: 5,350 mAh
Battery Type: Lithium Polymer 4S
Battery Power: 15.2V
Battery Weight: 462 gr
GPS Compatible: √
Altitude Hold: √
Position Hold: √
Flight Planning: √
Autonomous Flight: √
Target Tracking: √
Automatic Landing: √
Return To Home: √
Obstacle Avoidance: √
Headless Mode: √
Built-in Camera: √ (12.4 megapixels)

DYS

X-Ray V3

©banggood.com

Release Date: March 2016
Price: $170
Number of Rotors: 4
Control System: Remote Control
Applications: FPV Racing
Dimensions: 131 × 131 × 38 mm
Wingspan: 160 mm
Weight: 143 gr
Frame Material: Carbon Fiber
Rotor Material: Carbon Fiber
Motors: 1104-4000kv
Available Colors: Black
Controller Frequency: 5.8 GHz
Battery Capacity: 1,000 mAh
Battery Type: Lithium Polymer 3S
GPS Compatible: –
Altitude Hold: –
Position Hold: –
Flight Planning: –
Autonomous Flight: –
Target Tracking: –
Automatic Landing: –
Return To Home: –
Obstacle Avoidance: –
Headless Mode: –
Built-in Camera: √

Eachine

E10C Mini

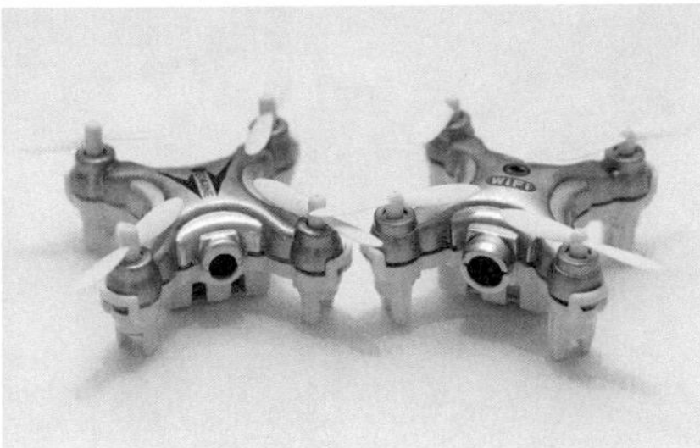

© firstquadcopter.com

Release Date: March 2016
Price: $35
Number of Rotors: 4
Control System: Remote Control
Applications: Recreation
Dimensions: 60 × 60 × 28 mm
Weight: 30 gr
Frame Material: Plastic
Rotor Material: Plastic
Available Colors: Pink, Silver
Max. Flight Time: 6 minutes
Max. Operating Range: 20 m
Controller Frequency: 2.4 GHz
Battery Capacity: 150 mAh
Battery Type: Lithium Polymer 1S
Battery Power: 3.7V
GPS Compatible: –
Altitude Hold: –
Position Hold: –
Flight Planning: –
Autonomous Flight: –
Target Tracking: –
Automatic Landing: –
Return To Home: –
Obstacle Avoidance: –
Headless Mode: –
Built-in Camera: √ (2 megapixels)

Eachine

E20 3D

© Helicomicro

Release Date: March 2016
Price: $25
Number of Rotors: 4
Control System: Remote Control
Applications: Recreation
Dimensions: 173 × 173 × 36 mm
Frame Material: Plastic
Rotor Material: Plastic
Available Colors: Black
Max. Flight Time: 9 minutes
Max. Operating Range: 60 m
Controller Frequency: 2.4 GHz
Battery Capacity: 400 mAh
Battery Type: Lithium Polymer 1S
Battery Power: 3.7V
GPS Compatible: –
Altitude Hold: –
Position Hold: –
Flight Planning: –
Autonomous Flight: –
Target Tracking: –
Automatic Landing: –
Return To Home: –
Obstacle Avoidance: –
Headless Mode: –
Built-in Camera: –

Foxtech

A4

© Foxtech

Release Date: March 2016
Price: $2,539
Number of Rotors: 4
Propellers: Foxtech Tough 2880
Control System: Remote Control, Flight Computer
Applications: Agriculture
Dimensions: 600 × 600 × 540 mm
Wingspan: 850 mm
Weight: 3.9 kg
Max. Payload: 8 kg
Frame Material: Carbon Fiber
Rotor Material: Carbon Fiber
Motors: Foxtech S8120 kV90
Available Colors: Black
Max. Flight Time: 25 minutes
Max. Flight Time with Payload: 15 minutes
Controller Frequency: 2.4 GHz
Battery Type: Lithium Polymer 6S
GPS Compatible: √
Altitude Hold: √
Position Hold: √
Flight Planning: √
Autonomous Flight: √
Target Tracking: –
Automatic Landing: –
Return To Home: √
Obstacle Avoidance: –
Headless Mode: –
Built-in Camera: –

FreeFly Systems

Alta 8

© FreeFly Systems

Release Date: March 2016
Price: $17,495
Number of Rotors: 8
Control System: Remote Control, Phone/Tablet
Applications: Film, Photography, FPV Racing
Wingspan: 1.32 m
Weight: 6.2 kg
Max. Payload: 9.1 kg
Frame Material: Carbon Fiber
Rotor Material: Carbon Fiber
Motors: Direct Drive 3-Phase PMAC Outrunner
Available Colors: Black
Min. Operating Temperature: -20°C
Max. Operating Temperature: 45°C
Controller Frequency: 2.4 GHz
Battery Capacity: 16,000 mAh
Battery Type: Lithium Polymer 6S
Battery Power: 22.2V
GPS Compatible: √
Altitude Hold: √
Position Hold: √
Flight Planning: √
Autonomous Flight: √
Target Tracking: –
Automatic Landing: √
Return To Home: √
Obstacle Avoidance: –
Headless Mode: –
Built-in Camera: –

Helic Max

Sky Vampire

© justdrones.com

Release Date: March 2016
Price: $60
Number of Rotors: 4
Control System: Remote Control
Applications: Film, Photography
Dimensions: 500 × 500 × 160 mm
Weight: 283 gr
Frame Material: Plastic
Rotor Material: Plastic
Available Colors: Black
Max. Flight Time: 10 minutes
Max. Operating Range: 150 m
Controller Frequency: 2.4 GHz
Battery Capacity: 1,200 mAh
Battery Type: Lithium Polymer
Battery Power: 7.4V
GPS Compatible: –
Altitude Hold: √
Position Hold: √
Flight Planning: –
Autonomous Flight: –
Target Tracking: –
Automatic Landing: –
Return To Home: √
Obstacle Avoidance: –
Headless Mode: –
Built-in Camera: –

Helic Max

Mini Sky Walker

© dhgate.com

Release Date: March 2016
Price: $25
Number of Rotors: 4
Control System: Remote Control
Applications: Recreation
Dimensions: 100 × 100 × 95 mm
Weight: 18 gr
Frame Material: Plastic
Rotor Material: Plastic
Available Colors: Green, Orange
Max. Flight Time: 6 minutes
Max. Operating Range: 50 m
Controller Frequency: 2.4 GHz
Battery Capacity: 200 mAh
Battery Type: Lithium Polymer
Battery Power: 3.7V
GPS Compatible: –
Altitude Hold: √
Position Hold: √
Flight Planning: –
Autonomous Flight: –
Target Tracking: –
Automatic Landing: –
Return To Home: –
Obstacle Avoidance: –
Headless Mode: –
Built-in Camera: –

Huajun Toys

W606-2 Move Wings

© Huajun Toys

Release Date: March 2016
Price: $55
Number of Rotors: 4
Control System: Remote Control
Applications: Recreation
Dimensions: 625 × 625 × 150 mm
Frame Material: Plastic
Rotor Material: Plastic
Available Colors: Black, Blue
Max. Flight Time: 13 minutes
Max. Operating Range: 300 m
Controller Frequency: 2.4 GHz
Battery Capacity: 1,200 mAh
Battery Type: Lithium Polymer
Battery Power: 7.4V
GPS Compatible: –
Altitude Hold: √
Position Hold: √
Flight Planning: –
Autonomous Flight: –
Target Tracking: –
Automatic Landing: √
Return To Home: √
Obstacle Avoidance: √
Headless Mode: √
Built-in Camera: √

Huajun Toys

W606-3 U-Fly

© gearbest.com

Release Date: March 2016
Price: $89
Number of Rotors: 4
Control System: Remote Control
Applications: Recreation
Dimensions: 420 × 320 × 150 mm
Weight: 600 gr
Frame Material: Plastic
Rotor Material: Plastic
Available Colors: Black
Max. Flight Time: 10 minutes
Max. Operating Range: 200 m
Controller Frequency: 2.4 GHz
Battery Capacity: 1,800 mAh
Battery Type: Lithium Polymer
Battery Power: 7.4V
GPS Compatible: –
Altitude Hold: √
Position Hold: √
Flight Planning: –
Autonomous Flight: –
Target Tracking: –
Automatic Landing: √
Return To Home: √
Obstacle Avoidance: –
Headless Mode: √
Built-in Camera: √ (2 megapixels)

Huajun Toys

W608-7

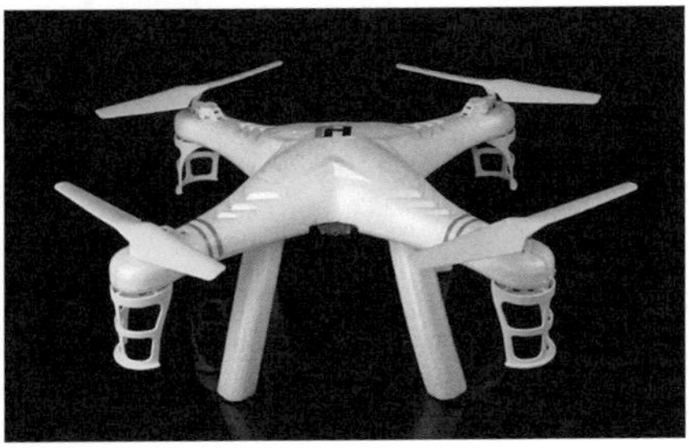

© Huajun Toys

Release Date: March 2016
Price: $89
Number of Rotors: 4
Control System: Remote Control
Applications: Recreation
Dimensions: 420 × 320 × 150 mm
Weight: 600 gr
Frame Material: Plastic
Rotor Material: Plastic
Available Colors: Black
Max. Flight Time: 10 minutes
Max. Operating Range: 200 m
Controller Frequency: 2.4 GHz
Battery Capacity: 1,800 mAh
Battery Type: Lithium Polymer
Battery Power: 7.4V
GPS Compatible: –
Altitude Hold: √
Position Hold: √
Flight Planning: –
Autonomous Flight: –
Target Tracking: –
Automatic Landing: –
Return To Home: √
Obstacle Avoidance: –
Headless Mode: –
Built-in Camera: √ (0.3 megapixels)

Hubsan

X4 Plus H107P

© Hubsan

Release Date: March 2016
Price: $45
Number of Rotors: 4
Control System: Remote Control
Applications: Film, Photography
Dimensions: 90 × 90 × 30 mm
Wingspan: 127 mm
Frame Material: Plastic
Rotor Material: Plastic
Available Colors: Black
Max. Flight Time: 10 minutes
Max. Operating Range: 100 m
Controller Frequency: 2.4 GHz
Battery Capacity: 520 mAh
Battery Type: Lithium Polymer 2S
Battery Power: 3.7V
GPS Compatible: √
Altitude Hold: √
Position Hold: √
Flight Planning: –
Autonomous Flight: –
Target Tracking: √
Automatic Landing: √
Return To Home: √
Obstacle Avoidance: –
Headless Mode: –
Built-in Camera: √

JJRC

H28

Release Date: March 2016
Price: $80
Number of Rotors: 4
Control System: Remote Control
Applications: Recreation
Dimensions: 470 × 470 × 190 mm
Wingspan: 560 mm
Weight: 390 gr
Frame Material: Plastic
Rotor Material: Plastic
Available Colors: Black
Max. Flight Time: 10 minutes
Max. Operating Range: 250 m
Controller Frequency: 2.4 GHz
Battery Capacity: 1,200 mAh
Battery Type: Lithium Polymer
Battery Power: 7.4V
GPS Compatible: –
Altitude Hold: √
Position Hold: √
Flight Planning: –
Autonomous Flight: –
Target Tracking: –
Automatic Landing: –
Return To Home: √
Obstacle Avoidance: –
Headless Mode: √
Built-in Camera: √ (2 megapixels)

MMC

Phoenix A6

Release Date: March 2016
Price: $2,800
Number of Rotors: 6
Control System: Remote Control
Applications: Film, Photography, Aerial Mapping, Inspection, Surveillance
Dimensions: 2.28 × 2.07 × 0.62 m
Weight: 11 kg
Max. Payload: 7 kg
Frame Material: Carbon Fiber
Rotor Material: Carbon Fiber
Motors: Q8L KV130
Available Colors: Red
Max. Flight Time: 75 minutes
Max. Speed: 32,4 km/h (20 mp/h)
Max. Flying Altitude: 4,500 m (14,764 ft)
Max. Ascent Speed: 2 m/s
Min. Operating Temperature: -10°C
Max. Operating Temperature: 50°C
Max. Operating Range: 5 km
Controller Frequency: 2.4 GHz
Battery Capacity: 32,000 mAh
Battery Type: Lithium Polymer 6S
Battery Power: 22.2V
GPS Compatible: √
Altitude Hold: √
Position Hold: √
Flight Planning: √
Autonomous Flight: √
Target Tracking: √
Automatic Landing: √
Return To Home: √
Obstacle Avoidance: –
Headless Mode: –
Built-in Camera: –

Revell Control

Steady Quad Cam

Release Date: March 2016
Price: $115
Number of Rotors: 4
Control System: Remote Control
Applications: Recreation
Dimensions: 106 × 106 × 35 mm
Weight: 808 gr
Frame Material: Plastic
Rotor Material: Plastic
Available Colors: White
Max. Flight Time: 5 minutes
Max. Operating Range: 50 m
Controller Frequency: 2.4 GHz
Battery Capacity: 520 mAh
Battery Type: Lithium Polymer
Battery Power: 3.7V
GPS Compatible: –
Altitude Hold: –
Position Hold: –
Flight Planning: –
Autonomous Flight: –
Target Tracking: –
Automatic Landing: –
Return To Home: –
Obstacle Avoidance: –
Headless Mode: √
Built-in Camera: √

Rise

RXD250

Release Date: March 2016
Price: $199
Number of Rotors: 4
Control System: Remote Control
Applications: FPV Racing
Dimensions: 250 mm
Weight: 300 gr
Frame Material: Plastic
Rotor Material: Plastic
Available Colors: Red
Max. Flight Time: 6 minutes
Controller Frequency: 2.4 GHz
Battery Capacity: 1,800 mAh
Battery Type: Lithium Polymer 3S
GPS Compatible: –
Altitude Hold: –
Position Hold: –
Flight Planning: –
Autonomous Flight: –
Target Tracking: –
Automatic Landing: –
Return To Home: –
Obstacle Avoidance: –
Headless Mode: –
Built-in Camera: –

Storm Racing Drone

SRD 195

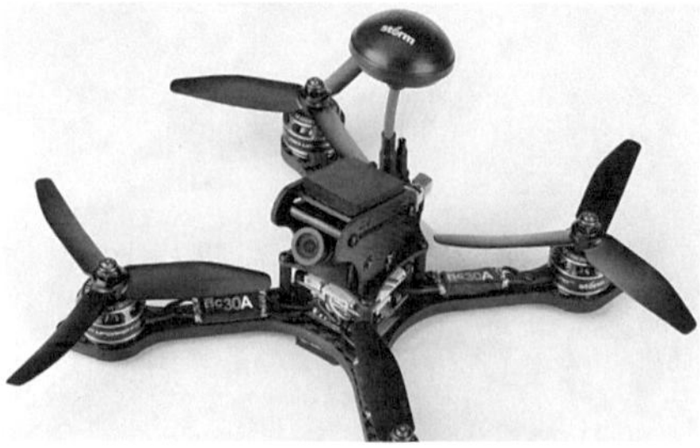

©SRD

Release Date: March 2016
Price: $469
Number of Rotors: 4
Control System: Remote Control
Applications: FPV Racing
Dimensions: 175 × 175 × 83 mm
Weight: 357 gr
Frame Material: Carbon Fiber
Rotor Material: Carbon Fiber
Motors: Storm M1306-3100kv
Available Colors: Black
Max. Operating Range: 1.5 km
Controller Frequency: 5.8 GHz
Battery Capacity: 1,300 mAh
Battery Type: Lithium Polymer
Battery Power: 14.8V
GPS Compatible: –
Altitude Hold: –
Position Hold: –
Flight Planning: –
Autonomous Flight: –
Target Tracking: –
Automatic Landing: –
Return To Home: –
Obstacle Avoidance: –
Headless Mode: –
Built-in Camera: √ (700TVL)

Storm Racing Drone

SRD 280 V3

©SRD

Release Date: March 2016
Price: $499
Number of Rotors: 4
Control System: Remote Control
Applications: FPV Racing
Dimensions: 230 × 245 × 110 mm
Weight: 516 gr
Frame Material: Carbon Fiber
Rotor Material: Carbon Fiber
Available Colors: Black, Green, Orange, Pink, White, Yellow
Max. Operating Range: 1.5 km
Controller Frequency: 5.8 GHz
Battery Capacity: 1,800 mAh
Battery Type: Lithium Polymer
Battery Power: 14.8V
GPS Compatible: –
Altitude Hold: –
Position Hold: –
Flight Planning: –
Autonomous Flight: –
Target Tracking: –
Automatic Landing: –
Return To Home: –
Obstacle Avoidance: –
Headless Mode: –
Built-in Camera: √ (700TVL)

Teamnet Group
Hirrus UAV

© Teamnet Group

Release Date: March 2016
Price: Unknown
Number of Rotors: 1 (Fixed Wing)
Control System: Ground Control Station
Applications: Military Purposes, Inspection, Reconnaissance, Surveillance
Dimensions: 1.2 × 3.3 m
Wingspan: 3.3 m
Weight: 9 kg
Max. Payload: 1 kg
Frame Material: Composite
Available Colors: White
Max. Flight Time: 180 minutes
Max. Speed: 110 km/h (68 mp/h)
Max. Operating Range: 20 km
GPS Compatible: √
Altitude Hold: √
Position Hold: –
Flight Planning: √
Autonomous Flight: √
Target Tracking: √
Automatic Landing: –
Return To Home: √
Obstacle Avoidance: √
Headless Mode: –
Built-in Camera: √ (24 megapixels)

UAV Solutions
Phoenix Ace LE

© UAV Solutions

Release Date: March 2016
Price: On Request
Number of Rotors: 4
Control System: Ground Control Station, Laptop, Tablet
Applications: Aerial Mapping, Reconnaissance, Surveillance
Dimensions: 1.62 × 1.62 m
Wingspan: 1.96 m
Weight: 18 kg
Max. Payload: 3.6 kg
Frame Material: Carbon Fiber, Kevlar
Available Colors: White
Max. Flight Time: 50 minutes
Max. Flying Altitude: 1,524 m (5,000 ft)
Max. Operating Range: 5 km
Controller Frequency: 2.4 GHz
Battery Type: Lithium Polymer
GPS Compatible: √
Altitude Hold: √
Position Hold: √
Flight Planning: √
Autonomous Flight: √
Target Tracking: –
Automatic Landing: √
Return To Home: √
Obstacle Avoidance: –
Headless Mode: –
Built-in Camera: –

UDIRC Toys

U845 Voyager 6

© yousky7.com

Release Date: March 2016
Price: $38
Number of Rotors: 6
Control System: Remote Control
Applications: Recreation
Dimensions: 210 × 210 × 82 mm
Wingspan: 298 mm
Weight: 112 gr
Frame Material: Plastic
Rotor Material: Plastic
Available Colors: Black
Max. Flight Time: 8 minutes
Max. Operating Range: 50 m
Controller Frequency: 2.4 GHz
Battery Capacity: 350 mAh
Battery Type: Lithium Polymer
Battery Power: 7V
GPS Compatible: –
Altitude Hold: √
Position Hold: √
Flight Planning: –
Autonomous Flight: –
Target Tracking: –
Automatic Landing: –
Return To Home: –
Obstacle Avoidance: –
Headless Mode: –
Built-in Camera: √

Xinlin Shiye

X165 mini Quad

© aliexpress.com

Release Date: March 2016
Price: $20
Number of Rotors: 4
Control System: Remote Control
Applications: Recreation
Dimensions: 70 × 70 × 25 mm
Frame Material: Plastic
Rotor Material: Plastic
Available Colors: Black, White
Max. Flight Time: 5 minutes
Max. Operating Range: 25 m
Controller Frequency: 2.4 GHz
Battery Capacity: 120 mAh
Battery Type: Lithium Polymer
Battery Power: 3.7V
GPS Compatible: –
Altitude Hold: √
Position Hold: √
Flight Planning: –
Autonomous Flight: –
Target Tracking: –
Automatic Landing: –
Return To Home: √
Obstacle Avoidance: –
Headless Mode: –
Built-in Camera: –

XK Innovations

Whirlwind X251

© droneflyers.com

Release Date: March 2016
Price: $135
Number of Rotors: 4
Control System: Remote Control
Applications: Film, Photography, FPV Racing
Dimensions: 177 × 177 × 83 mm
Weight: 205 gr
Frame Material: Plastic, Carbon Fiber
Rotor Material: Plastic, Carbon Fiber
Available Colors: White
Max. Flight Time: 12 minutes
Max. Speed: 75,6 km/h (47 mp/h)
Max. Operating Range: 300 m
Controller Frequency: 2.4 GHz
Battery Capacity: 950 mAh
Battery Type: Lithium Polymer 4S
Battery Power: 7.4V
GPS Compatible: –
Altitude Hold: –
Position Hold: –
Flight Planning: –
Autonomous Flight: –
Target Tracking: –
Automatic Landing: –
Return To Home: –
Obstacle Avoidance: –
Headless Mode: –
Built-in Camera: √

XK Innovations

X100 3D

© banggood.com

Release Date: March 2016
Price: $33
Number of Rotors: 4
Control System: Remote Control
Applications: Recreation
Dimensions: 140 × 140 × 40 mm
Weight: 37 gr
Frame Material: Plastic
Rotor Material: Plastic
Available Colors: White
Max. Flight Time: 5 minutes
Max. Operating Range: 100 m
Controller Frequency: 2.4 GHz
Battery Capacity: 250 mAh
Battery Type: Lithium Polymer
Battery Power: 3.7V
GPS Compatible: –
Altitude Hold: √
Position Hold: √
Flight Planning: –
Autonomous Flight: –
Target Tracking: –
Automatic Landing: –
Return To Home: –
Obstacle Avoidance: –
Headless Mode: –
Built-in Camera: –

Yi Zhan Toys

i4S

© topforphone.com

Release Date: March 2016
Price: $39
Number of Rotors: 4
Control System: Remote Control
Applications: Recreation
Dimensions: 135 × 145 × 220 mm
Weight: 46 gr
Frame Material: Plastic
Rotor Material: Plastic
Available Colors: Black
Max. Flight Time: 4 minutes
Max. Operating Range: 30 m
Controller Frequency: 2.4 GHz
Battery Capacity: 350 mAh
Battery Type: Lithium Polymer
Battery Power: 3.7V
GPS Compatible: –
Altitude Hold: √
Position Hold: √
Flight Planning: –
Autonomous Flight: –
Target Tracking: –
Automatic Landing: –
Return To Home: √
Obstacle Avoidance: –
Headless Mode: √
Built-in Camera: √ (2 megapixels)

Acme

Zoopa Q400 Hunter

© Acme

Release Date: April 2016
Price: $135
Number of Rotors: 4
Control System: Remote Control, Phone/Tablet
Applications: Recreation
Dimensions: 320 × 320 × 88 mm
Wingspan: 390 mm
Weight: 120 gr
Frame Material: Plastic
Rotor Material: Plastic
Available Colors: Black
Max. Flight Time: 9 minutes
Max. Operating Range: 100 m
Controller Frequency: 2.4 GHz
Battery Capacity: 600 mAh
Battery Type: Lithium Polymer 1S
Battery Power: 3.7V
GPS Compatible: –
Altitude Hold: –
Position Hold: –
Flight Planning: –
Autonomous Flight: –
Target Tracking: –
Automatic Landing: –
Return To Home: –
Obstacle Avoidance: –
Headless Mode: –
Built-in Camera: √

Aerix Drones

Black Talon 2.0

© Aerix Drones

Release Date: April 2016
Price: $115
Number of Rotors: 4
Control System: Remote Control
Applications: Recreation, FPV Racing
Dimensions: 100 × 100 × 25 mm
Wingspan: 140 mm
Weight: 68 gr
Frame Material: Plastic
Rotor Material: Plastic
Available Colors: Black
Max. Flight Time: 6 minutes
Max. Speed: 24 km/h (15 mp/h)
Max. Operating Range: 50 m
Controller Frequency: 2.4 GHz
Battery Capacity: 520 mAh
Battery Type: Lithium Polymer
Battery Power: 3.7V
GPS Compatible: –
Altitude Hold: √
Position Hold: √
Flight Planning: –
Autonomous Flight: –
Target Tracking: –
Automatic Landing: –
Return To Home: √
Obstacle Avoidance: –
Headless Mode: √
Built-in Camera: √ (2 megapixels)

Aeronavics

Icon

© Aeronavics

Release Date: April 2016
Price: $1,000
Number of Rotors: 6
Control System: Remote Control
Applications: Film, Photography, Aerial Mapping, Surveillance
Weight: 25 kg
Max. Payload: 15 kg
Frame Material: Carbon Fiber
Rotor Material: Carbon Fiber
Available Colors: White, Yellow
Max. Flight Time with Payload: 20 minutes
Max. Speed: 60 km/h (37 mp/h)
Controller Frequency: 2.4 GHz
Battery Type: Lithium Polymer 6S
GPS Compatible: √
Altitude Hold: √
Position Hold: √
Flight Planning: √
Autonomous Flight: √
Target Tracking: –
Automatic Landing: √
Return To Home: √
Obstacle Avoidance: –
Headless Mode: –
Built-in Camera: –

Altavian

Nova F7200 AT

© Altavian

Release Date: April 2016
Price: $15,000
Number of Rotors: 1 (Fixed Wing)
Control System: Ground Control Station
Applications: Aerial Mapping, Inspection, Maritime Patrol, Surveillance
Dimensions: 1.44 × 2.76 × 0.17 m
Weight: 7.8 kg
Max. Payload: 1.67 kg
Frame Material: Composite
Available Colors: White
Max. Flight Time: 90 minutes
Max. Flight Time with Payload: 90 minutes
Max. Speed: 107 km/h (67 mp/h)
Max. Flying Altitude: 3,000 m (9,842 ft)
GPS Compatible: √
Altitude Hold: √
Position Hold: –
Flight Planning: √
Autonomous Flight: √
Target Tracking: –
Automatic Landing: –
Return To Home: √
Obstacle Avoidance: √
Headless Mode: –
Built-in Camera: √

Cheerson

CX-10D Smart Q

© aliexpress.com

Release Date: April 2016
Price: $20
Number of Rotors: 4
Control System: Remote Control, Phone
Applications: Recreation
Dimensions: 45 × 45 × 22 mm
Weight: 15 gr
Frame Material: Plastic
Rotor Material: Plastic
Available Colors: Orange, Purple
Max. Flight Time: 5 minutes
Max. Operating Range: 25 m
Controller Frequency: 2.4 GHz
Battery Capacity: 150 mAh
Battery Type: Lithium Polymer
Battery Power: 3.7V
GPS Compatible: –
Altitude Hold: –
Position Hold: –
Flight Planning: –
Autonomous Flight: –
Target Tracking: –
Automatic Landing: –
Return To Home: –
Obstacle Avoidance: –
Headless Mode: –
Built-in Camera: –

Diatone Innovations

Tyrant 180

© Diatone Innovations

Release Date: April 2016
Price: $200
Number of Rotors: 4
Control System: Remote Control
Applications: FPV Racing
Dimensions: 184 × 143 × 60 mm
Weight: 320 gr
Frame Material: Carbon Fiber
Rotor Material: Plastic
Motors: BX2204-2300kv brushless
Available Colors: Black
Controller Frequency: 5.8 GHz
Battery Capacity: 1,300 mAh
Battery Type: Lithium Polymer 4S
Battery Power: 11.1V
Battery Weight: 280 gr
GPS Compatible: –
Altitude Hold: –
Position Hold: –
Flight Planning: –
Autonomous Flight: –
Target Tracking: –
Automatic Landing: –
Return To Home: –
Obstacle Avoidance: –
Headless Mode: –
Built-in Camera: √ (600TVL)

DJI

Matrice 600

© DJI

Release Date: April 2016
Price: $4,599
Number of Rotors: 6
Control System: Remote Control
Applications: Film, Photography, Aerial Mapping
Dimensions: 1.67 × 1.52 × 0.76 m
Wingspan: 1.13 m
Weight: 9.6 kg
Max. Payload: 6 kg
Frame Material: Carbon Fiber
Rotor Material: Plastic
Available Colors: Black
Max. Flight Time: 35 minutes
Max. Flight Time with Payload: 16 minutes
Max. Speed: 65 km/h (40 mp/h)
Max. Flying Altitude: 2,500 m (8,200 ft)
Max. Ascent Speed: 5 m/s
Max. Descent Speed: 3 m/s
Min. Operating Temperature: -10°C
Max. Operating Temperature: 40°C
Max. Operating Range: 5 km
Controller Frequency: 2.4 GHz
Battery Capacity: 4,500 mAh
Battery Type: Lithium Polymer 6S
Battery Power: 22.2V
Battery Weight: 595 gr
GPS Compatible: √
Altitude Hold: √
Position Hold: √
Flight Planning: –
Autonomous Flight: –
Target Tracking: √
Automatic Landing: √
Return To Home: √
Obstacle Avoidance: –
Headless Mode: –
Built-in Camera: –

Eachine

E10W

©helicomicro.com

Release Date: April 2016
Price: $35
Number of Rotors: 4
Control System: Remote Control
Applications: Recreation
Dimensions: 32 × 32 × 22 mm
Weight: 15 gr
Frame Material: Plastic
Rotor Material: Plastic
Available Colors: Blue
Max. Flight Time: 4 minutes
Max. Operating Range: 30 m
Controller Frequency: 2.4 GHz
Battery Capacity: 150 mAh
Battery Type: Lithium Polymer 1S
Battery Power: 3.7V
GPS Compatible: –
Altitude Hold: –
Position Hold: –
Flight Planning: –
Autonomous Flight: –
Target Tracking: –
Automatic Landing: –
Return To Home: –
Obstacle Avoidance: –
Headless Mode: –
Built-in Camera: √

Eachine

E30W

©banggood.com

Release Date: April 2016
Price: $45
Number of Rotors: 4
Control System: Remote Control
Applications: Recreation
Dimensions: 31,5 × 31,5 × 10,5 mm
Weight: 130 gr
Frame Material: Plastic
Rotor Material: Plastic
Available Colors: Black, White
Max. Flight Time: 7 minutes
Max. Operating Range: 70 m
Controller Frequency: 2.4 GHz
Battery Capacity: 550 mAh
Battery Type: Lithium Polymer 1S
Battery Power: 3.7V
GPS Compatible: –
Altitude Hold: –
Position Hold: –
Flight Planning: –
Autonomous Flight: –
Target Tracking: –
Automatic Landing: –
Return To Home: –
Obstacle Avoidance: –
Headless Mode: –
Built-in Camera: –

Eachine

Falcon 180

© banggood.com

Release Date: April 2016
Price: $230
Number of Rotors: 4
Control System: Remote Control
Applications: FPV Racing
Weight: 308 gr
Frame Material: Carbon Fiber
Rotor Material: Plastic
Motors: B1806-2300kv
Available Colors: Black
Max. Speed: 150 km/h (93 mp/h)
Controller Frequency: 5.8 GHz
Battery Capacity: 1,500 mAh
Battery Type: Lithium Polymer 3S
Battery Power: 11.1V
GPS Compatible: √
Altitude Hold: –
Position Hold: –
Flight Planning: –
Autonomous Flight: –
Target Tracking: –
Automatic Landing: –
Return To Home: √
Obstacle Avoidance: –
Headless Mode: –
Built-in Camera: √ (520TVL)

Fuqi Toys

FQ777-951W Mini

© aliexpress.com

Release Date: April 2016
Price: $32
Number of Rotors: 4
Control System: Remote Control
Applications: Recreation
Dimensions: 90 × 40 × 30 mm
Wingspan: 70 mm
Weight: 21 gr
Frame Material: Plastic
Rotor Material: Plastic
Available Colors: Black, Red, White
Max. Flight Time: 6 minutes
Max. Operating Range: 50 m
Controller Frequency: 2.4 GHz
Battery Capacity: 150 mAh
Battery Type: Lithium Polymer
Battery Power: 3.7V
GPS Compatible: –
Altitude Hold: –
Position Hold: –
Flight Planning: –
Autonomous Flight: –
Target Tracking: –
Automatic Landing: –
Return To Home: –
Obstacle Avoidance: –
Headless Mode: –
Built-in Camera: √ (0.3 megapixels)

Fuqi Toys

FQ777-954D Mini

© aliexpress.com

Release Date: April 2016
Price: $38
Number of Rotors: 4
Control System: Remote Control
Applications: Recreation
Dimensions: 60 × 60 × 26 mm
Weight: 17 gr
Frame Material: Plastic
Rotor Material: Plastic
Available Colors: Black, Red, White
Max. Flight Time: 7 minutes
Max. Operating Range: 30 m
Controller Frequency: 2.4 GHz
Battery Capacity: 150 mAh
Battery Type: Lithium Polymer
Battery Power: 3.7V
GPS Compatible: –
Altitude Hold: –
Position Hold: –
Flight Planning: –
Autonomous Flight: –
Target Tracking: –
Automatic Landing: –
Return To Home: –
Obstacle Avoidance: –
Headless Mode: –
Built-in Camera: √ (0.3 megapixels)

Gteng

T901C

© gearbest.com

Release Date: April 2016
Price: $40
Number of Rotors: 4
Control System: Remote Control
Applications: Recreation
Dimensions: 165 × 165 × 40 mm
Frame Material: Plastic
Rotor Material: Plastic
Available Colors: Blue
Max. Flight Time: 8 minutes
Max. Operating Range: 100 m
Controller Frequency: 2.4 GHz
Battery Capacity: 500 mAh
Battery Type: Lithium Polymer
Battery Power: 3.7V
GPS Compatible: –
Altitude Hold: –
Position Hold: –
Flight Planning: –
Autonomous Flight: –
Target Tracking: –
Automatic Landing: –
Return To Home: √
Obstacle Avoidance: –
Headless Mode: √
Built-in Camera: √ (720P HD)

Hesel EU

Splash Drone 2

© Hesel EU

Release Date: April 2016
Price: $900
Number of Rotors: 4
Control System: Remote Control, Phone/Tablet
Applications: Film, Photography
Wingspan: 500 mm
Weight: 1.15 kg
Max. Payload: 1.35 kg
Frame Material: Carbon Fiber
Rotor Material: Plastic
Motors: 4215-590kv
Available Colors: Orange, White
Max. Flight Time: 22 minutes
Max. Speed: 21,5 km/h (13,4 mp/h)
Max. Operating Range: 1 km
Controller Frequency: 2.4 GHz
Battery Capacity: 5,200 mAh
Battery Type: Lithium Polymer 4S
Battery Power: 14.8V
GPS Compatible: √
Altitude Hold: √
Position Hold: √
Flight Planning: √
Autonomous Flight: √
Target Tracking: √
Automatic Landing: √
Return To Home: √
Obstacle Avoidance: –
Headless Mode: –
Built-in Camera: –

Hubsan

Nano Q4 plus H111C

© Hubsan

Release Date: April 2016
Price: $38
Number of Rotors: 4
Control System: Remote Control
Applications: Recreation
Dimensions: 45 × 45 × 23 mm
Weight: 17 gr
Frame Material: Plastic
Rotor Material: Plastic
Available Colors: White
Max. Flight Time: 7 minutes
Max. Speed: 36 km/h (22,3 mp/h)
Max. Flying Altitude: 30 m (98 ft)
Controller Frequency: 2.4 GHz
Battery Capacity: 150 mAh
Battery Type: Lithium Polymer 1S
Battery Power: 3.7V
GPS Compatible: –
Altitude Hold: –
Position Hold: –
Flight Planning: –
Autonomous Flight: –
Target Tracking: –
Automatic Landing: –
Return To Home: –
Obstacle Avoidance: –
Headless Mode: –
Built-in Camera: √

Hubsan

Nano Q4 FPV H111D

© Hubsan

Release Date: April 2016
Price: $90
Number of Rotors: 4
Control System: Remote Control
Applications: Recreation
Dimensions: 68 × 68 × 25 mm
Weight: 18 gr
Frame Material: Plastic
Rotor Material: Plastic
Available Colors: White
Max. Flight Time: 6 minutes
Max. Speed: 36 km/h (22,3 mp/h)
Max. Flying Altitude: 50 m (164 ft)
Max. Operating Range: 100 m
Controller Frequency: 2.4 GHz
Battery Capacity: 180 mAh
Battery Type: Lithium Polymer 1S
Battery Power: 3.7V
GPS Compatible: –
Altitude Hold: –
Position Hold: –
Flight Planning: –
Autonomous Flight: –
Target Tracking: –
Automatic Landing: –
Return To Home: –
Obstacle Avoidance: –
Headless Mode: –
Built-in Camera: √

JJRC

H29C

© aliexpress.com

Release Date: April 2016
Price: $45
Number of Rotors: 4
Control System: Remote Control
Applications: Recreation
Dimensions: 285 × 280 × 95 mm
Weight: 100 gr
Frame Material: Plastic
Rotor Material: Plastic
Available Colors: Blue, Red
Max. Flight Time: 8 minutes
Max. Operating Range: 150 m
Controller Frequency: 2.4 GHz
Battery Capacity: 450 mAh
Battery Type: Lithium Polymer
Battery Power: 3.7V
GPS Compatible: –
Altitude Hold: √
Position Hold: √
Flight Planning: –
Autonomous Flight: –
Target Tracking: –
Automatic Landing: –
Return To Home: √
Obstacle Avoidance: –
Headless Mode: √
Built-in Camera: √ (2 megapixels)

MMC

HyDrone 1800

©MMC

Release Date: April 2016
Price: On Request
Number of Rotors: 6
Control System: Remote Control
Applications: Film, Photography, Inspection, Surveillance
Dimensions: 2.28 × 2.07 × 0.62 m
Weight: 12 kg
Max. Payload: 5 kg
Frame Material: Carbon Fiber
Rotor Material: Carbon Fiber
Motors: Q8L KV130 Hydrogen Fuel
Available Colors: Orange
Max. Flight Time: 270 minutes
Max. Speed: 32,4 km/h (20 mp/h)
Max. Flying Altitude: 4,500 m (14,764 ft)
Max. Ascent Speed: 2 m/s
Min. Operating Temperature: -10°C
Max. Operating Temperature: 50°C
Max. Operating Range: 100 km
Controller Frequency: 2.4 GHz
GPS Compatible: √
Altitude Hold: √
Position Hold: √
Flight Planning: √
Autonomous Flight: √
Target Tracking: √
Automatic Landing: √
Return To Home: √
Obstacle Avoidance: –
Headless Mode: –
Built-in Camera: –

MMC

HyDrone 1550

©MMC

Release Date: April 2016
Price: $1,900
Number of Rotors: 6
Control System: Remote Control
Applications: Film, Photography, Inspection, Surveillance
Dimensions: 1.55 × 1.34 × 0.61 m
Weight: 12 kg
Max. Payload: 1 kg
Frame Material: Carbon Fiber
Rotor Material: Carbon Fiber
Motors: Q8L KV130 Hydrogen Fuel
Available Colors: Red
Max. Flight Time: 150 minutes
Max. Speed: 28 km/h (18 mp/h)
Max. Flying Altitude: 3,000 m (9,842 ft)
Max. Ascent Speed: 2 m/s
Min. Operating Temperature: -5°C
Max. Operating Temperature: 40°C
Max. Operating Range: 5 km
Controller Frequency: 2.4 GHz
GPS Compatible: √
Altitude Hold: √
Position Hold: √
Flight Planning: √
Autonomous Flight: √
Target Tracking: √
Automatic Landing: √
Return To Home: √
Obstacle Avoidance: √
Headless Mode: √
Built-in Camera: √

SenseFly

Albris

Release Date: April 2016
Price: On Request
Number of Rotors: 4
Control System: Remote Control
Applications: Aerial Mapping, Inspection
Frame Material: Carbon Fiber
Rotor Material: Carbon Fiber
Available Colors: Black
Max. Flight Time: 22 minutes
Max. Speed: 43 km/h (27 mp/h)
Max. Ascent Speed: 7 m/s
Min. Operating Temperature: -10°C
Max. Operating Temperature: 40°C
Max. Operating Range: 1 km
Controller Frequency: 2.4 GHz
Battery Capacity: 10,000 mAh
Battery Type: Lithium Polymer 4S
GPS Compatible: √
Altitude Hold: √
Position Hold: √
Flight Planning: √
Autonomous Flight: √
Target Tracking: √
Automatic Landing: –
Return To Home: √
Obstacle Avoidance: –
Headless Mode: –
Built-in Camera: √ (38 megapixels)

Storm Racing Drone

SRD 240X

Release Date: April 2016
Price: $419
Number of Rotors: 6
Control System: Remote Control
Applications: FPV Racing
Dimensions: 200 × 200 × 60 mm
Weight: 414 gr
Frame Material: Carbon Fiber
Rotor Material: Carbon Fiber
Available Colors: Black
Max. Operating Range: 1.5 km
Controller Frequency: 5.8 GHz
Battery Capacity: 1,300 mAh
Battery Type: Lithium Polymer
Battery Power: 14.8V
GPS Compatible: –
Altitude Hold: –
Position Hold: –
Flight Planning: –
Autonomous Flight: –
Target Tracking: –
Automatic Landing: –
Return To Home: –
Obstacle Avoidance: –
Headless Mode: –
Built-in Camera: √ (700TVL)

Storm Racing Drone

SRD Y3

© SRD

Release Date: April 2016
Price: $489
Number of Rotors: 3
Control System: Remote Control
Applications: FPV Racing
Dimensions: 188 × 210 × 90 mm
Weight: 268 gr
Frame Material: Carbon Fiber
Rotor Material: Carbon Fiber
Motors: Storm M2204-UC 2300kv
Available Colors: White
Max. Operating Range: 1.5 km
Controller Frequency: 5.8 GHz
Battery Capacity: 1,000 mAh
Battery Type: Lithium Polymer
Battery Power: 11.1V
GPS Compatible: –
Altitude Hold: –
Position Hold: –
Flight Planning: –
Autonomous Flight: –
Target Tracking: –
Automatic Landing: –
Return To Home: –
Obstacle Avoidance: –
Headless Mode: –
Built-in Camera: √ (700TVL)

Tarot RC

130 FPV

© Tarot

Release Date: April 2016
Price: $150
Number of Rotors: 4
Control System: Remote Control
Applications: FPV Racing
Dimensions: 199 × 186 × 28 mm
Weight: 85 gr
Frame Material: Carbon Fiber
Rotor Material: Carbon Fiber
Motors: Tarot MT1104-4000kv brushless
Available Colors: Black
Controller Frequency: 5.8 GHz
Battery Capacity: 800 mAh
Battery Type: Lithium Polymer
Battery Power: 11.1V
GPS Compatible: –
Altitude Hold: –
Position Hold: –
Flight Planning: –
Autonomous Flight: –
Target Tracking: –
Automatic Landing: –
Return To Home: –
Obstacle Avoidance: –
Headless Mode: –
Built-in Camera: √ (300TVL)

Team Blacksheep

Vortex 285

© Team BlackSheep

Release Date: April 2016
Price: $339
Number of Rotors: 4
Control System: Remote Control
Applications: FPV Racing
Dimensions: 285 mm
Weight: 350 gr
Frame Material: Plastic, Carbon Fiber
Rotor Material: Carbon Fiber
Available Colors: Black
Max. Flight Time: 10 minutes
Controller Frequency: 5.8 GHz
Battery Capacity: 1,700 mAh
Battery Type: Lithium Polymer 3S
GPS Compatible: –
Altitude Hold: –
Position Hold: –
Flight Planning: –
Autonomous Flight: –
Target Tracking: –
Automatic Landing: –
Return To Home: –
Obstacle Avoidance: –
Headless Mode: –
Built-in Camera: √

Vision Aerial

SwitchBlade-Elite

© Vision Aerial

Release Date: April 2016
Price: $9,450
Number of Rotors: 3
Control System: Remote Control
Applications: Aerial Mapping, Inspection, Surveillance
Wingspan: 862 mm
Weight: 2.7 kg
Max. Payload: 2.3 kg
Frame Material: Aluminum, Carbon Fiber, Polycarbonate
Rotor Material: Carbon Fiber
Motors: 1125W, 360kv
Available Colors: Blue
Max. Flight Time: 40 minutes
Max. Speed: 100 km/h (62 mp/h)
Max. Flying Altitude: 4,500 m (14,764 ft)
Min. Operating Temperature: -15°C
Max. Operating Temperature: 45°C
Max. Operating Range: 16 km
Controller Frequency: 2.4 GHz
Battery Capacity: 13,000 mAh
Battery Type: Lithium Polymer 6S
GPS Compatible: √
Altitude Hold: √
Position Hold: √
Flight Planning: √
Autonomous Flight: √
Target Tracking: –
Automatic Landing: √
Return To Home: √
Obstacle Avoidance: –
Headless Mode: –
Built-in Camera: √

Volocopter GmbH

Volocopter VC200

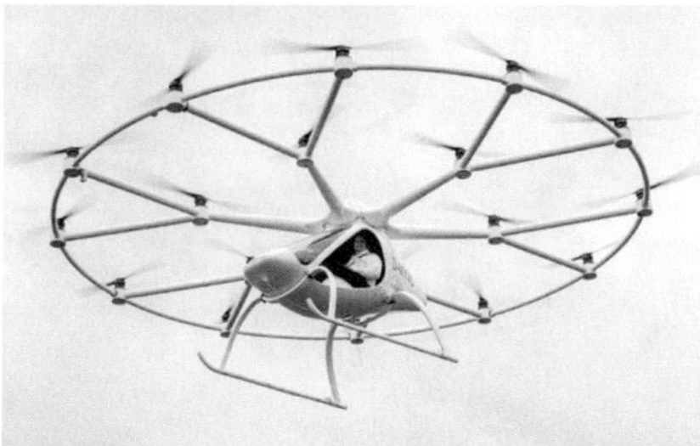

©Volocopter GmbH

Release Date: April 2016
Price: Unknown
Number of Rotors: 18
Control System: Ground Control Station
Applications: Passenger Transport
Dimensions: 2.9 × 9.15 × 2 m
Wingspan: 9.15 m
Weight: 300 kg
Max. Payload: 150 kg
Frame Material: Composite
Available Colors: White
Max. Flight Time: 20 minutes
Max. Speed: 100 km/h (62 mp/h)
Max. Flying Altitude: 1,981 m (6,500 ft)
Battery Type: Lithium Polymer
GPS Compatible: –
Altitude Hold: √
Position Hold: √
Flight Planning: –
Autonomous Flight: –
Target Tracking: –
Automatic Landing: √
Return To Home: √
Obstacle Avoidance: –
Headless Mode: –
Built-in Camera: √

Walkera

F210 3D

©gearbest.com

Release Date: April 2016
Price: $449
Number of Rotors: 4
Control System: Remote Control
Applications: FPV Racing
Dimensions: 182 × 182 × 103 mm
Wingspan: 312 mm
Weight: 370 gr
Frame Material: Carbon Fiber
Rotor Material: Plastic
Motors: WK-WS-28-014A 2500kv
Available Colors: Black
Max. Flight Time: 9 minutes
Min. Operating Temperature: -10°C
Max. Operating Temperature: 40°C
Max. Operating Range: 800 m
Controller Frequency: 2.4 GHz
Battery Capacity: 1,300 mAh
Battery Type: Lithium Polymer 4S
Battery Power: 14.8V
GPS Compatible: –
Altitude Hold: –
Position Hold: –
Flight Planning: –
Autonomous Flight: –
Target Tracking: –
Automatic Landing: –
Return To Home: –
Obstacle Avoidance: –
Headless Mode: –
Built-in Camera: √

Walkera

Rodeo 150

©geekbuying.com

Release Date: April 2016
Price: $229
Number of Rotors: 4
Control System: Remote Control
Applications: FPV Racing
Dimensions: 137 × 148 × 77 mm
Weight: 215 gr
Frame Material: Plastic, Carbon Fiber
Rotor Material: Plastic
Motors: WK-WS-28-014A 2500kv
Available Colors: Black, White
Max. Flight Time: 8 minutes
Min. Operating Temperature: -10°C
Max. Operating Temperature: 40°C
Max. Operating Range: 1 km
Controller Frequency: 2.4 GHz
Battery Capacity: 850 mAh
Battery Type: Lithium Polymer 2S
Battery Power: 7.4V
GPS Compatible: –
Altitude Hold: –
Position Hold: –
Flight Planning: –
Autonomous Flight: –
Target Tracking: –
Automatic Landing: –
Return To Home: –
Obstacle Avoidance: –
Headless Mode: –
Built-in Camera: √ (600TVL)

WLtoys

Q303C Spaceship

©WLtoys

Release Date: April 2016
Price: $90
Number of Rotors: 4
Control System: Remote Control
Applications: Recreation
Dimensions: 510 × 255 × 165 mm
Frame Material: Plastic
Rotor Material: Plastic
Available Colors: Black
Max. Flight Time: 9 minutes
Max. Operating Range: 150 m
Controller Frequency: 2.4 GHz
Battery Capacity: 1,500 mAh
Battery Type: Lithium Polymer
Battery Power: 7.4V
GPS Compatible: –
Altitude Hold: √
Position Hold: √
Flight Planning: –
Autonomous Flight: –
Target Tracking: –
Automatic Landing: –
Return To Home: √
Obstacle Avoidance: –
Headless Mode: √
Built-in Camera: √ (2 megapixels)

XAircraft America

Xplanet P20 V2

© Xaircraft

Release Date: April 2016
Price: On Request
Number of Rotors: 4
Control System: Remote Control, Phone/Tablet
Applications: Agriculture, Inspection
Dimensions: 1.21 × 1.21 × 0.36 m
Weight: 20 kg
Frame Material: Carbon Fiber
Rotor Material: Carbon Fiber
Available Colors: White, Red
Max. Flight Time: 15 minutes
Max. Speed: 28 km/h (18 mp/h)
Max. Flying Altitude: 3 m (10 ft)
Controller Frequency: 2.4 GHz
Battery Type: Lithium Polymer
GPS Compatible: √
Altitude Hold: √
Position Hold: √
Flight Planning: √
Autonomous Flight: √
Target Tracking: √
Automatic Landing: √
Return To Home: √
Obstacle Avoidance: –
Headless Mode: –
Built-in Camera: –

Xinlin Shiye

X181 FPV

© aliexpress.com

Release Date: April 2016
Price: $135
Number of Rotors: 4
Control System: Remote Control
Applications: Recreation, FPV Racing
Frame Material: Plastic
Rotor Material: Plastic
Available Colors: Black
Max. Flight Time: 7 minutes
Max. Operating Range: 100 m
Controller Frequency: 2.4 GHz
Battery Capacity: 1,500 mAh
Battery Type: Lithium Polymer
GPS Compatible: –
Altitude Hold: –
Position Hold: –
Flight Planning: –
Autonomous Flight: –
Target Tracking: –
Automatic Landing: –
Return To Home: √
Obstacle Avoidance: –
Headless Mode: √
Built-in Camera: √ (2 megapixels)

Yuneec

Typhoon H

© mlvdrone.fr

Release Date: April 2016
Price: $1,800
Number of Rotors: 6
Control System: Remote Control, Phone/Tablet
Applications: Film, Photography
Dimensions: 520 × 457 × 294 mm
Wingspan: 410 mm
Weight: 1.85 kg
Frame Material: Carbon Fiber
Rotor Material: Carbon Fiber
Available Colors: Black
Max. Flight Time: 25 minutes
Max. Speed: 36 km/h (22 mp/h)
Max. Flying Altitude: 122 m (400 ft)
Max. Ascent Speed: 5 m/s
Max. Descent Speed: 3 m/s
Min. Operating Temperature: 0°C
Max. Operating Temperature: 50°C
Max. Operating Range: 1 km
Controller Frequency: 2.4 GHz
Battery Capacity: 5,400 mAh
Battery Type: Lithium Polymer 4S
Battery Power: 14.8V
GPS Compatible: √
Altitude Hold: √
Position Hold: √
Flight Planning: √
Autonomous Flight: √
Target Tracking: √
Automatic Landing: –
Return To Home: √
Obstacle Avoidance: –
Headless Mode: –
Built-in Camera: √ (12 megapixels)

Airdog SIA

Airdog

© Airdog

Release Date: May 2016
Price: $1,599
Number of Rotors: 4
Control System: Remote Control
Applications: Film, Photography
Dimensions: 350 × 516 mm
Wingspan: 620 mm
Weight: 2 kg
Frame Material: Plastic
Rotor Material: Plastic
Available Colors: Yellow
Max. Flight Time: 18 minutes
Max. Speed: 67 km/h (42 mp/h)
Max. Flying Altitude: 3,500 m (11,483 ft)
Max. Operating Range: 250 m
Controller Frequency: 2.4 GHz
Battery Capacity: 5600 mAh
Battery Type: Lithium Polymer 6S
Battery Power: 14.8V
GPS Compatible: √
Altitude Hold: √
Position Hold: √
Flight Planning: √
Autonomous Flight: √
Target Tracking: √
Automatic Landing: √
Return To Home: √
Obstacle Avoidance: –
Headless Mode: –
Built-in Camera: –

Arris

F210

Release Date: May 2016
Price: $319
Number of Rotors: 4
Propellers: Arris 5045 Bullnose
Control System: Remote Control
Applications: FPV Racing
Weight: 390 gr
Frame Material: Carbon Fiber
Rotor Material: Plastic
Motors: Arris 2205-2600kv
Available Colors: Black
Max. Flight Time: 11 minutes
Controller Frequency: 5.8 GHz
Battery Capacity: 1,500 mAh
Battery Type: Lithium Polymer 4S
Battery Power: 11.1V
GPS Compatible: √
Altitude Hold: √
Position Hold: √
Flight Planning: √
Autonomous Flight: √
Target Tracking: –
Automatic Landing: √
Return To Home: –
Obstacle Avoidance: –
Headless Mode: –
Built-in Camera: √ (CMOS mini camera)

ATyges

RPAS FV8

©ATyges

Release Date: May 2016
Price: Unknown
Number of Rotors: 8
Control System: Flight Computer
Applications: Aerial Mapping, Agriculture, Inspection, Transport
Available Colors: Black
Max. Flight Time: 30 minutes
Battery Type: Lithium Polymer
GPS Compatible: √
Altitude Hold: √
Position Hold: √
Flight Planning: √
Autonomous Flight: √
Target Tracking: √
Automatic Landing: √
Return To Home: √
Obstacle Avoidance: √
Headless Mode: –
Built-in Camera: –

Cheerson

CX-36 Glider

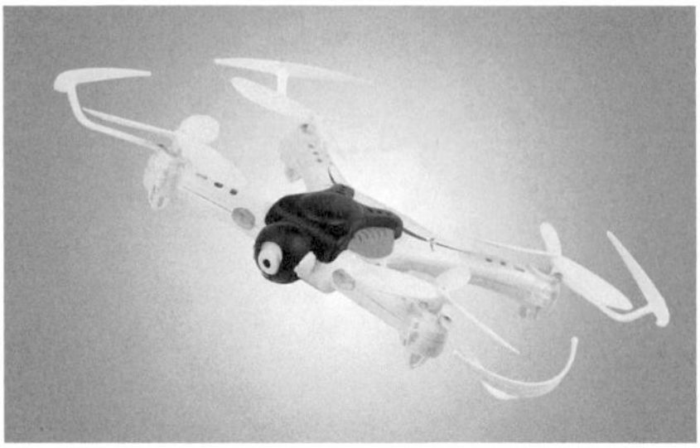

©Cheerson Hobby

Release Date: May 2016
Price: $35
Number of Rotors: 6
Control System: Remote Control, Phone/Tablet
Applications: Film, Photography, Recreation
Dimensions: 180 × 160 × 30 mm
Wingspan: 240 mm
Weight: 15 gr
Frame Material: Plastic
Rotor Material: Plastic
Available Colors: White
Max. Flight Time: 5 minutes
Max. Operating Range: 50 m
Controller Frequency: 2.4 GHz
Battery Capacity: 350 mAh
Battery Type: Lithium Polymer
Battery Power: 3.7V
GPS Compatible: –
Altitude Hold: √
Position Hold: √
Flight Planning: –
Autonomous Flight: –
Target Tracking: –
Automatic Landing: √
Return To Home: –
Obstacle Avoidance: –
Headless Mode: –
Built-in Camera: √ (0.3 megapixels)

Delta Drone

Delta X

©Delta Drone

Release Date: May 2016
Price: $28.000
Number of Rotors: 4
Control System: Flight Computer
Applications: Reconnaissance, Surveillance
Weight: 5.8 kg
Max. Payload: 1.5 kg
Frame Material: Carbon Fiber
Available Colors: Grey
Max. Flight Time: 20 minutes
Max. Speed: 65 km/h (40 mp/h)
GPS Compatible: √
Altitude Hold: √
Position Hold: √
Flight Planning: √
Autonomous Flight: √
Target Tracking: √
Automatic Landing: √
Return To Home: √
Obstacle Avoidance: –
Headless Mode: –
Built-in Camera: √

Eachine

Blade 185

Release Date: May 2016
Price: $300
Number of Rotors: 4
Control System: Remote Control
Applications: FPV Racing
Dimensions: 170 × 155 × 55 mm
Wingspan: 185 mm
Weight: 364 gr
Frame Material: Carbon Fiber
Rotor Material: Plastic
Motors: B1806-2300kv
Available Colors: Black
Controller Frequency: 5.8 GHz
Battery Capacity: 1,300 mAh
Battery Type: Lithium Polymer 3S
Battery Power: 11.1V
Battery Weight: 111 gr
GPS Compatible: √
Altitude Hold: –
Position Hold: –
Flight Planning: –
Autonomous Flight: –
Target Tracking: –
Automatic Landing: –
Return To Home: √
Obstacle Avoidance: –
Headless Mode: –
Built-in Camera: √ (520TVL)

Eachine

E40G

© Eachine

Release Date: May 2016
Price: $169
Number of Rotors: 4
Control System: Remote Control
Applications: FPV Racing
Dimensions: 150 × 150 × 38 mm
Weight: 71 gr
Frame Material: Plastic
Rotor Material: Plastic
Available Colors: Black
Max. Flight Time: 8 minutes
Max. Operating Range: 70 m
Controller Frequency: 5.8 GHz
Battery Capacity: 520 mAh
Battery Type: Lithium Polymer
Battery Power: 3.7V
GPS Compatible: –
Altitude Hold: √
Position Hold: √
Flight Planning: –
Autonomous Flight: –
Target Tracking: –
Automatic Landing: –
Return To Home: √
Obstacle Avoidance: –
Headless Mode: √
Built-in Camera: √ (2 megapixels)

Flypro

X-Eagle

©gearbest.com

Release Date: May 2016
Price: $699
Number of Rotors: 4
Control System: Remote Control, Phone/Tablet
Applications: Film, Photography
Dimensions: 290 × 290 × 180 mm
Wingspan: 350 mm
Weight: 1.26 kg
Frame Material: Plastic, Carbon Fiber
Rotor Material: Plastic, Carbon Fiber
Available Colors: Black
Max. Flight Time: 22 minutes
Max. Speed: 54 km/h (33,5 mp/h)
Max. Flying Altitude: 5,000 m (16,404 ft)
Max. Ascent Speed: 3 m/s
Max. Descent Speed: 1.5 m/s
Min. Operating Temperature: -10°C
Max. Operating Temperature: 40°C
Max. Operating Range: 300 m
Controller Frequency: 2.4 GHz
Battery Capacity: 5,200 mAh
Battery Type: Lithium Polymer 3S
Battery Power: 11V
Battery Weight: 370 gr
GPS Compatible: √
Altitude Hold: √
Position Hold: √
Flight Planning: –
Autonomous Flight: √
Target Tracking: √
Automatic Landing: √
Return To Home: √
Obstacle Avoidance: √
Headless Mode: –
Built-in Camera: –

Flytronic

Warmate

©wpn.pl

Release Date: May 2016
Price: Unknown
Number of Rotors: 1 (Fixed Wing)
Control System: Ground Control Station
Applications: Loitering munition
Dimensions: 1.1 × 1.4 m
Wingspan: 1.4 kg
Weight: 4 kg
Frame Material: Composite
Available Colors: Grey
Max. Flight Time: 50 minutes
Max. Speed: 80 km/h (49,7 mp/h)
Max. Flying Altitude: 500 m (1,640 ft)
Max. Operating Range: 40 km
GPS Compatible: √
Altitude Hold: √
Position Hold: –
Flight Planning: √
Autonomous Flight: √
Target Tracking: –
Automatic Landing: –
Return To Home: –
Obstacle Avoidance: –
Headless Mode: –
Built-in Camera: √

Gteng

T901F

Release Date: May 2016
Price: $50
Number of Rotors: 4
Control System: Remote Control
Applications: Recreation, FPV Racing
Dimensions: 165 × 165 × 40 mm
Frame Material: Plastic
Rotor Material: Plastic
Available Colors: Black
Max. Flight Time: 8 minutes
Max. Operating Range: 300 m
Controller Frequency: 5.8 GHz
Battery Capacity: 500 mAh
Battery Type: Lithium Polymer
Battery Power: 3.7V
GPS Compatible: –
Altitude Hold: –
Position Hold: –
Flight Planning: –
Autonomous Flight: –
Target Tracking: –
Automatic Landing: –
Return To Home: √
Obstacle Avoidance: –
Headless Mode: √
Built-in Camera: √ (2 megapixels)

Hobby Lord

Alpha H6

© rcuniverse.com

Release Date: May 2016
Price: $475
Number of Rotors: 6
Control System: Remote Control
Applications: Film, Photography
Wingspan: 1 m
Weight: 1.95 kg
Max. Payload: 2.5 kg
Frame Material: Carbon Fiber
Rotor Material: Carbon Fiber
Motors: ST4708 brushless
Available Colors: White, Yellow
Max. Flight Time: 25 minutes
Controller Frequency: 2.4 GHz
Battery Capacity: 15,000 mAh
Battery Type: Lithium Polymer 6S
Battery Power: 22.2V
GPS Compatible: √
Altitude Hold: √
Position Hold: √
Flight Planning: √
Autonomous Flight: √
Target Tracking: –
Automatic Landing: √
Return To Home: √
Obstacle Avoidance: –
Headless Mode: –
Built-in Camera: –

Huajun Toys

W606-5 Flanker

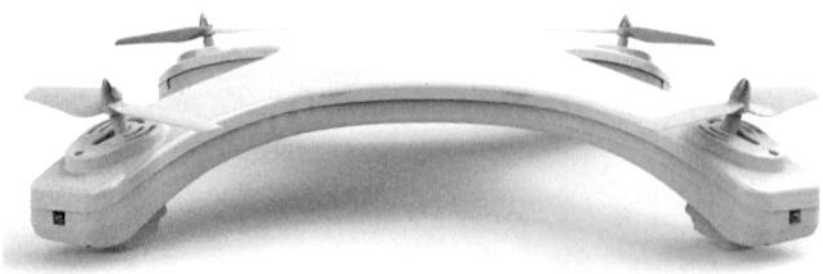

© Huajun Toys

Release Date: May 2016
Price: $80
Number of Rotors: 4
Control System: Remote Control
Applications: Recreation
Dimensions: 690 × 690 × 200 mm
Weight: 700 gr
Frame Material: Plastic
Rotor Material: Plastic
Available Colors: White
Max. Flight Time: 8 minutes
Max. Operating Range: 500 m
Controller Frequency: 2.4 GHz
Battery Capacity: 1,500 mAh
Battery Type: Lithium Polymer
Battery Power: 7.4V
GPS Compatible: –
Altitude Hold: √
Position Hold: √
Flight Planning: –
Autonomous Flight: √
Target Tracking: –
Automatic Landing: √
Return To Home: √
Obstacle Avoidance: –
Headless Mode: √
Built-in Camera: √

Idea Fly

Apollo

© Idea-Fly

Release Date: May 2016
Price: $560
Number of Rotors: 4
Control System: Remote Control
Applications: Film, Photography
Dimensions: 280 × 310 mm
Wingspan: 350 mm
Weight: 800 gr
Max. Payload: 500 gr
Frame Material: Plastic
Rotor Material: Carbon Fiber
Available Colors: Blue, Red, White
Max. Flight Time: 13 minutes
Max. Speed: 18 km/h (11 mp/h)
Max. Flying Altitude: 300 m (984 ft)
Max. Ascent Speed: 8 m/s
Max. Descent Speed: 4 m/s
Min. Operating Temperature: -20°C
Max. Operating Temperature: 50°C
Max. Operating Range: 500 m
Controller Frequency: 5.8 GHz
Battery Capacity: 2,200 mAh
Battery Type: Lithium Polymer 3S
Battery Power: 11.1V
Battery Weight: 300 gr
GPS Compatible: √
Altitude Hold: √
Position Hold: √
Flight Planning: √
Autonomous Flight: √
Target Tracking: –
Automatic Landing: √
Return To Home: √
Obstacle Avoidance: –
Headless Mode: –
Built-in Camera: –

Jin Xing Da
510G X-Predators

© thedronefiles.org

Release Date: May 2016
Price: $80
Number of Rotors: 4
Control System: Remote Control
Applications: Recreation, FPV Racing
Dimensions: 316 × 316 × 106 mm
Weight: 750 gr
Frame Material: Plastic
Rotor Material: Plastic
Available Colors: Black
Max. Flight Time: 10 minutes
Max. Operating Range: 100 m
Controller Frequency: 2.4 GHz
Battery Capacity: 600 mAh
Battery Type: Lithium Polymer
Battery Power: 3.7V
GPS Compatible: –
Altitude Hold: √
Position Hold: √
Flight Planning: –
Autonomous Flight: –
Target Tracking: –
Automatic Landing: √
Return To Home: √
Obstacle Avoidance: –
Headless Mode: √
Built-in Camera: √ (2 megapixels)

Jin Xing Da
512W

© aliexpress.com

Release Date: May 2016
Price: $30
Number of Rotors: 4
Control System: Remote Control
Applications: Recreation
Dimensions: 850 × 900 × 400 mm
Weight: 200 gr
Frame Material: Plastic
Rotor Material: Plastic
Available Colors: Blue, Yellow
Max. Flight Time: 8 minutes
Max. Operating Range: 30 m
Controller Frequency: 2.4 GHz
Battery Capacity: 180 mAh
Battery Type: Lithium Polymer
Battery Power: 3.7V
GPS Compatible: –
Altitude Hold: √
Position Hold: √
Flight Planning: √
Autonomous Flight: –
Target Tracking: –
Automatic Landing: –
Return To Home: √
Obstacle Avoidance: –
Headless Mode: –
Built-in Camera: √ (0.3 megapixels)

Kadet Defence Systems

JX2 NG

©Kadet Defence Systems

Release Date: May 2016
Price: Unknown
Number of Rotors: 1 (Fixed Wing)
Control System: Flight Computer
Applications: Military Purposes, Inspection, Reconnaissance, Surveillance
Dimensions: 2.6 × 2.5 m
Wingspan: 2.5 m
Weight: 35 kg
Frame Material: Composite
Motors: Two-stroke engine
Available Colors: Orange
Max. Flight Time: 45 minutes
Max. Speed: 216 km/h (134 mp/h)
Max. Operating Range: 40 km
GPS Compatible: √
Altitude Hold: √
Position Hold: –
Flight Planning: √
Autonomous Flight: √
Target Tracking: –
Automatic Landing: –
Return To Home: √
Obstacle Avoidance: √
Headless Mode: –
Built-in Camera: –

Lishitoys

L6056

©aliexpress.com

Release Date: May 2016
Price: $39
Number of Rotors: 4
Control System: Remote Control
Applications: Recreation
Dimensions: 220 × 220 × 60 mm
Weight: 500 gr
Frame Material: Plastic
Rotor Material: Plastic
Available Colors: Grey, White
Max. Flight Time: 8 minutes
Max. Operating Range: 100 m
Controller Frequency: 2.4 GHz
Battery Capacity: 380 mAh
Battery Type: Lithium Polymer
Battery Power: 3.7V
GPS Compatible: –
Altitude Hold: –
Position Hold: –
Flight Planning: –
Autonomous Flight: –
Target Tracking: √
Automatic Landing: –
Return To Home: √
Obstacle Avoidance: –
Headless Mode: √
Built-in Camera: √ (2 megapixels)

Meijiaxin Toys

X601H

©geekbuying.com

Release Date: May 2016
Price: $70
Number of Rotors: 6
Control System: Remote Control
Applications: Recreation, FPV Racing
Dimensions: 385 × 425 × 60 mm
Weight: 193 gr
Frame Material: Plastic
Rotor Material: Plastic
Available Colors: Black
Max. Flight Time: 8 minutes
Max. Operating Range: 70 m
Controller Frequency: 2.4 GHz
Battery Capacity: 700 mAh
Battery Type: Lithium Polymer
Battery Power: 3.7V
GPS Compatible: –
Altitude Hold: √
Position Hold: √
Flight Planning: –
Autonomous Flight: √
Target Tracking: –
Automatic Landing: –
Return To Home: √
Obstacle Avoidance: –
Headless Mode: √
Built-in Camera: √ (0.3 megapixels)

Meteomatics

Meteodrone

©Meteomatics

Release Date: May 2016
Price: On Request
Number of Rotors: 4
Control System: Flight Computer
Applications: Meteorological Research
Dimensions: 700 × 700 mm
Weight: 1.5 kg
Frame Material: Plastic, Carbon Fiber
Rotor Material: Carbon Fiber
Available Colors: White
Max. Flight Time: 20 minutes
Max. Speed: 60 km/h (37 mp/h)
Max. Flying Altitude: 1,500 m (4,921 ft)
Max. Ascent Speed: 10 m/s
GPS Compatible: √
Altitude Hold: √
Position Hold: √
Flight Planning: √
Autonomous Flight: –
Target Tracking: –
Automatic Landing: –
Return To Home: √
Obstacle Avoidance: –
Headless Mode: –
Built-in Camera: –

Meteomatics

Meteodrone XL

© Meteomatics

Release Date: May 2016
Price: On Request
Number of Rotors: 4
Control System: Flight Computer
Applications: Meteorological Research
Dimensions: 900 × 900 mm
Weight: 5 kg
Max. Payload: 1 kg
Frame Material: Plastic, Carbon Fiber
Rotor Material: Carbon Fiber
Available Colors: White
Max. Flight Time: 40 minutes
Max. Speed: 40 km/h (25 mp/h)
Max. Flying Altitude: 3,000 m (9,842 ft)
Max. Ascent Speed: 6 m/s
GPS Compatible: √
Altitude Hold: √
Position Hold: √
Flight Planning: √
Autonomous Flight: –
Target Tracking: –
Automatic Landing: –
Return To Home: √
Obstacle Avoidance: –
Headless Mode: –
Built-in Camera: –

Meteomatics

Meteodrone SSE

© Meteomatics

Release Date: May 2016
Price: On Request
Number of Rotors: 6
Control System: Flight Computer
Applications: Meteorological Research
Dimensions: 600 × 600 mm
Weight: 700 gr
Frame Material: Plastic, Carbon Fiber
Rotor Material: Plastic
Available Colors: White
Max. Flight Time: 12 minutes
Max. Speed: 75 km/h (46,6 mp/h)
Max. Flying Altitude: 1,500 m (4,921 ft)
Max. Ascent Speed: 19 m/s
GPS Compatible: √
Altitude Hold: √
Position Hold: √
Flight Planning: √
Autonomous Flight: –
Target Tracking: –
Automatic Landing: –
Return To Home: √
Obstacle Avoidance: –
Headless Mode: –
Built-in Camera: –

Microdrones.com

MD4-3000

Release Date: May 2016
Price: Appr. $50,000
Number of Rotors: 4
Control System: Remote Control
Applications: Film, Photography, Inspection, Surveillance
Dimensions: 2.05 × 1.89 × 0.36 m
Weight: 10.4 kg
Max. Payload: 4 kg
Frame Material: Carbon Fiber
Rotor Material: Carbon Fiber
Available Colors: White
Max. Flight Time: 45 minutes
Max. Speed: 57,6 km/h (35,7 mp/h)
Max. Flying Altitude: 4,000 m (13,123 ft)
Max. Ascent Speed: 10 m/s
Min. Operating Temperature: -10°C
Max. Operating Temperature: 50°C
Max. Operating Range: 50 km
Controller Frequency: 2.4 GHz
Battery Capacity: 13,000 mAh
Battery Type: Lithium Polymer 6S
Battery Power: 22.2V
GPS Compatible: √
Altitude Hold: √
Position Hold: √
Flight Planning: √
Autonomous Flight: –
Target Tracking: √
Automatic Landing: √
Return To Home: √
Obstacle Avoidance: –
Headless Mode: –
Built-in Camera: –

Monstertronic

Sky Force

Release Date: May 2016
Price: $155
Number of Rotors: 4
Control System: Remote Control
Applications: Recreation
Dimensions: 520 × 520 × 70 mm
Frame Material: Plastic
Rotor Material: Plastic
Available Colors: Black, White
Controller Frequency: 2.4 GHz
Battery Type: Lithium Polymer 2S
GPS Compatible: –
Altitude Hold: √
Position Hold: √
Flight Planning: –
Autonomous Flight: –
Target Tracking: –
Automatic Landing: –
Return To Home: √
Obstacle Avoidance: –
Headless Mode: –
Built-in Camera: √

RC Logger

RC Eye One Navigator 250

© droenart.com

Release Date: May 2016
Price: $250
Number of Rotors: 4
Control System: Remote Control
Applications: FPV Racing
Wingspan: 250 mm
Frame Material: Plastic
Rotor Material: Plastic
Available Colors: Black
Max. Flight Time: 10 minutes
Controller Frequency: 2.4 GHz
Battery Capacity: 2,400 mAh
Battery Type: Lithium Polymer 3S
Battery Power: 7V
GPS Compatible: –
Altitude Hold: √
Position Hold: √
Flight Planning: –
Autonomous Flight: –
Target Tracking: –
Automatic Landing: –
Return To Home: –
Obstacle Avoidance: –
Headless Mode: –
Built-in Camera: √

Robota

Eclipse

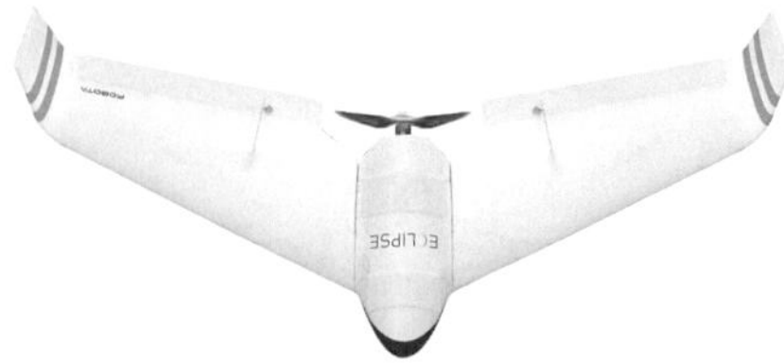

© Robota

Release Date: May 2016
Price: On Request
Number of Rotors: 1 (Fixed Wing)
Control System: Flight Computer
Applications: Aerial Mapping, Agriculture, Inspection
Dimensions: 350 × 1370 mm
Wingspan: 1370 mm
Weight: 1 kg
Max. Payload: 360 gr
Frame Material: EPP Foam
Available Colors: White
Max. Flight Time: 50 minutes
Max. Speed: 72 km/h (44,7 mp/h)
Min. Operating Temperature: -10°C
Max. Operating Temperature: 38°C
Max. Operating Range: 4.8 km
Controller Frequency: 2.4 GHz
Battery Type: Lithium Polymer
GPS Compatible: √
Altitude Hold: √
Position Hold: –
Flight Planning: √
Autonomous Flight: √
Target Tracking: –
Automatic Landing: –
Return To Home: √
Obstacle Avoidance: –
Headless Mode: –
Built-in Camera: √

Sci.Aero

TopoDrone 4Scight

© Sci.Aero

Release Date: May 2016
Price: Unknown
Number of Rotors: 4
Control System: Remote Control, Phone/Tablet
Applications: Film, Photography, Agriculture, Inspection, Surveillance
Dimensions: 560 × 640 × 250 mm
Wingspan: 850 mm
Weight: 1.5 kg
Max. Payload: 500 gr
Frame Material: Plastic
Rotor Material: Plastic
Available Colors: Orange
Max. Flight Time: 22 minutes
Max. Speed: 60 km/h (37 mp/h)
Max. Operating Range: 500 m
Controller Frequency: 2.4 GHz
GPS Compatible: √
Altitude Hold: √
Position Hold: √
Flight Planning: √
Autonomous Flight: √
Target Tracking: –
Automatic Landing: –
Return To Home: √
Obstacle Avoidance: –
Headless Mode: –
Built-in Camera: √

Sky Hero

OB1

© Sky Hero

Release Date: May 2016
Price: $299
Number of Rotors: 4
Control System: Remote Control
Applications: FPV Racing
Dimensions: 135 × 135 mm
Wingspan: 202 mm
Weight: 116 gr
Frame Material: Plastic, Carbon Fiber
Rotor Material: Plastic
Available Colors: Blue, Green, Red, White
Max. Flight Time: 10 minutes
Controller Frequency: 2.4 GHz
Battery Type: Lithium Polymer 3S
GPS Compatible: –
Altitude Hold: –
Position Hold: –
Flight Planning: –
Autonomous Flight: –
Target Tracking: –
Automatic Landing: –
Return To Home: –
Obstacle Avoidance: –
Headless Mode: –
Built-in Camera: √

Threod Systems

EOS mini UAS

© Threod Systems

Release Date: May 2016
Price: Unknown
Number of Rotors: 1 (Fixed Wing)
Control System: Ground Control Station
Applications: Aerial Mapping, Inspection, Surveillance
Dimensions: 1.6 × 3.5 m
Wingspan: 3.5 m
Weight: 7 kg
Frame Material: Composite
Available Colors: Grey
Max. Flight Time: 2 hours
Max. Speed: 100 km/h (62 mp/h)
Max. Flying Altitude: 1,500 m (4,921 ft)
Max. Operating Range: 20 km
GPS Compatible: √
Altitude Hold: √
Position Hold: –
Flight Planning: √
Autonomous Flight: √
Target Tracking: –
Automatic Landing: √
Return To Home: √
Obstacle Avoidance: –
Headless Mode: –
Built-in Camera: √

TRNDlabs

SKEYE Nano HD

© TRNDlabs

Release Date: May 2016
Price: $69
Number of Rotors: 4
Control System: Remote Control
Applications: Recreation
Dimensions: 40 × 40 × 22 mm
Weight: 14 gr
Frame Material: Plastic
Rotor Material: Plastic
Available Colors: Orange
Max. Flight Time: 4 minutes
Max. Operating Range: 50 m
Controller Frequency: 2.4 GHz
Battery Capacity: 120 mAh
Battery Type: Lithium Polymer
Battery Power: 3.7V
GPS Compatible: –
Altitude Hold: –
Position Hold: –
Flight Planning: –
Autonomous Flight: –
Target Tracking: –
Automatic Landing: –
Return To Home: –
Obstacle Avoidance: –
Headless Mode: –
Built-in Camera: √ (3 megapixels)

Vulcan

Airlift

© Vulcan

Release Date: May 2016
Price: On Request
Number of Rotors: 8
Control System: Remote Control
Applications: Transport
Frame Material: Carbon Fiber
Rotor Material: Carbon Fiber
Available Colors: Black
GPS Compatible: √
Altitude Hold: √
Position Hold: √
Flight Planning: √
Autonomous Flight: √
Target Tracking: √
Automatic Landing: √
Return To Home: √
Obstacle Avoidance: –
Headless Mode: –
Built-in Camera: –

XK Innovations

X252 Shuttle

© banggood.com

Release Date: May 2016
Price: $149
Number of Rotors: 4
Control System: Remote Control
Applications: Recreation
Dimensions: 400 × 400 × 190 mm
Weight: 230 gr
Frame Material: Plastic
Rotor Material: Plastic
Available Colors: Black, White
Max. Flight Time: 12 minutes
Max. Operating Range: 300 m
Controller Frequency: 2.4 GHz
Battery Capacity: 950 mAh
Battery Type: Lithium Polymer
Battery Power: 7.4V
GPS Compatible: –
Altitude Hold: √
Position Hold: √
Flight Planning: –
Autonomous Flight: –
Target Tracking: –
Automatic Landing: –
Return To Home: –
Obstacle Avoidance: –
Headless Mode: –
Built-in Camera: –

Zerotech

Dobby

© bhphotovideo.com

Release Date: May 2016
Price: $390
Number of Rotors: 4
Control System: Remote Control, Phone/Tablet
Applications: Film, Photography
Dimensions: 135 × 145 × 37 mm
Weight: 200 gr
Frame Material: Carbon Fiber
Rotor Material: Carbon Fiber
Motors: 6210-380kv
Available Colors: White
Max. Flight Time: 9 minutes
Max. Flying Altitude: 3,000 m (9,842 ft)
Min. Operating Temperature: 0°C
Max. Operating Temperature: 40°C
Max. Operating Range: 100 m
Controller Frequency: 2.4 GHz
Battery Capacity: 970 mAh
Battery Type: Lithium Polymer 2S
Battery Power: 7.6V
GPS Compatible: √
Altitude Hold: √
Position Hold: √
Flight Planning: √
Autonomous Flight: –
Target Tracking: √
Automatic Landing: √
Return To Home: √
Obstacle Avoidance: –
Headless Mode: –
Built-in Camera: √ (13 megapixels)

Airhogs

Connect Mission Drone

© Airhogs

Release Date: June 2016
Price: $150
Number of Rotors: 4
Control System: Phone/Tablet
Applications: Recreation
Frame Material: Styrofoam
Available Colors: Black
Max. Flight Time: 30 minutes
Controller Frequency: 2.4 GHz
Battery Type: Lithium Polymer
GPS Compatible: –
Altitude Hold: –
Position Hold: –
Flight Planning: –
Autonomous Flight: –
Target Tracking: –
Automatic Landing: –
Return To Home: –
Obstacle Avoidance: –
Headless Mode: –
Built-in Camera: –

Arris
X180

©Arris Hobby

Release Date: June 2016
Price: $255
Number of Rotors: 4
Propellers: Arris 4050
Control System: Remote Control
Applications: FPV Racing
Dimensions: 254 × 221 × 178 mm
Weight: 1.3 kg
Frame Material: Carbon Fiber
Rotor Material: Plastic
Motors: Arris 2205-2600kv
Available Colors: Black
Max. Flight Time: 10 minutes
Max. Operating Range: 1 km
Controller Frequency: 5.8 GHz
Battery Capacity: 1,500 mAh
Battery Type: Lithium Polymer 4S
Battery Power: 11.1V
GPS Compatible: √
Altitude Hold: √
Position Hold: √
Flight Planning: √
Autonomous Flight: √
Target Tracking: –
Automatic Landing: √
Return To Home: –
Obstacle Avoidance: –
Headless Mode: –
Built-in Camera: √ (Sony 700TVL)

Cheerson
CX-37

©Cheerson Hobby

Release Date: June 2016
Price: $35
Number of Rotors: 6
Control System: Remote Control
Applications: Film, Photography, Recreation
Dimensions: 100 × 100 × 30 mm
Wingspan: 140 mm
Weight: 35 gr
Frame Material: Plastic
Rotor Material: Plastic
Available Colors: Blue, White
Max. Flight Time: 5 minutes
Max. Operating Range: 30 m
Controller Frequency: 2.4 GHz
Battery Capacity: 350 mAh
Battery Type: Lithium Polymer
Battery Power: 3.7V
GPS Compatible: –
Altitude Hold: √
Position Hold: √
Flight Planning: –
Autonomous Flight: –
Target Tracking: –
Automatic Landing: √
Return To Home: –
Obstacle Avoidance: –
Headless Mode: –
Built-in Camera: √ (0.3 megapixels)

Diatone Innovations

Crusader 190

© Diatone Innovations

Release Date: June 2016
Price: $245
Number of Rotors: 4
Control System: Remote Control
Applications: FPV Racing
Weight: 330 gr
Frame Material: Carbon Fiber
Rotor Material: Plastic
Motors: BX2206-2000kv brushless
Available Colors: Black
Controller Frequency: 5.8 GHz
Battery Capacity: 3,000 mAh
Battery Type: Lithium Polymer 2S
Battery Power: 7.4V
GPS Compatible: –
Altitude Hold: –
Position Hold: –
Flight Planning: –
Autonomous Flight: –
Target Tracking: –
Automatic Landing: –
Return To Home: –
Obstacle Avoidance: –
Headless Mode: –
Built-in Camera: √ (800TVL)

Draganfly

Draganflyer Commander

© Draganfly

Release Date: June 2016
Price: On Request
Number of Rotors: 4
Control System: Remote Control
Applications: Film, Photography, Aerial Mapping, Surveillance
Dimensions: 870 × 870 × 295 mm
Wingspan: 1070 mm
Weight: 2.75 kg
Max. Payload: 1 kg
Frame Material: Carbon Fiber
Rotor Material: Carbon Fiber
Available Colors: Black
Max. Flight Time: 40 minutes
Max. Speed: 50 km/h (31 mp/h)
Max. Flying Altitude: 2,400 m (7,874 ft)
Max. Ascent Speed: 2 m/s
Max. Descent Speed: 2 m/s
Max. Operating Range: 200 m
Controller Frequency: 2.4 GHz
Battery Type: Lithium Polymer
GPS Compatible: √
Altitude Hold: √
Position Hold: √
Flight Planning: –
Autonomous Flight: –
Target Tracking: –
Automatic Landing: √
Return To Home: √
Obstacle Avoidance: –
Headless Mode: –
Built-in Camera: –

DreamQii

PlexiDrone

Release Date: June 2016
Price: $1,329
Number of Rotors: 8
Control System: Remote Control
Applications: Film, Photography
Dimensions: 380 × 380 × 130 mm
Wingspan: 540 mm
Weight: 1.3 kg
Max. Payload: 1 kg
Frame Material: Carbon Fiber
Rotor Material: Carbon Fiber
Available Colors: Black
Max. Flight Time: 30 minutes
Max. Speed: 70 km/h (44 mp/h)
Max. Operating Range: 1 km
Controller Frequency: 2.4 GHz
Battery Capacity: 5,000 mAh
Battery Type: Lithium Polymer 3S
Battery Power: 11.1V
GPS Compatible: √
Altitude Hold: √
Position Hold: √
Flight Planning: –
Autonomous Flight: –
Target Tracking: –
Automatic Landing: √
Return To Home: √
Obstacle Avoidance: –
Headless Mode: –
Built-in Camera: –

Dromida

Kodo HD

Release Date: June 2016
Price: $59
Number of Rotors: 4
Control System: Remote Control
Applications: FPV Racing
Wingspan: 106 mm
Weight: 40 gr
Frame Material: Plastic
Rotor Material: Plastic
Available Colors: Black
Max. Flight Time: 5 minutes
Controller Frequency: 2.4 GHz
Battery Capacity: 390 mAh
Battery Type: Lithium Polymer 1S
Battery Power: 3.7V
GPS Compatible: –
Altitude Hold: –
Position Hold: –
Flight Planning: –
Autonomous Flight: –
Target Tracking: –
Automatic Landing: –
Return To Home: –
Obstacle Avoidance: –
Headless Mode: –
Built-in Camera: √

DYS

Lightning X220

©DYS

Release Date: June 2016
Price: $280
Number of Rotors: 4
Control System: Remote Control
Applications: FPV Racing
Wingspan: 220 mm
Weight: 340 gr
Frame Material: Carbon Fiber
Rotor Material: Carbon Fiber
Motors: 2205-2300kv
Available Colors: Black
Controller Frequency: 5.8 GHz
Battery Capacity: 2,200 mAh
Battery Type: Lithium Polymer 4S
GPS Compatible: √
Altitude Hold: –
Position Hold: –
Flight Planning: –
Autonomous Flight: –
Target Tracking: –
Automatic Landing: –
Return To Home: –
Obstacle Avoidance: –
Headless Mode: –
Built-in Camera: √

Flyability

Elios

©Flyability

Release Date: June 2016
Price: On Request
Number of Rotors: 4
Control System: Remote Control, Phone/Tablet
Applications: Inspection, Surveillance
Dimensions: 400 mm
Weight: 700 gr
Frame Material: Aluminum, Carbon Fiber, Plastic
Rotor Material: Plastic
Available Colors: Black
Max. Flight Time: 10 minutes
Max. Speed: 25 km/h (15,5 mp/h)
Max. Ascent Speed: 2.5 m/s
Min. Operating Temperature: -10°C
Max. Operating Temperature: 35°C
Max. Operating Range: 5 km
Controller Frequency: 2.4 GHz
Battery Capacity: 2800 mAh
Battery Type: Lithium Polymer 3S
GPS Compatible: √
Altitude Hold: √
Position Hold: √
Flight Planning: √
Autonomous Flight: √
Target Tracking: √
Automatic Landing: √
Return To Home: √
Obstacle Avoidance: –
Headless Mode: –
Built-in Camera: √

Foxtech

Screamer 250

©Foxtech

Release Date: June 2016
Price: $270
Number of Rotors: 5
Control System: Remote Control
Applications: FPV Racing
Dimensions: 250 mm
Weight: 610 gr
Frame Material: Carbon Fiber
Rotor Material: Carbon Fiber
Motors: Foxtech 2208-2500kv
Available Colors: Black
Max. Flight Time: 5 minutes
Max. Speed: 130 km/h (80 mp/h)
Controller Frequency: 5.8 GHz
Battery Capacity: 1,500 mAh
Battery Type: Lithium Polymer 4S
GPS Compatible: –
Altitude Hold: –
Position Hold: –
Flight Planning: –
Autonomous Flight: –
Target Tracking: –
Automatic Landing: –
Return To Home: –
Obstacle Avoidance: –
Headless Mode: –
Built-in Camera: √ (600TVL)

Gear2Play

Condor

©droneshop.nl

Release Date: June 2016
Price: $115
Number of Rotors: 4
Control System: Remote Control
Applications: Recreation
Frame Material: Plastic
Rotor Material: Plastic
Available Colors: Black
Max. Flight Time: 8 minutes
Max. Operating Range: 30 m
Controller Frequency: 2.4 GHz
Battery Capacity: 380 mAh
Battery Type: Lithium Polymer
Battery Power: 3.7V
GPS Compatible: –
Altitude Hold: –
Position Hold: –
Flight Planning: –
Autonomous Flight: –
Target Tracking: –
Automatic Landing: –
Return To Home: √
Obstacle Avoidance: –
Headless Mode: √
Built-in Camera: √

Gteng

T902C

© deviceranking.cz

Release Date: June 2016
Price: $25
Number of Rotors: 4
Control System: Remote Control
Applications: Recreation
Dimensions: 120 × 120 × 64 mm
Frame Material: Plastic
Rotor Material: Plastic
Available Colors: White
Max. Flight Time: 8 minutes
Max. Operating Range: 100 m
Controller Frequency: 2.4 GHz
Battery Capacity: 500 mAh
Battery Type: Lithium Polymer
Battery Power: 3.7V
GPS Compatible: –
Altitude Hold: –
Position Hold: –
Flight Planning: –
Autonomous Flight: –
Target Tracking: –
Automatic Landing: –
Return To Home: √
Obstacle Avoidance: –
Headless Mode: √
Built-in Camera: √ (720P HD)

IoT Group

ROAM-e

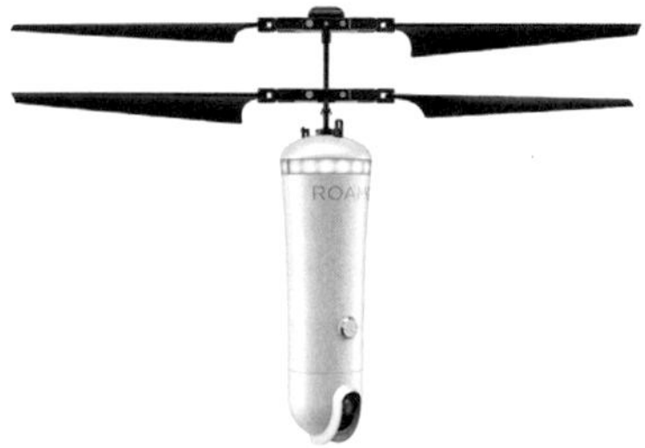

© IoT Group

Release Date: June 2016
Price: $399
Number of Rotors: 2
Control System: Phone/Tablet
Applications: Film, Photography
Dimensions: 85 mm
Frame Material: Plastic
Rotor Material: Plastic
Available Colors: White
Max. Flight Time: 20 minutes
Max. Operating Range: 25 m
Controller Frequency: 2.4 GHz
GPS Compatible: √
Altitude Hold: √
Position Hold: √
Flight Planning: –
Autonomous Flight: √
Target Tracking: √
Automatic Landing: –
Return To Home: √
Obstacle Avoidance: –
Headless Mode: –
Built-in Camera: √ (5 megapixels)

JJRC

JJ500

© JJRC

Release Date: June 2016
Price: $54
Number of Rotors: 4
Control System: Remote Control
Applications: Recreation
Dimensions: 234 × 234 × 58 mm
Weight: 75 gr
Frame Material: Plastic
Rotor Material: Plastic
Available Colors: Blue, Red
Max. Flight Time: 6 minutes
Max. Operating Range: 100 m
Controller Frequency: 2.4 GHz
Battery Capacity: 500 mAh
Battery Type: Lithium Polymer
Battery Power: 3.7V
GPS Compatible: –
Altitude Hold: √
Position Hold: √
Flight Planning: –
Autonomous Flight: –
Target Tracking: –
Automatic Landing: –
Return To Home: –
Obstacle Avoidance: –
Headless Mode: –
Built-in Camera: √ (0.3 megapixels)

Lishitoys

L6058

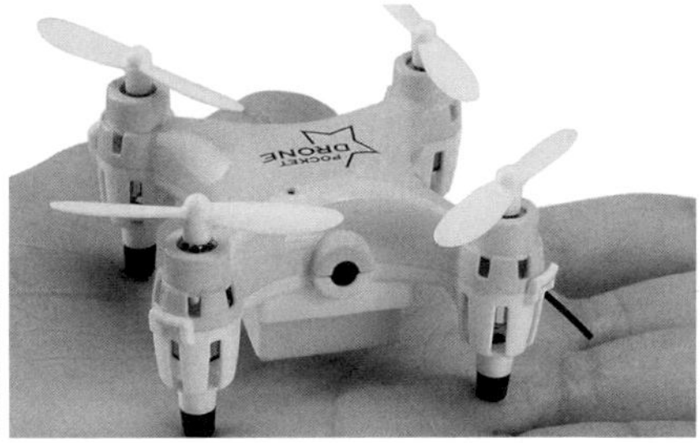

© aliexpress.com

Release Date: June 2016
Price: $18
Number of Rotors: 4
Control System: Remote Control
Applications: Recreation
Dimensions: 45 × 45 × 30 mm
Frame Material: Plastic
Rotor Material: Plastic
Available Colors: Blue
Max. Flight Time: 8 minutes
Max. Operating Range: 30 m
Controller Frequency: 2.4 GHz
Battery Capacity: 120 mAh
Battery Type: Lithium Polymer
Battery Power: 3.7V
GPS Compatible: –
Altitude Hold: –
Position Hold: –
Flight Planning: –
Autonomous Flight: –
Target Tracking: –
Automatic Landing: –
Return To Home: √
Obstacle Avoidance: –
Headless Mode: –
Built-in Camera: –

Meijiaxin Toys

X904

© 2kita.by

Release Date: June 2016
Price: $29
Number of Rotors: 4
Control System: Remote Control
Applications: Recreation
Dimensions: 126 × 126 × 35 mm
Frame Material: Plastic
Rotor Material: Plastic
Available Colors: Grey
Max. Flight Time: 6 minutes
Max. Operating Range: 50 m
Controller Frequency: 2.4 GHz
Battery Capacity: 380 mAh
Battery Type: Lithium Polymer
Battery Power: 3.7V
GPS Compatible: –
Altitude Hold: –
Position Hold: –
Flight Planning: –
Autonomous Flight: –
Target Tracking: –
Automatic Landing: –
Return To Home: –
Obstacle Avoidance: –
Headless Mode: –
Built-in Camera: –

Meijiaxin Toys

X905C

© aliexpress.com

Release Date: June 2016
Price: $20
Number of Rotors: 4
Control System: Remote Control
Applications: Recreation
Dimensions: 62 × 62 × 20 mm
Frame Material: Plastic
Rotor Material: Plastic
Available Colors: White
Max. Flight Time: 6 minutes
Max. Operating Range: 30 m
Controller Frequency: 2.4 GHz
Battery Capacity: 120 mAh
Battery Type: Lithium Polymer
Battery Power: 3.7V
GPS Compatible: –
Altitude Hold: –
Position Hold: –
Flight Planning: –
Autonomous Flight: –
Target Tracking: –
Automatic Landing: –
Return To Home: √
Obstacle Avoidance: –
Headless Mode: √
Built-in Camera: √ (2 megapixels)

ONAGOfly

Smart Nano Drone

©ONAGOfly

Release Date: June 2016
Price: $259
Number of Rotors: 4
Control System: Remote Control
Applications: Recreation
Dimensions: 100 × 100 × 45 mm
Weight: 135 gr
Frame Material: Plastic
Rotor Material: Plastic
Available Colors: White
Controller Frequency: 2.4 GHz
Battery Capacity: 800 mAh
Battery Type: Lithium Polymer
Battery Power: 7.4V
GPS Compatible: –
Altitude Hold: –
Position Hold: –
Flight Planning: –
Autonomous Flight: –
Target Tracking: –
Automatic Landing: –
Return To Home: –
Obstacle Avoidance: –
Headless Mode: –
Built-in Camera: √ (15 megapixels)

Onyxstar

Xena 8F

©Onyxstar

Release Date: June 2016
Price: Unknown
Number of Rotors: 8
Control System: Remote Control
Applications: Film, Photography, Aerial Mapping, Agriculture, Inspection, Surveillance, Transport
Wingspan: 720 mm
Weight: 3.5 kg
Max. Payload: 2 kg
Frame Material: Carbon Fiber, Composite
Rotor Material: Carbon Fiber
Available Colors: Red, White
Max. Flight Time: 37 minutes
Max. Flying Altitude: 5,000 m (16,404 ft)
Min. Operating Temperature: -15°C
Max. Operating Temperature: 40°C
Controller Frequency: 2.4 GHz
Battery Capacity: 10,000 mAh
Battery Type: Lithium Polymer 6S
GPS Compatible: √
Altitude Hold: √
Position Hold: √
Flight Planning: √
Autonomous Flight: √
Target Tracking: –
Automatic Landing: √
Return To Home: √
Obstacle Avoidance: –
Headless Mode: –
Built-in Camera: –

PrecisionHawk

Lancaster 5

© PrecisionHawk

Release Date: June 2016
Price: On Request
Number of Rotors: 1 (Fixed Wing)
Control System: Remote Control
Applications: Aerial Mapping, Agriculture, Inspection, Surveillance
Wingspan: 1.5 m
Weight: 2.4 kg
Max. Payload: 1.15 kg
Frame Material: Plastic, Carbon Fiber
Rotor Material: Plastic
Available Colors: White
Max. Flight Time: 45 minutes
Max. Speed: 79 km/h (49 mp/h)
Max. Flying Altitude: 2,500 m (8,200 ft)
Max. Operating Range: 2 km
GPS Compatible: √
Altitude Hold: √
Position Hold: √
Flight Planning: √
Autonomous Flight: √
Target Tracking: √
Automatic Landing: √
Return To Home: √
Obstacle Avoidance: √
Headless Mode: –
Built-in Camera: √

Rise

RXS270

© Rise

Release Date: June 2016
Price: $249
Number of Rotors: 4
Control System: Remote Control
Applications: FPV Racing
Wingspan: 270 mm
Weight: 460 gr
Frame Material: Plastic
Rotor Material: Plastic
Available Colors: Black
Max. Flight Time: 8 minutes
Controller Frequency: 2.4 GHz
Battery Capacity: 2,200 mAh
Battery Type: Lithium Polymer 3S
GPS Compatible: –
Altitude Hold: –
Position Hold: –
Flight Planning: –
Autonomous Flight: –
Target Tracking: –
Automatic Landing: –
Return To Home: –
Obstacle Avoidance: –
Headless Mode: –
Built-in Camera: –

Storm Racing Drone

SRD NR8

©SRD

Release Date: June 2016
Price: $399
Number of Rotors: 4
Control System: Remote Control
Applications: FPV Racing
Dimensions: 143 × 159 × 75 mm
Weight: 149 gr
Frame Material: Carbon Fiber
Rotor Material: Carbon Fiber
Available Colors: White
Max. Operating Range: 1.5 km
Controller Frequency: 5.8 GHz
Battery Capacity: 600 mAh
Battery Type: Lithium Polymer
Battery Power: 14.8V
GPS Compatible: –
Altitude Hold: –
Position Hold: –
Flight Planning: –
Autonomous Flight: –
Target Tracking: –
Automatic Landing: –
Return To Home: –
Obstacle Avoidance: –
Headless Mode: –
Built-in Camera: √ (520TVL)

Storm Racing Drone

SRD 180 V3

©hostingpics.net

Release Date: June 2016
Price: $429
Number of Rotors: 4
Control System: Remote Control
Applications: FPV Racing
Dimensions: 170 × 190 × 60 mm
Weight: 350 gr
Frame Material: Carbon Fiber
Rotor Material: Carbon Fiber
Motors: M1306-3100kv
Available Colors: Black
Max. Operating Range: 1.5 km
Controller Frequency: 5.8 GHz
Battery Capacity: 1,000 mAh
Battery Type: Lithium Polymer
Battery Power: 14.8V
GPS Compatible: –
Altitude Hold: –
Position Hold: –
Flight Planning: –
Autonomous Flight: –
Target Tracking: –
Automatic Landing: –
Return To Home: –
Obstacle Avoidance: –
Headless Mode: –
Built-in Camera: √ (700TVL)

Tarot RC

120 FPV

©Tarot

Release Date: June 2016
Price: $150
Number of Rotors: 4
Control System: Remote Control
Applications: Recreation
Dimensions: 199 × 186 × 28 mm
Weight: 85 gr
Frame Material: Carbon Fiber
Rotor Material: Carbon Fiber
Motors: MT1104 4000kv brushless
Available Colors: Black
Controller Frequency: 5.8 GHz
Battery Capacity: 800 mAh
Battery Type: Lithium Polymer
Battery Power: 11.1V
GPS Compatible: –
Altitude Hold: –
Position Hold: –
Flight Planning: –
Autonomous Flight: –
Target Tracking: –
Automatic Landing: –
Return To Home: –
Obstacle Avoidance: –
Headless Mode: –
Built-in Camera: √ (300TVL)

Tarot RC

150 FPV

©Tarot

Release Date: June 2016
Price: $170
Number of Rotors: 4
Control System: Remote Control
Applications: FPV Racing
Dimensions: 215 × 189 × 28 mm
Weight: 152 gr
Frame Material: Carbon Fiber
Rotor Material: Carbon Fiber
Motors: MT1104 4000kv brushless
Available Colors: Black
Controller Frequency: 5.8 GHz
Battery Capacity: 800 mAh
Battery Type: Lithium Polymer
Battery Power: 11.1V
GPS Compatible: –
Altitude Hold: –
Position Hold: –
Flight Planning: –
Autonomous Flight: –
Target Tracking: –
Automatic Landing: –
Return To Home: –
Obstacle Avoidance: –
Headless Mode: –
Built-in Camera: √ (300TVL)

Tekever

AR4 Light Ray Compact

Release Date: June 2016
Price: On Request
Number of Rotors: 1 (Fixed Wing)
Control System: Ground Control Station, Laptop
Applications: Military Purposes, Inspection, Reconnaissance, Surveillance, Target Acquisition
Dimensions: 1.2 × 1.8 m
Wingspan: 1.8 m
Weight: 5 kg
Max. Payload: 1 kg
Frame Material: Composite
Available Colors: Grey
Max. Flight Time: 2 hours
Max. Speed: 80 km/h (49,7 mp/h)
Max. Operating Range: 20 km
GPS Compatible: √
Altitude Hold: √
Position Hold: –
Flight Planning: √
Autonomous Flight: √
Target Tracking: –
Automatic Landing: –
Return To Home: √
Obstacle Avoidance: √
Headless Mode: –
Built-in Camera: √

WLtoys

Q343 Mini Quadcopter

Release Date: June 2016
Price: $25
Number of Rotors: 4
Control System: Remote Control
Applications: Recreation
Dimensions: 69 × 69 × 27 mm
Weight: 14 gr
Frame Material: Plastic
Rotor Material: Plastic
Available Colors: Blue, Green
Max. Flight Time: 6 minutes
Max. Operating Range: 30 m
Controller Frequency: 2.4 GHz
Battery Capacity: 150 mAh
Battery Type: Lithium Polymer
Battery Power: 3.7V
GPS Compatible: –
Altitude Hold: –
Position Hold: –
Flight Planning: –
Autonomous Flight: –
Target Tracking: –
Automatic Landing: –
Return To Home: –
Obstacle Avoidance: –
Headless Mode: –
Built-in Camera: √

Airhogs

Star Trek U.S.S. Enterprise

© spinmaster.com

Release Date: July 2016
Price: $119
Number of Rotors: 4
Control System: Remote Control
Applications: Recreation
Dimensions: 457 × 305 × 132 mm
Weight: 500 gr
Frame Material: EPP Foam, Plastic
Rotor Material: Plastic
Available Colors: White
Controller Frequency: 2.4 GHz
Battery Capacity: 650 mAh
Battery Type: Lithium Polymer
Battery Power: 3.7V
GPS Compatible: –
Altitude Hold: –
Position Hold: –
Flight Planning: –
Autonomous Flight: –
Target Tracking: –
Automatic Landing: –
Return To Home: –
Obstacle Avoidance: –
Headless Mode: –
Built-in Camera: –

Altavian

Galaxy R8700

© Altavian

Release Date: July 2016
Price: $15,000
Number of Rotors: 8
Control System: Flight Computer
Applications: Aerial Mapping, Agriculture, Inspection
Dimensions: 1.27 × 1.27 × 0.22 m
Wingspan: 1.2 m
Weight: 11.1 kg
Max. Payload: 1.36 kg
Frame Material: Carbon Fiber
Available Colors: White
Max. Flight Time: 25 minutes
Max. Speed: 54 km/h (33,5 mp/h)
Max. Flying Altitude: 3,048 m (10,000 ft)
Max. Ascent Speed: 1.5 m/s
Max. Descent Speed: 2.75 m/s
Max. Operating Range: 4.8 km
Controller Frequency: 2.4 GHz
Battery Type: Lithium Polymer
GPS Compatible: √
Altitude Hold: √
Position Hold: √
Flight Planning: √
Autonomous Flight: √
Target Tracking: –
Automatic Landing: √
Return To Home: √
Obstacle Avoidance: –
Headless Mode: –
Built-in Camera: –

BirdsEyeView Aerobotics

FireFly6 Pro

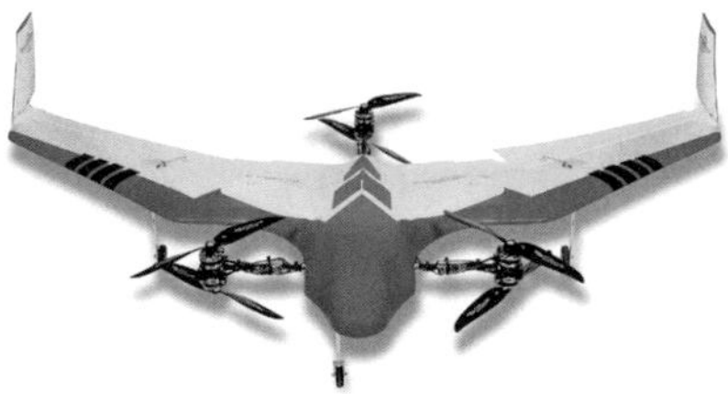

© BirdsEyeView

Release Date: July 2016
Price: $5,999
Number of Rotors: 6 (Fixed Wing)
Control System: Ground Control Station
Applications: Film, Photography, Aerial Mapping, Agriculture, Inspection
Dimensions: 0.95 × 1.52 m
Wingspan: 1.52 m
Weight: 4.5 kg
Max. Payload: 1.1 kg
Frame Material: EPO Foam (carbon fiber reinforced)
Available Colors: White/Red
Max. Flight Time: 50 minutes
Max. Speed: 105 km/h (65 mp/h)
Max. Operating Range: 53 km
Controller Frequency: 2.4 GHz
Battery Capacity: 5,200 mAh
Battery Type: Lithium Polymer 3S
GPS Compatible: √
Altitude Hold: √
Position Hold: √
Flight Planning: √
Autonomous Flight: √
Target Tracking: √
Automatic Landing: √
Return To Home: √
Obstacle Avoidance: √
Headless Mode: –
Built-in Camera: –

Cheerson

CX-10WD

©unmannedtechshop.co.uk

Release Date: July 2016
Price: $43
Number of Rotors: 4
Control System: Remote Control, Phone
Applications: Recreation
Dimensions: 42 × 42 × 25 mm
Frame Material: Plastic
Rotor Material: Plastic
Available Colors: Gold, Pink, Red, Silver
Max. Flight Time: 4 minutes
Max. Operating Range: 40 m
Controller Frequency: 2.4 GHz
Battery Capacity: 150 mAh
Battery Type: Lithium Polymer
Battery Power: 3.7V
GPS Compatible: –
Altitude Hold: √
Position Hold: –
Flight Planning: –
Autonomous Flight: –
Target Tracking: –
Automatic Landing: √
Return To Home: √
Obstacle Avoidance: –
Headless Mode: √
Built-in Camera: √ (0.3 megapixels)

Facebook

Aquila

© Facebook

Release Date: July 2016
Price: Unknown
Number of Rotors: 4 (Fixed Wing)
Control System: Flight Computer
Applications: Internet Beaming
Wingspan: 45 m
Weight: 453 kg
Frame Material: Carbon Fiber
Motors: Solar Powered
Available Colors: Black
Max. Speed: 129 km/h (80 mp/h)
GPS Compatible: √
Altitude Hold: √
Position Hold: √
Flight Planning: √
Autonomous Flight: √
Target Tracking: √
Automatic Landing: √
Return To Home: √
Obstacle Avoidance: √
Headless Mode: –
Built-in Camera: –

Fuqi Toys

FQ777-01

© helicomicro.com

Release Date: July 2016
Price: $90
Number of Rotors: 4
Control System: Remote Control
Applications: Recreation, FPV Racing
Dimensions: 340 × 340 × 24 mm
Weight: 129 gr
Frame Material: Plastic
Rotor Material: Plastic
Available Colors: Black, White
Max. Flight Time: 8 minutes
Max. Operating Range: 120 m
Controller Frequency: 2.4 GHz
Battery Capacity: 700 mAh
Battery Type: Lithium Polymer
Battery Power: 3.7V
GPS Compatible: –
Altitude Hold: √
Position Hold: √
Flight Planning: –
Autonomous Flight: –
Target Tracking: –
Automatic Landing: –
Return To Home: √
Obstacle Avoidance: –
Headless Mode: √
Built-in Camera: √ (0.3 megapixels)

Guizhou Aviation Industry Group (CAIG)

Soar Dragon

Release Date: July 2016
Price: Unknown
Number of Rotors: – (Fixed Wing)
Control System: Ground Control Station
Applications: Military Purposes, Inspection, Armed Reconnaissance, Surveillance
Dimensions: 14.33 × 24.86 × 5.41 m
Wingspan: 24.86 m
Motors: 1 × Guizhou WP-13 turbojet
Available Colors: Black
Max. Flight Time: 10 hours
Max. Speed: 750 km/h (466 mp/h)
Max. Flying Altitude: 18,000 m (59,055 ft)
Max. Operating Range: 7,000 km
GPS Compatible: √
Altitude Hold: √
Position Hold: –
Flight Planning: √
Autonomous Flight: √
Target Tracking: √
Automatic Landing: √
Return To Home: √
Obstacle Avoidance: √
Headless Mode: –
Built-in Camera: √

HuaXiang

8971V

©gearbest.com

Release Date: July 2016
Price: $40
Number of Rotors: 4
Control System: Remote Control
Applications: Film, Photography, Recreation
Dimensions: 175 × 175 × 60 mm
Frame Material: Plastic
Rotor Material: Plastic
Motors: Brushed
Available Colors: Black, White
Max. Flight Time: 6 minutes
Max. Operating Range: 80 m
Controller Frequency: 2.4 GHz
Battery Capacity: 500 mAh
Battery Type: Lithium Polymer
Battery Power: 3.7V
GPS Compatible: –
Altitude Hold: √
Position Hold: √
Flight Planning: –
Autonomous Flight: –
Target Tracking: –
Automatic Landing: –
Return To Home: –
Obstacle Avoidance: –
Headless Mode: √
Built-in Camera: √ (2 megapixels)

Jin Xing Da

511V Pterosaurs

©aliexpress.com

Release Date: July 2016
Price: $22
Number of Rotors: 4
Control System: Remote Control
Applications: Recreation
Dimensions: 236 × 346 × 65 mm
Wingspan: 346 mm
Frame Material: Plastic
Rotor Material: Plastic
Motors: Brushed
Available Colors: Blue, Silver
Max. Flight Time: 6 minutes
Controller Frequency: 2.4 GHz
Battery Capacity: 400 mAh
Battery Type: Lithium Polymer
Battery Power: 7.4V
GPS Compatible: –
Altitude Hold: –
Position Hold: –
Flight Planning: –
Autonomous Flight: –
Target Tracking: –
Automatic Landing: –
Return To Home: –
Obstacle Avoidance: –
Headless Mode: –
Built-in Camera: –

JJRC

H31

©geekbuying.com

Release Date: July 2016
Price: $32
Number of Rotors: 4
Control System: Remote Control
Applications: Recreation
Dimensions: 290 × 290 × 90 mm
Weight: 73 gr
Frame Material: Plastic
Rotor Material: Plastic
Motors: Brushed
Available Colors: White
Max. Flight Time: 10 minutes
Max. Operating Range: 80 m
Controller Frequency: 2.4 GHz
Battery Capacity: 400 mAh
Battery Type: Lithium Polymer
Battery Power: 3.7V
GPS Compatible: –
Altitude Hold: –
Position Hold: –
Flight Planning: –
Autonomous Flight: –
Target Tracking: –
Automatic Landing: –
Return To Home: √
Obstacle Avoidance: –
Headless Mode: √
Built-in Camera: –

Martin UAV

V-Bat

Release Date: July 2016
Price: On Request
Number of Rotors: 1 (Fixed Wing)
Control System: Flight Computer
Applications: Aerial Mapping, Border Patrol, Inspection, Surveillance
Dimensions: 2.4 × 2.74 m
Wingspan: 2.74
Weight: 31 kg
Max. Payload: 2.3 kg
Frame Material: Aluminum, Carbon Fiber, Kevlar
Motors: 150cc two-stroke engine
Available Colors: White
Max. Flight Time: 8 hours
Max. Speed: 166 km/h (103 mp/h)
Max. Flying Altitude: 4,572 m (15,000 ft)
Max. Operating Range: 560 km
Controller Frequency: 2.4 GHz
GPS Compatible: √
Altitude Hold: √
Position Hold: √
Flight Planning: √
Autonomous Flight: √
Target Tracking: √
Automatic Landing: √
Return To Home: √
Obstacle Avoidance: √
Headless Mode: –
Built-in Camera: √

Monstertronic

FPV Fighter

© Monstertronic

Release Date: July 2016
Price: $190
Number of Rotors: 4
Control System: Remote Control
Applications: FPV Racing
Dimensions: 190 × 190 × 50 mm
Frame Material: Plastic
Rotor Material: Plastic
Available Colors: Black
Controller Frequency: 5.8 GHz
Battery Capacity: 780 mAh
Battery Type: Lithium Polymer 1S
Battery Power: 3.7V
GPS Compatible: –
Altitude Hold: √
Position Hold: √
Flight Planning: –
Autonomous Flight: –
Target Tracking: –
Automatic Landing: –
Return To Home: √
Obstacle Avoidance: –
Headless Mode: –
Built-in Camera: √

Perspective Robotics GmbH

Fotokite Phi

© fotokite.com

Release Date: July 2016
Price: $500
Number of Rotors: 4
Control System: Remote Control
Applications: Film, Photography
Wingspan: 350 mm
Weight: 350 gr
Max. Payload: 50 gr
Frame Material: Carbon Fiber
Rotor Material: Plastic
Available Colors: Blue
Max. Flight Time: 10 minutes
Max. Speed: – (tethered cable)
Max. Operating Range: 8 m
Controller Frequency: 2.4 GHz
Battery Type: Lithium Polymer 3S
Battery Power: 11V
GPS Compatible: –
Altitude Hold: √
Position Hold: √
Flight Planning: –
Autonomous Flight: –
Target Tracking: √
Automatic Landing: –
Return To Home: –
Obstacle Avoidance: –
Headless Mode: –
Built-in Camera: –

Sistemas de Control Remoto S.L.

Atlantic

© hispaviacion.es

Release Date: July 2016
Price: Unknown
Number of Rotors: 1 (Fixed Wing)
Control System: Ground Control Station
Applications: Military Purposes, Inspection, Reconnaissance, Surveillance
Dimensions: 2.8 × 3.8 m
Wingspan: 3.8 m
Weight: 45 kg
Max. Payload: 7 kg
Frame Material: Composite
Rotor Material: Wood
Motors: Two-stroke, two-cylinders petrol engine
Available Colors: Grey
Max. Flight Time: 5 hours
Max. Speed: 150 km/h (93 mp/h)
Max. Flying Altitude: 3,000 m (9,842 ft)
Max. Operating Range: 100 km
GPS Compatible: √
Altitude Hold: √
Position Hold: –
Flight Planning: √
Autonomous Flight: √
Target Tracking: –
Automatic Landing: √
Return To Home: √
Obstacle Avoidance: √
Headless Mode: –
Built-in Camera: √

Storm Racing Drone

SRD130 Loki-X3

Release Date: July 2016
Price: $429
Number of Rotors: 4
Control System: Remote Control
Applications: FPV Racing
Dimensions: 105 × 132 × 34 mm
Weight: 172 gr
Frame Material: Carbon Fiber
Rotor Material: Carbon Fiber
Motors: M1306-3100kv
Available Colors: Black
Max. Operating Range: 1.5 km
Controller Frequency: 5.8 GHz
Battery Capacity: 600 mAh
Battery Type: Lithium Polymer
Battery Power: 14.8V
GPS Compatible: –
Altitude Hold: –
Position Hold: –
Flight Planning: –
Autonomous Flight: –
Target Tracking: –
Automatic Landing: –
Return To Home: –
Obstacle Avoidance: –
Headless Mode: –
Built-in Camera: √ (600TVL)

Storm Racing Drone

SRD163 Loki-X4

Release Date: July 2016
Price: $439
Number of Rotors: 4
Control System: Remote Control
Applications: FPV Racing
Dimensions: 135 × 175 × 34 mm
Weight: 216 gr
Frame Material: Carbon Fiber
Rotor Material: Carbon Fiber
Motors: M1806 2300kv
Available Colors: Black
Max. Operating Range: 1.5 km
Controller Frequency: 5.8 GHz
Battery Capacity: 850 mAh
Battery Type: Lithium Polymer
Battery Power: 14.8V
GPS Compatible: –
Altitude Hold: –
Position Hold: –
Flight Planning: –
Autonomous Flight: –
Target Tracking: –
Automatic Landing: –
Return To Home: –
Obstacle Avoidance: –
Headless Mode: –
Built-in Camera: √ (600TVL)

Storm Racing Drone

SRD199 Loki-X5

© helipal.com

Release Date: July 2016
Price: $469
Number of Rotors: 4
Control System: Remote Control
Applications: FPV Racing
Dimensions: 163 × 200 × 34 mm
Weight: 282 gr
Frame Material: Carbon Fiber
Rotor Material: Carbon Fiber
Motors: M2204-UC 2300kv
Available Colors: Black
Max. Operating Range: 1.5 km
Controller Frequency: 5.8 GHz
Battery Capacity: 1,300 mAh
Battery Type: Lithium Polymer
Battery Power: 14.8V
GPS Compatible: –
Altitude Hold: –
Position Hold: –
Flight Planning: –
Autonomous Flight: –
Target Tracking: –
Automatic Landing: –
Return To Home: –
Obstacle Avoidance: –
Headless Mode: –
Built-in Camera: √ (600TVL)

Storm Racing Drone

SRD 200X

© SRD

Release Date: July 2016
Price: $399
Number of Rotors: 4
Control System: Remote Control
Applications: FPV Racing
Dimensions: 188 × 180 × 60 mm
Weight: 350 gr
Frame Material: Carbon Fiber
Rotor Material: Carbon Fiber
Available Colors: Black
Max. Operating Range: 1.5 km
Controller Frequency: 5.8 GHz
Battery Capacity: 1,300 mAh
Battery Type: Lithium Polymer
Battery Power: 14.8V
GPS Compatible: –
Altitude Hold: –
Position Hold: –
Flight Planning: –
Autonomous Flight: –
Target Tracking: –
Automatic Landing: –
Return To Home: –
Obstacle Avoidance: –
Headless Mode: –
Built-in Camera: √ (700TVL)

Storm Racing Drone

SRD FlexRC Owl Storm Edition

©SRD

Release Date: July 2016
Price: $399
Number of Rotors: 4
Control System: Remote Control
Applications: FPV Racing
Dimensions: 178 × 173 × 33 mm
Weight: 175 gr
Frame Material: Carbon Fiber
Rotor Material: Carbon Fiber
Motors: M1306-3100kv
Available Colors: Black
Max. Operating Range: 1.5 km
Controller Frequency: 5.8 GHz
Battery Capacity: 600 mAh
Battery Type: Lithium Polymer
Battery Power: 14.8V
GPS Compatible: –
Altitude Hold: –
Position Hold: –
Flight Planning: –
Autonomous Flight: –
Target Tracking: –
Automatic Landing: –
Return To Home: –
Obstacle Avoidance: –
Headless Mode: –
Built-in Camera: √ (600TVL)

Storm Racing Drone

Storm Drone 6 V3

©SRD

Release Date: July 2016
Price: $669
Number of Rotors: 6
Propellers: Gemfan 8045C
Control System: Remote Control
Applications: Film, Photography
Dimensions: 450 × 375 × 215 mm
Weight: 870 gr
Frame Material: Carbon Fiber
Rotor Material: Carbon Fiber
Motors: M2208 1260kv
Available Colors: Black
Max. Flight Time: 17 minutes
Max. Operating Range: 1.5 km
Controller Frequency: 2.4 GHz
Battery Capacity: 5,500 mAh
Battery Type: Lithium Polymer
Battery Power: 11.1V
GPS Compatible: √
Altitude Hold: √
Position Hold: √
Flight Planning: √
Autonomous Flight: √
Target Tracking: √
Automatic Landing: √
Return To Home: √
Obstacle Avoidance: –
Headless Mode: –
Built-in Camera: –

Storm Racing Drone

SRD TBS Vendetta

Release Date: July 2016
Price: $779
Number of Rotors: 4
Control System: Remote Control
Applications: FPV Racing
Dimensions: 230 × 220 × 50 mm
Weight: 415 gr
Frame Material: Carbon Fiber
Rotor Material: Carbon Fiber
Motors: M1306-3100kv
Available Colors: Black
Max. Operating Range: 1.5 km
Controller Frequency: 5.8 GHz
Battery Capacity: 1,300 mAh
Battery Type: Lithium Polymer
Battery Power: 14.8V
GPS Compatible: –
Altitude Hold: –
Position Hold: –
Flight Planning: –
Autonomous Flight: –
Target Tracking: –
Automatic Landing: –
Return To Home: –
Obstacle Avoidance: –
Headless Mode: –
Built-in Camera: √ (600TVL)

Thales

Fulmar X

Release Date: July 2016
Price: $1.1 million
Number of Rotors: 1 (Fixed Wing)
Control System: Ground Control Station
Applications: Military Purposes, Reconnaissance, Surveillance
Dimensions: 1.2 × 3 × 0.5 m
Wingspan: 3 m
Weight: 20 kg
Max. Payload: 8 kg
Frame Material: Carbon Fiber
Available Colors: Grey, Yellow
Max. Flight Time: 12 hours
Max. Flying Altitude: 3,000 m (9,842 ft)
Max. Operating Range: 800 km
GPS Compatible: √
Altitude Hold: √
Position Hold: –
Flight Planning: √
Autonomous Flight: √
Target Tracking: √
Automatic Landing: –
Return To Home: √
Obstacle Avoidance: –
Headless Mode: –
Built-in Camera: √

Yi Zhan Toys

i4W

© aliexpress.com

Release Date: July 2016
Price: $55
Number of Rotors: 4
Control System: Remote Control
Applications: Recreation
Dimensions: 135 × 145 × 250 mm
Weight: 42 gr
Frame Material: Plastic
Rotor Material: Plastic
Available Colors: Black
Max. Flight Time: 5 minutes
Max. Operating Range: 50 m
Controller Frequency: 2.4 GHz
Battery Capacity: 450 mAh
Battery Type: Lithium Polymer
Battery Power: 3.7V
GPS Compatible: –
Altitude Hold: √
Position Hold: √
Flight Planning: –
Autonomous Flight: –
Target Tracking: –
Automatic Landing: √
Return To Home: √
Obstacle Avoidance: –
Headless Mode: √
Built-in Camera: √ (2 megapixels)

Aeronautics Ltd.

Orbiter 1K

© Wikipedia/Alexmilt

Release Date: May 2015
Price: $175,000
Number of Rotors: 1 (Fixed Wing)
Control System: Remote Control, Laptop
Applications: Loitering Munition, Reconnaissance
Dimensions: 1 × 2.2 × 0.3 m
Wingspan: 2.2 m
Weight: 7 kg
Max. Payload: 1.5 kg
Available Colors: Grey
Max. Flight Time: 90 minutes
Max. Speed: 105 km/h (65 mp/h)
Max. Operating Range: 15 km
GPS Compatible: √
Altitude Hold: √
Position Hold: –
Flight Planning: √
Autonomous Flight: √
Target Tracking: √
Automatic Landing: √
Return To Home: √
Obstacle Avoidance: √
Headless Mode: –
Built-in Camera: √

Arris

X210

© Arris Hobby

Release Date: August 2016
Price: $289
Number of Rotors: 4
Control System: Remote Control
Applications: FPV Racing
Wingspan: 210 mm
Weight: 84 gr
Frame Material: Carbon Fiber
Motors: 2205-2300kv brushless
Available Colors: Red
Max. Operating Range: 1 km
Controller Frequency: 5.8 GHz
Battery Capacity: 1,500 mAh
Battery Type: Lithium Polymer 4S
GPS Compatible: –
Altitude Hold: √
Position Hold: √
Flight Planning: –
Autonomous Flight: –
Target Tracking: –
Automatic Landing: –
Return To Home: √
Obstacle Avoidance: –
Headless Mode: –
Built-in Camera: √

Ascent AeroSystems

Sprite

© Hobbyking

Release Date: August 2016
Price: $849
Number of Rotors: 2
Control System: Remote Control, Phone/Tablet
Applications: Film, Photography
Dimensions: 350 mm
Wingspan: 100 mm
Weight: 1.2 kg
Frame Material: Plastic
Rotor Material: Plastic
Available Colors: White
Max. Flight Time: 12 minutes
Max. Speed: 36 km/h (22 mp/h)
Max. Ascent Speed: 10 m/s
Max. Operating Range: 6 km
Controller Frequency: 2.4 GHz
Battery Type: Lithium Polymer
Battery Power: 11.1V
GPS Compatible: √
Altitude Hold: √
Position Hold: √
Flight Planning: √
Autonomous Flight: √
Target Tracking: √
Automatic Landing: √
Return To Home: √
Obstacle Avoidance: –
Headless Mode: –
Built-in Camera: –

Dromida

XL 370 FPV

© bhphotovideo.com

Release Date: August 2016
Price: $160
Number of Rotors: 4
Control System: Remote Control
Applications: Recreation
Dimensions: 533 × 342 × 140 mm
Weight: 1 kg
Frame Material: Plastic
Rotor Material: Plastic
Available Colors: Blue, Red
Max. Flight Time: 12 minutes
Controller Frequency: 2.4 GHz
Battery Type: Lithium Polymer 2S
GPS Compatible: –
Altitude Hold: √
Position Hold: √
Flight Planning: –
Autonomous Flight: –
Target Tracking: –
Automatic Landing: √
Return To Home: √
Obstacle Avoidance: –
Headless Mode: –
Built-in Camera: √ (1080p HD)

DYS

XDR220

© fpvmodel.com

Release Date: August 2016
Price: $330
Number of Rotors: 4
Propellers: XT50403 Tri-prop
Control System: Remote Control
Applications: FPV Racing
Dimensions: 190 × 190 × 53 mm
Wingspan: 220 mm
Weight: 310 gr
Frame Material: Carbon Fiber, Plastic
Rotor Material: Plastic
Motors: SE2205-2300kv
Available Colors: Blue, Red, White
Controller Frequency: 5.8 GHz
Battery Capacity: 1,500 mAh
Battery Type: Lithium Polymer 4S
GPS Compatible: –
Altitude Hold: –
Position Hold: –
Flight Planning: –
Autonomous Flight: –
Target Tracking: –
Automatic Landing: –
Return To Home: –
Obstacle Avoidance: –
Headless Mode: –
Built-in Camera: √

Eachine

QX90

Release Date: August 2016
Price: $119
Number of Rotors: 4
Control System: Remote Control
Applications: FPV Racing
Dimensions: 118 × 118 × 70 mm
Weight: 39 gr
Frame Material: Carbon Fiber
Rotor Material: Plastic
Motors: 8520 brush motor
Available Colors: Black
Max. Flight Time: 6 minutes
Controller Frequency: 5.8 GHz
Battery Capacity: 600 mAh
Battery Type: Lithium Polymer
Battery Power: 3.7V
GPS Compatible: –
Altitude Hold: –
Position Hold: –
Flight Planning: –
Autonomous Flight: –
Target Tracking: –
Automatic Landing: –
Return To Home: –
Obstacle Avoidance: –
Headless Mode: –
Built-in Camera: √ (520TVL)

Eachine

Wizard X220

Release Date: August 2016
Price: $199
Number of Rotors: 4
Control System: Remote Control
Applications: FPV Racing
Wingspan: 220 mm
Weight: 535 gr
Frame Material: Carbon Fiber
Rotor Material: Plastic
Motors: MN2205 2300kv
Available Colors: Black
Max. Flight Time: 12 minutes
Controller Frequency: 5.8 GHz
Battery Type: Lithium Polymer
GPS Compatible: –
Altitude Hold: –
Position Hold: –
Flight Planning: –
Autonomous Flight: –
Target Tracking: –
Automatic Landing: –
Return To Home: –
Obstacle Avoidance: –
Headless Mode: –
Built-in Camera: √ (700TVL)

Elbit Systems

Skylark C

© Elbit Systems

Release Date: August 2016
Price: Unknown
Number of Rotors: 1
Control System: Flight Computer
Applications: Military Purposes, Maritime Patrol, Reconnaissance, Surveillance
Wingspan: 3.6 m
Weight: 15 kg
Available Colors: Grey
Max. Flight Time: 7 hours
Max. Flying Altitude: 5,000 m (16,404 ft)
Max. Operating Range: 40 km
GPS Compatible: √
Altitude Hold: √
Position Hold: –
Flight Planning: √
Autonomous Flight: √
Target Tracking: –
Automatic Landing: √
Return To Home: √
Obstacle Avoidance: √
Headless Mode: –
Built-in Camera: √

Estes Rockets

Proto-Z

© Estes Rockets

Release Date: August 2016
Price: $30
Number of Rotors: 4
Control System: Remote Control
Applications: Recreation
Wingspan: 100 mm
Weight: 17 gr
Frame Material: Plastic
Rotor Material: Plastic
Available Colors: Blue
Max. Flight Time: 5 minutes
Controller Frequency: 2.4 GHz
Battery Capacity: 110 mAh
Battery Type: Lithium Polymer
Battery Power: 3.7V
GPS Compatible: –
Altitude Hold: –
Position Hold: –
Flight Planning: –
Autonomous Flight: –
Target Tracking: –
Automatic Landing: –
Return To Home: –
Obstacle Avoidance: –
Headless Mode: –
Built-in Camera: –

Fleye

Flee

© Fleye

Release Date: August 2016
Price: $1,250
Number of Rotors: 1
Control System: Remote Control, Phone/Tablet
Applications: Film, Photography
Dimensions: 230 mm
Weight: 450 gr
Frame Material: Plastic
Rotor Material: Plastic
Available Colors: White
Max. Flight Time: 10 minutes
Max. Speed: 15 km/h (10 mp/h)
Max. Operating Range: 100 m
Controller Frequency: 2.4 GHz
Battery Capacity: 1,500 mAh
Battery Type: Lithium Polymer 3S
GPS Compatible: √
Altitude Hold: √
Position Hold: √
Flight Planning: √
Autonomous Flight: √
Target Tracking: √
Automatic Landing: –
Return To Home: √
Obstacle Avoidance: –
Headless Mode: –
Built-in Camera: √

GoolRC

X5C

© rcmoment.com

Release Date: August 2016
Price: $40
Number of Rotors: 4
Control System: Remote Control
Applications: Recreation, FPV Racing
Dimensions: 315 × 315 × 85 mm
Weight: 110 gr
Frame Material: Plastic
Rotor Material: Plastic
Available Colors: Red
Max. Flight Time: 5 minutes
Max. Operating Range: 50 m
Controller Frequency: 2.4 GHz
Battery Capacity: 500 mAh
Battery Type: Lithium Polymer
Battery Power: 3.7V
GPS Compatible: –
Altitude Hold: –
Position Hold: –
Flight Planning: –
Autonomous Flight: –
Target Tracking: –
Automatic Landing: –
Return To Home: √
Obstacle Avoidance: –
Headless Mode: √
Built-in Camera: √ (2 megapixels)

Gteng

T903

© hightechplace.com

Release Date: August 2016
Price: $32
Number of Rotors: 4
Control System: Remote Control
Applications: Recreation
Dimensions: 150 × 150 × 40 mm
Frame Material: Plastic
Rotor Material: Plastic
Available Colors: Green
Max. Flight Time: 8 minutes
Max. Operating Range: 100 m
Controller Frequency: 2.4 GHz
Battery Capacity: 350 mAh
Battery Type: Lithium Polymer
Battery Power: 3.7V
GPS Compatible: –
Altitude Hold: –
Position Hold: –
Flight Planning: –
Autonomous Flight: –
Target Tracking: –
Automatic Landing: –
Return To Home: √
Obstacle Avoidance: –
Headless Mode: √
Built-in Camera: –

Gteng

T905F

© rcmoment.com

Release Date: August 2016
Price: $85
Number of Rotors: 4
Control System: Remote Control
Applications: Recreation, FPV Racing
Dimensions: 321 × 313 × 100 mm
Frame Material: Plastic
Rotor Material: Plastic
Available Colors: Black
Max. Flight Time: 8 minutes
Max. Operating Range: 150 m
Controller Frequency: 2.4 GHz
Battery Capacity: 350 mAh
Battery Type: Lithium Polymer
Battery Power: 7.4V
GPS Compatible: √
Altitude Hold: √
Position Hold: √
Flight Planning: √
Autonomous Flight: √
Target Tracking: √
Automatic Landing: √
Return To Home: √
Obstacle Avoidance: √
Headless Mode: –
Built-in Camera: √ (720P HD)

Idea-Fly

Grasshopper F210

©dhgate.com

Release Date: August 2016
Price: $359
Number of Rotors: 4
Control System: Remote Control
Applications: FPV Racing
Dimensions: 205 × 240 × 90 mm
Weight: 344 gr
Frame Material: Carbon Fiber
Rotor Material: Plastic
Motors: 2205-2300kv
Available Colors: Yellow
Max. Flight Time: 16 minutes
Max. Speed: 230 km/h (143 mp/h)
Max. Flying Altitude: 800 m (2,624 ft)
Controller Frequency: 2.4 GHz
Battery Capacity: 1,300 mAh
Battery Type: Lithium Polymer
Battery Power: 14.8V
Battery Weight: 148 gr
GPS Compatible: √
Altitude Hold: √
Position Hold: –
Flight Planning: –
Autonomous Flight: –
Target Tracking: –
Automatic Landing: –
Return To Home: –
Obstacle Avoidance: –
Headless Mode: –
Built-in Camera: √ (700TVL)

JJRC

H11WH

©rcmoment.com

Release Date: August 2016
Price: $65
Number of Rotors: 4
Control System: Remote Control, Phone/Tablet
Applications: Recreation, FPV Racing
Dimensions: 200 × 200 × 47 mm
Wingspan: 270 mm
Weight: 153 gr
Frame Material: Plastic
Rotor Material: Plastic
Motors: Brushed
Available Colors: Black
Max. Flight Time: 7 minutes
Max. Operating Range: 60 m
Controller Frequency: 2.4 GHz
Battery Capacity: 1,100 mAh
Battery Type: Lithium Polymer
Battery Power: 3.7V
Battery Weight: 23 gr
GPS Compatible: –
Altitude Hold: √
Position Hold: √
Flight Planning: –
Autonomous Flight: –
Target Tracking: –
Automatic Landing: –
Return To Home: √
Obstacle Avoidance: –
Headless Mode: √
Built-in Camera: √ (2 megapixels)

JJRC

H23

© rcmoment.com

Release Date: August 2016
Price: $35
Number of Rotors: 4
Control System: Remote Control
Applications: Recreation
Dimensions: 230 × 230 × 70 mm
Weight: 114 gr
Frame Material: Plastic
Rotor Material: Plastic
Available Colors: Blue, Yellow
Max. Flight Time: 7 minutes
Controller Frequency: 2.4 GHz
Battery Capacity: 750 mAh
Battery Type: Lithium Polymer
Battery Power: 3.7V
GPS Compatible: –
Altitude Hold: –
Position Hold: –
Flight Planning: –
Autonomous Flight: –
Target Tracking: –
Automatic Landing: –
Return To Home: √
Obstacle Avoidance: –
Headless Mode: √
Built-in Camera: –

JTT UAV

T60 V1

© alibaba.com

Release Date: August 2016
Price: from $25,999
Number of Rotors: 6
Control System: Flight Computer
Applications: Film, Photography, Inspection, Surveillance
Wingspan: 1.2 m
Weight: 10.7 kg
Max. Payload: 4.5 kg
Available Colors: White
Max. Flight Time: 55 minutes
Max. Speed: 65 km/h (40 mp/h)
Max. Flying Altitude: 1,500 m (4,921 ft)
Max. Operating Range: 8 km
Controller Frequency: 2.4 GHz
Battery Capacity: 44,000 mAh
Battery Type: Lithium Polymer
GPS Compatible: √
Altitude Hold: √
Position Hold: √
Flight Planning: √
Autonomous Flight: √
Target Tracking: –
Automatic Landing: √
Return To Home: √
Obstacle Avoidance: √
Headless Mode: √
Built-in Camera: –

Kespry

Kespry Drone 2.0

©Kespry

Release Date: August 2016
Price: On Request
Number of Rotors: 4
Control System: Phone/Tablet
Applications: Film, Photography, Aerial Mapping, Surveillance
Weight: 2 kg
Frame Material: Plastic, Carbon Fiber
Rotor Material: Carbon Fiber
Available Colors: White
Max. Flight Time: 30 minutes
Max. Speed: 40 km/h (25 mp/h)
GPS Compatible: √
Altitude Hold: –
Position Hold: –
Flight Planning: –
Autonomous Flight: √
Target Tracking: –
Automatic Landing: √
Return To Home: –
Obstacle Avoidance: √
Headless Mode: –
Built-in Camera: √

L3 Unmanned Systems

HQ-60

©unmanned-aerial.com

Release Date: August 2016
Price: Unknown
Number of Rotors: 5 (Fixed Wing)
Control System: Ground Control Station
Applications: Military Purposes, Reconnaissance, Surveillance
Dimensions: 2.44 × 3.81 × 0.56 m
Wingspan: 3.81 m
Weight: 43 kg
Max. Payload: 5.4 kg
Frame Material: Composite
Rotor Material: Carbon Fiber
Available Colors: Grey
Max. Flight Time: 15 hours
Max. Speed: 75 km/h (46,6 mp/h)
Max. Flying Altitude: 4,267 m (14,000 ft)
Max. Operating Range: 1,100 km
GPS Compatible: √
Altitude Hold: √
Position Hold: √
Flight Planning: √
Autonomous Flight: √
Target Tracking: √
Automatic Landing: √
Return To Home: √
Obstacle Avoidance: √
Headless Mode: –
Built-in Camera: √

Meijiaxin Toys
X104T

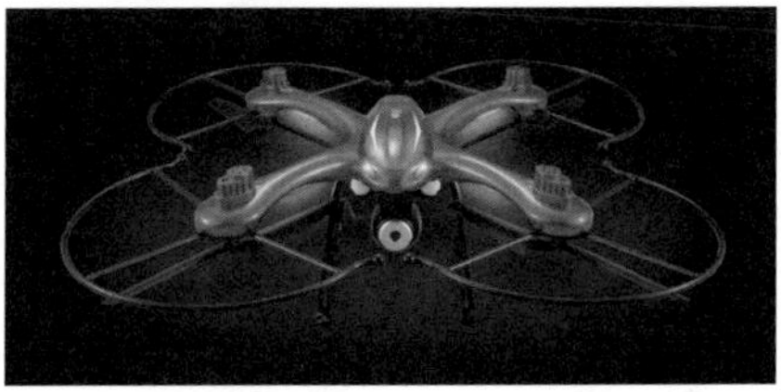

©gearbest.com

Release Date: August 2016
Price: $100
Number of Rotors: 4
Control System: Remote Control
Applications: Recreation, FPV Racing
Dimensions: 500 × 500 × 180 mm
Frame Material: Plastic
Rotor Material: Plastic
Available Colors: Red
Max. Flight Time: 15 minutes
Max. Operating Range: 100 m
Controller Frequency: 5.8 GHz
Battery Capacity: 1,200 mAh
Battery Type: Lithium Polymer
Battery Power: 7.4V
GPS Compatible: –
Altitude Hold: √
Position Hold: √
Flight Planning: –
Autonomous Flight: –
Target Tracking: –
Automatic Landing: –
Return To Home: √
Obstacle Avoidance: –
Headless Mode: √
Built-in Camera: √

Meijiaxin Toys
X301H

©hobbycenter.ru

Release Date: August 2016
Price: $55
Number of Rotors: 4
Control System: Remote Control
Applications: Recreation, FPV Racing
Dimensions: 310 × 310 × 58 mm
Frame Material: Plastic
Rotor Material: Plastic
Available Colors: Black
Max. Flight Time: 7 minutes
Max. Operating Range: 70 m
Controller Frequency: 2.4 GHz
Battery Capacity: 600 mAh
Battery Type: Lithium Polymer
Battery Power: 3.7V
GPS Compatible: –
Altitude Hold: √
Position Hold: √
Flight Planning: –
Autonomous Flight: –
Target Tracking: –
Automatic Landing: –
Return To Home: √
Obstacle Avoidance: –
Headless Mode: √
Built-in Camera: √

Nano-Racing

Nano Racer

© Olivier Regneault

Release Date: August 2016
Price: $315
Number of Rotors: 4
Control System: Remote Control
Applications: Recreation, FPV Racing
Wingspan: 120 mm
Frame Material: Carbon Fiber, Plastic
Available Colors: Blue, Grey, Red
Controller Frequency: 2.4 GHz
Battery Capacity: 700 mAh
Battery Type: Lithium Polymer 1S
Battery Power: 3.7V
GPS Compatible: √
Altitude Hold: √
Position Hold: √
Flight Planning: √
Autonomous Flight: √
Target Tracking: √
Automatic Landing: √
Return To Home: √
Obstacle Avoidance: √
Headless Mode: √
Built-in Camera: √

Nihui Model Technology

NH-010

© aliexpress.com

Release Date: August 2016
Price: $12
Number of Rotors: 4
Control System: Remote Control
Applications: Recreation
Dimensions: 80 × 80 × 30 mm
Frame Material: Plastic
Rotor Material: Plastic
Motors: Brushed
Available Colors: Red
Max. Flight Time: 5 minutes
Max. Operating Range: 30 m
Controller Frequency: 2.4 GHz
Battery Capacity: 150 mAh
Battery Type: Lithium Polymer
Battery Power: 3.7V
GPS Compatible: –
Altitude Hold: –
Position Hold: –
Flight Planning: –
Autonomous Flight: –
Target Tracking: –
Automatic Landing: –
Return To Home: √
Obstacle Avoidance: –
Headless Mode: √
Built-in Camera: –

Parrot

Disco

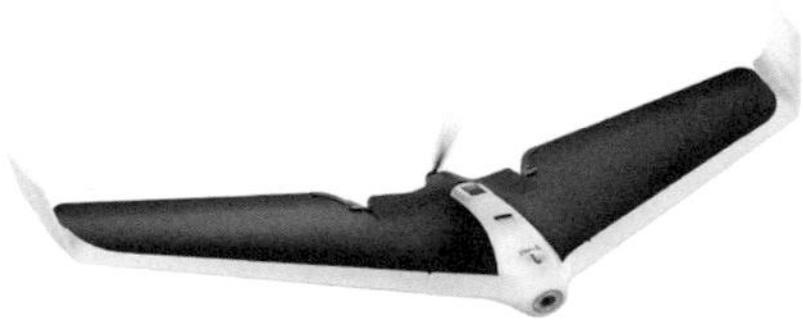

© bhphotovideo.com

Release Date: August 2016
Price: $1299
Number of Rotors: 1 (Fixed Wing)
Control System: Remote Control, Phone/Tablet
Applications: Film, Photography, FPV Racing
Dimensions: 0.58 × 1.15 × 0.12 m
Wingspan: 1.15 m
Frame Material: EPP Foam, Carbon Fiber
Rotor Material: Plastic
Motors: 1280kv
Available Colors: Black
Max. Flight Time: 45 minutes
Max. Speed: 80 km/h (49,7 mp/h)
Max. Operating Range: 2 km
Controller Frequency: 2.4 GHz
Battery Capacity: 2,700 mAh
Battery Type: Lithium Polymer 3S
GPS Compatible: √
Altitude Hold: √
Position Hold: –
Flight Planning: √
Autonomous Flight: √
Target Tracking: √
Automatic Landing: –
Return To Home: √
Obstacle Avoidance: –
Headless Mode: –
Built-in Camera: √

Parrot

Mambo

© apple.com

Release Date: August 2016
Price: $125
Number of Rotors: 4
Control System: Remote Control, Phone/Tablet
Applications: Film, Photography
Dimensions: 180 × 180 mm
Weight: 63 gr
Frame Material: Plastic
Rotor Material: Plastic
Available Colors: White
Max. Flight Time: 9 minutes
Max. Operating Range: 100 m
Controller Frequency: 2.4 GHz
Battery Capacity: 550 mAh
Battery Type: Lithium Polymer
GPS Compatible: –
Altitude Hold: –
Position Hold: –
Flight Planning: –
Autonomous Flight: –
Target Tracking: –
Automatic Landing: –
Return To Home: –
Obstacle Avoidance: –
Headless Mode: –
Built-in Camera: √

PowerVision

PowerEgg

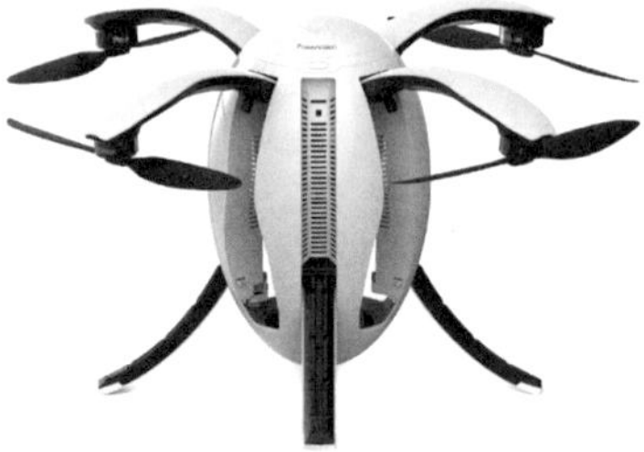

© muziker.lu

Release Date: August 2016
Price: $1,875
Number of Rotors: 4
Control System: Remote Control
Applications: Film, Photography, Recreation
Dimensions: 272 × 476 mm
Wingspan: 448 mm
Weight: 2.1 kg
Frame Material: Plastic
Rotor Material: Plastic
Available Colors: White
Max. Flight Time: 23 minutes
Max. Speed: 47 km/h (29 mp/h)
Max. Flying Altitude: 4,000 m (13,123 ft)
Max. Ascent Speed: 5 m/s
Max. Descent Speed: 2 m/s
Min. Operating Temperature: 0°C
Max. Operating Temperature: 40°C
Max. Operating Range: 5 km
Controller Frequency: 2.4 GHz
Battery Capacity: 6,400 mAh
Battery Type: Lithium Polymer
Battery Power: 14.8V
Battery Weight: 580 gr
GPS Compatible: √
Altitude Hold: √
Position Hold: √
Flight Planning: √
Autonomous Flight: √
Target Tracking: √
Automatic Landing: √
Return To Home: √
Obstacle Avoidance: –
Headless Mode: √
Built-in Camera: √

Prodrone

PD6B-AW-ARM

© Prodrone

Release Date: August 2016
Price: Unknown
Number of Rotors: 6
Control System: Remote Control
Applications: Transport
Wingspan: 1450 mm
Max. Payload: 20 kg
Frame Material: Plastic, Carbon Fiber
Rotor Material: Carbon Fiber
Available Colors: Orange
Max. Flight Time: 30 minutes
Max. Speed: 60 km/h (37 mp/h)
Max. Flying Altitude: 5,000 m (16,404 ft)
Controller Frequency: 2.4 GHz
Battery Capacity: 16,000 mAh
Battery Type: Lithium Polymer
Battery Power: 22.2V
GPS Compatible: √
Altitude Hold: √
Position Hold: √
Flight Planning: √
Autonomous Flight: –
Target Tracking: √
Automatic Landing: √
Return To Home: √
Obstacle Avoidance: –
Headless Mode: –
Built-in Camera: –

SMD Smart Drone

Mirage Pro

©SMD

Release Date: August 2016
Price: $1,249
Number of Rotors: 4
Control System: Remote Control, Phone/Tablet
Applications: Film, Photography, Aerial Mapping, Agriculture, Inspection
Frame Material: Plastic
Rotor Material: Plastic
Available Colors: Grey
Max. Flight Time: 30 minutes
Controller Frequency: 2.4 GHz
Battery Capacity: 10,000 mAh
Battery Type: Lithium Polymer
GPS Compatible: √
Altitude Hold: √
Position Hold: √
Flight Planning: –
Autonomous Flight: –
Target Tracking: –
Automatic Landing: √
Return To Home: √
Obstacle Avoidance: √
Headless Mode: –
Built-in Camera: √

Syma Toys

X8HC

©hightechtoys.nl

Release Date: August 2016
Price: $85
Number of Rotors: 4
Control System: Remote Control
Applications: Film, Photography, Recreation
Dimensions: 500 × 500 × 190 mm
Frame Material: Plastic
Rotor Material: Plastic
Available Colors: Gold
Max. Flight Time: 7 minutes
Max. Operating Range: 70 m
Controller Frequency: 2.4 GHz
Battery Capacity: 2,000 mAh
Battery Type: Lithium Polymer
Battery Power: 7.4V
GPS Compatible: –
Altitude Hold: –
Position Hold: –
Flight Planning: –
Autonomous Flight: –
Target Tracking: –
Automatic Landing: –
Return To Home: √
Obstacle Avoidance: –
Headless Mode: √
Built-in Camera: √ (2 megapixels)

Vantage Robotics
Snap

©wired.com

Release Date: August 2016
Price: $1,295
Number of Rotors: 4
Control System: Remote Control, Phone/Tablet
Applications: Film, Photography
Dimensions: 317 × 335 × 63 mm
Weight: 570 gr
Max. Payload: 200 gr
Frame Material: Plastic
Rotor Material: Plastic
Available Colors: White
Max. Flight Time: 20 minutes
Max. Speed: 54 km/h (33,5 mp/h)
Max. Flying Altitude: 5,000 m (16,404 ft)
Max. Operating Range: 1.5 km
Controller Frequency: 2.4 GHz
Battery Capacity: 3,000 mAh
Battery Type: Lithium Polymer 4S
Battery Power: 14.8V
Battery Weight: 220 gr
GPS Compatible: √
Altitude Hold: √
Position Hold: √
Flight Planning: √
Autonomous Flight: –
Target Tracking: √
Automatic Landing: √
Return To Home: √
Obstacle Avoidance: √
Headless Mode: –
Built-in Camera: √ (Sony Exmor IMX 377)

Wltoys
Q333

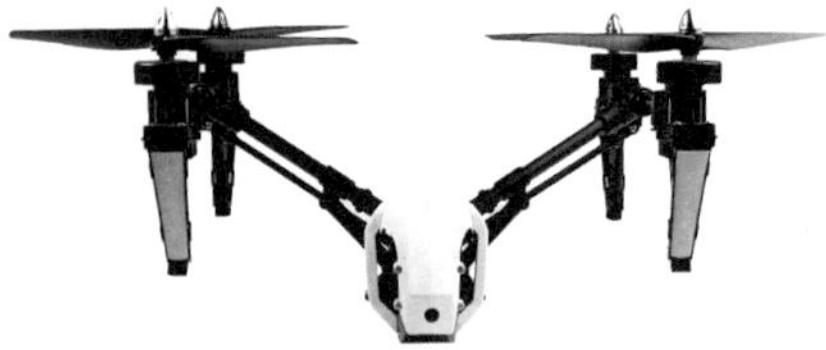

©aliexpress.com

Release Date: August 2016
Price: $115
Number of Rotors: 4
Control System: Remote Control
Applications: FPV Racing
Dimensions: 450 × 450 × 140 mm
Weight: 1.6 kg
Frame Material: Plastic
Rotor Material: Plastic
Available Colors: White
Max. Flight Time: 8 minutes
Max. Operating Range: 100 m
Controller Frequency: 5.8 GHz
Battery Capacity: 2,000 mAh
Battery Type: Lithium Polymer
Battery Power: 7.4V
GPS Compatible: –
Altitude Hold: √
Position Hold: √
Flight Planning: –
Autonomous Flight: –
Target Tracking: –
Automatic Landing: –
Return To Home: √
Obstacle Avoidance: –
Headless Mode: √
Built-in Camera: √ (0.3 megapixels)

Airhogs

Helix Sentinel

©Airhogs

Release Date: September 2016
Price: $200
Number of Rotors: 4
Control System: Remote Control, Phone/Tablet
Applications: Recreation
Frame Material: EPP Foam
Rotor Material: Plastic
Available Colors: Black
Controller Frequency: 2.4 GHz
Battery Type: Lithium Polymer
GPS Compatible: –
Altitude Hold: –
Position Hold: –
Flight Planning: –
Autonomous Flight: –
Target Tracking: –
Automatic Landing: –
Return To Home: √
Obstacle Avoidance: √
Headless Mode: –
Built-in Camera: √ (720P HD)

Applied Aeronautics

Albatross

©prweb.com

Release Date: September 2016
Price: $6,100
Number of Rotors: 1 (Fixed Wing)
Control System: Ground Control Station
Applications: Aerial Mapping, Inspection, Reconnaissance, Surveillance
Wingspan: 3 m
Weight: 6 kg
Max. Payload: 4.4 kg
Frame Material: Composite
Available Colors: White
Max. Flight Time: 5 hours
Max. Speed: 145 km/h (90 mp/h)
Max. Operating Range: 100 km
GPS Compatible: √
Altitude Hold: √
Position Hold: –
Flight Planning: √
Autonomous Flight: √
Target Tracking: –
Automatic Landing: √
Return To Home: √
Obstacle Avoidance: –
Headless Mode: –
Built-in Camera: √

Blade

Conspiracy 220

© Horizon Hobby

Release Date: September 2016
Price: $399
Number of Rotors: 4
Control System: Remote Control
Applications: Recreation, FPV Racing
Dimensions: 180 × 180 mm
Weight: 550 gr
Frame Material: Carbon Fiber
Rotor Material: Plastic
Available Colors: Green
Max. Flight Time: 5 minutes
Controller Frequency: 5.8 GHz
Battery Capacity: 1,300 mAh
Battery Type: Lithium Polymer 4S
GPS Compatible: –
Altitude Hold: –
Position Hold: –
Flight Planning: –
Autonomous Flight: –
Target Tracking: –
Automatic Landing: –
Return To Home: –
Obstacle Avoidance: –
Headless Mode: –
Built-in Camera: √ (650TVL)

Carrera RC

Mario Copter

© ociostock.com

Release Date: September 2016
Price: $99
Number of Rotors: 4
Control System: Remote Control
Applications: Recreation
Dimensions: 380 × 230 × 510 mm
Weight: 82 gr
Max. Payload: 9 gr
Frame Material: Plastic
Rotor Material: Plastic
Available Colors: Black
Controller Frequency: 2.4 GHz
Battery Capacity: 1,000 mAh
Battery Type: Lithium Polymer
Battery Power: 3.7V
GPS Compatible: –
Altitude Hold: –
Position Hold: √
Flight Planning: √
Autonomous Flight: –
Target Tracking: –
Automatic Landing: –
Return To Home: √
Obstacle Avoidance: –
Headless Mode: –
Built-in Camera: –

DJI

Mavic Pro

© DJI

Release Date: September 2016
Price: $999
Number of Rotors: 4
Control System: Remote Control, Phone/Tablet
Applications: Film, Photography
Wingspan: 335 mm
Weight: 743 gr
Frame Material: Aluminum, Fiberglass, Plastic, Nylon
Rotor Material: Carbon Fiber
Available Colors: Grey
Max. Flight Time: 27 minutes
Max. Speed: 65 km/h (40 mp/h)
Max. Flying Altitude: 5,000 m (16,404 ft)
Max. Ascent Speed: 5 m/s
Max. Descent Speed: 3 m/s
Min. Operating Temperature: 0°C
Max. Operating Temperature: 40°C
Max. Operating Range: 7 km
Controller Frequency: 2.4 GHz
Battery Capacity: 3,830 mAh
Battery Type: Lithium Polymer 3S
Battery Power: 11.4V
Battery Weight: 240 gr
GPS Compatible: √
Altitude Hold: √
Position Hold: √
Flight Planning: √
Autonomous Flight: √
Target Tracking: √
Automatic Landing: √
Return To Home: √
Obstacle Avoidance: √
Headless Mode: √
Built-in Camera: √ (12 megapixels)

Eachine

E010 mini

© Phaser Computers

Release Date: September 2016
Price: $25
Number of Rotors: 4
Control System: Remote Control
Applications: FPV Racing
Dimensions: 80 × 80 × 50 mm
Frame Material: Plastic
Rotor Material: Plastic
Available Colors: Green, Red
Max. Flight Time: 5 minutes
Max. Operating Range: 30 m
Controller Frequency: 2.4 GHz
Battery Capacity: 150 mAh
Battery Type: Lithium Polymer
Battery Power: 3.7V
GPS Compatible: –
Altitude Hold: –
Position Hold: –
Flight Planning: –
Autonomous Flight: –
Target Tracking: –
Automatic Landing: –
Return To Home: √
Obstacle Avoidance: –
Headless Mode: √
Built-in Camera: –

Eachine

Falcon 250 Pro

© banggood.com

Release Date: September 2016
Price: $299
Number of Rotors: 4
Control System: Remote Control
Applications: FPV Racing
Dimensions: 240 × 190 × 70 mm
Weight: 535 gr
Frame Material: Carbon Fiber
Rotor Material: Plastic
Available Colors: Black
Controller Frequency: 2.4 GHz
Battery Capacity: 1,500 mAh
Battery Type: Lithium Polymer 3S
Battery Power: 11.1V
Battery Weight: 7 gr
GPS Compatible: –
Altitude Hold: –
Position Hold: –
Flight Planning: –
Autonomous Flight: –
Target Tracking: –
Automatic Landing: –
Return To Home: –
Obstacle Avoidance: –
Headless Mode: –
Built-in Camera: √ (700TVL)

GoolRC

T5G

© aliexpress.com

Release Date: September 2016
Price: $53
Number of Rotors: 4
Control System: Remote Control
Applications: Recreation, FPV Racing
Dimensions: 285 × 285 × 95 mm
Weight: 96 gr
Frame Material: Plastic
Rotor Material: Plastic
Available Colors: Blue, Red
Max. Flight Time: 5 minutes
Max. Operating Range: 50 m
Controller Frequency: 2.4 GHz
Battery Capacity: 450 mAh
Battery Type: Lithium Polymer
Battery Power: 3.7V
GPS Compatible: –
Altitude Hold: –
Position Hold: –
Flight Planning: –
Autonomous Flight: –
Target Tracking: –
Automatic Landing: –
Return To Home: √
Obstacle Avoidance: –
Headless Mode: √
Built-in Camera: √ (2 megapixels)

HSE

AG-V6A Plus

©HSE

Release Date: September 2016
Price: On Request
Number of Rotors: 6
Control System: Remote Control
Applications: Agriculture
Wingspan: 1.63 m
Weight: 34.6 kg
Frame Material: Carbon Fiber
Available Colors: Black
Max. Flight Time: 50 minutes
Max. Speed: 33 km/h (20 mp/h)
Controller Frequency: 2.4 GHz
Battery Type: Lithium Polymer
GPS Compatible: √
Altitude Hold: √
Position Hold: √
Flight Planning: √
Autonomous Flight: √
Target Tracking: –
Automatic Landing: √
Return To Home: √
Obstacle Avoidance: –
Headless Mode: –
Built-in Camera: –

Jin Xing Da

506G/W

©aliexpress.com

Release Date: September 2016
Price: $75
Number of Rotors: 4
Control System: Remote Control
Applications: Recreation
Dimensions: 390 × 390 × 150 mm
Weight: 550 gr
Frame Material: Plastic
Rotor Material: Plastic
Available Colors: Black
Max. Operating Range: 300 m
Controller Frequency: 5.8 GHz
Battery Capacity: 2,000 mAh
Battery Type: Lithium Polymer
Battery Power: 7.4V
GPS Compatible: –
Altitude Hold: √
Position Hold: √
Flight Planning: –
Autonomous Flight: √
Target Tracking: –
Automatic Landing: √
Return To Home: √
Obstacle Avoidance: –
Headless Mode: √
Built-in Camera: √ (2 megapixels)

JJRC

H36

Release Date: September 2016
Price: $17
Number of Rotors: 4
Control System: Remote Control
Applications: Recreation
Dimensions: 82 × 82 × 30 mm
Weight: 22 gr
Frame Material: Plastic
Rotor Material: Plastic
Available Colors: Black, Blue
Max. Flight Time: 6 minutes
Max. Operating Range: 30 m
Controller Frequency: 2.4 GHz
Battery Capacity: 150 mAh
Battery Type: Lithium Polymer
Battery Power: 3.7V
GPS Compatible: –
Altitude Hold: –
Position Hold: –
Flight Planning: –
Autonomous Flight: –
Target Tracking: –
Automatic Landing: –
Return To Home: √
Obstacle Avoidance: –
Headless Mode: √
Built-in Camera: –

Matternet

Matternet Two

Release Date: September 2016
Price: On Request
Number of Rotors: 4
Control System: Remote Control
Applications: Delivery, Transport
Dimensions: 800 × 800 × 260 mm
Weight: 9.5 kg
Max. Payload: 2 kg
Available Colors: White
Max. Speed: 36 km/h (22 mp/h)
Min. Operating Temperature: -10°C
Max. Operating Temperature: 40°C
Max. Operating Range: 20 km
Controller Frequency: 2.4 GHz
Battery Type: Lithium Polymer
GPS Compatible: √
Altitude Hold: √
Position Hold: √
Flight Planning: √
Autonomous Flight: √
Target Tracking: √
Automatic Landing: √
Return To Home: √
Obstacle Avoidance: –
Headless Mode: –
Built-in Camera: –

Meijiaxin Toys

X906T

©chinabrands.com

Release Date: September 2016
Price: $79
Number of Rotors: 4
Control System: Remote Control
Applications: Recreation, FPV Racing
Dimensions: 137 × 137 × 35 mm
Frame Material: Plastic
Rotor Material: Plastic
Available Colors: Black
Max. Flight Time: 6 minutes
Max. Operating Range: 100 m
Controller Frequency: 2.4 GHz
Battery Capacity: 380 mAh
Battery Type: Lithium Polymer
Battery Power: 3.7V
GPS Compatible: –
Altitude Hold: –
Position Hold: –
Flight Planning: –
Autonomous Flight: –
Target Tracking: –
Automatic Landing: –
Return To Home: √
Obstacle Avoidance: –
Headless Mode: –
Built-in Camera: √ (720P HD)

Meijiaxin Toys

X908T

©helicomicro.com

Release Date: September 2016
Price: $70
Number of Rotors: 4
Control System: Remote Control
Applications: Recreation, FPV Racing
Dimensions: 128 × 128 × 41 mm
Frame Material: Plastic
Rotor Material: Plastic
Available Colors: Black
Max. Flight Time: 6 minutes
Max. Operating Range: 100 m
Controller Frequency: 2.4 GHz
Battery Capacity: 380 mAh
Battery Type: Lithium Polymer
Battery Power: 3.7V
GPS Compatible: –
Altitude Hold: –
Position Hold: –
Flight Planning: –
Autonomous Flight: –
Target Tracking: –
Automatic Landing: –
Return To Home: √
Obstacle Avoidance: –
Headless Mode: –
Built-in Camera: √ (720P HD)

Novadem

U130

© Novadem

Release Date: September 2016
Price: On Request
Number of Rotors: 4
Control System: Remote Control
Applications: Aerial Mapping, Inspection, Thermography
Dimensions: 1.3 × 1.3 × 0.2 m
Weight: 2.2 kg
Max. Payload: 800 gr
Frame Material: Carbon Fiber, Plastic
Rotor Material: Carbon Fiber
Available Colors: Yellow
Max. Flight Time: 20 minutes
Max. Speed: 43 km/h (27 mp/h)
Max. Flying Altitude: 2,200 m (7,218 ft)
Controller Frequency: 2.4 GHz
Battery Type: Lithium Polymer
GPS Compatible: √
Altitude Hold: √
Position Hold: √
Flight Planning: √
Autonomous Flight: √
Target Tracking: –
Automatic Landing: √
Return To Home: √
Obstacle Avoidance: √
Headless Mode: –
Built-in Camera: √

PowerVision

PowerEye

© bhphotovideo.com

Release Date: September 2016
Price: $2,700
Number of Rotors: 4
Control System: Remote Control, Phone
Applications: Film, Photography, Aerial Mapping
Dimensions: 513 × 513 × 310 mm
Wingspan: 650 mm
Weight: 3.95 kg
Frame Material: Carbon Fiber, Plastic
Rotor Material: Plastic
Available Colors: Black
Max. Flight Time: 30 minutes
Max. Speed: 65 km/h (40 mp/h)
Max. Flying Altitude: 4,000 m (13,123 ft)
Max. Descent Speed: 5 m/s
Min. Operating Temperature: -10°C
Max. Operating Temperature: 40°C
Max. Operating Range: 5 km
Controller Frequency: 2.4 GHz
Battery Capacity: 9,000 mAh
Battery Type: Lithium Polymer
Battery Power: 22.2V
Battery Weight: 1.23 kg
GPS Compatible: √
Altitude Hold: √
Position Hold: √
Flight Planning: √
Autonomous Flight: √
Target Tracking: √
Automatic Landing: √
Return To Home: √
Obstacle Avoidance: –
Headless Mode: √
Built-in Camera: –

Prodrone

PD6-CI-L

©Prodrone

Release Date: September 2016
Price: On Request
Number of Rotors: 6
Control System: Remote Control
Applications: Inspection
Wingspan: 600 mm
Weight: 6 kg
Max. Payload: 2 kg
Available Colors: Black
Max. Flight Time: 10 minutes
Max. Speed: 20 km/h (12 mp/h)
Controller Frequency: 2.4 GHz
Battery Capacity: 6,000 mAh
Battery Type: Lithium Polymer
GPS Compatible: –
Altitude Hold: √
Position Hold: √
Flight Planning: –
Autonomous Flight: –
Target Tracking: –
Automatic Landing: –
Return To Home: √
Obstacle Avoidance: √
Headless Mode: –
Built-in Camera: √

QuadH2o

SarahH2o

©QuadH2o

Release Date: September 2016
Price: $4,795
Number of Rotors: 6
Control System: Remote Control
Applications: Inspection, Search & Rescue
Dimensions: 650 × 740 × 240 mm
Weight: 4.8 kg
Max. Payload: 1 kg
Frame Material: Composite
Available Colors: White
Max. Flight Time: 25 minutes
Max. Speed: 56 km/h (35 mp/h)
Max. Ascent Speed: 6 m/s
Max. Descent Speed: 4.5 m/s
Min. Operating Temperature: -10°C
Max. Operating Temperature: 50°C
Max. Operating Range: 1 km
Controller Frequency: 2.4 GHz
Battery Capacity: 7,000 mAh
Battery Type: Lithium Polymer 6S
GPS Compatible: √
Altitude Hold: √
Position Hold: √
Flight Planning: √
Autonomous Flight: √
Target Tracking: –
Automatic Landing: √
Return To Home: √
Obstacle Avoidance: –
Headless Mode: –
Built-in Camera: –

Rotor X

Atom RX122

© Rotor X

Release Date: September 2016
Price: $750
Number of Rotors: 4
Control System: Remote Control
Applications: FPV Racing
Wingspan: 122 mm
Weight: 250 gr
Frame Material: Carbon Fiber, Plastic
Rotor Material: Plastic
Available Colors: Black, Blue, Grey, Red, White, Yellow
Max. Speed: 100 km/h (62 mp/h)
Controller Frequency: 5.8 GHz
Battery Capacity: 6,400 mAh
Battery Type: Lithium Polymer
GPS Compatible: –
Altitude Hold: –
Position Hold: –
Flight Planning: –
Autonomous Flight: –
Target Tracking: –
Automatic Landing: –
Return To Home: –
Obstacle Avoidance: –
Headless Mode: –
Built-in Camera: √ (600TVL)

SMD Smart Drone

Red Arrow

© droneflyers.wordpress.com

Release Date: September 2016
Price: $495
Number of Rotors: 4
Control System: Remote Control, Phone/Tablet
Applications: FPV Racing
Wingspan: 250 mm
Weight: 550 gr
Frame Material: Plastic
Rotor Material: Plastic
Motors: 2204-2200kv brushless
Available Colors: Red
Max. Flight Time: 15 minutes
Max. Speed: 72 km/h (44,7 mp/h)
Max. Flying Altitude: 100 m (328 ft)
Max. Ascent Speed: 3 m/s
Max. Descent Speed: 2 m/s
Min. Operating Temperature: -10°C
Max. Operating Temperature: 45°C
Max. Operating Range: 300 m
Controller Frequency: 2.4 GHz
Battery Capacity: 2,200 mAh
Battery Type: Lithium Polymer 3S
Battery Power: 12V
GPS Compatible: –
Altitude Hold: –
Position Hold: –
Flight Planning: –
Autonomous Flight: –
Target Tracking: –
Automatic Landing: –
Return To Home: –
Obstacle Avoidance: –
Headless Mode: –
Built-in Camera: √ (720P HD)

UDIRC Toys

U45 Raven

©usatoyz.com

Release Date: September 2016
Price: $60
Number of Rotors: 4
Control System: Remote Control
Applications: Recreation, FPV Racing
Dimensions: 400 × 260 × 89 mm
Weight: 120 gr
Frame Material: Plastic
Rotor Material: Plastic
Available Colors: Black
Max. Flight Time: 6 minutes
Controller Frequency: 2.4 GHz
Battery Type: Lithium Polymer 2S
GPS Compatible: –
Altitude Hold: √
Position Hold: √
Flight Planning: –
Autonomous Flight: –
Target Tracking: –
Automatic Landing: √
Return To Home: √
Obstacle Avoidance: –
Headless Mode: √
Built-in Camera: √

Vertical Technologies

DeltaQuad One

©Vertical Technologies

Release Date: September 2016
Price: $8,695
Number of Rotors: 5 (Fixed Wing)
Control System: Remote Control, Ground Control Station
Applications: Aerial Mapping, Inspection, Surveillance
Dimensions: 0.9 × 2.35 m
Wingspan: 2.35 m
Weight: 4.9 kg
Max. Payload: 1 kg
Frame Material: EPP Foam
Available Colors: Black
Max. Flight Time: 2 hours
Max. Speed: 100 km/h (62 mp/h)
Min. Operating Temperature: -20°C
Max. Operating Temperature: 40°C
Max. Operating Range: 120 km
Controller Frequency: 2.4 GHz
Battery Capacity: 23,000 mAh
Battery Type: Lithium Polymer 4S
GPS Compatible: √
Altitude Hold: √
Position Hold: √
Flight Planning: √
Autonomous Flight: √
Target Tracking: √
Automatic Landing: √
Return To Home: √
Obstacle Avoidance: √
Headless Mode: √
Built-in Camera: √

Yi Zhan Toys

i3

© amazon.com

Release Date: September 2016
Price: $30
Number of Rotors: 4
Control System: Remote Control
Applications: Recreation, FPV Racing
Dimensions: 165 × 160 × 38 mm
Frame Material: Plastic
Rotor Material: Plastic
Available Colors: Black
Max. Flight Time: 7 minutes
Max. Operating Range: 50 m
Controller Frequency: 2.4 GHz
Battery Capacity: 350 mAh
Battery Type: Lithium Polymer
Battery Power: 3.7V
GPS Compatible: –
Altitude Hold: –
Position Hold: –
Flight Planning: –
Autonomous Flight: –
Target Tracking: –
Automatic Landing: –
Return To Home: √
Obstacle Avoidance: –
Headless Mode: √
Built-in Camera: √

Airhogs

Hyper Stunt

© Airhogs

Release Date: October 2016
Price: $35
Number of Rotors: 4
Control System: Remote Control
Applications: Recreation
Dimensions: 80 × 80 × 75 mm
Weight: 250 gr
Frame Material: Plastic
Rotor Material: Plastic
Available Colors: Red, Yellow
Max. Flight Time: 5 minutes
Max. Operating Range: 50 m
Controller Frequency: 2.4 GHz
Battery Capacity: 140 mAh
Battery Type: Lithium Polymer
Battery Power: 3.7V
GPS Compatible: –
Altitude Hold: –
Position Hold: –
Flight Planning: –
Autonomous Flight: –
Target Tracking: –
Automatic Landing: –
Return To Home: –
Obstacle Avoidance: –
Headless Mode: –
Built-in Camera: –

AltiUAS

Transition

© AltiUAS

Release Date: October 2016
Price: On Request
Number of Rotors: 5 (Fixed Wing)
Control System: Ground Control Station
Applications: Recreation
Dimensions: 2.3 × 3 × 0.49 m
Wingspan: 3 m
Weight: 12 kg
Max. Payload: 1 kg
Frame Material: Carbon Fiber
Available Colors: Grey, Orange
Max. Flight Time: 10 hours
GPS Compatible: √
Altitude Hold: √
Position Hold: √
Flight Planning: √
Autonomous Flight: √
Target Tracking: –
Automatic Landing: √
Return To Home: √
Obstacle Avoidance: –
Headless Mode: –
Built-in Camera: √

Blade

Inductrix FPV

© Horizon Hobby

Release Date: October 2016
Price: $115
Number of Rotors: 4
Control System: Remote Control
Applications: Recreation, FPV Racing
Dimensions: 83 × 83 × 28 mm
Weight: 24 gr
Frame Material: Plastic
Rotor Material: Plastic
Available Colors: Yellow
Max. Flight Time: 4 minutes
Max. Operating Range: 150 m
Controller Frequency: 5.8 GHz
Battery Capacity: 200 mAh
Battery Type: Lithium Polymer 1S
Battery Power: 3.7V
GPS Compatible: –
Altitude Hold: –
Position Hold: –
Flight Planning: –
Autonomous Flight: –
Target Tracking: –
Automatic Landing: –
Return To Home: –
Obstacle Avoidance: –
Headless Mode: –
Built-in Camera: √

Blade

Theory W

© Horizon Hobby

Release Date: October 2016
Price: $250
Number of Rotors: 1 (Fixed Wing)
Control System: Remote Control
Applications: Recreation, FPV Racing
Dimensions: 360 × 760 mm
Wingspan: 760 mm
Weight: 480 gr
Frame Material: Z-Foam
Rotor Material: Plastic
Motors: 2205-2350kv
Available Colors: White
Max. Flight Time: 7 minutes
Controller Frequency: 5.8 GHz
Battery Capacity: 1,300 mAh
Battery Type: Lithium Polymer 4S
GPS Compatible: –
Altitude Hold: –
Position Hold: –
Flight Planning: –
Autonomous Flight: –
Target Tracking: –
Automatic Landing: –
Return To Home: –
Obstacle Avoidance: –
Headless Mode: –
Built-in Camera: √

Dromida

Hovershot

© Amain Hobbies

Release Date: October 2016
Price: $100
Number of Rotors: 4
Control System: Remote Control, Phone
Applications: Recreation, FPV Racing
Wingspan: 120 mm
Weight: 72 gr
Frame Material: Plastic
Rotor Material: Plastic
Available Colors: Grey
Max. Flight Time: 10 minutes
Controller Frequency: 2.4 GHz
Battery Capacity: 650 mAh
Battery Type: Lithium Polymer
Battery Power: 3.7V
GPS Compatible: –
Altitude Hold: √
Position Hold: √
Flight Planning: –
Autonomous Flight: –
Target Tracking: –
Automatic Landing: √
Return To Home: √
Obstacle Avoidance: –
Headless Mode: –
Built-in Camera: √ (1 megapixel)

Dronetools

DroneQuad

© Dronetools

Release Date: October 2016
Price: On Request
Number of Rotors: 4
Control System: Flight Computer, Laptop
Applications: Film, Photography
Frame Material: Carbon Fiber
Rotor Material: Carbon Fiber
Available Colors: White
Max. Flight Time: 6 hours
Max. Operating Range: 1.7 km
Controller Frequency: 5.8 GHz
Battery Capacity: 7,000 mAh
Battery Type: Lithium Polymer
GPS Compatible: √
Altitude Hold: √
Position Hold: √
Flight Planning: √
Autonomous Flight: √
Target Tracking: –
Automatic Landing: √
Return To Home: √
Obstacle Avoidance: –
Headless Mode: –
Built-in Camera: –

Eachine

QX95

© studiosport.fr

Release Date: October 2016
Price: $100
Number of Rotors: 4
Control System: Remote Control
Applications: FPV Racing
Dimensions: 125 × 85 × 55 mm
Weight: 56 gr
Frame Material: Carbon Fiber
Rotor Material: Plastic
Motors: 8520 brush motor
Available Colors: Black
Max. Flight Time: 6 minutes
Controller Frequency: 5.8 GHz
Battery Capacity: 600 mAh
Battery Type: Lithium Polymer
Battery Power: 3.7V
GPS Compatible: –
Altitude Hold: –
Position Hold: –
Flight Planning: –
Autonomous Flight: –
Target Tracking: –
Automatic Landing: –
Return To Home: –
Obstacle Avoidance: –
Headless Mode: –
Built-in Camera: √ (520P HD)

Eznov

Ez Manta

© Eznov

Release Date: October 2016
Price: Unknown
Number of Rotors: 3 (Fixed Wing)
Control System: Remote Control, Phone/Tablet
Applications: Aerial Mapping, Agriculture, Inspection, Surveillance
Frame Material: Carbon Fiber, EPP Foam
Available Colors: Black
Battery Type: Lithium Polymer
GPS Compatible: –
Altitude Hold: √
Position Hold: √
Flight Planning: √
Autonomous Flight: √
Target Tracking: √
Automatic Landing: √
Return To Home: √
Obstacle Avoidance: –
Headless Mode: –
Built-in Camera: –

Fuqi Toys

FQ777-03 Tiny Buddy

© geekbuying.com

Release Date: October 2016
Price: $20
Number of Rotors: 4
Control System: Remote Control
Applications: Recreation
Dimensions: 220 × 220 × 210 mm
Weight: 8 gr
Frame Material: Plastic
Rotor Material: Plastic
Available Colors: White
Max. Flight Time: 5 minutes
Max. Operating Range: 50 m
Controller Frequency: 2.4 GHz
Battery Capacity: 80 mAh
Battery Type: Lithium Polymer
Battery Power: 3.7V
GPS Compatible: –
Altitude Hold: –
Position Hold: –
Flight Planning: –
Autonomous Flight: –
Target Tracking: –
Automatic Landing: –
Return To Home: –
Obstacle Avoidance: –
Headless Mode: √
Built-in Camera: –

General Atomics

Avenger ER

Release Date: October 2016
Price: Unknown
Number of Rotors: – (Fixed Wing)
Control System: Ground Control Station
Applications: Military Purposes, Armed Reconnaissance, Surveillance, Target Acquisition
Dimensions: 13.4 × 23.2 m
Wingspan: 23.2 m
Weight: 8,800 kg
Frame Material: Composite
Motors: 1 × Pratt & Whitney Canada PW545B turbofan
Available Colors: Grey
Max. Flight Time: 20 hours
Max. Speed: 740 km/h (460 mp/h)
Max. Flying Altitude: 15,240 m (50,000 ft)
GPS Compatible: √
Altitude Hold: √
Position Hold: –
Flight Planning: √
Autonomous Flight: √
Target Tracking: √
Automatic Landing: √
Return To Home: √
Obstacle Avoidance: √
Headless Mode: –
Built-in Camera: √

General Atomics

MQ-1C Gray Eagle ER

Release Date: October 2016
Price: Unknown
Number of Rotors: – (Fixed Wing)
Control System: Ground Control Station
Applications: Military Purposes, Armed Reconnaissance, Surveillance, Target Acquisition
Dimensions: 8.53 × 17 × 2.1 m
Wingspan: 17 m
Weight: 1,905 kg
Frame Material: Composite
Motors: 1 × Thielert Centurion 1.7 heavy-fuel engine
Available Colors: Grey
Max. Flight Time: 42 hours
Max. Speed: 309 km/h (192 mp/h)
Max. Flying Altitude: 8,840 m (29,000 ft)
GPS Compatible: √
Altitude Hold: √
Position Hold: –
Flight Planning: √
Autonomous Flight: √
Target Tracking: √
Automatic Landing: √
Return To Home: √
Obstacle Avoidance: √
Headless Mode: –
Built-in Camera: √

Héliceo

Fox4

© Hélicéo

Release Date: October 2016
Price: On Request
Number of Rotors: 4
Control System: Flight Computer
Applications: Film, Photography, Aerial Mapping, Inspection
Dimensions: 860 × 1050 × 530 mm
Weight: 4 kg
Max. Payload: 300 gr
Frame Material: Aluminum, Carbon Fiber
Available Colors: White
Max. Flight Time: 25 minutes
Max. Speed: 50 km/h (31 mp/h)
Max. Flying Altitude: 2,000 m (6,560 ft)
Min. Operating Temperature: -10°C
Max. Operating Temperature: 45°C
Max. Operating Range: 10 km
Controller Frequency: 2.4 GHz
Battery Capacity: 5,000 mAh
Battery Type: Lithium Polymer
GPS Compatible: √
Altitude Hold: √
Position Hold: √
Flight Planning: √
Autonomous Flight: √
Target Tracking: √
Automatic Landing: √
Return To Home: √
Obstacle Avoidance: –
Headless Mode: –
Built-in Camera: √

Hélicéo

FoxyPro

© steemitimages.com

Release Date: October 2016
Price: On Request
Number of Rotors: 1 (Fixed Wing)
Control System: Ground Control Station
Applications: Aerial Mapping, Inspection
Dimensions: 1.4 × 3.5 m
Wingspan: 3.5 m
Weight: 9 kg
Max. Payload: 2 kg
Frame Material: Composite
Available Colors: Orange
Max. Flight Time: 60 minutes
Max. Speed: 140 km/h (87 mp/h)
Max. Flying Altitude: 5,000 m (16,404 ft)
Min. Operating Temperature: -10°C
Max. Operating Temperature: 45°C
Max. Operating Range: 60 km
Controller Frequency: 2.4 GHz
Battery Capacity: 10,000 mAh
Battery Type: Lithium Polymer
GPS Compatible: √
Altitude Hold: √
Position Hold: –
Flight Planning: √
Autonomous Flight: √
Target Tracking: √
Automatic Landing: √
Return To Home: √
Obstacle Avoidance: –
Headless Mode: –
Built-in Camera: √

Idea-Fly

Poseidon 480

© rcmoment.com

Release Date: October 2016
Price: $1,565
Number of Rotors: 4
Control System: Remote Control
Applications: Film, Photography, FPV Racing
Dimensions: 400 × 400 × 290 mm
Weight: 2.74 kg
Max. Payload: 1.2 kg
Frame Material: Carbon Fiber, Plastic
Rotor Material: Carbon Fiber
Available Colors: White
Controller Frequency: 2.4 GHz
Battery Capacity: 4,500 mAh
Battery Type: Lithium Polymer 6S
Battery Power: 22.2V
GPS Compatible: √
Altitude Hold: √
Position Hold: √
Flight Planning: √
Autonomous Flight: √
Target Tracking: –
Automatic Landing: √
Return To Home: √
Obstacle Avoidance: –
Headless Mode: –
Built-in Camera: √ (700TVL)

Intel

Falcon 8+

© Intel

Release Date: October 2016
Price: $29,999
Number of Rotors: 8
Control System: Flight Computer, Remote Control
Applications: Aerial Mapping, Inspection, Surveillance
Dimensions: 768 × 817 × 160 mm
Weight: 1.2 kg
Max. Payload: 800 gr
Frame Material: Carbon Fiber
Available Colors: Yellow
Max. Flight Time: 26 minutes
Max. Speed: 65 km/h (40 mp/h)
Max. Flying Altitude: 4,000 m (13,123 ft)
Min. Operating Temperature: -5°C
Max. Operating Temperature: 40°C
Max. Operating Range: 1 km
Controller Frequency: 2.4 GHz
Battery Type: Lithium Polymer
GPS Compatible: √
Altitude Hold: √
Position Hold: √
Flight Planning: √
Autonomous Flight: √
Target Tracking: –
Automatic Landing: √
Return To Home: √
Obstacle Avoidance: –
Headless Mode: –
Built-in Camera: √

JJRC

JJPRO P130

©banggood.com

Release Date: October 2016
Price: $130
Number of Rotors: 4
Control System: Remote Control
Applications: FPV Racing
Dimensions: 160 × 180 × 75 mm
Weight: 197 gr
Frame Material: Carbon Fiber
Rotor Material: Plastic
Motors: 1806-2300kv
Available Colors: Green, Purple, Yellow
Controller Frequency: 5.8 GHz
Battery Capacity: 850 mAh
Battery Type: Lithium Polymer 3S
Battery Power: 11.1V
GPS Compatible: –
Altitude Hold: –
Position Hold: –
Flight Planning: –
Autonomous Flight: –
Target Tracking: –
Automatic Landing: –
Return To Home: –
Obstacle Avoidance: –
Headless Mode: –
Built-in Camera: √ (800TVL)

JJRC

JJPRO X1G

©aliexpress.com

Release Date: October 2016
Price: $150
Number of Rotors: 4
Control System: Remote Control
Applications: FPV Racing
Dimensions: 340 × 340 × 115 mm
Weight: 254 gr
Frame Material: Plastic
Rotor Material: Plastic
Motors: 1806-2300kv
Available Colors: Black
Max. Flight Time: 15 minutes
Max. Operating Range: 200 m
Controller Frequency: 2.4 GHz
Battery Capacity: 1,300 mAh
Battery Type: Lithium Polymer
Battery Power: 7.4V
GPS Compatible: –
Altitude Hold: –
Position Hold: –
Flight Planning: –
Autonomous Flight: –
Target Tracking: –
Automatic Landing: –
Return To Home: –
Obstacle Avoidance: –
Headless Mode: –
Built-in Camera: √ (600TVL)

RC Logger

RC Eye Twist 300

©droneart.com

Release Date: October 2016
Price: $888
Number of Rotors: 4
Control System: Remote Control
Applications: Recreation
Wingspan: 320 mm
Frame Material: Carbon Fiber
Rotor Material: Plastic
Motors: 2212-1200kv brushless
Available Colors: Black
Max. Flight Time: 10 minutes
Controller Frequency: 2.4 GHz
Battery Capacity: 1,800 mAh
Battery Type: Lithium Polymer 4S
Battery Power: 14.8V
GPS Compatible: –
Altitude Hold: –
Position Hold: –
Flight Planning: –
Autonomous Flight: –
Target Tracking: –
Automatic Landing: –
Return To Home: –
Obstacle Avoidance: –
Headless Mode: √
Built-in Camera: –

Revell Control

Pulse FPV

©conrad.com

Release Date: October 2016
Price: $279
Number of Rotors: 4
Control System: Remote Control
Applications: Recreation, FPV Racing
Dimensions: 190 × 190 × 50 mm
Weight: 120 gr
Frame Material: Plastic
Rotor Material: Plastic
Available Colors: White
Max. Flight Time: 10 minutes
Controller Frequency: 2.4 GHz
Battery Capacity: 600 mAh
Battery Type: Lithium Polymer 2S
GPS Compatible: √
Altitude Hold: √
Position Hold: √
Flight Planning: –
Autonomous Flight: –
Target Tracking: √
Automatic Landing: –
Return To Home: √
Obstacle Avoidance: –
Headless Mode: √
Built-in Camera: √

Revell Control

VR Shot

©ercmarket.com

Release Date: October 2016
Price: $149
Number of Rotors: 4
Control System: Remote Control
Applications: Recreation
Dimensions: 550 × 350 × 130 mm
Weight: 921 gr
Frame Material: Plastic
Rotor Material: Plastic
Available Colors: Blue
Max. Flight Time: 7 minutes
Controller Frequency: 2.4 GHz
Battery Type: Lithium Polymer
GPS Compatible: –
Altitude Hold: √
Position Hold: √
Flight Planning: –
Autonomous Flight: –
Target Tracking: –
Automatic Landing: –
Return To Home: √
Obstacle Avoidance: –
Headless Mode: √
Built-in Camera: √

Rise

Vusion 250

©getfpv.com

Release Date: October 2016
Price: $219
Number of Rotors: 4
Control System: Remote Control
Applications: FPV Racing
Wingspan: 250 mm
Weight: 502 gr
Frame Material: Plastic
Rotor Material: Plastic
Motors: 2280kv brushless
Available Colors: Black
Controller Frequency: 5.8 GHz
Battery Capacity: 1,500 mAh
Battery Type: Lithium Polymer 3S
Battery Power: 11.1V
GPS Compatible: –
Altitude Hold: –
Position Hold: –
Flight Planning: –
Autonomous Flight: –
Target Tracking: –
Automatic Landing: –
Return To Home: –
Obstacle Avoidance: –
Headless Mode: –
Built-in Camera: √ (600TVL)

Robodrone

Hornet

©Robodrone Industries

Release Date: October 2016
Price: $625
Number of Rotors: 4
Control System: Remote Control
Applications: FPV Racing
Dimensions: 310 × 310 mm
Frame Material: Carbon Fiber, Plastic
Rotor Material: Plastic
Available Colors: Blue, Green, Grey, Black
Battery Type: Lithium Polymer
GPS Compatible: √
Altitude Hold: √
Position Hold: √
Flight Planning: √
Autonomous Flight: –
Target Tracking: –
Automatic Landing: –
Return To Home: √
Obstacle Avoidance: –
Headless Mode: –
Built-in Camera: √

Ruiven Technologies

Kudolo Wing

©kudolo.com

Release Date: October 2016
Price: Unknown
Number of Rotors: 4
Control System: Remote Control, Phone/Tablet
Applications: Recreation
Dimensions: 448 × 420 mm
Weight: 730 gr
Frame Material: Plastic
Rotor Material: Plastic
Available Colors: Blue, Orange, Purple, Yellow
Max. Flight Time: 20 minutes
Max. Flying Altitude: 2,500 m (8,200 ft)
Min. Operating Temperature: 0°C
Max. Operating Temperature: 40°C
Max. Operating Range: 2 km
Controller Frequency: 2.4 GHz
Battery Capacity: 2,200 mAh
Battery Type: Lithium Polymer 3S
Battery Power: 11.1V
GPS Compatible: √
Altitude Hold: √
Position Hold: √
Flight Planning: √
Autonomous Flight: –
Target Tracking: √
Automatic Landing: √
Return To Home: √
Obstacle Avoidance: √
Headless Mode: √
Built-in Camera: √

SenseFly

Ebee Plus

© SenseFly

Release Date: October 2016
Price: $17,990
Number of Rotors: 1 (Fixed Wing)
Control System: Remote Control
Applications: Aerial Mapping, Surveillance
Wingspan: 1.1 m
Weight: 690 gr
Frame Material: EPP Foam, Carbon Fiber
Rotor Material: Carbon Fiber
Available Colors: Black
Max. Flight Time: 60 minutes
Max. Speed: 90 km/h (56 mp/h)
Max. Flying Altitude: 122 m (400 ft)
Min. Operating Temperature: -10°C
Max. Operating Temperature: 40°C
Max. Operating Range: 3 km
Controller Frequency: 2.4 GHz
Battery Capacity: 2,150 mAh
Battery Type: Lithium Polymer
Battery Power: 11.1V
GPS Compatible: √
Altitude Hold: √
Position Hold: –
Flight Planning: √
Autonomous Flight: √
Target Tracking: √
Automatic Landing: –
Return To Home: √
Obstacle Avoidance: –
Headless Mode: –
Built-in Camera: √

Syma Toys

X5UC

© aliexpress.com

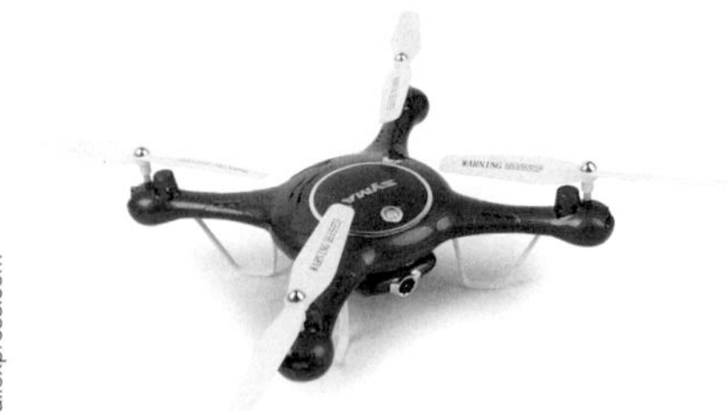

Release Date: October 2016
Price: $50
Number of Rotors: 4
Control System: Remote Control
Applications: Film, Photography, Recreation
Dimensions: 320 × 320 × 70 mm
Frame Material: Plastic
Rotor Material: Plastic
Available Colors: White
Max. Flight Time: 10 minutes
Max. Operating Range: 70 m
Controller Frequency: 2.4 GHz
Battery Capacity: 500 mAh
Battery Type: Lithium Polymer
Battery Power: 3.7V
GPS Compatible: –
Altitude Hold: –
Position Hold: –
Flight Planning: –
Autonomous Flight: –
Target Tracking: –
Automatic Landing: √
Return To Home: √
Obstacle Avoidance: –
Headless Mode: √
Built-in Camera: √ (2 megapixels)

Tarot RC

Peeper Long Flight Time TL750S1

© aliexpress.com

Release Date: October 2016
Price: $689
Number of Rotors: 4
Control System: Remote Control
Applications: Recreation
Wingspan: 750 mm
Weight: 1.56 kg
Frame Material: Aluminum, Carbon Fiber
Rotor Material: Carbon Fiber
Motors: 4008 brushless
Available Colors: Red
Max. Flight Time: 47 minutes
Controller Frequency: 5.8 GHz
Battery Capacity: 10,000 mAh
Battery Type: Lithium Polymer
Battery Power: 22.2V
GPS Compatible: –
Altitude Hold: √
Position Hold: √
Flight Planning: –
Autonomous Flight: –
Target Tracking: –
Automatic Landing: √
Return To Home: √
Obstacle Avoidance: –
Headless Mode: –
Built-in Camera: –

UDIRC Toys

U28-1 kestrel

© ripmax.com

Release Date: October 2016
Price: $195
Number of Rotors: 4
Control System: Remote Control
Applications: Film, Photography
Dimensions: 156 × 165 × 64 mm
Weight: 69 gr
Frame Material: Plastic
Rotor Material: Plastic
Available Colors: White
Max. Flight Time: 6 minutes
Min. Operating Temperature: -10°C
Max. Operating Temperature: 40°C
Max. Operating Range: 50 m
Controller Frequency: 2.4 GHz
Battery Capacity: 500 mAh
Battery Type: Lithium Polymer
Battery Power: 3.7V
GPS Compatible: –
Altitude Hold: √
Position Hold: √
Flight Planning: –
Autonomous Flight: –
Target Tracking: √
Automatic Landing: √
Return To Home: √
Obstacle Avoidance: –
Headless Mode: –
Built-in Camera: √

Walkera

AiBao

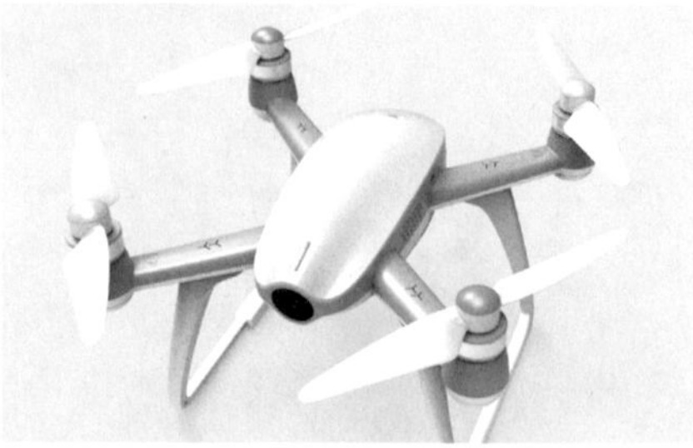

© banggood.com

Release Date: October 2016
Price: $399
Number of Rotors: 4
Control System: Remote Control
Applications: Recreation, Gaming, FPV Racing
Wingspan: 280 mm
Weight: 570 gr
Frame Material: Plastic
Rotor Material: Plastic
Available Colors: Gold, Pink, White
Max. Flight Time: 20 minutes
Max. Speed: 72 km/h (44,7 mp/h)
Max. Ascent Speed: 5 m/s
Max. Descent Speed: 3 m/s
Min. Operating Temperature: -25°C
Max. Operating Temperature: 60°C
Max. Operating Range: 1.5 km
Controller Frequency: 2.4 GHz
Battery Capacity: 5,200 mAh
Battery Type: Lithium Polymer
Battery Power: 7.6V
GPS Compatible: √
Altitude Hold: √
Position Hold: √
Flight Planning: √
Autonomous Flight: √
Target Tracking: √
Automatic Landing: √
Return To Home: √
Obstacle Avoidance: –
Headless Mode: √
Built-in Camera: √ (4K HD)

Xcraft UAV

X Plus One

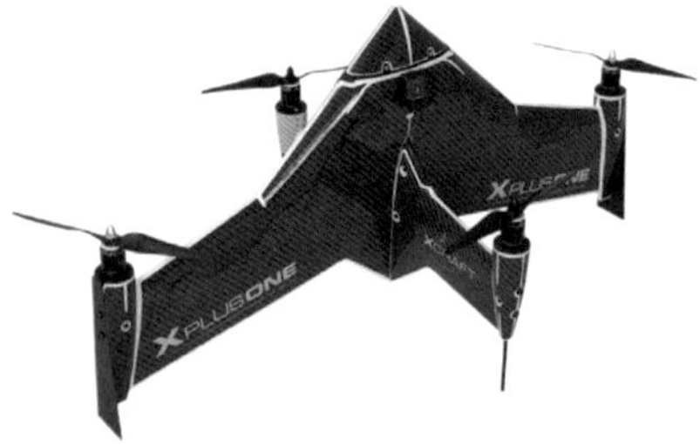

© xCraft

Release Date: October 2016
Price: $795
Number of Rotors: 4
Control System: Phone/Tablet
Applications: Film, Photography, FPV Racing
Dimensions: 825 × 482 × 355 mm
Weight: 1.3 kg
Frame Material: Carbon Fiber
Available Colors: Black
Max. Flight Time: 20 minutes
Max. Speed: 96,5 km/h (60 mp/h)
Controller Frequency: 2.4 GHz
Battery Type: Lithium Polymer
Battery Power: 11.1V
GPS Compatible: –
Altitude Hold: √
Position Hold: √
Flight Planning: √
Autonomous Flight: √
Target Tracking: √
Automatic Landing: √
Return To Home: √
Obstacle Avoidance: √
Headless Mode: –
Built-in Camera: –

Zero-Zero Robotics

Hover Camera

© Zero-Zero Robotics

Release Date: October 2016
Price: $399
Number of Rotors: 4
Control System: Phone/Tablet
Applications: Film, Photography, Recreation
Dimensions: 182 × 232 × 33 mm
Weight: 242 gr
Frame Material: Carbon Fiber
Rotor Material: Plastic
Available Colors: Black
Max. Flight Time: 10 minutes
Max. Speed: 29 km/h (18 mp/h)
Max. Flying Altitude: 2,000 m (6,560 ft)
Min. Operating Temperature: 5°C
Max. Operating Temperature: 35°C
Controller Frequency: 2.4 GHz
Battery Capacity: 1,360 mAh
Battery Type: Lithium Polymer 2S
Battery Power: 7.6V
Battery Weight: 71 gr
GPS Compatible: √
Altitude Hold: √
Position Hold: √
Flight Planning: √
Autonomous Flight: √
Target Tracking: √
Automatic Landing: –
Return To Home: √
Obstacle Avoidance: √
Headless Mode: √
Built-in Camera: √ (13 megapixels)

Yuneec

Breeze

© tested.com

Release Date: October 2016
Price: $299
Number of Rotors: 4
Control System: Remote Control, Phone
Applications: Film, Photography, FPV Racing
Dimensions: 196 × 196 × 65 mm
Wingspan: 240 mm
Weight: 385 gr
Frame Material: Plastic
Rotor Material: Plastic
Motors: 2206 brushless
Available Colors: White
Max. Flight Time: 12 minutes
Max. Speed: 18 km/h (11 mp/h)
Max. Flying Altitude: 80 m (262 ft)
Max. Ascent Speed: 1 m/s
Min. Operating Temperature: 0°C
Max. Operating Temperature: 40°C
Controller Frequency: 2.4 GHz
Battery Capacity: 1,150 mAh
Battery Type: Lithium Polymer 3S
Battery Power: 11.1V
GPS Compatible: √
Altitude Hold: √
Position Hold: √
Flight Planning: √
Autonomous Flight: √
Target Tracking: √
Automatic Landing: √
Return To Home: √
Obstacle Avoidance: –
Headless Mode: √
Built-in Camera: √ (13 megapixels)

AeroVironment

Quantix

© AeroVironment

Release Date: November 2016
Price: $5,500
Number of Rotors: 4 (Fixed Wing, VTOL)
Control System: Flight Computer, Tablet
Applications: Aerial Mapping, Agriculture, Inspection
Available Colors: White
Max. Flight Time: 45 minutes
Max. Speed: 64 km/h (40 mp/h)
Max. Flying Altitude: 122 m (400 ft)
Controller Frequency: 2.4 GHz
Battery Type: Lithium Polymer
GPS Compatible: √
Altitude Hold: √
Position Hold: √
Flight Planning: √
Autonomous Flight: √
Target Tracking: –
Automatic Landing: √
Return To Home: √
Obstacle Avoidance: –
Headless Mode: –
Built-in Camera: √ (18 megapixels)

Airhogs

X-Stream Video

© Airhogs

Release Date: November 2016
Price: $50
Number of Rotors: 4
Control System: Phone
Applications: Recreation
Frame Material: Plastic
Rotor Material: Plastic
Available Colors: Grey
Controller Frequency: 2.4 GHz
Battery Type: Lithium Polymer
GPS Compatible: –
Altitude Hold: √
Position Hold: √
Flight Planning: √
Autonomous Flight: –
Target Tracking: √
Automatic Landing: –
Return To Home: √
Obstacle Avoidance: –
Headless Mode: √
Built-in Camera: √ (720P HD)

Aosenma

CG035

©geekbuying.com

Release Date: November 2016
Price: $299
Number of Rotors: 4
Control System: Remote Control, Phone
Applications: Film, Photography, FPV Racing
Dimensions: 330 × 330 × 137 mm
Frame Material: Plastic
Rotor Material: Plastic
Available Colors: White
Max. Flight Time: 20 minutes
Max. Operating Range: 300 m
Controller Frequency: 2.4 GHz
Battery Capacity: 3,200 mAh
Battery Type: Lithium Polymer
Battery Power: 7.4V
GPS Compatible: –
Altitude Hold: √
Position Hold: √
Flight Planning: √
Autonomous Flight: –
Target Tracking: –
Automatic Landing: √
Return To Home: √
Obstacle Avoidance: –
Headless Mode: √
Built-in Camera: √ (1080P HD)

Ares RC

Recon FPV

© Amain Hobbies

Release Date: November 2016
Price: $169
Number of Rotors: 4
Control System: Remote Control
Applications: Recreation, FPV Racing
Dimensions: 157 × 157 × 41 mm
Weight: 69 gr
Frame Material: Plastic
Rotor Material: Plastic
Available Colors: Blue
Controller Frequency: 2.4 GHz
Battery Capacity: 500 mAh
Battery Type: Lithium Polymer 1S
Battery Power: 3.7V
GPS Compatible: –
Altitude Hold: –
Position Hold: –
Flight Planning: –
Autonomous Flight: –
Target Tracking: –
Automatic Landing: –
Return To Home: –
Obstacle Avoidance: –
Headless Mode: –
Built-in Camera: √ (720P HD)

Ares RC

Quantum FPV

© Amain Hobbies

Release Date: November 2016
Price: $80
Number of Rotors: 4
Control System: Remote Control
Applications: FPV Racing
Dimensions: 312 × 323 × 92 mm
Weight: 167 gr
Frame Material: Plastic
Rotor Material: Plastic
Available Colors: Black
Controller Frequency: 2.4 GHz
Battery Capacity: 350 mAh
Battery Type: Lithium Polymer 2S
Battery Power: 7.4V
GPS Compatible: –
Altitude Hold: –
Position Hold: –
Flight Planning: –
Autonomous Flight: –
Target Tracking: –
Automatic Landing: –
Return To Home: –
Obstacle Avoidance: –
Headless Mode: –
Built-in Camera: √ (720P HD)

C-Astral

Bramor ppX

© C-Astral

Release Date: November 2016
Price: On Request
Number of Rotors: 1 (Fixed Wing)
Control System: Ground Control Station
Applications: Aerial Mapping, Agriculture, Inspection, Surveillance
Dimensions: 0.96 × 2.3 m
Wingspan: 2.3 m
Weight: 4.7 kg
Max. Payload: 1 kg
Frame Material: Carbon Fiber, Kevlar
Available Colors: White
Max. Flight Time: 3.5 hours
Max. Speed: 79 km/h (49 mp/h)
Max. Flying Altitude: 5,000 m (16,404 ft)
Max. Operating Range: 150 km
Controller Frequency: 2.4 GHz
Battery Type: Lithium Polymer
GPS Compatible: √
Altitude Hold: √
Position Hold: –
Flight Planning: √
Autonomous Flight: √
Target Tracking: –
Automatic Landing: –
Return To Home: √
Obstacle Avoidance: √
Headless Mode: –
Built-in Camera: √ (42 megapixels)

Chengdu Aircraft Industry Group (CAIG)

Cloud Shadow

©Chris Pocock

Release Date: November 2016
Price: Unknown
Number of Rotors: 1 (Fixed Wing)
Control System: Ground Control Station
Applications: Military Purposes, Armed Reconnaissance, Surveillance, Target Acquisition
Weight: 2,300 kg
Frame Material: Composite
Motors: 1 × WP-11C turbojet engine
Available Colors: Grey
Max. Speed: 620 km/h (385 mp/h)
Max. Flying Altitude: 15,000 m (49,213 ft)
Max. Operating Range: 290 km
GPS Compatible: √
Altitude Hold: √
Position Hold: –
Flight Planning: √
Autonomous Flight: √
Target Tracking: √
Automatic Landing: √
Return To Home: √
Obstacle Avoidance: √
Headless Mode: –
Built-in Camera: √

China Aerospace Science and Technology Corporation (CASC)

CH-5 Rainbow

©eastpendulum.com

Release Date: November 2016
Price: Unknown
Number of Rotors: 1 (Fixed Wing)
Control System: Ground Control Station
Applications: Military Purposes, Armed Reconnaissance, Surveillance, Target Acquisition
Dimensions: 11.2 × 21 × 3.8 m
Wingspan: 21 m
Weight: 2,225 kg
Max. Payload: 1,000 kg
Frame Material: Composite
Available Colors: Grey
Max. Flight Time: 60 hours
Max. Speed: 480 km/h (298 mp/h)
Max. Flying Altitude: 9,000 m (29,527 ft)
Max. Operating Range: 10,000 km
GPS Compatible: √
Altitude Hold: √
Position Hold: –
Flight Planning: √
Autonomous Flight: √
Target Tracking: √
Automatic Landing: √
Return To Home: √
Obstacle Avoidance: √
Headless Mode: –
Built-in Camera: √

Defence Research Development Organisation of India (DRDO)

Rustom 2

©zonefiveaviation.com/Angad Singh

Release Date: November 2016
Price: Unknown
Number of Rotors: 2 (Fixed Wing)
Control System: Ground Control Station
Applications: Military Purposes, Armed Reconnaissance, Surveillance, Target Acquisition
Dimensions: 9.5 × 20.6 × 2.4 m
Wingspan: 20.6 m
Weight: 1,800 kg
Max. Payload: 350 kg
Frame Material: Composite
Motors: 2 × NPO-Saturn 36MT engine, wing-mounted turboprop
Available Colors: Grey
Max. Flight Time: 24 hours
Max. Speed: 225 km/h (140 mp/h)
Max. Flying Altitude: 9,144 m (30,000 ft)
Max. Operating Range: 350 km
GPS Compatible: √
Altitude Hold: √
Position Hold: –
Flight Planning: √
Autonomous Flight: √
Target Tracking: √
Automatic Landing: √
Return To Home: √
Obstacle Avoidance: √
Headless Mode: –
Built-in Camera: √

DJI

Inspire 2

©droneshop.com

Release Date: November 2016
Price: $2,999
Number of Rotors: 4
Control System: Remote Control, Phone/Tablet
Applications: Film, Photography
Dimensions: 480 × 470 × 320 mm
Weight: 3.44 kg
Max. Payload: 810 gr
Frame Material: Carbon Fiber, Plastic
Rotor Material: Carbon Fiber
Available Colors: Grey
Max. Flight Time: 27 minutes
Max. Speed: 94 km/h (58,4 mp/h)
Max. Flying Altitude: 5,000 m (16,404 ft)
Max. Ascent Speed: 6 m/s
Max. Descent Speed: 4 m/s
Min. Operating Temperature: -20°C
Max. Operating Temperature: 40°C
Max. Operating Range: 7 km
Controller Frequency: 2.4 GHz
Battery Capacity: 4,280 mAh
Battery Type: Lithium Polymer 6S
Battery Power: 22.8V
Battery Weight: 515 gr
GPS Compatible: √
Altitude Hold: √
Position Hold: √
Flight Planning: √
Autonomous Flight: √
Target Tracking: √
Automatic Landing: √
Return To Home: √
Obstacle Avoidance: √
Headless Mode: √
Built-in Camera: –

DJI

Phantom 4 Pro

© DJI

Release Date: November 2016
Price: $1,499
Number of Rotors: 4
Control System: Remote Control, Phone/Tablet
Applications: Film, Photography
Wingspan: 350 mm
Weight: 1.38 kg
Max. Payload: 462 gr
Frame Material: Plastic, Magnesium
Rotor Material: Carbon Fiber
Available Colors: Black, White
Max. Flight Time: 30 minutes
Max. Speed: 72 km/h (45 mp/h)
Max. Flying Altitude: 6,000 m (19,685 ft)
Max. Ascent Speed: 6 m/s
Max. Descent Speed: 5 m/s
Min. Operating Temperature: 0°C
Max. Operating Temperature: 40°C
Max. Operating Range: 7 km
Controller Frequency: 2.4 GHz
Battery Capacity: 5,870 mAh
Battery Type: Lithium Polymer 4S
Battery Power: 15.2V
Battery Weight: 468 gr
GPS Compatible: √
Altitude Hold: √
Position Hold: √
Flight Planning: √
Autonomous Flight: √
Target Tracking: √
Automatic Landing: √
Return To Home: √
Obstacle Avoidance: √
Headless Mode: √
Built-in Camera: √ (20 megapixels)

Drone America

NavX

© Drone America

Release Date: November 2016
Price: On Request
Number of Rotors: 4
Control System: Ground Control Station, Laptop, Tablet
Applications: Aerial Mapping, Agriculture, Inspection, Surveillance
Wingspan: 800 mm
Weight: 6.8 kg
Max. Payload: 4.5 kg
Frame Material: Carbon Fiber
Available Colors: White
Max. Flight Time: 25 minutes
Max. Speed: 54 km/h (33,5 mp/h)
Max. Flying Altitude: 3,657 m (12,000 ft)
Battery Type: Lithium Polymer
GPS Compatible: √
Altitude Hold: √
Position Hold: √
Flight Planning: √
Autonomous Flight: √
Target Tracking: –
Automatic Landing: √
Return To Home: √
Obstacle Avoidance: √
Headless Mode: –
Built-in Camera: –

Eachine

EX110

© aliexpress.com

Release Date: November 2016
Price: $100
Number of Rotors: 4
Control System: Remote Control
Applications: Recreation, FPV Racing
Dimensions: 92 × 92 mm
Wingspan: 110 mm
Weight: 43 gr
Frame Material: Carbon Fiber
Rotor Material: Plastic
Motors: Chaoli 8520
Available Colors: Black
Max. Flight Time: 6 minutes
Controller Frequency: 5.8 GHz
Battery Capacity: 600 mAh
Battery Type: Lithium Polymer
Battery Power: 3.7V
GPS Compatible: –
Altitude Hold: –
Position Hold: –
Flight Planning: –
Autonomous Flight: –
Target Tracking: –
Automatic Landing: –
Return To Home: –
Obstacle Avoidance: –
Headless Mode: –
Built-in Camera: √ (800TVL)

Flypro

X-Jaguar

© Flypro

Release Date: November 2016
Price: $295
Number of Rotors: 4
Control System: Remote Control
Applications: FPV Racing
Dimensions: 166 × 166 × 88 mm
Wingspan: 190 mm
Weight: 510 gr
Frame Material: Aluminum, Carbon Fiber
Rotor Material: Plastic
Available Colors: Red
Max. Flight Time: 9 minutes
Min. Operating Temperature: -10°C
Max. Operating Temperature: 50°C
Controller Frequency: 5.8 GHz
Battery Capacity: 1,300 mAh
Battery Type: Lithium Polymer 4S
GPS Compatible: –
Altitude Hold: –
Position Hold: –
Flight Planning: –
Autonomous Flight: –
Target Tracking: –
Automatic Landing: –
Return To Home: –
Obstacle Avoidance: –
Headless Mode: –
Built-in Camera: √ (600TVL)

Gear2Play

Focus

©blokker.nl

Release Date: November 2016
Price: $55
Number of Rotors: 4
Control System: Remote Control
Applications: Recreation
Dimensions: 260 × 260 mm
Frame Material: Plastic
Rotor Material: Plastic
Available Colors: White
Max. Flight Time: 15 minutes
Max. Operating Range: 30 m
Controller Frequency: 2.4 GHz
Battery Capacity: 380 mAh
Battery Type: Lithium Polymer
Battery Power: 3.7V
GPS Compatible: –
Altitude Hold: –
Position Hold: –
Flight Planning: –
Autonomous Flight: –
Target Tracking: –
Automatic Landing: –
Return To Home: √
Obstacle Avoidance: –
Headless Mode: √
Built-in Camera: √ (1.2 megapixels)

Gteng

T906W

©gearbest.com

Release Date: November 2016
Price: $30
Number of Rotors: 4
Control System: Remote Control
Applications: Recreation
Dimensions: 59 × 59 × 30 mm
Weight: 200 gr
Frame Material: Plastic
Rotor Material: Plastic
Available Colors: Black
Max. Flight Time: 7 minutes
Max. Operating Range: 25 m
Controller Frequency: 2.4 GHz
Battery Capacity: 200 mAh
Battery Type: Lithium Polymer
Battery Power: 3.7V
GPS Compatible: –
Altitude Hold: –
Position Hold: –
Flight Planning: √
Autonomous Flight: –
Target Tracking: –
Automatic Landing: –
Return To Home: √
Obstacle Avoidance: –
Headless Mode: √
Built-in Camera: √

Gteng

T907W

© aliexpress.com

Release Date: November 2016
Price: $45
Number of Rotors: 4
Control System: Remote Control
Applications: Recreation
Dimensions: 170 × 170 × 70 mm
Weight: 45 gr
Frame Material: Plastic
Rotor Material: Plastic
Available Colors: Orange
Max. Flight Time: 6 minutes
Max. Operating Range: 50 m
Controller Frequency: 2.4 GHz
Battery Capacity: 500 mAh
Battery Type: Lithium Polymer
GPS Compatible: –
Altitude Hold: √
Position Hold: –
Flight Planning: –
Autonomous Flight: –
Target Tracking: –
Automatic Landing: √
Return To Home: √
Obstacle Avoidance: –
Headless Mode: √
Built-in Camera: √ (0.3 megapixels)

Huajun Toys

W606-6

© banggood.com

Release Date: November 2016
Price: $53
Number of Rotors: 4
Control System: Remote Control
Applications: Recreation, FPV Racing
Dimensions: 400 × 400 × 70 mm
Weight: 120 gr
Frame Material: Plastic
Rotor Material: Plastic
Available Colors: Black, Blue, Pink
Max. Flight Time: 11 minutes
Max. Operating Range: 80 m
Controller Frequency: 2.4 GHz
Battery Capacity: 650 mAh
Battery Type: Lithium Polymer
Battery Power: 3.7V
GPS Compatible: –
Altitude Hold: √
Position Hold: √
Flight Planning: –
Autonomous Flight: –
Target Tracking: –
Automatic Landing: √
Return To Home: √
Obstacle Avoidance: –
Headless Mode: √
Built-in Camera: √ (720P HD)

JJRC

H37

©geekbuying.com

Release Date: November 2016
Price: $20
Number of Rotors: 4
Control System: Remote Control
Applications: Recreation, FPV Racing
Dimensions: 135 × 160 × 25 mm
Weight: 73 gr
Frame Material: Plastic
Rotor Material: Plastic
Available Colors: Black, Pink
Max. Flight Time: 8 minutes
Max. Operating Range: 40 m
Controller Frequency: 2.4 GHz
Battery Capacity: 500 mAh
Battery Type: Lithium Polymer
Battery Power: 3.7V
GPS Compatible: –
Altitude Hold: √
Position Hold: √
Flight Planning: –
Autonomous Flight: –
Target Tracking: –
Automatic Landing: –
Return To Home: √
Obstacle Avoidance: –
Headless Mode: √
Built-in Camera: √ (2 megapixels)

Kaiser Baas

Gamma

©Kaiser Baas

Release Date: November 2016
Price: $149
Number of Rotors: 4
Control System: Remote Control
Applications: Recreation, FPV Racing
Dimensions: 330 × 330 × 68 mm
Weight: 120 gr
Frame Material: Plastic
Rotor Material: Plastic
Available Colors: Black
Max. Flight Time: 6 minutes
Max. Operating Range: 80 m
Controller Frequency: 2.4 GHz
Battery Capacity: 80 mAh
Battery Type: Lithium Polymer
Battery Power: 3.7V
GPS Compatible: –
Altitude Hold: √
Position Hold: √
Flight Planning: √
Autonomous Flight: √
Target Tracking: –
Automatic Landing: –
Return To Home: √
Obstacle Avoidance: –
Headless Mode: √
Built-in Camera: √ (720P HD)

Kaiser Baas

Theta

© Kaiser Baas

Release Date: November 2016
Price: $105
Number of Rotors: 4
Control System: Remote Control
Applications: Recreation, FPV Racing
Dimensions: 170 × 165 × 45 mm
Weight: 70 gr
Frame Material: Plastic
Rotor Material: Plastic
Available Colors: Black
Max. Flight Time: 5 minutes
Max. Speed: 16,5 km/h (10 mp/h)
Max. Ascent Speed: 5 m/s
Min. Operating Temperature: 0°C
Max. Operating Temperature: 40°C
Max. Operating Range: 60 m
Controller Frequency: 2.4 GHz
Battery Capacity: 520 mAh
Battery Type: Lithium Polymer
Battery Power: 3.7V
GPS Compatible: –
Altitude Hold: √
Position Hold: √
Flight Planning: √
Autonomous Flight: √
Target Tracking: –
Automatic Landing: –
Return To Home: √
Obstacle Avoidance: –
Headless Mode: √
Built-in Camera: √ (720P HD)

Meijiaxin Toys

X909T

© banggood.com

Release Date: November 2016
Price: $50
Number of Rotors: 4
Control System: Remote Control
Applications: Recreation, FPV Racing
Dimensions: 77 × 76 mm
Frame Material: Plastic
Rotor Material: Plastic
Available Colors: Black, White
Max. Flight Time: 6 minutes
Max. Operating Range: 50 m
Controller Frequency: 2.4 GHz
Battery Capacity: 800 mAh
Battery Type: Lithium Polymer
Battery Power: 3.7V
GPS Compatible: –
Altitude Hold: –
Position Hold: –
Flight Planning: –
Autonomous Flight: –
Target Tracking: –
Automatic Landing: –
Return To Home: √
Obstacle Avoidance: –
Headless Mode: –
Built-in Camera: √

PowerUp

Power Up FPV

© PowerUp

Release Date: November 2016
Price: $140
Number of Rotors: 1 (Fixed Wing)
Control System: Phone/Tablet
Applications: Recreation, FPV Racing
Frame Material: Carbon Fiber, Paper, Plastic
Rotor Material: Plastic
Available Colors: Red, White
Max. Flight Time: 10 minutes
Max. Speed: 33 km/h (20 mp/h)
Controller Frequency: 2.4 GHz
Battery Type: Lithium Polymer
GPS Compatible: –
Altitude Hold: √
Position Hold: –
Flight Planning: √
Autonomous Flight: √
Target Tracking: –
Automatic Landing: √
Return To Home: √
Obstacle Avoidance: –
Headless Mode: –
Built-in Camera: √

Revell Control

Navigator

© Revell Control

Release Date: November 2016
Price: $399
Number of Rotors: 4
Control System: Remote Control
Applications: Recreation, FPV Racing
Dimensions: 250 × 250 × 80 mm
Weight: 900 gr
Frame Material: Plastic
Rotor Material: Plastic
Available Colors: Black
Max. Operating Range: 200 m
Controller Frequency: 2.4 GHz
Battery Type: Lithium Polymer
GPS Compatible: √
Altitude Hold: √
Position Hold: √
Flight Planning: √
Autonomous Flight: –
Target Tracking: √
Automatic Landing: –
Return To Home: √
Obstacle Avoidance: –
Headless Mode: √
Built-in Camera: √

Revell Control

Quadrotox

Release Date: November 2016
Price: $89
Number of Rotors: 4
Control System: Remote Control
Applications: Recreation, FPV Racing
Dimensions: 200 × 200 × 50 mm
Weight: 450 gr
Frame Material: Plastic
Rotor Material: Plastic
Available Colors: Black
Controller Frequency: 2.4 GHz
Battery Capacity: 780 mAh
Battery Type: Lithium Polymer 1S
GPS Compatible: –
Altitude Hold: –
Position Hold: –
Flight Planning: –
Autonomous Flight: –
Target Tracking: –
Automatic Landing: –
Return To Home: √
Obstacle Avoidance: –
Headless Mode: √
Built-in Camera: –

Revell Control

Sky Spider

Release Date: November 2016
Price: $125
Number of Rotors: 4
Control System: Remote Control
Applications: Recreation
Dimensions: 510 × 510 × 100 mm
Weight: 300 gr
Frame Material: EPP Foam
Rotor Material: Plastic
Available Colors: Yellow
Controller Frequency: 2.4 GHz
Battery Type: Lithium Polymer
GPS Compatible: –
Altitude Hold: –
Position Hold: –
Flight Planning: –
Autonomous Flight: –
Target Tracking: –
Automatic Landing: –
Return To Home: √
Obstacle Avoidance: –
Headless Mode: –
Built-in Camera: –

Storm Racing Drone

SRD 101

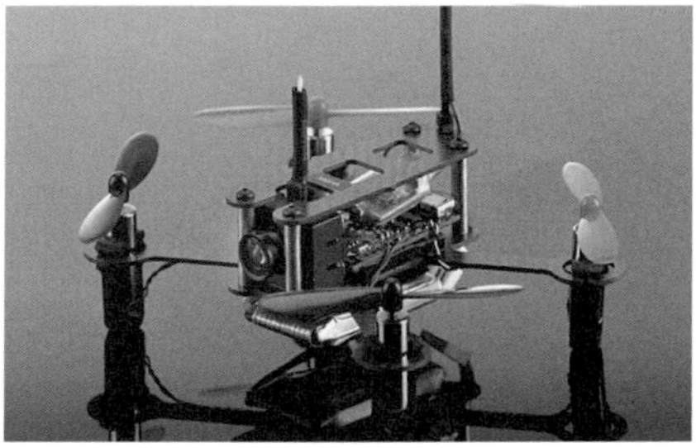

© helipal.com

Release Date: November 2016
Price: $99
Number of Rotors: 4
Control System: Remote Control
Applications: FPV Racing
Dimensions: 98 × 77 × 39 mm
Wingspan: 101 mm
Weight: 44 gr
Frame Material: Carbon Fiber
Rotor Material: Plastic
Motors: BM720 coreless
Available Colors: Black
Controller Frequency: 5.8 GHz
Battery Capacity: 600 mAh
Battery Type: Lithium Polymer 1S
GPS Compatible: –
Altitude Hold: –
Position Hold: –
Flight Planning: –
Autonomous Flight: –
Target Tracking: –
Automatic Landing: –
Return To Home: –
Obstacle Avoidance: –
Headless Mode: –
Built-in Camera: √ (600TVL)

Tobyrich

Moskito

© Tobyrich

Release Date: November 2016
Price: $60
Number of Rotors: 1 (Fixed Wing)
Control System: Phone/Tablet
Applications: Recreation
Dimensions: 220 × 220 × 80 mm
Wingspan: 220 mm
Weight: 18 gr
Frame Material: EPP Foam
Rotor Material: Plastic
Available Colors: White
Max. Flight Time: 12 minutes
Max. Operating Range: 60 m
Controller Frequency: 2.4 GHz
Battery Capacity: 80 mAh
Battery Type: Lithium Polymer
GPS Compatible: –
Altitude Hold: –
Position Hold: –
Flight Planning: –
Autonomous Flight: –
Target Tracking: –
Automatic Landing: –
Return To Home: –
Obstacle Avoidance: –
Headless Mode: –
Built-in Camera: –

Walkera

Voyager 4

© conrad.com

Release Date: November 2016
Price: $3,099
Number of Rotors: 4
Control System: Remote Control, Phone/Tablet
Applications: Film, Photography
Dimensions: 465 × 465 × 352 mm
Wingspan: 382 mm
Weight: 3.25 kg
Motors: WK-WS-42-002A brushless
Available Colors: White
Max. Flight Time: 20 minutes
Max. Speed: 72 km/h (44,7 mp/h)
Max. Flying Altitude: 1,000 m (3,280 ft)
Max. Ascent Speed: 5 m/s
Max. Descent Speed: 4 m/s
Min. Operating Temperature: -10°C
Max. Operating Temperature: 40°C
Max. Operating Range: 1.5 km
Controller Frequency: 2.4 GHz
Battery Capacity: 4,500 mAh
Battery Type: Lithium Polymer 6S
Battery Power: 22.2V
GPS Compatible: √
Altitude Hold: √
Position Hold: √
Flight Planning: √
Autonomous Flight: √
Target Tracking: √
Automatic Landing: √
Return To Home: √
Obstacle Avoidance: √
Headless Mode: –
Built-in Camera: √

Wingsland Technology

S6

© reichelt.de

Release Date: November 2016
Price: $160
Number of Rotors: 4
Control System: Remote Control
Applications: Film, Photography
Dimensions: 138 × 169 × 33 mm
Weight: 260 gr
Frame Material: Plastic
Rotor Material: Plastic
Available Colors: Black, Green, Orange, Silver, Yellow
Max. Flight Time: 10 minutes
Max. Speed: 29 km/h (18 mp/h)
Max. Ascent Speed: 8 m/s
Max. Descent Speed: 8 m/s
Min. Operating Temperature: 0°C
Max. Operating Temperature: 40°C
Max. Operating Range: 100 m
Controller Frequency: 2.4 GHz
Battery Capacity: 1,400 mAh
Battery Type: Lithium Polymer 2S
Battery Power: 7.6V
GPS Compatible: √
Altitude Hold: √
Position Hold: √
Flight Planning: √
Autonomous Flight: √
Target Tracking: √
Automatic Landing: √
Return To Home: √
Obstacle Avoidance: √
Headless Mode: √
Built-in Camera: √ (13 megapixels)

Ares RC

Xview FPV

©radioshack.com

Release Date: December 2016
Price: $90
Number of Rotors: 4
Control System: Remote Control, Phone
Applications: FPV Racing
Dimensions: 137 × 137 × 38 mm
Weight: 42 gr
Frame Material: Plastic
Rotor Material: Plastic
Available Colors: Black
Controller Frequency: 2.4 GHz
Battery Capacity: 300 mAh
Battery Type: Lithium Polymer 1S
Battery Power: 3.7V
GPS Compatible: –
Altitude Hold: –
Position Hold: –
Flight Planning: –
Autonomous Flight: –
Target Tracking: –
Automatic Landing: –
Return To Home: –
Obstacle Avoidance: –
Headless Mode: –
Built-in Camera: √

Arris

C250 FPV

©hobby-wing.com

Release Date: December 2016
Price: $599
Number of Rotors: 4
Control System: Remote Control
Applications: FPV Racing
Dimensions: 195 × 230 × 42 mm
Weight: 350 gr
Frame Material: Carbon Fiber
Rotor Material: Plastic
Motors: Arris S2205-2300kv
Available Colors: Black
Max. Flight Time: 10 minutes
Controller Frequency: 5.8 GHz
Battery Capacity: 1,500 mAh
Battery Type: Lithium Polymer 4S
GPS Compatible: –
Altitude Hold: √
Position Hold: –
Flight Planning: –
Autonomous Flight: –
Target Tracking: –
Automatic Landing: –
Return To Home: –
Obstacle Avoidance: –
Headless Mode: –
Built-in Camera: √

ASUAV

Youbi XV-130

©gearbest.com

Release Date: December 2016
Price: $135
Number of Rotors: 4
Control System: Remote Control
Applications: FPV Racing
Dimensions: 115 × 150 × 35 mm
Wingspan: 150 mm
Weight: 115 gr
Frame Material: Carbon Fiber, Plastic
Rotor Material: Plastic
Motors: 1306-3600kv
Available Colors: Black, Grey, White
Controller Frequency: 5.8 GHz
Battery Type: Lithium Polymer
GPS Compatible: –
Altitude Hold: –
Position Hold: –
Flight Planning: –
Autonomous Flight: –
Target Tracking: –
Automatic Landing: –
Return To Home: –
Obstacle Avoidance: –
Headless Mode: –
Built-in Camera: √ (600TVL)

Cheerson

CX-117

©Cheerson Hobby

Release Date: December 2016
Price: $69
Number of Rotors: 4
Control System: Remote Control
Applications: FPV Racing
Dimensions: 115 × 115 mm
Weight: 59 gr
Frame Material: Carbon Fiber
Available Colors: Black
Max. Flight Time: 8 minutes
Battery Capacity: 600 mAh
Battery Type: Lithium Polymer 2S
GPS Compatible: –
Altitude Hold: –
Position Hold: –
Flight Planning: –
Autonomous Flight: –
Target Tracking: –
Automatic Landing: –
Return To Home: –
Obstacle Avoidance: –
Headless Mode: –
Built-in Camera: √ (1 megapixel)

Cheerson

CX-93S Thorn Jump

© Cheerson Hobby

Release Date: December 2016
Price: $105
Number of Rotors: 4
Control System: Remote Control
Applications: Recreation
Dimensions: 165 × 165 × 45 mm
Weight: 100 gr
Frame Material: Plastic
Rotor Material: Plastic
Available Colors: Black
Max. Flight Time: 6 minutes
Max. Speed: 36 km/h (22 mp/h)
Min. Operating Temperature: 0°C
Max. Operating Temperature: 40°C
Max. Operating Range: 100 m
Controller Frequency: 2.4 GHz
Battery Capacity: 520 mAh
Battery Type: Lithium Polymer
Battery Power: 7.4V
GPS Compatible: –
Altitude Hold: –
Position Hold: –
Flight Planning: –
Autonomous Flight: –
Target Tracking: –
Automatic Landing: –
Return To Home: –
Obstacle Avoidance: –
Headless Mode: –
Built-in Camera: √ (2 megapixels)

DaMing

DM009 Conqueror

© justdrones.com.au

Release Date: December 2016
Price: $60
Number of Rotors: 4
Control System: Remote Control
Applications: Recreation, FPV Racing
Dimensions: 315 × 315 × 100 mm
Weight: 135 gr
Frame Material: Plastic
Rotor Material: Plastic
Available Colors: Black
Max. Flight Time: 13 minutes
Max. Operating Range: 150 m
Controller Frequency: 2.4 GHz
Battery Capacity: 600 mAh
Battery Type: Lithium Polymer
Battery Power: 7.4V
GPS Compatible: –
Altitude Hold: √
Position Hold: √
Flight Planning: √
Autonomous Flight: √
Target Tracking: –
Automatic Landing: √
Return To Home: √
Obstacle Avoidance: –
Headless Mode: –
Built-in Camera: √ (5 megapixels)

Eachine

Fatbee FB90

© Eachine

Release Date: December 2016
Price: $120
Number of Rotors: 4
Control System: Remote Control
Applications: Recreation, FPV Racing
Dimensions: 118 × 118 × 70 mm
Wingspan: 90 mm
Weight: 63 gr
Frame Material: Plastic
Rotor Material: Plastic
Motors: 8520 2S Coreless
Available Colors: Yellow
Controller Frequency: 5.8 GHz
Battery Capacity: 400 mAh
Battery Type: Lithium Polymer
Battery Power: 7.4V
GPS Compatible: –
Altitude Hold: –
Position Hold: –
Flight Planning: –
Autonomous Flight: –
Target Tracking: –
Automatic Landing: –
Return To Home: –
Obstacle Avoidance: –
Headless Mode: –
Built-in Camera: √ (520TVL)

Eachine

E50

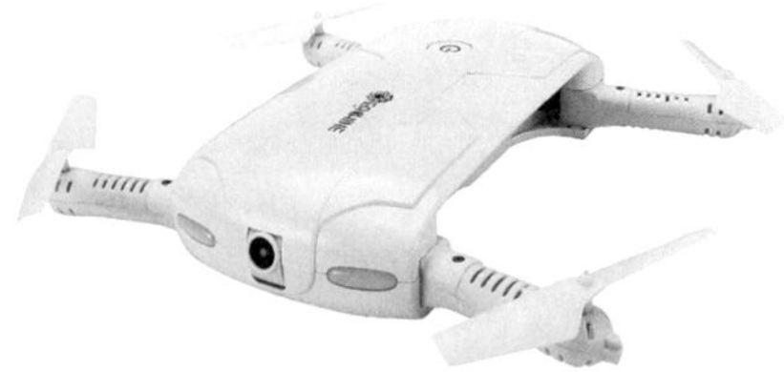

© helicomicro.com

Release Date: December 2016
Price: $100
Number of Rotors: 4
Control System: Remote Control
Applications: Recreation, FPV Racing
Dimensions: 135 × 160 × 25 mm
Frame Material: Plastic
Rotor Material: Plastic
Available Colors: Pink, White
Max. Flight Time: 8 minutes
Max. Operating Range: 40 m
Controller Frequency: 2.4 GHz
Battery Capacity: 500 mAh
Battery Type: Lithium Polymer
Battery Power: 3.7V
GPS Compatible: –
Altitude Hold: √
Position Hold: √
Flight Planning: –
Autonomous Flight: –
Target Tracking: –
Automatic Landing: √
Return To Home: √
Obstacle Avoidance: –
Headless Mode: √
Built-in Camera: √ (2 megapixels)

Embention

M600 multicopter

© Embention

Release Date: December 2016
Price: Unknown
Number of Rotors: 6
Control System: Remote Control
Applications: Film, Photography, Aerial Mapping, Agriculture, Inspection
Dimensions: 900 × 900 × 300 mm
Weight: 6.5 kg
Max. Payload: 1 kg
Frame Material: Carbon Fiber
Motors: Disk 4822 390kv
Available Colors: Black
Max. Flight Time: 40 minutes
Max. Speed: 80 km/h (49,7 mp/h)
Max. Operating Range: 12 km
Controller Frequency: 2.4 GHz
Battery Capacity: 16,000 mAh
Battery Type: Lithium Polymer 6S
GPS Compatible: √
Altitude Hold: √
Position Hold: √
Flight Planning: √
Autonomous Flight: √
Target Tracking: –
Automatic Landing: √
Return To Home: √
Obstacle Avoidance: –
Headless Mode: –
Built-in Camera: –

Flytronic

FT-5 LOS

© WB Group

Release Date: December 2016
Price: Unknown
Number of Rotors: 2 (Fixed Wing)
Control System: Ground Control Station
Applications: Inspection, Surveillance, Reconnaissance
Dimensions: 3.1 × 6.4 m
Wingspan: 6.4 m
Weight: 55 kg
Max. Payload: 30 kg
Frame Material: Composite
Available Colors: Grey
Max. Flight Time: 12 hours
Max. Speed: 180 km/h (112 mp/h)
Max. Flying Altitude: 5,000 m (16,404 ft)
Max. Operating Range: 180 km
GPS Compatible: √
Altitude Hold: √
Position Hold: √
Flight Planning: √
Autonomous Flight: √
Target Tracking: –
Automatic Landing: √
Return To Home: √
Obstacle Avoidance: –
Headless Mode: –
Built-in Camera: √

HiSystems GmbH

Mikrokopter 8-3500

© mikrokopter.de

Release Date: December 2016
Price: $16,000
Number of Rotors: 8
Control System: Flight Computer
Applications: Aerial Mapping, Agriculture, Inspection, Surveillance
Dimensions: 1.08 × 1.16 × 0.45 m
Weight: 3.35 kg
Max. Payload: 5.5 kg
Frame Material: Carbon Fiber
Available Colors: White
Max. Flight Time: 40 minutes
Max. Speed: 54 km/h (33,5 mp/h)
Max. Flying Altitude: 5,000 m (16,404 ft)
Min. Operating Temperature: -5°C
Max. Operating Temperature: 40°C
Max. Operating Range: 4 km
Controller Frequency: 2.4 GHz
Battery Capacity: 4,500 mAh
Battery Type: Lithium Polymer 6S
GPS Compatible: √
Altitude Hold: √
Position Hold: √
Flight Planning: √
Autonomous Flight: √
Target Tracking: –
Automatic Landing: √
Return To Home: √
Obstacle Avoidance: –
Headless Mode: –
Built-in Camera: –

Hubsan

X4 Star Pro H507A

© Hubsan

Release Date: December 2016
Price: $64
Number of Rotors: 4
Control System: Remote Control
Applications: Recreation, FPV Racing
Dimensions: 225 × 225 × 60 mm
Weight: 160 gr
Frame Material: Plastic
Rotor Material: Plastic
Available Colors: Black
Max. Flight Time: 10 minutes
Max. Operating Range: 100 m
Controller Frequency: 2.4 GHz
Battery Capacity: 550 mAh
Battery Type: Lithium Polymer
Battery Power: 7.6V
Battery Weight: 24 gr
GPS Compatible: √
Altitude Hold: √
Position Hold: √
Flight Planning: √
Autonomous Flight: √
Target Tracking: √
Automatic Landing: √
Return To Home: √
Obstacle Avoidance: –
Headless Mode: √
Built-in Camera: √ (720P HD)

Immersion RC

Vortex 150 mini

©helipal.com

Release Date: December 2016
Price: $319
Number of Rotors: 4
Control System: Remote Control
Applications: FPV Racing
Weight: 250 gr
Frame Material: Carbon Fiber
Rotor Material: Plastic
Motors: 1306-3200kv
Available Colors: Green, Red
Max. Flight Time: 4 minutes
Controller Frequency: 5.8 GHz
Battery Capacity: 500 mAh
Battery Type: Lithium Polymer 4S
GPS Compatible: –
Altitude Hold: –
Position Hold: –
Flight Planning: –
Autonomous Flight: –
Target Tracking: –
Automatic Landing: –
Return To Home: –
Obstacle Avoidance: –
Headless Mode: –
Built-in Camera: √ (600TVL)

Makerfire

Micro FPV Lite Quad

©rcdronearena.com

Release Date: December 2016
Price: $52
Number of Rotors: 4
Control System: Remote Control
Applications: Recreation, FPV Racing
Dimensions: 64 × 64 mm
Weight: 22 gr
Frame Material: Plastic
Rotor Material: Plastic
Motors: 6 × 15 coreless
Available Colors: Green, Red
Max. Flight Time: 4 minutes
Max. Operating Range: 300 m
Controller Frequency: 2.4 GHz
Battery Capacity: 200 mAh
Battery Type: Lithium Polymer
Battery Power: 3.7V
GPS Compatible: –
Altitude Hold: –
Position Hold: –
Flight Planning: –
Autonomous Flight: –
Target Tracking: –
Automatic Landing: –
Return To Home: –
Obstacle Avoidance: –
Headless Mode: –
Built-in Camera: √ (520TVL)

Meijiaxin Toys

Bugs 3

©justdrones.com.au

Release Date: December 2016
Price: $109
Number of Rotors: 4
Control System: Remote Control
Applications: Recreation, FPV Racing
Dimensions: 440 × 440 × 150 mm
Wingspan: 310 mm
Weight: 460 gr
Frame Material: Plastic
Rotor Material: Plastic
Motors: Brushless
Available Colors: Black
Max. Flight Time: 20 minutes
Max. Operating Range: 500 m
Controller Frequency: 2.4 GHz
Battery Capacity: 1,800 mAh
Battery Type: Lithium Polymer
Battery Power: 7.4V
GPS Compatible: –
Altitude Hold: √
Position Hold: √
Flight Planning: –
Autonomous Flight: –
Target Tracking: –
Automatic Landing: –
Return To Home: –
Obstacle Avoidance: –
Headless Mode: –
Built-in Camera: –

Muvi

Q1 drone

©toyscrates.com

Release Date: December 2016
Price: $699
Number of Rotors: 4
Control System: Remote Control
Applications: Film, Photography
Wingspan: 450 mm
Frame Material: Plastic
Rotor Material: Plastic
Available Colors: Black
Max. Flight Time: 20 minutes
Max. Speed: 72 km/h (44,7 mp/h)
Max. Ascent Speed: 6 m/s
Max. Descent Speed: 4 m/s
Max. Operating Range: 700 m
Controller Frequency: 2.4 GHz
Battery Capacity: 6,800 mAh
Battery Type: Lithium Polymer
GPS Compatible: √
Altitude Hold: √
Position Hold: √
Flight Planning: √
Autonomous Flight: √
Target Tracking: √
Automatic Landing: √
Return To Home: √
Obstacle Avoidance: –
Headless Mode: √
Built-in Camera: –

Tobyrich

SmartPlane Pro

© Tobyrich

Release Date: December 2016
Price: $175
Number of Rotors: 1 (Fixed Wing)
Control System: Phone/Tablet
Applications: Recreation, FPV Racing
Dimensions: 330 × 300 × 80 mm
Wingspan: 300 mm
Weight: 34 gr
Frame Material: Carbon Fiber, Durinum
Rotor Material: Plastic
Available Colors: Black
Max. Flight Time: 12 minutes
Max. Operating Range: 100 m
Controller Frequency: 2.4 GHz
Battery Capacity: 185 mAh
Battery Type: Lithium Polymer
GPS Compatible: –
Altitude Hold: –
Position Hold: –
Flight Planning: –
Autonomous Flight: –
Target Tracking: –
Automatic Landing: –
Return To Home: –
Obstacle Avoidance: –
Headless Mode: –
Built-in Camera: –

Wingcopter

Wingcopter 178 Heavy Lift

© Wingcopter

Release Date: December 2016
Price: On Request
Number of Rotors: 4 (Fixed Wing)
Control System: Flight Computer, Tablet
Applications: Delivery, Transport
Dimensions: 1.32 × 1.78 × 0.52 m
Wingspan: 1.78 m
Weight: 6 kg
Max. Payload: 6 kg
Available Colors: White, Yellow
Max. Speed: 150 km/h (93 mp/h)
Max. Flying Altitude: 5,000 m (16,404 ft)
Max. Ascent Speed: 6 m/s
Max. Operating Range: 100 km
GPS Compatible: √
Altitude Hold: √
Position Hold: √
Flight Planning: √
Autonomous Flight: √
Target Tracking: √
Automatic Landing: √
Return To Home: √
Obstacle Avoidance: –
Headless Mode: –
Built-in Camera: –

Index

ARGENTINA

Aerodreams
Buenos Aires
22 UAV Strix

Ejército Argentino
Buenos Aires
www.argentina.gob.ar/ejercito
80 Lipán M3

Nostromo Defensa
Córdoba
www.nostromo-group.com
26 Yarara 01-B
81 Cabure III

AUSTRALIA

IoT Group
Sydney
www.myroam-e.com
520 ROAM-e

Sci.Aero
Balcatta
www.sci.aero/4scight-3
353 CyberQuad Maxi
511 TopoDrone 4Scight

Kaiser Baas
Rowville
www.kaiserbaas.com
318 Alpha
462 Delta
591 Gamma
592 Theta

Xtreem
(By Swann productions)
Port Melbourne
www.swann.com
239 Atom Quadcopter
383 Mini Stealth Drone
414 Atom II Lightning Fast Mini
413 Electro-Max Eye
413 Gravity Pursuit
414 Quadforce

AUSTRIA

Airborne Robotics
Klagenfurt am Wörthersee
www.airborne-robotics.com
274 Air 6
275 Air 8

Carrera RC
(By Stadlbauer Marketing & Vertrieb GmbH)
Salzburg
www.carrera-rc.com
243 Quadrocopter X1
374 Quadrocopter Video Next
556 Mario Copter

BELGIUM

AltiGator
Waterloo
www.altigator.com
84 ALG-EOS

Fleye
Liège
www.gofleye.com
544 Flee

Onyxstar
(by Altigator)
Waterloo
www.onyxstar.net
181 Fox-C8 HD
302 Hydra12
407 Fox-C8 XT RTF
523 Xena 8F

Sky-Hero
Brussels
www.sky-hero.com
168 Spy
182 Spyder 6
249 Little Spyder
305 Little Six
353 Anakin
464 Anakin Club Racer
511 OB1

Trimble
(A Service-Drone Company)
Ghent
www.trimble.com
370 ZX5 multirotor

BRAZIL

SantosLab
Rio de Janeiro
www.santoslab.com
47 Carcará I
127 Carcará II

Xmobots
São Paulo
www.xmobots.com.br
60 Apoena 1000
108 Nauru 500A
239 Echar 20A

CANADA

4 front Robotics
Calgary, Alberta
www.4frontrobotics.com
226 Navig8 Electric

Aeromao
Mississauga, Ontario
www.aeromao.com/
184 Aeromapper Talon
211 Aeromapper 300
283 Aeromapper EV2

Aeryon Labs Inc.
Waterloo, Ontario
www.aeryon.com
62 Scout
128 SkyRanger R60

Airhogs
(a Spin Master company)
Toronto, Ontario
www.airhogs.com
134 Helix X4 Stunt
372 Helix Video Drone
373 Millenium Falcon Quad
397 Helix Ion
451 Helix Race
514 Connect Mission Drone
528 Star Trek U.S.S. Enterprise
555 Helix Sentinel
566 Hyper Stunt
582 X-Stream Video

Draganfly
Saskatoon, Saskatchewan
www.draganfly.com
99 Draganflyer X4-P
244 Draganflyer Guardian
516 Draganflyer Commander

DreamQii
Toronto, Ontario
www.dreamqii.com
517 PlexiDrone

ING Robotic Aviation
Ottawa, Ontario
www.ingrobotic.com
147 Serenity

Rotor X
(By Hexwing Industries)
Laval, Quebec
www.go.rotorxracing.com
564 Atom RX122

CHILE

Idetec
Santiago
www.idetec-uas.com
77 Stardust
246 AG-Wing

R.M.S.
Quilpué
www.rms.co.cl
72 Mantarraya

CHINA

AEE Technology Co., Ltd.
Shenzhen, Guangdong Province
www.en.aee.com
171 F50
219 Toruk AP10
297 F100
343 Toruk AP9
385 Toruk AP11

AeroSky
(By Xheli)
www.xheli.com
125 550
158 C17
164 P12
266 C6
284 QAV 250
284 QAV ZMR250
309 QAV 280
344 X350

All Tech
(a Keweitai brand)
Shenzhen, Guangdong Province
195 KWT-Z4M
468 KWT-X6L

Aosenma
Shantou, Guangdong Province
211 CG022
583 CG035

Arris
Chengdu, Sichuan Province
www.arrishobby.com
285 X-Speed 250
415 X-Speed 280
430 M700
499 F210
515 X180
540 X210
597 C250 FPV

Art-Tech
Shenzhen, Guangdong Province
www.art-tech.com
138 Mini X6
256 AT-100

Assault
(by Hobbyking)
Hong Kong
www.hobbyking.com
256 Reaper 500 3D

ASUAV (Awesome)
Dongguan, Guangdong Province
www.awesomeuav.com
598 Youbi XV-130

Cheerson
Shantou, Guangdong Province
www.cheersonhobby.com
165 CX-20
186 CX-10
187 CX-11
212 CX-30
310 CX-22
311 CX-31
374 CX-10C
387 CX-33S
415 CX-32S Falcon
416 CX-Stars
432 CX-10W
432 CX-35 Leopard
453 CX-91 Jumper
484 CX-10D
500 CX-36
515 CX-37
529 CX-10WD
598 CX-117
599 CX-93S Thorn Jump

Chengdu Aircraft Industry Group (CAIG)
Chengdu, Sichuan Province
www.cac.avic.com
54 Sky Wing III
62 Pterodactyl I
101 VMA-01
149 VD200
375 Wing Loong II
585 Cloud Shadow

China Aerospace Science and Industry Corporation (CASIC)
Beijing
www.cccme.org.cn
51 Sky Hawk 1
68 HW-300 Blade
69 HW-310
70 WJ 600 Sky Hawk
75 HW-210
113 SF-460 Blade
243 HW-510

China Aerospace Science and Technology Corporation (CASC)
Beijing
www.english.spacechina.com
49 CH-802
86 CH-803
113 CH-92
220 CH-4 Rainbow
358 CH-3 Rainbow
375 CH-91
585 CH-5 Rainbow

DaMing (DM)
Handan City, Hebei Province
311 Spy 007
398 Spy 003
599 Conqueror

Diatone Innovations
Donggang, Guangdong Province
www.diatone.hk

485 Tyrant 180
516 Crusader 190

DJI
Shenzhen, Guangdong Province
www.dji.com
121 Phantom 1
135 Spreading Wings S800
146 Phantom 2 Vision
155 Phantom 2
159 Spreading Wings S1000
178 Phantom 2 Vision+
212 Spreading Wings S900
231 Spreading Wings S1000+
244 Inspire 1 V2.0
300 Phantom 3
325 Matrice 100
433 Agras MG-1
470 Phantom 4
485 Matrice 600
557 Mavic Pro
586 Inspire 2
587 Phantom 4 Pro

DYS
Huizhou, Guangdong Province
www.dys.hk/
221 D800 X4
232 D800 V6
233 D800 X8
245 D1100 V8
312 BG-250
399 X230 FPV
470 X-Ray V3
518 Lightning X220
541 XDR220

Eachine
Beijing
www.eachine.com
203 H1 Sky Walker
213 X6
245 CG023
287 Q200
313 H8 Mini
326 3D X4
359 H7
359 H8 3D Mini
388 Racer 250
400 H8C Mini
434 Assassin 180 FPV
434 Falcon 250
435 H99D
454 H99W
471 E10C Mini
471 E20 3D
486 E10W
486 E30W
487 Falcon 180
501 Blade 185
501 E40G
542 QX90
542 Wizard X220
557 E010 mini
558 Falcon 250 Pro
569 QX95
588 EX110
600 E50
600 Fatbee FB90

EHANG
Guangzhou, Guangdong Province
www.ehang.com
400 Ghost 2.0

eTurbine
Hong Kong
360 TB 250

Ewatt Aerospace
Wuhan, Hubei Province
www.ewattuav.com
140 EWZ-S8

Fineco
Guangdong Province
188 FX1 Nano
313 FX-4V
327 FX-3
346 FX-2
346 FX-7Ci
361 FX-21
436 FX-6Ci

Flypro
Shenzhen, Guangdong Province
www.flypro.com
189 X400
196 X600
268 PX400
502 X-Eagle
588 X-Jaguar

Fly Sight
Shenzhen, Guangdong Province
www.flysight.com
455 F250

Foxtech
Tianjin
www.foxtechfpv.com
347 Devourer130 V2
472 A4
519 Screamer 250

FUAV
Shenzen, Guangdong Province
www.f-uav.com
347 Seraphi Pro

Fuqi Toys co.,Ltd
Shantou, Guangdong Province
www.fuqitoys.com
328 FQ777-124 Pocket Drone
362 FQ777-956C Sirius
377 FQ777-AF957 The Gazer
389 FQ777-955C Scorpius
436 FQ777-126 Spider
437 FQ777-954W FPV
487 FQ777-951W Mini
488 FQ777-954D Mini
530 FQ777-01
570 FQ777-03 Tiny Buddy

GoolRC
Hong Kong
www.goolrc.com
544 X5C
558 T5G

Gteng
Shantou, Guangdong Province
www.en.gtengtoys.com
488 T901C
503 T901F
520 T902C
545 T903
545 T905F
589 T906W
590 T907W

Guizhou Aviation Industry Group (CAIG)
Guiyang, Guizhou province
www.gaiec.cn/en/

13 WZ-2000
125 Harrier Hawk I
531 Soar Dragon

Harbin Aviation Industrial Group (HAIG)
Harbin, Heilongjiang Province
28 BZK-005 Giant Eagle

Helic Max
(a hengdi digital technology company)
Shenzhen, Guangdong Province
www.helicmax.com
277 Sky Hawkeye
329 Sky Cruiser
348 Sky Roller
455 Sky Phantom
473 Sky Vampire
473 Mini Sky Walker

HiSky Hobby
(By Chiyuan Electronics Co. Ltd)
Guanghzou, Guangdong Province
189 HMX120
330 HMX280
417 Talon 280

Hobby King
Hong Kong
www.hobbyking.com
214 Super-X
363 Black Widow 260 FPV
456 KingKong 260 FPV

Hobby Lord
Shenzhen, Guangdong Province
www.hobbylord.com
100 ST550 Bumblebee
143 ST800
314 Alfa Y3
503 Alfa H6

Hongdu Avation Industry Group (HAIG)
Nanchang, Jiangxi Province
www.hongdu.com.cn
117 Sky Crossbow (Tian-Nu)
151 Sharp Sword (Li-Jian)

Huajun Toys
Shantou, Guangdong Province
www.huajuntoys.com
173 W608-1
222 W608-2
246 W608-8
301 W609-7
315 W609-8 Pathfinder
315 W609-9
316 W609-10
390 W609-5
474 W606-2 Move Wings
474 W606-3 U-Fly
475 W608-7
504 W606-5 Flanker
590 W606-6

HuaXiang
Shantou, Guangdong Province
www.cn.huaxiangtoys.com
197 8953 Camera
198 8957 Raider
269 8983 Sky Walker
288 8927
316 8923 Smart
330 8933V Exceed
349 8963 Camera
437 8925 Explorer
531 8971V

Hubsan
Shenzhen, Guangdong Province
www.hubsan.com
135 Sky Cruiser H302F
141 X4 H107C
160 X4 H107L
204 Nano Q4 H111
402 X4 Pro FPV H109S
403 X4 Plus H107D+
438 X4 Cam Plus H107C+
438 X4 Cam H501C
456 X4 Air H501S
457 X4 Desire H502S
475 X4 Plus H107P
489 Nano Q4 plus H111C
490 Nano Q4 FPV H111D
602 X4 Star Pro H507A

Idea-Fly
Shenzhen, Guangdong Province
www.idea-fly.com
198 Storm-800
234 Hero-550
390 Mars-350
504 Apollo
546 Grasshopper F210
573 Poseidon 480

Immersion RC
Hong Kong
www.immersionrc.com
259 XuGong V2
301 Vortex 285
403 Vortex 250 Pro
603 Vortex 150 mini

Jin Xing Da
Shantou, Guangdong Province
www.jxdtoys.com
130 385
174 388
174 392
179 389
190 391 Space Trek Galaxy
190 393 Space Trek Defiant
191 395 Airbus
331 398
349 502
377 509G
404 509V
417 509W
505 510G X-Predators
505 512W
532 511V Pterosaurs
559 506G/W

JJRC
Shantou, Guangdong Province
www.jjrctoy.com
199 F182
234 JJ600A
247 H3
247 H8C
259 H9D
260 H5C
260 H6C
261 H7
269 JJ800
277 JJ810
278 V686
289 JJ820
302 H12C
317 H8D
317 JJ1000
331 H15
332 H13C

332 JJ669
333 JJ830
350 H8 Mini
350 JJ850
363 H20
364 H6D
364 H10
365 H18
365 JJ200
378 H5P
391 H11D
392 H12W
392 H21
404 H22
418 H11C
418 H26D
419 H5M
419 X1
420 H26
420 H26W
439 H6W
439 H25G
460 H20W
460 H29
461 H30C
461 H29G
476 H28
490 H29C
521 JJ500
532 H31
546 H11WH
547 H23
560 H36
574 JJPRO P130
574 JJPRO X1G
591 H37

JTT UAV
Shenzhen, Guangdong Province
www.en.jttuav.com
405 T50
547 T60 V1

Kongcopter
Shanghai
205 FH-800
235 FQ-700

Lishitoys
Shantou, Guangdong Province
www.cn.lishitoys.com/eng
278 L6052
290 L6039
366 L6053
440 L6055
506 L6056
521 L6058

Meijiaxin Toys (MJX)
Hong Kong
www.mjxtoys.com
261 X700C
318 X400 V2
333 X300C
334 X500
334 X600
335 X800
351 X705C
351 X900
379 X101
440 X902
507 X601H
522 X904
522 X905C
549 X104T
549 X301H
561 X906T
561 X908T
592 X909T
604 Bugs 3

MMC
Shenzen, Guangdong Province
www.mmcuav.com
441 F4 Precision Agriculture Drone
441 F6 Precision Agriculture Drone
476 Phoenix A6
491 HyDrone 1800
491 HyDrone 1550

MOLA
(By Nine Eagles)
Shanghai
www.i-mola.com/
367 Tourist 1
379 Fanatic 1

Nihui Model Technology
Shantou, Guangdong Province
214 U207 Intruder
279 U807
550 NH-010

Nine Eagles
Shanghai
www.nineeagle.com
206 Galaxy Visitor 3
235 Galaxy Visitor 6
421 Galaxy Visitor 7

PowerVision
Beijing
www.powervision.me
552 PowerEgg
562 PowerEye

Quanum
(by Hobbyking)
www.hobbyking.com
Hong Kong
199 Nova
236 Venture

RC Logger
Hong Kong
www.rclogger.com
162 RC Eye One Xtreme
319 RC Eye One NovaX 350
510 RC Eye One Navigator 250
575 RC Eye Twist 300

Ruiven Technologies
Fuzhou, Fujian Province
www.en.ruiven.cc
577 Kudolo Wing

Skyrc
Shenzhen, Guangdong Province
www.skyrc.com
445 Sokar FPV

Speedwolf
Shenzhen, Guangdong Province
www.speedwolf.net
250 Vajra 80

SMD Smart Drone
Shenzhen, Guangdong Province
www.smd-uav.com/en/
553 Mirage Pro
564 Red Arrow

SRD (Storm Racing Drone)
(A Helipal company)
Hong Kong

www.helipal.com
270 Storm Drone 4 mini
271 Storm Drone 4 V3
337 SRD 260 Pro V2
338 Storm Drone 8
410 SRD Lumenier QAV210
422 SRD 370 Naza V2
464 SRD 210
465 SRD Ninja 250
465 SRD 250 V3
478 SRD 195
478 SRD 280 V3
492 SRD 240X
493 SRD Y3
525 SRD 180 V3
525 SRD NR8
535 SRD130 Loki-X3
535 SRD163 Loki-X4
536 SRD199 Loki-X5
536 SRD 200X
537 Storm Drone 6 V3
537 SRD FlexRC Owl Storm Edition
538 SRD TBS Vendetta
595 SRD 101

Syma Toys
Shantou, Guangdong Province
www.symatoys.com
116 X1
131 X3 Pioneer
163 X4 Assault
192 X5C Explorer
192 X6 Super Ship
207 X2
251 X11 Hornet
251 X12S Nano
305 X5S
306 X8C Venture
306 X13 Storm
319 X5SW
369 X8G
410 X18
466 X54HC
553 X8HC
578 X5UC

Tarot RC
Wenzhou, Zhejiang Province
www.tarot-rc.com
281 200 FPV Mini
380 250 FPV
493 130 FPV
526 120 FPV
526 150 FPV
579 Peeper Long Flight Time TL750S1

Team BlackSheep
Hong Kong
www.team-blacksheep.com
103 Discovery
132 Caipirinha
223 Gemini FPV Racer
466 Vendetta
494 Vortex 285

UDIRC Toys Industrial Co., Ltd.
Shantou, Guangdong Province
www.udirc.com
169 U830 Demon
237 U841-1 HD
272 U840 mini
294 U842-1 Lark FPV
307 U27 Free Loop
307 U818A-1 Discovery
320 U839 Nano
355 U842 Falcon
355 U843 multi-skywalker
446 U818A Discovery WiFi
480 U845 Voyager 6
565 U45 Raven
579 U28-1 kestrel

Walkera
Guangzhou, Guangdong Province
www.walkera.com
170 QR X350 Pro
200 QR Y100
217 Tali H500
224 Scout X4
295 Voyager 3
320 QR X350 Premium
356 QR X900
371 Runner 250R
394 Runner 250 Pro
447 F210
447 Furious 320
495 F210 3D
496 Rodeo 150
580 AiBao
596 Voyager 4

Shenzhen Wingsland Technology Co., Ltd.
Shenzhen, Guangdong Province
www.szsungreen.com
308 Scarlet minivet
448 K3
448 M5
596 S6

WLtoys
Guangdong Province
www.wltoys.eu/en
130 V959 Pro Battleship
176 V303 Seeker
177 V323 Skywalker
183 V252 Skylark
201 V333N Ufo
208 V282/V292 mini Skylark Egg
209 V333 Ufo
209 V353 Galaxy
238 V636 Skylark
238 V393 Brushless Explorer
253 V646 mini Ufo
254 V383 500 3D
254 V676 mini Ufo
272 V686G Explore
321 Q202 Aircraft Carrier
321 V606 Explorers
371 Q212 Spaceship
383 Q282J Spaceship
395 Q222 Spaceship
412 DQ242G
423 Flying 3D X6 Plus
496 Q303C Spaceship
527 Q343 Mini Quadcopter
554 Q333

XAircraft America
Guangzhou, Guangdong Province
www.xaircraft.cn/en
295 Xmission
497 Xplanet P20 V2

Xi'an Aisheng Technology Group
Xi'an, Shaanxi Province
www.aisheng.nwpu.edu.cn
64 ASN-207
64 ASN-209
78 ASN-102
79 ASN-216

79 ASN-229
104 ASN-218

Xinlin Shiye
Shantou, Guangdong Province
www.xinlinshiye.com
395 X118 FPV
480 X165 mini Quad
497 X181 FPV

Xiro
(By Zero UAV Tech.Co.,LTD)
Shenzhen, Guangdong Province
282 Xplorer
449 Xplorer 2
449 Xplorer Mini

XK Innovations
Shenzhen, Guangdong Province
www.xk-innovations.com
322 Air Dancer X350
322 Alien X250
323 Detect X380
396 X260
424 Aircam X500
481 Whirlwind X251
481 X100 3D
513 X252 Shuttle

Yi Zhan Toys
Shantou, Guangdong Province
www.yizhantoys.com
218 X4
342 Tarantula X6
425 i6S
450 i7H
450 i8H
482 i4S
539 i4W
566 i3

Yuneec International
Jinxi, Jiangsu Province
www.yuneec.com
171 Flying Eyes HX3
264 Typhoon Q500
357 Typhoon Q500 4K
384 Tornado H920
425 Typhoon G
498 Typhoon H
581 Breeze

Zerotech
Beijing
www.zerotech.com
210 HighOne
225 E-Epic
426 Guardian Z10
514 Dobby

Zero-Zero Robotics
Shenzhen, Guangdong Province
gethover.com
581 Hover Camera

CZECH REPUBLIC

Robodrone
Brno
www.robodrone.com
304 Kingfisher
352 Sparrow
577 Hornet

DENMARK

Sky-watch
Aalborg
www.sky-watch.com
144 Huginn X1D
263 Cumulus V1

ENGLAND

BAE Systems
Farnborough
baesystems.com
40 HERTI
86 Telemos

Extreme Fliers
London
www.extremefliers.com/
156 Microdrone 2.0
361 Microdrone 3.0

Flyingwings
Newport Pagnell
www.flyingwings.co.uk
287 Venturi Evo

Muvi
(By Veho Global Group)
Eastleigh, UK
368 X drone
604 Q1 drone

QuadH2o
(By XtremeVision360 Ltd.)
Worthing
137 QuadH2o
279 HexH2o
563 SarahH2o

QuestUAV
Northumberland
www.questuav.com
181 Q-200 Surveyor Pro
451 Q-100 Datahawk

Skyartec
Eaton Socon
www.skyartec.co.uk
142 Butterfly

Vulcan
Mitcheldean, UK
www.vulcanuav.com
163 Black Widow
217 Raven
513 Airlift

ESTONIA

Threod Systems
Viimsi
threod.com
512 EOS mini UAS

EUROPEAN UNION

Airbus
(former European Aeronautic Defence and Space Company, partnered by France, Germany and Spain)
Toulouse (France),
Leiden (Netherlands)
www.airbus.com
22 KZO
45 Harfang
50 Barracuda
133 Atlante
325 Talarion

Dassault Aviation
(Partnered by France, Greece, Italy, Spain, Sweden and Switzerland)
Paris (France)
117 nEUROn

MDBA
(Partnered by England, France, Germany and Italy)
Stevenage (UK)
www.mbda-systems.com
97 Fire Shadow

FINLAND

Avartek
Masala
www.avartek.fi
37 AT-04

FRANCE

Delair-Tech
Toulouse
www.delair-tech.com
195 DT18
312 DT26X

Delta Drone
Dardilly
www.deltadrone.com
257 Delta Y
500 Delta X

EscaDrone
Grenoble
www.escadrone.com
360 Squad3

Eznov
Metz
www.ezmanta.eznov.com
570 Ez Manta

Hélicéo
Nantes
www.heliceo.com
572 Fox4
572 FoxyPro

HESEL EU
St Pierre de Maillé
www.helsel.eu
329 Splash Drone
489 Splash Drone 2

Lehmann Aviation
Blois
www.lehmannaviation.com
35 LP960
58 LV580
71 LM450
115 LA100
136 LA200
136 LA300

Nano-Racing
Paris
550 Nano Racer

Novadem
Aix-en-Provence
www.novadem.fr
407 NX110m
562 U130

Parrot
Paris
www.parrot.com
75 AR.Drone
95 AR.Drone 2.0
206 Rolling Spider
248 Bebop
336 Airborne Cargo
336 Airborne Night
352 Hydrofoil
421 Bebop 2
551 Disco
551 Mambo

Sagem
Paris
www.safran-electronics-defense.com
12 Sperwer MK-II

Squadrone Systems
Grenoble
www.hexoplus.com
422 Hexo Plus

Survey-Copter
(by Airbus Group)
Pierrelatte
www.survey-copter.com/en
182 DVF2000

Thales
Paris
www.thalesgroup.com/en
216 Watchkeeper WK450
411 Spy'Ranger
538 Fulmar X

Xamen Technologies
Pau
www.xamen.fr
210 BE 4000
356 LE 4-8 X Dual

Xtim
Marseille
www.mybionicbird.com
263 Bionic Bird

GERMANY

Acme
Westenholz
www.acme-online.de
194 Zoopa Q165
194 Zoopa Q410 Movie
227 Zoopa Q650
266 Zoopa Q55 Zepto
273 Zoopa Q155 Roonin
296 Zoopa Q420 Cruiser
323 Zoopa Q Crosswave 290
324 Zoopa Q165 Riot
343 Zoopa Q550 Evolution
372 Zoopa Q650 Razor
384 Zoopa Q600 Mantis
482 Zoopa Q400 Hunter

Aibotix
Kassel
www.aibotix.com
164 Aibot X6 V2

Aidrones
Koenigs Wusterhausen
www.aidrones.de
285 AiD-MC8

AirRobot
Arnsberg
www.airrobot.de
91 AR 100-B
105 AR 200
172 AR 180

AscTec
(Ascending Technologies GmbH)
Krailling
www.asctec.de

57 Falcon 8
73 Pelican
91 Hummingbird
229 Firefly

Bormatec
Ravensburg
www.bormatec.com
74 Maja
96 Cam Flyer Q
99 Explorer
310 Ninox

EMT-Penzberg
Penzberg
www.emt-penzberg.de
15 UAV X-13
50 Fancopter
63 Aladin

Flairics
(by Spectairgroup)
Meerbusch
www.flairics.com
213 Journalist
401 Inspector S

Heigth Tech
Meerbusch
www.heighttech.com
288 HT-8 C180

HiSystems GmbH
Moormerland
www.mikrokopter.de
129 Mikrokopter 6S12
602 Mikrokopter 8-3500

Microdrones.com
Siegen
www.microdrones.com
65 MD4-1000
70 MD4-200
509 MD4-3000

Monstertronic
Berlin
www.monstertronic.org
161 Sky Agent V2
335 Invader V2
367 Sirius Quattro
405 Sky Eagle
406 Sky Bot
406 Sky Hawk
462 Sky Bolt
463 Sky Runner
509 Sky Force
533 FPV Fighter

Revell Control
Bünde
www.revell-control.de
137 Atomium
215 Nano Quad Pro
249 Hexatron
270 XS Nano Quad
291 Nano Quad Cam
291 Rayvore
292 Spot
303 X-Spy 2.0
304 Nano Hex
369 Proto Quad
394 Nano Pocket
408 Backflip 3D
409 Formula Q
409 Funtic
445 Orbix
477 Steady Quad Cam
575 Pulse FPV
576 VR Shot
593 Navigator
594 Quadrotox
594 Sky Spider

ScaraBot Technologies
Regensburg
www.scarabot.de
153 Scarabot X8

Service-drone.com
Berlin
www.service-drone.com
237 G4 Surveying Robot
280 G4 Blackbird SR
280 G4 Eagle V2 Cargo
281 G4 Skycrane
292 G4 Recon One

Tobyrich
Bremen
www.tobyrich.com
381 X-spy
595 Moskito
605 SmartPlane Pro

Volocopter GmbH
Bruchsal
www.volocopter.com/en/
495 Volocopter VC200

Wingcopter
Darmstadt
www.wingcopter.com
605 Wingcopter 178 Heavy Lift

GREECE

BSK Defense
Chania
www.bsk-defense.com
299 Phaethon G

Hellenic Aerospace Industry
Tanagra
www.haicorp.com/en/
20 E1-79 Pegasus II

HUNGARY

BHE Robotics
Budapest
www.bherobotics.com
185 BXAP15

INDIA

Aurora Integrated Systems
Bengaluru
www.aurora-is.com
76 Altius MK-II
83 Urban View

Defence Research Development Organisation of India (DRDO)
New Delhi
www.drdo.gov.in
63 Rustom 1
94 Imperial Eagle
94 Netra
122 Abhyas
586 Rustom 2

Kadet Defence Systems
Kolkata
www.kadet-uav.com
506 JX2 NG

National Aerospace Laboratories (NAL)
Bangalore

www.nal.res.in/en
92 Slybird
118 Golden Hawk

OM UAV
New Delhi
www.omuavsystems.com
175 Guardian Fixedwing
191 Baaz Mini

INDONESIA

PT Dirgantara Indonesia
Bandung
www.indonesian-aerospace.com
167 Puna Wulung

IRAN

Iran Aircraft Manufacturing Industrial Company
Shahin Shahr
25 HESA Ababil 3
74 HESA Karrar
97 Shahed 129
126 HESA H110 Sarir

Iran Aviation Industries Organisation (IAIO)
Tehran
129 Hamaseh
143 Yasir
151 Fotros

Qods Aviation Industries
Tehran
215 Mohajer 4B
248 Mohajer 2N

IRELAND

Versadrones
Cork
www.versadrones.com
132 Versa X6

ISRAEL

Aeronautics Ltd.
Yavne
www.aeronautics-sys.com
4 Aerostar (RQ-26A)
19 Orbiter 2
85 Orbiter 3
90 Dominator XP
539 Orbiter 1K

Bluebird Aero Systems Ltd.
Kadima Zoran
www.bluebird-uav.com
202 Micro B

Elbit Systems
Haifa
www.elbitsystems.com
27 Skylark II
61 Hermes 90
65 Hermes 900
80 Skylark I-LE
326 Skylark I-LEX
454 Skylark 3
543 Skylark C

Flytrex
Tel Aviv
www.flytrex.com
327 Sky

Innocon Ltd.
Holon
www.innoconltd.com
17 Mini Falcon
102 Micro Falcon
121 Spider

Israel Aerospace Industries (IAI)
Tel Aviv
www.iai.co.il
15 Bird-Eye 500
21 Bird-Eye 400
21 Heron 1/Shovel
27 I-View MK 50
33 Heron TP/Eitan
37 I-View MK 150
38 I-View MK 250
53 Harop
61 Mosquito
67 Bird-Eye 650
67 Searcher III
77 Panther
78 Mini Panther
83 Ghost
92 ETOP
165 Super Heron
204 Bird-Eye 650D
458 FE-Panther
458 Harpy NG
459 Green Dragon
459 Rotem-L

Rafael
Haifa
www.rafael.co.il
29 Skylite B

Uvision
Kochav Yair–Tzur Yigal
www.uvisionuav.com
87 Sparrow
339 Hero-30
339 Hero-70
340 Hero-120
340 Hero-250
341 Hero-400
341 Hero-900

ITALY

Aermatica
Colverde
www.aermatica.com
124 Anteos

IDS Corporation
Pisa
www.idscorporation.com
146 IA-3 Colibri
147 IA-17 manta

Leonardo S.p.A.
Rome
www.leonardocompany.com/en
7 DRS RQ-15 Neptune
20 Sky-X
31 Falco
35 Sky-Y

Piaggio Aerospace
Villanova D'Albenga
www.p1hh.piaggioaerospace.it
262 P.1HH HammerHead

Virtualrobotix
Calcio
www.virtualrobotix.it
183 VR Tiger Shark X8
294 VR Spark 350
467 VR Mapper 300

JAPAN

Autonomous Control Systems Laboratory (ACSL)
Chiba City

www.en.acsl.co.jp
282 Mini Surveyor MS-06LA

Prodrone
Nagoya City
www.prodrone.jp
408 Byrd
552 PD6B-AW-ARM
563 PD6-CI-L

LATVIA

Airdog SIA
Riga
www.airdog.com
498 Airdog

UAV Factory
Marupe
www.uavfactory.com
142 Penguin B

MALAYSIA

Composite Technology Research Malaysia (CTRM)
Malacca
www.ctrm.com
31 Aludra MK1
32 Aludra SR-08

MEXICO

Arca Space Corporation
Las Cruces
www.arcaspace.com
240 AirStrato Accelerator
241 AirStrato Explorer
241 AirStrato Pioneer

Hydra Technologies
Jalisco
www.hydra-technologies.com
46 E1 Gavilan
46 S4 Ehécatl

NETHERLANDS

Acecore Technologies
Oss
www.acecoretechnologies.com
296 NEO

Aerialtronics
The Hague
www.aerialtronics.com
274 Altura Zenith

Gear2Play
(By Spectron Toys)
Deventer
www.spectrontoys.com
389 Nano Smart Drone
402 Sky
519 Condor
589 Focus

TRNDlabs
Enschede
www.trndlabs.com
252 SKEYE Nano
354 SKEYE Mini
382 SKEYE Pico Drone
446 SKEYE Hexa Drone
512 SKEYE Nano HD

Vertical Technologies
Badhoevendorp
www.verticaltechnologies.com
565 DeltaQuad One

NEW ZEALAND

Aeronavics
Raglan
www.aeronavics.com
172 Skyjib X4
184 Bot
185 XM6
255 Skyjib X8
467 Navi
483 Icon

Skycam UAV
Palmerston North
www.kahunet.co.nz
103 Kahu
107 Swamp Fox

NORWAY

Northern Research Institute (NORUT)
Tromsø
www.norut.no/en/
57 CryoWing

Prox Dynamics
Oslo
www.proxdynamics.com
123 PD-100 Black hornet Nano

Robot Aviation
Hønefoss
www.robotaviation.com
463 Skyrobot FX450

PAKISTAN

Global Industrial and Defefense Solutions (GIDS)
Rawalpindi
www.gids.com.pk
87 UQAB-II
114 Shahpar
258 Scout Mini

Integrated Dynamics
Karachi
www.idaerospace.com
13 Border Eagle Mk-II
23 Desert Hawk
34 Vision MK-II
53 Shadow MK-I
108 Pride
114 Nishan MK-II
115 Skycam UAV

National Engineering and Scientific Commission (NESCOM)
Islamabad
152 Burraq

SATUMA
Islamabad
www.satuma.com.pk
17 Jasoos II/Bravo+
48 Flamingo
55 Muhkbar
88 Thunder SR
119 Mini UAV Salaar

POLAND

Flytronic
(a WB Group company)
Ożarów Mazowiecki
www.wbgroup.pl
123 Fly Eye
179 Flysar tactical UAV
376 Manta VTOL

502 Warmate
601 FT-5 LOS

Rzeszów University of Technology
Rzeszów
www.w.prz.edu.pl
76 PR-5 Wiewiór Plus

Trigger Composites
Grodzisko Dolne
www.trigger.pl
216 Pteryx

PORTUGAL

Tekever
Lisbon
www.tekever.com/en
106 AR4 Light Ray Evolution
138 AR1 Blue Ray
207 AR2 Carcará
293 VR1 Colibri
354 AR3 Net Ray
381 AR5 Life Ray Evolution
527 AR4 Light Ray Compact

ROMANIA

Teamnet Group
Bucharest
www.teamnet.ro
479 Hirrus UAV

RUSSIA

AFM-Servers Ltd.
Moscow
297 Ptero-G0

Enics
Kazan
www.enics.ru
43 DSAS E25
156 Eleron 3SV
160 Eleron 10SV

Irkut Corporation
Irkutsk
www.irkut.com
72 Irkut-200
112 Irkut-3

Izhmash Unmanned Systems
(by the Kalashnikov Group)
Izhevsk
www.kalashnikov.com
81 Granat-4
145 Granat-2
161 Granat-3
289 Takhion Mini-UAV
391 Granat-1

Kronshtadt Group
Saint Petersburg
www.kronshtadt.ru
60 Dozor-600

Popov Omsk Radio Factory
Omsk
www.relero.ru
118 Iskatel

Special Technological Centre
Saint Petersburg
www.rfntr.neva.ru/eng
250 Orlan-10

UAV Systems Group
(By the Kalashnikov Group)
Izhevsk
www.unmanned.ru/en/
200 Supercam S350

Vega Radio Engineering Corporation
Moscow
www.vega.su
49 Tipchak

ZALA Aero Group
(By the Kalashnikov Group)
Izhevsk
www.kalashnikov.com/brand/zala-aero/products
48 Zala 421-8
59 Zala 421-12
71 Zala 421-21
89 Zala 421-16E
100 Zala 421-16EM
127 Zala 421-22

SERBIA

Utva Aviation Industry
Pančevo
44 Vrabac
148 Pegaz 011

SINGAPORE

Makerfire
www.makerfire.com
603 Micro FPV Lite Quad

ST Aerospace
www.stengg.com/en/aerospace
85 Skyblade III
96 Skyblade 360
109 Skyblade IV
168 Ustar-Y

SLOVENIA

Airnamics
Medvode
www.airnamics.com
386 R5

C-Astral
Ajdovščina
www.c-astral.com
140 Bramor C4EYE
231 Bramor rTK
373 Atlas UAS
584 Bramor ppX

SPAIN

ATyges
Malaga
www.atyges.es
298 RPAS FV6
452 RPAS FV1
499 RPAS FV8

CATUAV
Barcelona
www.catuav.com
106 Argos
267 Mineos

Dronetools
Sevilla
www.dronetools.es
110 DroneHexa
569 DroneQuad

Embention
Alicante
www.embention.com
111 Flamingo Firefighting
601 M600 multicopter

Sistemas de Control Remoto S.L.
Madrid
www.scrtargets.es
55 SCRAB I
56 SCRAB II
534 Atlantic

SOUTH AFRICA

AltiUAS
(former SteadiDrone)
Knysna
www.altiuas.com
567 Transition

Denel Dynamics
Centurion
www.deneldynamics.co.za
110 Hungwe
276 Seeker 200
276 Seeker 400

SteadiDrone
(Now AltiUAS)
Knysna
www.steadidrone.com
109 EI8HT
122 QU4D 2013
169 QU4D 2014
223 Mavrik
293 Flare
338 Vader

SOUTH KOREA

Korea Aerospace Industries (KAI)
Sacheon
www.koreaaero.com/english/
4 RQ-101 Night Intruder 300
166 Devil Killer

Ucon System
Daejeon
www.uconsystem.com
382 Remo M-002

SWEDEN

Intuitive Aerial
Linköping
www.intuitiveaerial.com
457 Aerigon MK II

SWITZERLAND

Autonomous Systems Lab
Zurich
www.atlantiksolar.ethz.ch/
128 Atlantiksolar

Flyability
Lausanne
www.flyability.com/elios/
518 Elios

Meteomatics
St Gallen
www.meteomatics.com
507 Meteodrone
508 Meteodrone XL
508 Meteodrone SSE

Perspective Robotics GmbH
Zürich
www.fotokite.com
303 Fotokite Pro
534 Fotokite Phi

SenseFly
(A Parrot Company)
Lausanne
www.sensefly.com
73 Swinglet
167 Ebee
492 Albris
578 Ebee Plus

UMS AERO Group
Baar
www.umsskeldar.aero
170 F-720 (Discoverer II)
193 F-330

TAIWAN

Align
Taichung
www.align.com
154 M424
219 M480L
227 M690L
298 M470
397 MR25

National Chung-Shan Institute of Science and Technology (NCSIST)
Taoyuan City
www.ncsist.org.tw/eng
40 Chung Shyang II
262 Cardinal II

Thunder Tiger Robotix
Taichung City
www.thundertiger.com
208 Super Hornet X650
400 Ghost +

TURKEY

Kale-Baykar
Istanbul
www.baykarmakina.com
28 Bayraktar Mini
180 Bayraktar TB2

Turkish Aerospace Industries (TAI)
Ankara
www.tai.com.tr/en
5 Keklik
12 Pelikan/IHA-X2
33 Gözcü
44 Turna-G
107 Şimşek
271 Anka-B

UNITED ARAB EMIRATES

ADCOM Systems
Abu-Dhabi
www.adcom-systems.com
104 Yabhon Smart Eye
124 Yabhon U-40
273 Yabhon R

UNITED STATES

3D Robotics
Berkeley, California
www.3dr.com
133 Y6 RTF
218 Iris+
225 Aero-M Fixed Wing
226 X8-M
240 X8+
308 Solo

Action Drone
Chula Vista, California
www.actiondroneusa.com
154 AD1
428 AD2

Adorama
New York City, New York
www.adorama.com
255 Aries Blackbird X10

Aerial Technology International
Bend, Oregon
www.aerialtechnology.com
283 AgBOT
428 Thor X4

Aerix Drones
Rochester, New York
www.aerixdrones.com
324 Nano Drone
385 Turbo-X
396 Aerius
429 Vidius
483 Black Talon 2.0

Aerovel Corporation
White Salmon, Washington
www.aerovel.com
69 Flexrotor

AeroVironment
Monrovia, California
www.avinc.com
8 RQ-14 Dragon Eye
11 RQ-11 Raven
38 Wasp III
45 RQ-20 Puma
88 Shrike VTOL
98 RQ-12A (Wasp AE)
98 Switchblade
582 Quantix

AgEagle
Neodesha, Kansas
www.ageagle.com
468 RX60

Agribotix
Boulder, Colorado
www.agribotix.com
201 Hornet

Allied Drones
Huntington Beach, California
www.allieddrones.com
267 EF44 Atlas

Altavian
Gainesville, Florida
www.altavian.com
173 Nova F6500
484 Nova F7200 AT
528 Galaxy R8700

Applied Aeronautics
Austin, Texas
www.appliedaeronautics.com
555 Albatross

Applied Research Associates (ARA)
Albuquerque, New-Mexico
www.ara.com
39 Nighthawk IV

Arcturus UAV
Rohnert Park, California
www.arcturus-uav.com
90 T-20
177 Jump 15
275 Jump 20

Ares RC
(by Firelands Group)
Champaign, Illinois
www.ares-rc.com
149 Ethos QX 130
228 Ethos FPV
228 Ethos HD
242 Spectre X
357 Spidex
386 Spidex 3D
429 Shadow 240
452 Crossfire Racer
469 X:Bolt 250
583 Recon FPV
584 Quantum FPV
597 Xview FPV

Ascent AeroSystems
Tempe, Arizona
www.ascentaerosystems.com
540 Sprite

Aurora Flight Sciences
Manassas, Virginia
www.aurora.aero
139 Orion

Autel Robotics
Bothell, Washington
344 X-Star
430 X-Star Premium

Bee Rotor
Scottsdale, Arizona
www.beerotor.com
453 Victory 230

BirdsEyeView Aerobotics
Andover, New Hampshire
www.birdseyeview.aero
186 FireFly6
529 FireFly6 Pro

Blade
(by Horizon Hobby)
Champaign, Illinois
www.bladehelis.com
134 Nano QX
139 350 QX
145 180 QX
158 200 QX
159 350 QX2
229 Nano QX FPV
230 Pico QX
242 350 QX3
286 Glimpse
299 Mach25 FPV Racer
309 Chroma
345 Inductrix
431 Zeyrok
431 Inductrix 200
469 Nano QX2 FPV
556 Conspiracy 220
567 Inductrix FPV
568 Theory W

Boeing Insitu
Bingen, Washington
www.insitu.com
7 X-45A
19 ScanEagle
25 Dominator
36 Integrator
36 X48-B
101 Phantom Eye
178 RQ-21A Blackjack
230 ScanEagle 2

Brock Technologies
(Now Blackbar Engineering)
Tucson, Arizona
www.blackbarengineering.com
89 Havoc UAS

Composite Engineering Inc.
(Now Kratos Defense & Security Solutions)
San Diego, California
www.kratosusd.com
43 BQM-167 Skeeter

CyPhy Works, Inc.
Danvers, Massachusetts
www.cyphyworks.com
220 Pocket Flyer

Dromida
(a Hobbico company)
Champaign, Illinois
www.dromida.com
221 Omnius
232 Kodo
286 Omnius FPV
358 Vista UAV
376 Vista FPV
433 Verso
517 Kodo HD
541 XL 370 FPV
568 Hovershot

Drone America
Reno, Nevada
www.droneamerica.com
155 Phoebus
202 DAx8
203 HuginnAir
387 Savant
587 NavX

Dronetech UAV
San Antonio, Texas
www.dronetechuav.com
187 AV-1 Albatross
300 AV-2 Pelican

Easy Aerial
Brooklyn, New York
www.easyaerial.com
435 Easy Drone XL

EMAX
Brea, California
www.emaxmodel.com
345 Nighthawk 280 Pro

Estes Rockets
(a Hobbico company)
Penrose, Colorado
www.estesrockets.com
150 Proto-X
257 Proto-X FPV
258 Proto-X SLT
388 Proto-N
543 Proto-Z

Event38 Unmanned Systems
Akron, Ohio
www.event38.com
188 E384
401 E386

Facebook
Menlo Park, California
www.facebook.com
530 Aquila

FreeFly Systems
Redmond, Washington
www.freeflysystems.com
328 Alta 6
472 Alta 8

General Atomics
San Diego, California
www.ga.com
34 MQ-9 Reaper
56 Avenger
58 MQ-1C Gray Eagle
196 Predator XP
571 Avenger ER
571 MQ-1C Gray Eagle ER

Griffon Aerospace
Madison, Alabama
www.griffonaerospace.com
16 MQM-170A Outlaw G1
47 MQM-171 BroadSword

HeliMax
Champaign, Illinois
www.helimaxrc.com
111 1SQ RTF
150 1Si
197 230Si
362 Voltage 500 3D
416 Form 500

Hobbyzone
(By Horizon hobby)
Champaign, Illinois
www.horizonhobby.com
233 Faze
348 Zugo

HoneyComb
Wilsonville, Oregon
www.honeycombcorp.com/
268 AgDrone

Honeywell Aerospace
Phoenix, Arizona
www.aerospace.honeywell.com
42 RQ-16 T-Hawk

HSE (Homeland Surveillance & Electronics)
Casselberry, Florida
www.hse-uav.com
314 AG-V8A Plus
559 AG-V6A Plus

Intel
Santa Clara, California
www.intel.com
573 Falcon 8+

Kespry
Menlo Park, California
www.kespry.com
548 Kespry Drone 2.0

Keyshare US
Los Angeles, California
www.keyshareus.com
366 Glint

L3 Unmanned Systems
San Diego, California
www2.l3t.com/uas/index.htm
148 Cutlass
548 HQ-60

Lockheed Martin
Bethesda, Maryland

www.lockheedmartin.com
26 Desert Hawk III
39 RQ-170 Sentinel
141 Stalker XE
180 Indago

Lynxmotion
Swanton, Vermont
www.lynxmotion.com
105 Hunter VTail 500
116 Hunter VTail 400
126 Crazy2Fly

Martin UAV
Plano, Texas
www.martinuav.com
378 Super Bat DA-50
533 V-Bat

Matternet
Menlo Park, California
www.mttr.net
290 Matternet One
560 Matternet Two

Monarch Inc.
(Formerly CAL UAS)
Ridgecrest, California
www.monarchmakers.com
175 AG

Mota Group Inc
San Jose, California
mota.com
442 Giga 6000
442 Giga 8000
443 Jetjat Live-W
443 Jetjat Nano-C
444 Pro Live 4000
444 Pro Live 5000

NASA (National Aeronautics and Space Administration)
Edwards, California
www.nasa.gov
393 Prandtl-D No.3

Navmar Applied Sciences Corporation (NASC)
Warminster, Pennsylvania
www.nasc.com
205 TigerShark XP

Northrop Grumman
West Falls Church, Virginia
www.northropgrumman.com
5 RQ-4A Global Hawk
11 X-47A Pegasus
32 RQ-4B Global Hawk
68 Bat
84 X-47B
102 MQ-4C Triton

ONAGOfly
Delaware City, Delaware
www.onagofly.com
523 Smart Nano Drone

Physical Sciences Inc.
Andover, Massachusetts
www.psicorp.com
131 Instant Eye Mk-2 Gen3-A0

PMG Multi-Rotors
Corona, California
www.pmgmultirotors.com
162 Recon

PowerUp
New York City, New York
www.poweruptoys.com
144 Power Up 2.0
166 Power Up 3.0
593 Power Up FPV

Precision Drone
Noblesville, Indiana
www.precisiondrone.com
236 Pacesetter
368 Scout

PrecisionHawk
Raleigh, New York
www.precisionhawk.com
524 Lancaster 5

Prioria Robotics
Gainesville, Florida
www.prioria.com
42 Maveric
152 Hex
153 Hex Mini

Rapid Composites
Sarasota, Florida
www.rapidcomposites.com
393 Bullray Heavy Aero

Raytheon
Waltham, Massachusetts
wwwraytheon.com
8 Silver Fox
59 JLENS
222 Coyote

Rise
(a Hobbico company)
Champaign, Illinois
www.explore-rise.com/drones
477 RXD250
524 RXS270
576 Vusion 250

RoboFlight
(By AgPixel)
Denver, Colorado
www.agpixel.com
337 RF-70

Robota
Dallas, Texas
www.robota.us
95 Triton UAV
510 Eclipse

SeaHex
Newport, Rhode Island
176 SeaHex

Silent Falcon UAS
(An Aerospace company)
Alberquerque, New Mexico
www.silentfalconuas.com
380 Silent Falcon

Textron Systems
Providence, Rhode Island
www.textronsystems.com
9 RQ-7A Shadow
16 RQ-7B Shadow
54 Aerosonde Mark 4.7

Top Flight Technologies
Malden, Massachusetts
www.tflighttech.com
252 Airborg H6 1500

Traxxas
McKinny, Texas
www.traxxas.com
411 Aton
412 Aton Plus

Trident LLC
Columbus, Ohio
www.trident-design.com
423 Carbon Flyer

TurboAce
Orange, California
www.turboace.com
224 Infinity 6

UAV Solutions
Jessup, Maryland
www.uav-solutions.com
370 Phoenix 30
479 Phoenix Ace LE

Vantage Robotics
San Leandro, California
www.vantagerobotics.com
554 Snap

Vision Aerial
Bozeman, Montana
www.visionaerial.com
494 SwitchBlade-Elite

Volt Robotics
St-Louis, Missouri
www.voltrobotics.org
112 WaveSight
342 Venture

XactSense
Providence, Rhode Island
www.xactsense.com
193 Titan1

Xcraft UAV
Coeur d'Alene, Idaho
www.xcraft.io
580 X Plus One

xFold
San Francisco, California
www.xfoldrig.com
424 Dragon X12

THE DRONE CHRONICLES

Concept and design: Rob van Leijsen
Authors: Joerg Bader, Nicolas Nova, Brice Pauset, Rob van Leijsen
Publisher: Joerg Bader
Distribution: Spector Books, Leipzig / les presses du réel, Dijon
Editorial coordination: Tiphaine Carrère, Olivia Fahmy, Maud Pollien, Natalia Reichert, Alexandra Theiler
Translation: Judith Hayward, Natalia Reichert
Copy-editing and proofreading: Natalia Reichert

Printing and binding: Musumeci S.p.A., Quart
ISBN: 978-3-95905-310-5

Thanks to:
Markus Farner, Cintia Stucker, Veronica Tracchia

éditions Centre de la photographie Genève
www.centrephotogeneve.ch

Spector Books
www.spectorbooks.com

The Centre de la photographie Genève is supported by Ville de Genève

With the support of:

VILLE DE NYON

Région de Nyon

FONDATION LEENAARDS